Emotions, Qualia, and Consciousness

Istituto Italiano per gli Studi Filosofici

Series on Biophysics and Biocybernetics

Coordinating Editor: Cloe Taddei-Ferretti

Vol. 1: Biophysics of Photoreception: Molecular and Phototransductive Events
edited by: C. Taddei-Ferretti

Vol. 2: Biocybernetics of Vision: Integrative Mechanisms and Cognitive Processes
edited by: C. Taddei-Ferretti

Vol. 3: High-Dilution Effects on Cells and Integrated Systems
edited by: C. Taddei-Ferretti and P. Marotta

Vol. 4: Macromolecular Interplay in Brain Associative Mechanisms
edited by: A. Neugebauer

Vol. 5: From Structure to Information in Sensory Systems
edited by: C. Taddei-Ferretti and C. Musio

Vol. 6: Downward Processes in the Perception Representation Mechanisms
edited by: C. Taddei-Ferretti and C. Musio

Vol. 7: Chaos and Noise in Biology and Medicine
edited by: M. Barbi and S. Chillemi

Vol. 8: Neuronal Bases and Psychological Aspects of Consciousness
edited by: C. Taddei-Ferretti and C. Musio

Vol. 9: Neuronal Coding of Perceptual Systems
edited by: W. Backhaus

Vol. 11: Vision: The Approach of Biophysics and Neurosciences
edited by: C. Musio

Vol. 12: Memory and Emotions
edited by: P. Calabrese and A. Neugebauer

ISTITUTO ITALIANO PER GLI STUDI FILOSOFICI
SERIES ON BIOPHYSICS AND BIOCYBERNETICS
Vol. 10 – Biocybernetics

Emotions, Qualia, and Consciousness

Proceedings of the International School of Biocybernetics
Casamicciola, Napoli, Italy, 19–24 October 1998

Edited by

Alfred Kaszniak

Center for Consciousness Studies
Departments of Psychology, Neurology, and Psychiatry
University of Arizona, Tucson, Arizona, USA

World Scientific
Singapore • New Jersey • London • Hong Kong

Published by

World Scientific Publishing Co. Pte. Ltd.

P O Box 128, Farrer Road, Singapore 912805

USA office: Suite 1B, 1060 Main Street, River Edge, NJ 07661

UK office: 57 Shelton Street, Covent Garden, London WC2H 9HE

British Library Cataloguing-in-Publication Data
A catalogue record for this book is available from the British Library.

EMOTIONS, QUALIA, AND CONSCIOUSNESS

ISBN 981-02-4165-8

Printed in Singapore by World Scientific Printers (S) Pte Ltd

PREFACE

This is the tenth volume of the Series on Biophysics and Biocybernetics promoted by the Istituto Italiano per gli Studi Filosofici, Naples, Italy.

It appears as the Proceedings of the Course of the International School of Biocybernetics on "Emotion, Qualia, and Consciousness", which was inaugurated at the site of the Institute, Palazzo Serra di Cassano, Naples, Italy, and was held at the Hotel Gran Paradiso, Casamicciola, isle of Ischia, Italy, from 19 to 24 October, 1998, under direction of this volume's editor.

The School, directed by Cloe Taddei-Ferretti, is promoted and supported by the Istituto Italiano per gli Studi Filosofici. The organization of this Course was carried forward by the Center for Consciousness Studies, University of Arizona, Tucson, Arizona, U.S.A.

The experience of emotion is a ubiquitous component of the stream of consciousness, and emotional qualia (i.e., the "feeling" of emotion) appear to interact with other contents and processes of consciousness in complex ways. Further, recent research has suported the hypothesis that important functional aspects of emotion can operate outside of conscious awareness, begging questions about what particular neural structures, processes, stimulus conditions, and environmental contexts are necessary for the conscious experience of emotion. A full understanding of human emotion does not seem possible without an exploration of the nature and correlates of those consciously experienced qualia of emotion. Many scholars and scientists also now believe that no scientific or philosophic account of consciousness can be complete without an understanding of the role of emotion. Some have even argued for a necessary role of emotion in the genesis of consicousness itself. In an effort to advance understanding of these complex issues, this Course on "Emotion, Qualia, and Consciousness" was organized. Following an introductory lecture by the editor of this volume, invited faculty spoke on philosophical perspectives, perspectives from evolutionary biology, neuroscience, neuropsychology, and psychophysiology, psychological perspectives, as well as cognitive, social, and clinical perspectives and robotics. A common thread characterizing several of these lectures was the sense that theory and empirical studies in the domains of emotion research and consciousness studies should be mutually informative and interactive. The contributors to the present volume collectively make a persuasive argument for continued vigorous investigation of the interrelationships of emotion, qualia, and consciousness.

I would like to thank the members of the Course Advisory Board for their helpful advice and suggestions: R. Lane (USA), J. LeDoux (USA), J. Panksepp (USA), T. Radil (CZ), C. Taddei-Ferretti (I).

I would also like to acknowledge the partial financial support of the Center for Consciousness Studies (funded through a grant from the Fetzer Institute, Kalamazoo, Michigan, USA) at the University of Arizona (USA).

Due to the help of the Istituto per gli Studi Filosofici, it was possible to award grants to several deserving participants in the Course, especially those coming from countries needing financial help.

Sincere thanks go to all the scientists and scholars who agreed to lecture, who contributed to the Course with their comments and discussions, and who fostered, together with all the participants, a positive atmosphere in a highly stimulating intellectual milieu.

Finally, I wish to thank Jim Laukes (Center for Consciousness Studies, University of Arizona, USA) who served as the Course Organizer, Antonio Cotugno (Istituto di Cibernetica, CNR, Arco Felice, Naples, Italy) who was the Local Staff, and Nunzia Aprile (Istituto Italiano per gli studi Filosofici) who was the Administrative Advisor. Without their hard work and dedication the Course could not have been realized, nor its cordial atmosphere maintained. The beauty of the isle of Ischia and the courtesy of the staff of the Hotel Gran Paradiso at Casamicciola also contributed enormously to success of the Course.

Alfred W. Kaszniak

CONTENTS

BIOLOGICAL PERSPECTIVES

CONCLUSION

PARTICIPANTS

INTRODUCTORY LECTURE

EMOTION AND CONSCIOUSNESS: CURRENT RESEARCH AND CONTROVERSIES

ALFRED W. KASZNIAK
Center for Consciousness Studies, Departments of Psychology, Neurology & Psychiatry, University of Arizona, 1503 E. University, Tucson, Arizona 85721, U.S.A.

ABSTRACT

The experience of emotion is a nearly constant aspect of the stream of consciousness, and emotional experience appears to interact with other contents and processes of consciousness in complex ways. Recent research has supported the hypothesis that important functional aspects of emotion can operate outside of conscious awareness. Such research raises questions about what conditions are necessary and sufficient for the conscious experience of emotion. Conversely, many scholars and scientists now believe that no scientific or philosophic account of consciousness can be complete without an understanding of the role of emotion. This paper briefly reviews select recent contributions to an understanding of the relationship between emotion and consciousness. Experimental psychological, neuropsychological, and neuroscientific approaches are highlighted, and the author's own recent studies of conscious emotional experience and emotion physiology in different neurological disorders are described. The paper closes with some speculations, derived from this research, concerning the role of frontal brain systems in the conscious experience of emotion.

1. Introduction

1.1 Why is Emotion Important in Understanding Consciousness, and why is Consciousness Important in Understanding Emotion?

Recently, an increasing number of scholars and scientists have recognized the importance of emotion theory and research for the study of consciousness. Most all would agree that casual introspection reveals emotion to be a nearly constant aspect of the human phenomenal experience. Some consciousness theorists (e.g., Damasio, 1999; Watt, 1999) have posited emotion to play a necessary role in the genesis of all conscious experience. Empirical evidence has been interpreted as consistent with the hypothesis that emotions play a necessary role in such cognitive processes as reasoning (Churchland, 1998; Damasio 1994) and creativity (Csikszentmihalyi, 1990; Nielsen 1998), and neural network models of consciousness have included emotion as a necessary component (Levine 1998, Taylor 1992). Others (e.g., DeLancey, 1996) have argued that the qualia of emotional experience provide a critical example for any philosophical speculation concerning the functional role of phenomenal conscious states. Conversely, emotion researchers have become concerned with understanding the role of conscious experience in emotion (e.g., LeDoux, 1996; Panksepp, 1998). Recent developments in basic and clinical neuroscience have resulted in rapid progress toward understanding the neural bases of emotion (Lane, Nadel, Allen, & Kaszniak, 2000; LeDoux, 1996; Panksepp, 1998). These developments also have encouraged theoretic speculation and empirical research on the neural correlates of conscious emotional experience (e.g., Kaszniak, Reminger, Rapcsak, & Glisky, 1999; Lane, 2000). The identification of neural systems critical for the conscious experience of emotion may also provide important clues in the search for neural circuitry upon which other domains of conscious experience are dependent.

1.2 Components of Emotion and their Interrelationships

Despite disagreements regarding the necessary and sufficient conditions for identifying an emotion (for review, see Griffiths, 1997), analyses often describe five components: (1) Physiological (CNS and autonomic) arousal, (2) cognitive appraisal, (3) conscious experience (the qualia or "feeling" of emotion), (4) action tendency, and (5) expressive behavior (including facial expression). These components are complexly intercorrelated. Various investigators have shown differential relationships between dimensions of self-reported emotional experience, physiology, and expression. As reviewed by Lang and colleagues (Lang, Greenwald, Bradley, & Hamm, 1993), self-reported emotional valence (pleasant to unpleasant) experience has been found to co-vary with facial (zygomatic and corrugator muscle regions) electromyography (EMG), heart rate, and startle reflex magnitude. In contrast, self-reported emotional arousal (intensity) experience has been shown to co-vary with skin conductance response (SCR), particular aspects of electroencephalographic (EEG) activity, and measures of functional brain activation.

For both the general public and many scientists, emotion is identified with feeling, and is thus inextricably linked to consciousness. Clore (1994), for example, views feeling as a necessary condition for emotion, and sees conscious cognitive appraisal as preceding other emotional reactions. Some cognitive appraisal (in terms of positive or negative valuation in relation to personal goals, need states, of self-preservation) may indeed be a necessary condition for the activation of the various other emotion components (Lazarus, 1991). However, such appraisals need not necessarily be conscious (Frijda, 1993). In recent years, empirical evidence has made it clear that the conscious experience of emotion is not a necessary contributor to all emotion components. In humans, autonomic physiological and expressive motoric aspects of emotion can occur in response to an emotional stimulus that is below threshold for recognition. In studies employing backward masking of visual emotional stimuli (preventing these stimuli from being consciously recognized), both SCR (Öhman & Soares, 1994; Öhman,

Flykt, & Lundqvist, 2000) and facial EMG responses (Dimberg, Thunberg, & Elmehed, 2000) have been demonstrated.

Other evidence that is consistent with the interpretation that important functions of emotion occur nonconsciously comes from nonhuman animal research. As LeDoux's (1996, 2000) studies of fear in rodents have shown, the amygdala is a key structure in both the stimulus evaluation of threatening events and the production of defensive responses. These defensive responses appear to be evolutionarily selected, involuntary, automatic consequences of an initial rapid evaluation of stimulus significance, and do not require cortical mediation. Thus, LeDoux argues that subcortical (e.g., thalamic-amygdala circuitry) emotion systems are involved in fast, evolutionarily selected, and likely nonconscious (to the extent that consciousness requires cortical activity) aspects of emotion. Recent evidence from neuroimaging studies is consistent with LeDoux's hypothesis that the amygdala plays a role in emotion that does not require conscious perception of the eliciting stimulus. Functional magnetic resonance imaging (fMRI) has demonstrated amygdala activation in response to emotional stimuli (facial expressions) even when conscious awareness of the stimuli is prevented by backward masking (Whalen, Rauch, Etcoff, McInerny, Lee, & Jenike, 1998).

2. Searching for the Neural Correlates of Conscious Emotional Experience: General Considerations

Clore and Ortony (2000) have argued that consciously experienced emotions are always "about something." By this they mean that emotions are:

> ...affective (i.e., positively or negatively valenced) states that have objects (what philosophers call "intentional" states), which is why not every occurrence of an affective feeling constitutes an emotion. For example, to the extent that "fear" refers to an affective state directed toward a specific object, it qualifies as an emotion, and to the extent that "anxiety" refers to an affective state without an object, it does not qualify as an emotion. Thus, when one is afraid, the fear is crucially about something in particular, but when one feels anxious, the anxiety is not focally about anything in particular. (pp. 26-27)

If we accept Clore and Ortony's definition, then it would appear from the evidence reviewed above that conscious emotion can be both conceptually and experimentally dissociated from other emotion components (i.e., nonconscious aspects of stimulus appraisal, autonomic response, facial expression). What then might be the neural structures and processes that are necessary for the conscious experience of emotion? As noted above, LeDoux (1996) has argued that important processes in emotion take place nonconsciously, with some of the outputs of these processes delivered to consciousness. LeDoux specifically posits the following cortical inputs to be hypothetically necessary for the conscious experience of emotion: (1) direct cortical inputs from the amygdala, (2) inputs from amygdala to nonspecific arousal systems, and (3) feedback to the amygdala and cortical areas from the bodily changes of emotion,. What approaches are available to test such hypotheses? Invasive nonhuman animal studies, which allow for the most direct experimental manipulation of neural structures and processes, are unlikely to be very helpful. This is because the third-person assessment of conscious emotional experience is generally dependent upon verbal reportability, thus necessitating the study of humans where invasive procedures (for the purpose of experimentation alone) are not ethically permissible. Direct stimulation or recording from the brain of awake persons who are undergoing neurosurgery for a specific clinical disorder (e.g., epilepsy or a brain tumor) is certainly possible. However, interpretation of resultant data is often clouded by the fact that the clinical disorder in question itself results in alteration of brain function, potentially including aspects relevant to emotion. An alternative approach to searching for the neural correlates of conscious emotional experience assesses possible dissociations between conscious and nonconscious emotion components in persons with neurological disorders involving theoretically relevant brain systems. Another approach involves the application of functional neuroimaging technology (e.g., fMRI or positron emission tomography; PET) within experimental paradigms that attempt to elicit conscious emotional experience. A selective review of examples of these two (i.e., neuropsychological and neuroimaging) approaches will comprise the remainder of the present paper.

3. A Neuropsychological Approach to Testing the Facial Feedback Hypothesis of Conscious Emotional Experience

One of the hypothetical neural ingredients of conscious emotional experience proposed by LeDoux (1996) is feedback to cortical areas and the amygdala from the peripheral bodily components of emotional response. This hypothesis is based on the original peripheral feedback theory of William James (1884/1922). James argued that emotions are experienced as a direct consequence of the bodily (e.g., visceral, motoric) feedback we receive when confronted with an emotional stimulus. Walter Cannon (1927), who argued that visceral changes are too slow to effectively produce the seemingly immediate experience of emotion, critiqued James' theory. Cannon offered an alternative hypothesis that activation of subcortical brain centers by external stimulation produces both emotional experience and the muscular and visceral changes of emotion. Nonetheless, Cannon did not argue against the idea that bodily feedback does contribute in some way to the overall experience of emotion. A more recent variant of the bodily feedback theory of conscious emotional experience is the facial feedback hypothesis of Tomkins (1962, 1963). This hypothesis posits that facial expression provides cutaneous and proprioceptive feedback that is a necessary contributor to emotional experience. Tomkins' facial feedback theory placed primary importance on the facial musculature as a necessary condition for emotional experience. He argued that a number of facial cues, including the activity of the tongue and facial muscles, provide essential emotional information to the brain. The face, because of its fine nerve and muscle differentiation, is more capable of rapid, flexible, and sensitive responses than the slower-moving viscera (thus addressing Cannon's original objection to James' peripheral feedback theory).

Several empirical studies have attempted to test Tomkins' theory. These studies have typically employed direct manipulation of facial expression to determine its impact on conscious emotion experience. Results of these studies have been interpreted as consistent with the facial feedback theory. For example, Ekman and colleagues (Ekman, Levenson, & Friesen, 1983; Levenson Ekman & Friesen, 1990) found specific patterns of autonomic nervous system activity to result from posed facial muscle contractions that are characteristic of specific emotional expressions. These studies also found participants' verbal reports of emotional experience to correspond to the emotion-specific muscle contraction patterns. Similar results have been found in studies that ask participants to inhibit or exaggerate facial expressions (for review, see Camras, Holland, & Patterson 1993). In these studies, participants reported lowered intensity of emotional experience in the inhibition conditions as compared to the exaggeration conditions. However, more recent work (Gross & Levenson, 1993; 1997) has been unable to replicate this finding. Studies employing directed facial action have thus provided mixed results regarding the question of whether facial feedback is sufficient to produce conscious emotional experience. A more direct test of whether or not facial action is *necessary* for the experience of emotion would involve the study of individuals with little or no capacity for facial expression. If such individuals reported experiencing emotions in the same manner and with the same intensity as normal individuals, then it would be difficult to claim that facial expression is necessary for the experience of emotion.

3.1 Testing the Facial Feedback Hypotheses in Persons with Parkinson's Disease

Persons with Parkinson's disease (PD), due to brain nigro-striatal pathway dysfunction, may show markedly reduced spontaneous facial expression (termed "masked facies"), despite intact voluntary (pyramidal) control of facial muscle movement (Smith, Smith, & Ellgring, 1996). Persons with PD who exhibit masked facies would lack the feedback from facial expression that Tomkins posited to be a strong contributor to the conscious experience of emotion. The facial feedback hypothesis thus would predict that those individuals with PD who show masked facies should report proportionately reduced emotional experience in response to salient stimuli.

Contrary to this prediction, empirical studies have failed to find evidence for reduced emotional experience reports in individuals with PD. A study by Katsikitis and Pilowsky (1991) showed persons with PD and healthy control participants to report similar ratings of amusement to cartoon slides, despite the fact that those with PD smiled less often during slide viewing. Similarly, a study by Smith *et al.* (1996) showed that reduced spontaneous facial expression in PD was not associated with decreased self-ratings of emotional experience while viewing emotionally arousing video clips. Although these studies provide some evidence that the reduced facial expression of PD is not associated with comparable reductions in self-reported emotional experience, further research appeared needed to adequately explore this issue. Changes in facial expression cannot always be clearly observed. Hence, the use of more sensitive measures of facial muscle activity might help to better understand the true relationship between facial feedback and conscious emotional experience.

In addition to relying upon observational measures of facial expression, previous investigations did not address possible effects of antiparkinsonian medication on emotional expression and experience. Participants in the Katsikitis and Pilowsky (1991) and Smith *et al.* (1996) studies were receiving standard dopamine replacement therapy. Since dopaminergic medication could conceivably affect the facial expressiveness of persons with PD (Hunker, Abbs, & Barlow, 1982), the experimental manipulation of medication status could potentially shed light on the relationship between facial expression and emotional experience.

3.2 Further Testing of the Facial Feedback Hypothesis in PD Using Facial EMG

Dalby, Reminger, Kaszniak, and Montgomery (in preparation) evaluated the relationship between facial muscle activation in response to emotional stimuli and the self-reported experience of emotion in persons with idiopathic PD. Nineteen nondemented, nondepressed persons with PD were compared to 19 age-,sex-, and education-matched healthy normal control (NC) participants. A neurologist with expertise in movement disorders made the diagnoses of PD, and all patients had recent complete physical and neurological examinations to rule out any other concomitant illness. The participants with PD were selected on the basis of all having evidence of masked facies, as determined by standardized examination procedures, and all showed bilateral limb movement involvement (indicative of disease progression beyond the initial stages of PD).

Data were collected in one morning and one afternoon session, separated by several days. This two-day data collection format was employed so that the participants with PD could be tested both "on" and "off" of their levodopa medication. With the supervision of their attending neurologist, PD participants completed one "off" drug condition, in which they abstained from levodopa medication for at least 12 hours prior to testing, and one "on" drug condition, in which they ingested levodopa medication about one hour prior to testing. By necessity, the "off" drug condition had to be a morning session, since PD participants would have had great difficulty abstaining from medication in the daytime (due to the increase in their movement impairment resulting from medication abstinence). For NC participants, these two conditions merely corresponded to one morning session and one afternoon session. Session order was counter-balanced across participants within both groups.

The International Affective Picture System (IAPS) photographic slides, developed by Peter Lang and colleagues (Center for the Study of Emotion and Attention, 1999), were employed as the emotional stimuli. The IAPS slides vary in content (e.g., animate/inanimate, human/animal, etc.) complexity, luminance, and central image size. These slides are standardized according to how pleasant/unpleasant (valence) and calm/excited (arousal) each slide makes an individual feel (using a self-report format described below). The focus on these two dimensions of emotional experience is justified by factor-analytic studies of evaluative responses to a large variety of emotional stimuli, in which valence and arousal dimensions were found to account for most of the variance (for review see Russell, 1980). Lang, Bradley, and Cuthbert (1999) provide extensive normative data on the IAPS pictures, with young adult consensus ratings available for emotional valence and arousal experienced in response to the stimuli. In development of this normative data base, Lang and colleagues had participants report their experience of emotion using the Self-Assessment Manikin (SAM - Lang, 1980), which employs a cartoon-like visual analog scale designed to minimize the effects that language may have in reporting response to emotionally-arousing stimuli. Both the valence and arousal dimensions are ordinally scaled with five figures, and the option is given to make ratings between two figures, providing a scale ranging from 1 to 9 for each dimension (see Fig. 1).

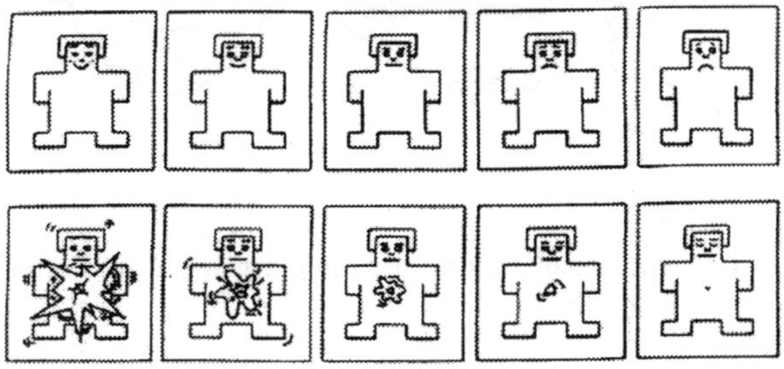

Figure 1. The Self-Assessment Manikin (SAM), used to obtain quantitative ratings of emotional valence (top row) and arousal (bottom row) experienced in response to standardized emotional stimuli.

Normative data was used to create two valence- and arousal-matched sets of slide images, selected from the complete IAPS set. Two separate slide sets were needed to accommodate the drug "on" vs. drug "off" testing conditions. Eighteen of the most valent and most arousing slides (9 pleasant, 9 unpleasant), as rated by both male and female college students (Lang & Greenwald, 1988) were chosen and matched to 18 additional slides by valence and arousal ratings, and by slide content. Thus, each slide set consisted of 27 slides (9 pleasant/positive, 9 neutral, and 9 unpleasant/negative), block-randomized in sets of three to distribute the slide valence types evenly throughout the presentation. Slides with content of a highly distressing nature (e.g., extremely bloody scenes) or of a potentially more shocking nature (e.g., scenes involving full nudity) were not included in the sets. Because the original IAPS set was normed on college students, an attempt was also made to include some pictures relevant to older adults (e.g., surgery), given that the average age for the PD and NC groups was approximately 70 years.

Facial EMG signals were recorded for five seconds before slide onset and for six seconds during slide presentation, using standard electrode preparation, amplification, time sampling, and signal filtering procedures. Surface facial EMG activity was recorded bilaterally over the corrugator and zygomatic major muscle regions (Fig. 2). EMG change scores were calculated separately at each muscle site by subtracting the EMG activity during the second immediately preceding slide onset from the average response during the 6-second slide-viewing interval.

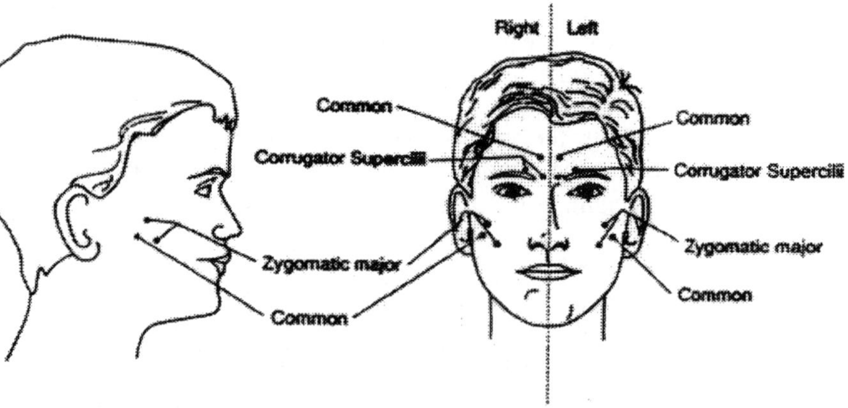

Figure 2. EMG electrode placement sites for recording zygomatic and corrugator facial muscle activity in response to emotional stimuli

The use of EMG allows for measurement of muscle activity even in the absence of overt facial actions (Tassinary & Cacioppo, 1992). Previous research has shown that, during negative emotional experiences elicited by either emotional imagery or visual scenes, greater EMG activity occurs in the region of the corrugator supercilium muscle, which draws the eyebrows together. During positive emotional experiences, greater EMG activity occurs in the region of the zygomatic major muscle which draws the ends of the mouth up and back (e.g., Brown & Schwartz, 1980; Dimberg, 1990; Lang *et al.*, 1993). Even when measurable degrees of visible facial expression are absent, facial EMG activity has differentiated the valence (positive vs. negative) and intensity of facial expressive response to emotional scenes (Cacioppo, Petty, Losch, & Kim, 1986).

Participants were told that they would be viewing pictures differing in emotional content, and that each picture should be attended to for the entire time it is shown. After picture offset, a blank screen was shown during which valence and arousal ratings were made. Computer digitized auditory instructions requested that the participant rate the slide on both dimensions (valence and arousal) by pushing one of nine buttons placed adjacent to or between the five SAM figures. A variable slide interval of about 20-35 seconds occurred between each slide presentation, allowing time for each participant to rate his or her emotional experience, and also for EMG activity to return to baseline.

For the NC participants, the expected (based on previous facial EMG and emotion studies) pattern of greatest zygomatic EMG change in response to the positively valent slides and greatest corrugator EMG change in response to the negatively valent slides was observed. Also as expected, we found significantly reduced facial EMG change scores for both the bilateral zygomatic (in response to the positively valent slides) and corrugator regions (in response to the negatively valent slides) for the PD group in comparison to the healthy controls (Fig. 3). However, the SAM self-reports of experienced emotional valence and arousal were virtually identical for the two groups (Fig. 4). The drug "on" versus "off" condition did not significantly affect either the facial EMG change scores or the SAM valence or arousal ratings for the PD group.

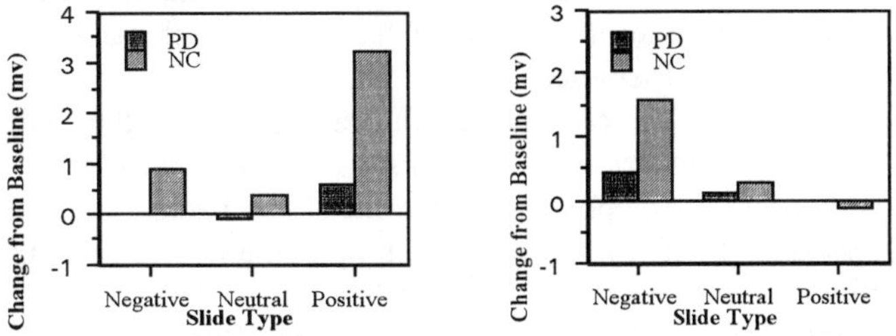

Figure 3. Zygomatic (left graph) and corrugator (right graph) facial EMG change in response to emotional slides for Parkinson's (PD) and normal control (NC) groups.

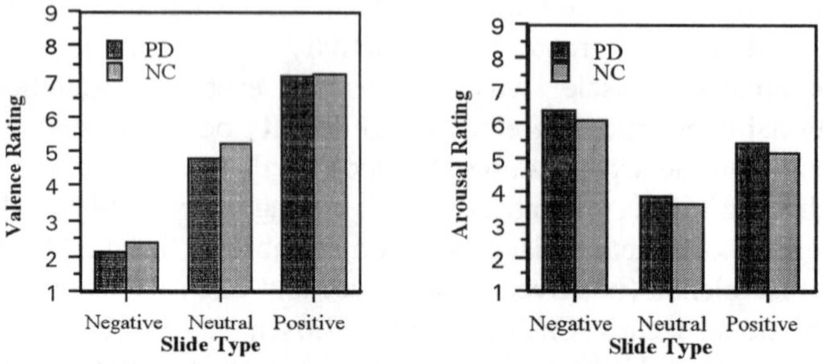

Figure 4. Emotional experience self-report ratings (Self-Assessment Manikin; SAM) on valence (left graph) and arousal (right graph) dimensions for Parkinson's (PD) and normal control (NC) groups.

Each participant's EMG change scores for both the corrugator and zygomatic muscle regions were correlated with his/her SAM valence ratings for each slide. Linear correlations were computed for corrugator data and quadratic correlations were computed for zygomatic data, based on previous research (Lang *et al.*, 1993) indicating that such models would best explain the relationships between valence ratings and these particular muscle groups. All resulting correlation coefficients were statistically significant, with corrugator EMG activity showing a negative linear correlation with valence ratings, and zygomatic EMG activity showing a positive quadratic correlation with valence ratings. In order to determine whether the PD group had a significantly lower correlation of facial muscle activity and SAM self-report of emotional valence experience in comparison to the control group, average correlations for the two groups were compared using a significance test described by Cohen and Cohen (1983). Significant differences were found for both the corrugator and zygomatic muscle regions. Thus, the correlation of EMG changes scores with valence ratings was reliably lower for the PD group when compared to the healthy controls. Given that the PD and NC groups did not differ in their subjective ratings of experience in response to the IAPS slides, this is consistent with the interpretation that facial feedback is not making a necessary contribution to emotional experience.

Although these results are contrary to the predictions of the facial feedback hypothesis, interpretive limitations of the study must be considered. This study cannot definitively rule out the possibility that facial feedback, via mechanisms not adequately assessed by EMG measures, is making some necessary contribution to conscious emotional experience. First, other facial factors such as facial temperature, blood flow to the face, audition of one's own voice, or breathing patterns could be peripheral contributors to the conscious experience of emotion. Further research will be necessary to determine the importance of such factors. In addition, some emotion researchers (Damasio, 1994; Matsumoto & Lee, 1993) have proposed that facial feedback may contribute to conscious emotional experience via subcortical or cortical activation of neuronal systems involved in facial motor control rather than through feedback from changes in the facial musculature itself. Because our study was limited to measurement of peripheral facial muscle activation, and since it is possible that persons with PD may have normal initial activation (although not output) of subcortical or cortical facial motor control systems, the validity of this iteration of the feedback hypothesis cannot be addressed. Another possibility that will require further investigation concerns changes that could occur in the subjective scaling of experienced emotion in persons with PD. PD is a progressive illness with onset in adulthood, after individuals have had a presumably normal development of emotional experience and facial expression. It remains possible that facial feedback does make a significant contribution to emotional experience, but that persons with PD gradually "rescale" their subjective ratings across a now-reduced range of emotional experience. Such a possibility could result in normal-appearing ratings of emotional valence and arousal experience, even though the range of actual experience might be reduced. The study of individuals with life-long facial paralysis (e.g., Möbius' syndrome; see Cole, 1997; Goldblatt & Williams, 1986), and the future application of multidimensional scaling techniques to the ratings of conscious emotional experience, might be capable of resolving this question.

Thus, our study cannot definitively rule out the possibility that some aspect of facial feedback makes a necessary contribution to the conscious experience of emotion. It does, however, illustrate the way in which the combination of neuropsychological and psychophysiological research methodologies can be used to test specific hypotheses derived from theories about the neural correlates of conscious emotion.

4. Are There Particular Brain Structures Necessary for the Conscious Experience of Emotion?

Another example of the potential of combining neuropsychological and psychophysiological research methodologies comes from the study of persons with damage to the ventromedial frontal area, including the anterior cingulate gyrus (see Fig. 5). Damage to the anterior cingulate (e.g., in the case of surgical lesions for treatment of intractable pain) has been reported to result in altered emotional experience (Devinsky, Morrell, & Vogt, 1995). The rostral anterior cingulate has dense connections with the orbital frontal cortex, the amygdala, and other paralimbic structures (e.g., the insular cortex), all of which are believed to be involved in emotional processes (for review see Price, Carmichael, & Drevets, 1996). Lane (2000) has hypothesized that the anterior cingulate cortex functions to provide conscious working memory for “interoceptive emotional information,” analogous to the role of nearby dorsolateral prefrontal cortex in cognitive working memory (Goldman-Rakic, 1992).

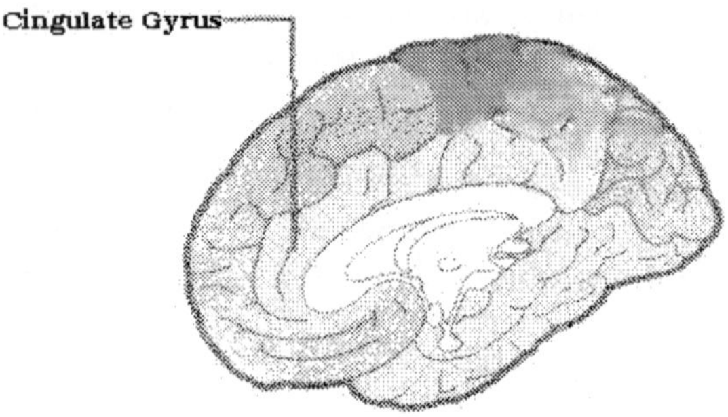

Figure 5. Medial surface of the right cerebral hemisphere, showing location of the cingulate gyrus.

4.1 Evidence from the Study of Patients with Ventromedial Frontal Lobe Damage

In a series of studies by Damasio and colleagues (for review, see Bechara, Damasio & Damasio, 2000; Adolphs, Tranel, Bechara, Damasio & Damasio, 1996), it has been shown that patients with ventromedial frontal damage do not show the expected SCR response to emotionally significant visual scenes. Previous research has shown larger SCRs in response to emotionally salient versus nonemotional stimuli (for review see Andreassi, 1995). Further, SCRs have been found to be positively correlated with self-rated emotional arousal (Lang, *et al.*, 1993). Damasio (1994) hypothesized that frontally damaged patients who do not show SCRs to emotional stimuli will not have the "conscious body state characteristic of an emotion" (p. 209). Damasio (1994) provided anecdotal examples of frontally-damaged patients who did not show SCRs to strongly negative emotional scenes and who also reported that they did not feel the usually expected negative emotion. Among patients with damage to the ventromedial frontal region, those with extensive anterior cingulate gyrus damage often show the most severely defective SCRs in response to emotional visual scenes (Tranel, 2000).

In an attempt to further test Damasio's hypothesis, Kaszniak, Reminger, Rapcsak, and Glisky (1999) studied 7 patients with ventromedial frontal lobe damage (including the anterior cingulate in all cases). We employed the same IAPS emotional slide methodology as that described above in the Dalby *et al.* study of PD patients. These frontally damaged patients showed overall smaller SCRs and a lack of SCR differentiation in response to neutral versus negatively or positively arousing visual stimuli. In contrast, healthy controls showed the expected larger SCRs in response to positively or negatively valent scenes in comparison to neutral slides. These same patients also showed a corresponding lack of differentiation in their self-reported (using the SAM response method) emotional arousal experience. Despite their abnormal SCR differentiation and subjective arousal ratings in response to the emotional scenes, these patients indicated a normal appreciation of the general meaning of emotionally-salient stimuli, as reflected in valence experience ratings (again, using the SAM response method) which were nearly identical to those of normal controls. Patients with ventromedial frontal damage thus provide evidence consistent with the interpretation that the anterior cingulate (and possibly related frontal structures) plays an important role in both the conscious experience of emotional arousal and the differentiation of autonomic physiologic response to emotional stimuli.

4.2 Evidence from Neuroimaging Studies

The search for the neural correlates of conscious emotional experience has also been advanced by experiments which attempt to experimentally manipulate conscious versus nonconscious aspects of emotion while participants' regional brain activation is measured by functional neuroimaging techniques. In a PET experiment, Lane, Fink, Chau, and Dolan (1997) found specifically increased activity in the rostral anterior cingulate cortex (Broadman's area 32) and medial prefrontal cortex during attention to subjective emotional experience elicited by a set of 12 consecutive IAPS emotional scenes, in comparison with a condition directing attention to the spatial context of the scenes. This result is consistent with the above-described observations from studies of patients with ventromedial frontal lobe damage, indicating a specific relationship between anterior cingulate activation and conscious emotional experience.

Another neuroimaging approach examines patients showing an apparent dissociation between conscious and nonconscious components of emotion (e.g., impairment in the self-reported conscious experience of emotion in the context of behavior suggesting that nonconscious aspects of emotional response are intact). Alexithymia is a term originally proposed to describe people who seem to lack an adequate language for their emotional experience. Richard Lane and colleagues (Lane, Sechrest, Riedel, Weldon, Kaszniak, & Schwartz, 1996) have presented evidence showing that alexithymia is more properly conceptualized as a general problem in the conscious experience and discrimination of emotion (i.e., alexithymic patients showed impairment in both nonverbal and verbal emotion stimulus perception tasks). Lane, Ahern, Schwartz, and Kaszniak (1997) have argued that alexithymia can be conceptualized as an emotional equivalent of blindsight (i.e., the state where a person's primary visual cortex is damaged but the capacities for visual localization, discrimination, and acuity are demonstrably preserved despite phenomenal blindness - for review see Kentridge & Heywood, 1999). If this characterization of alexithymia is correct, then it should be possible to compare the brain physiology of alexithymic individuals to those of nonalexithymic persons, with results having implications for the search for neural correlates of conscious emotion. Such an approach would be similar to that in which experiments involving blindsight patients have been used to explore the neural correlates of visual consciousness. These kinds of studies, with patients who meet clinical criteria for the diagnosis of alexithymia, are not yet available. However, Lane and colleagues (1998) have used PET measurement of regional brain activation to examine the neural correlates of varying degrees of "emotional awareness" among normal volunteers. Differences between these volunteers in degree of emotional awareness were measured with the Levels of Emotional Awareness Scale (LEAS, Lane, *et al.*, 1990), which has been shown to correlate with verbal and nonverbal emotion stimulus recognition/discrimination accuracy (Lane, *et al.*, 1996). Lane *et al.* (1998) found higher scores on the LEAS to be associated with greater blood flow in a supra-callosal region of the anterior cingulate cortex (Broadman's area 24), for both film-elicited and recall-elicited emotion conditions.

Based upon the differences in regions of the anterior cingulate that were activated in different PET studies, Lane (2000) has speculated that the primary or phenomenal conscious experience of emotion may be dependent upon the dorsal anterior cingulate. In contrast, secondary or reflective emotional consciousness (involving the evaluation, representation, and reflection upon primary emotional experience) may depend more upon the rostral anterior cingulate/medial prefrontal cortex..

4.3 The Frontal Lobes and Conscious Emotion: An Integrative Speculation

In an attempt to integrate the various sources of evidence concerning the role of the frontal lobes in conscious emotional experience, the following speculative proposal is offered (modified from Kaszniak *et al.*, 1999): Following (Lane, 2000), it is hypothesized that the dorsal anterior cingulate gyrus serves as a working memory "convergence area" for bodily aspects of emotion (processed within the anterior insula) to gain access to primary or phenomenal conscious representation. The rostral anterior cingulate was suggested to serve a similar role in reflective emotional consciousness. In contrast, as suggested by Levine (1998), the orbital frontal areas may be involved (perhaps both consciously and nonconsciously) in limiting choices, based upon emotional factors, among courses of action. These choices themselves may be generated by dorsolateral frontal cortices (see Dehaene & Changeux, 1991). This model would be consistent with clinical descriptions (e.g., Damasio, 1994) of patients with orbital frontal damage who often develop numerous alternative strategies for solving problems but seem unable to decide between them and take appropriate action. The orbital frontal cortex, with its connections to the amygdala, also appears to be importantly involved in the inhibition of emotional impulses, given the marked disinhibition that is seen in patients with orbital frontal damage (e.g., Damasio, Grabowski, Frank, Galaburda, & Damasio, 1994). Connections between the orbital frontal cortex and the anterior cingulate may allow conscious representations of emotional bodily states (integrated with other conscious representations, such as plans and contextual judgments) to influence impulse control in the service of adaptive behavior regulation.

Testing of this speculative account will clearly require further research. Potentially informative experiments could examine (with observational, self-report, and psychophysiological assessments) persons with circumscribed damage to the relevant frontal lobe regions. The additional inclusion of experimental paradigms in which emotionally-salient stimuli would be prevented from being consciously perceived (e.g., through backward masking) might also be able to help in answering questions about the functional roles played by apparently dissociable conscious and nonconscious emotion components.

References

Adolphs, R., D. Tranel, A. Bechara, H. Damasio and A.R. Damasio (1996) "Neuropsychological approaches to reasoning and decision-making", in: *Neurobiology of Decision-Making,* A.R. Damasio *et al.*, eds, Berlin: Springer-Verlag, pp. 159-179.

Andreassi, J.L. (1995) *Psychophysiology: Human Behavior and Physiological Response*, Hillsdale, NJ: Lawrence Erlbaum Associates.

Bechara, A., H. Damasio and A.R. Damasio (2000) "Emotion, decision making and the Orbitofrontal cortex", *Cerebral Cortex* 10:295-307.

Brown, S.L. and G.E. Schwartz (1980) "Relationships between facial electromyography and subjective experience during affective imagery", *Biological Psychology* 11:49-62.

Camras, L.A., E.A. Holland and M.J. Patterson (1993) "Facial expression", in: *Handbook of Emotions*, M. Lewis and J.M. Haviland, eds, New York: Guilford Press, pp. 199-208.

Cannon, W.B. (1927) "The James-Lange theory of emotions: A critical examination and an alternative theory", *American Journal of Psychology* 39:106-124.

Cacioppo, J.T., R.E. Petty, M.E. Losch and H.S. Kim (1986) "Electromyographic activity over facial muscle regions can differentiate the valence and intensity of affective reactions", *Journal of Personality and Social Psychology* 50:260-268.

Center for the Study of Emotion and Attention [CSEA-NIMH] (1999) *The international affective picture system: Digitized photographs*, Gainesville, Florida: The Center for Research in Psychophysiology, University of Florida.

Churchland, P.S. (1998) "Feeling reasons", in: *On the Contrary: Critical essays, 1987-1997*, P.M. Churchland and P.S. Churchland, eds, Cambridge, MA: MIT Press, pp. 231-254.

Clore, G.L. (1994) "Why emotions are never unconscious", in: *The Nature of Emotion: Fundamental Questions*, P. Ekman and R.J. Davidson, eds, New York: Oxford University Press, pp. 285-290.

Clore, G.L. and A. Ortony (2000) "Cognition in emotion: Always, sometimes, or never?", in: *Cognitive Neuroscience of Emotion*, R.D. Lane, L. Nadel, G.L. Ahern, J.J.B. Allen, A.W. Kaszniak, S.Z. Rapcsak and G.E. Schwartz, eds, New York: Oxford University Press, pp. 24-61.

Cohen, J. and P. Cohen (1983) *Applied Multiple Regression/Correlation Analysis for the Behavioral Sciences (2nd ed.)*, Hillsdale, NJ: Lawrence Erlbaum Associates.

Cole, J. (1997) *About Face*, Cambridge, Massachusetts: MIT Press.

Csikszentmihalyi, M. (1990) *Flow: The Psychology of Optimal Experience*, New York: Harper Perennial.

Dalby, P.R., S.L. Reminger, A.W. Kaszniak and E.B. Montgomery (in preparation) "Expressive and subjective features of emotion in idiopathic Parkinson's disease: A test of the facial feedback hypothesis".

Damasio, A.R. (1994) *Descartes' Error: Emotion, Reason and the Human Brain*, New York: G.P. Putnam.

Damasio, A.R. (1999) *The Feeling of What Happens: Body and Emotion in the Making of Consciousness*, New York: Harcourt Brace and Company.

Damasio, H., T. Grabowski, R. Frank, A.M. Galaburda and A.R. Damasio (1994) "The return of Phineas Gage: clues about the brain from the skull of a famous patient", *Science* 264:1102-1105.

Dehaene, S. and J.-P. Changeux (1991) "The Wisconsin Card Sorting Test: Theoretical analysis and modeling in a neuronal network", *Cerebral Cortex* 1:62-79.

DeLancey, C. (1996) "Emotion and the function of consciousness", *Journal of Consciousness Studies* 3:492-499.

Devinsky, O., M.J. Morrell and B.A. Vogt (1995) "Contributions of anterior cingulate cortex to behavior", *Brain* 11: 279-306.

Dimberg, U. (1990) "Facial electromyography and emotional reactions", *Psychophysiology* 27:481-494.

Dimberg, U., M. Thunberg and K. Elmehed (2000) "Unconscious facial reactions to emotional facial expressions", *Psychological Science* 11:86-89.

Ekman, P., R.W. Levenson and W. V. Friesen (1983) "Autonomic nervous system activity distinguishes among emotions", *Science* 221:1208-1210.

Frijda, N.H. (1993) "The place of appraisal in emotion", *Cognition and Emotion* 7:357-387.

Goldblatt, D. and D. Williams (1986) "I an sniling: Möbius' syndrome inside and out", *Journal of Child Neurology* 1:71-78.

Goldman-Rakic, P.S. (1992) "Working memory and the mind", *Scientific American* 267 (3):111-117.

Griffiths, P.E. (1997) *What Emotions Really Are: The Problem of Psychological Categories*, Chicago: University of Chicago Press.

Gross, J.J. and R.W. Levenson (1993) "Emotional suppression: physiology, self-report, and expressive behavior", *Journal of Personality and Social Psychology* 64:970-986.

Gross, J.J. and R.W. Levenson (1997) "Hiding feelings: the acute effects of inhibiting negative and positive emotion", *Journal of Abnormal Psychology* 106:95-103.

Hunker, C.J., J.A. Abbs and S.M. Barlow (1982) "The relationship between Parkinson rigidity and hypokinesia in the orofacial system: A quantitative analysis", *Neurology* 32:749-754.

James, W. (1922) "What is an emotion?", in: *Psychology classics: A series of reprints and translations*, K. Dunlap, ed, Baltimore: Wilkins and Wilkins. (Reprinted from *Mind*, 1884, 188-205).

Kaszniak, A.W., S.L. Reminger, S.Z. Rapcsak and E.L. Glisky (1999) "Conscious experience and autonomic response to emotional stimuli following frontal lobe damage", in: *Toward a Science of Consciousness III: The Third Tucson Discussions and Debates*, S.R. Hameroff, A.W. Kaszniak and D.J. Chalmers, eds, Cambridge, Massachusetts: MIT Press, pp.201-213.

Katsikitis, M. and I. Pilowsky (1991) "A controlled quantitative study of facial expression in Parkinson's disease and depression", *Journal of Nervous and Mental Disease* 179:683-688.

Kentridge, R.W. and C.A. Heywood (1999) "The status of blindsight: Near-threshold vision, islands of cortex and the Riddoch phenomenon", *Journal of Consciousness Studies* 6:3-11.

Lane, R.D. (2000) "Neural correlates of conscious emotional experience", in: *Cognitive Neuroscience of Emotion*, R.D. Lane, L. Nadel, G.L. Ahern, J.J.B. Allen, A.W. Kaszniak, S.Z. Rapcsak and G.E. Schwartz, eds, New York: Oxford University Press, pp. 345-370.

Lane, R.D., G.L. Ahern, G.E. Schwartz and A.W. Kaszniak (1997) "Is alexithymia the emotional equivalent of blindsight?", *Biological Psychiatry* 42:834-844.

Lane, R.D., G.R. Fink, P.M.L. Chau and R.J. Dolan (1997) "Neural activation during selective attention to subjective emotional responses", *Neuroreport* 8:3969-3972.

Lane, R.D., L. Nadel, J.J.B. Allen and A.W. Kaszniak (2000) "The study of emotion from the perspective of cognitive neuroscience", in: *Cognitive Neuroscience of Emotion*, R.D. Lane, L. Nadel, G.L. Ahern, J.J.B. Allen,

A.W. Kaszniak, S.Z. Rapcsak and G.E. Schwartz, eds, New York: Oxford University Press, pp. 3-11.

Lane, R.D., D. Quinlan, G. Schwartz, P. Walker, and S. Zeitlin (1990) "The levels of emotional awareness scale: A cognitive-developmental measure of emotion", *Journal of Personality Assessment* 55:124-134.

Lane, R.D., E.M. Reiman, B. Alexelrod, L.-S. Yun, A. Holmes and G.E. Schwartz (1998) "Neural correlates of levels of emotional awareness: evidence of an interaction between emotion and attention in the anterior cingulate cortex", *Journal of Cognitive Neuroscience* 10:525-535.

Lane, R.D., L. Sechrest, R. Riedel, V. Weldon, A.W. Kaszniak and G.E Schwartz (1996) "Impaired verbal and nonverbal emotion recognition in alexithymia", *Psychosomatic Medicine* 58:203-210.

Lang, P.J., M.M. Bradley and B.N. Cuthbert (1999) *International affective picture system (IAPS): Technical manual and affective ratings*, Gainesville, FL: The Center for Research in Psychophysiology, University of Florida.

Lang, P.J. and M.K. Greenwald (1988) *The international affective picture system standardization procedure and initial group results for affective judgements: Technical report 1A*, Gainesville, Florida: The Center for Research in Psychophysiology, University of Florida.

Lang, P.J., M.K. Greenwald, M.M. Bradley and A.O. Hamm (1993) "Looking at pictures: Affective, facial, visceral, and behavioral reactions", *Psychophysiology* 30: 261-273.

Lazarus, R.S. (1991) *Emotion and Adaptation*, New York: Oxford University Press.

Levenson, R.W., P. Ekman and W.V. Friesen, (1990) "Voluntary facial action generates emotion-specific autonomic nervous system activity", *Psychophysiology* 27:363-384.

LeDoux, J. (1996) *The Emotional Brain: The Mysterious Underpinnings of Emotional Life*, New York: Simon and Schuster.

LeDoux, J. (2000) "Cognitive-emotional interactions: Listen to the brain", in: *Cognitive Neuroscience of Emotion*, R.D. Lane, L. Nadel, G.L. Ahern, J.J.B. Allen, A.W. Kaszniak, S.Z. Rapcsak and G.E. Schwartz, eds, New York: Oxford University Press, pp. 129-155.

Levine, D.S. (1998) "Emotion and consciousness: A shotgun marriage?", in: *Toward a Science of Consciousness II: The Second Tucson Discussions and Debates*, S.R. Hameroff, A.W. Kaszniak and A.C. Scott, eds, Cambridge, Massachusetts: MIT Press, pp. 513-520.

Matsumoto, D. and M. Lee (1993) "Consciousness, volition, and the neuropsychology of facial expressions of emotion", *Consciousness and Cognition* 2:237-254.

Nielsen, L. (1998) "Modeling creativity: Taking the evidence seriously", in: *Toward a Science of Consciousness II: The Second Tucson Discussions and*

Debates , S.R. Hameroff, A.W. Kaszniak, and A.C. Scott, eds, Cambridge, Massachusetts: MIT Press, pp. 717-724.

Öhman, A., A. Flykt and D. Lundqvist (2000) "Unconscious emotion: Evolutionary perspectives, psychophysiological data and neuropsychological mechanisms", in: *Cognitive Neuroscience of Emotion*, R.D. Lane, L. Nadel, G.L. Ahern, J.J.B. Allen, A.W. Kaszniak, S.Z. Rapcsak and G.E. Schwartz, eds, New York: Oxford University Press, pp. 296-327.

Öhman, A. and J.J.F. Soares (1994) "Unconscious anxiety: Phobic responses to masked stimuli", *Journal of Abnormal Psychology* 103:231-240.

Panksepp, J. (1998) *Affective Neuroscience*, New York: Oxford University Press.

Price, J.L., S.T. Carmichael and W.C. Drevets (1996) "Networks related to the orbital and medial prefrontal cortex: A substrate for emotional behavior?", *Progress in Brain Research* 107:523-536.

Russell, J.A. (1980) "A circumplex model of affect", *Journal of Personality and Social Psychology* 39:1161-1178.

Smith, M.C., M.K. Smith and H. Ellgring (1996) "Spontaneous and posed facial expression in Parkinson's disease", *Journal of the International Neuropsychological Society* 2:383-391.

Tassinari, L.G., and J. T. Cacioppo (1992) "Unobservable facial actions and emotion", *Psychological Science* 3:28-33.

Taylor, J.G. (1992) "Towards a neural network model of the mind", *Neural Network World* 6:797-812.

Tomkins, S.S. (1962) *Affect, Imagery, Consciousness: Vol. I. The Positive Affects.* New York: Springer.

Tomkins, S.S. (1963) *Affect, Imagery, Consciousness: Vol. II. The Negative Affects*, New York: Springer.

Tranel, D. (2000) "Electrodermal activity in cognitive neuroscience: Neuroanatomical and neuropsychological correlates", in: *Cognitive Neuroscience of Emotion*, R.D. Lane, L. Nadel, G.L. Ahern, J.J.B. Allen, A.W. Kaszniak, S.Z. Rapcsak and G.E. Schwartz, eds, New York: Oxford University Press, pp. 192-224.

Watt, D.F. (1999) "At the intersection of emotion and consciousness: Affective neuroscience and extended reticular thalamic activating system (ERTAS) theories of consciousness", in: *Toward a Science of Consciousness III: The Third Tucson Discussions and Debates*, S.R. Hameroff, A.W. Kaszniak and D.J. Chalmers, eds, Cambridge, Massachusetts: MIT Press, pp. 215-229.

Whalen, P.J., S.L. Rauch, N.L. Etcoff, S. McInerny, M.B. Lee and M.A. Jenike (1998) "Masked presentations of emotional facial expressions modulate amygdala activity without explicit knowledge", *Journal of Neuroscience* 18:411-418.

PHILOSOPHICAL PERSPECTIVES

INTRODUCTION: PHILOSOPHICAL PERSPECTIVES

ALFRED W. KASZNIAK
Center for Consciousness Studies, Departments of Psychology, Neurology & Psychiatry, University of Arizona, 1503 E. University, Tucson, Arizona 85721, U.S.A.

Since at least the time of Socrates, Western philosophy has been concerned with the question "What is an emotion?" However, as Solomon (2000) has pointed out, throughout most of the history of philosophical discussion, emotion has been seen as a slave to the master of reason:

> First and foremost, there is the inferior role of emotion - the idea that emotion is as such more primitive, less intelligent, more bestial, less dependable, and more dangerous than reason, and thus needs to be controlled by reason... Second, and more profoundly, there is the reason-emotion distinction itself - as if we were dealing with two different natural kinds, two conflicting and antagonistic aspects of the soul. Even those philosophers who sought to integrate them and reduce one to the other (typically reducing emotion to an inferior genus of reason, a "confused perception" or "distorted judgment") maintained the distinction and continued to insist on the superiority of reason. (Solomon, 2000, p. 3)

In recent decades, this view has been shifting, in part responsive to empirical observations concerning emotion-cognition interrelationships from the fields of basic neuroscience (LeDoux, 1996), behavioral neurology (Adolphs, Tranel, Bechara, Damasio, & Damasio, 1996; Damasio, 1994), cognitive neuroscience (Lane, Nadel, Allen, & Kaszniak, 2000), and experimental social psychology (Schwartz & Clore, 1996). There is currently a renewed philosophical interest in emotion reflected in books and papers that address (among other topics): (1) an understanding emotions as psychological categories (Griffiths, 1997); (2) the defining characteristics of typical emotions (Ben-Ze'ev, 1997); (3) the implications of an analysis of emotion qualia for functionalist accounts of consciousness (DeLancey, 1996); (4) the rationality of emotions (de Sousa, 1989; Greenspan, 1988); and, (5) the roles of emotion in reasoning and rational choice (Churchland, 1998; Elster, 1998, 1999; Stocker, 1996).

Contributors to this first section of the present volume provide current philosophical perspectives on these and other complex issues concerning emotion and its relation to other aspects of mind. The questions addressed by these authors are basic to any comprehensive consideration of emotion and consciousness: By what criteria should emotions be categorized and classified? What are the basic common characteristics of typical emotions? What are the experiential features that contribute to our concept of emotional intensity? What do thought-experiments, such as the inversion of emotional qualia, tell us about the validity of functionalist accounts of emotion? How might phenomenological analysis help to link biological processes and beliefs or other world-views in our understanding of emotion? What is the relationship between emotion, values, and cognitive revision? How can we best account for the intersubjective perception of emotion?

As the contributions of each of the authors in this section makes clear, careful philosophical analysis provides a necessary foundation for empirical research on emotion and consciousness Such analysis also provides a framework within which to integrate empirical observations and examine their implications for fundamental questions about human nature. As a "slave-master" (Solomon, 2000) metaphor for emotion and reason is gradually replaced by conceptualizations that recognize the subtlety, complexity, and interdependence of these aspects of mind, a more integrative (and hopefully accurate) view of ourselves may result.

References

Adolphs, R., D. Tranel, A. Bechara, H. Damasio and A.R. Damasio (1996), "Neuropsychological approaches to reasoning and decision-making", in: *Neurobiology of Decision-Making,* A.R. Damasio *et al.*, eds, Berlin: Springer-Verlag, pp. 159-179.

Ben-Ze'ev, A. (1997) "The affective realm", *New Ideas in Psychology* 15:247-259.

Churchland, P.S. (1998) "Feeling reasons", in: *On the contrary: Critical essays, 1987-1997*, P.M. Churchland and P.S. Churchland, eds, Cambridge, Massachusetts: MIT Press, pp. 231-254.

Damasio, A.R. (1994) *Descartes' Error: Emotion, Reason and the Human Brain*, New York: G.P. Putnam.

DeLancey, C. (1996) "Emotion and the function of consciousness", *Journal of Consciousness Studies* 3:492-499.

de Sousa, R. (1989) *The Rationality of Emotion*, Cambridge, Massachusetts: MIT Press.

Elster, J. (1998) *Alchemies of the Mind: Rationality and the Emotions*, Cambridge, England: Cambridge University Press.

Elster, J. (1999) *Strong Feelings: Emotion, Addiction, and Human Behavior*, Cambridge, Massachusetts: MIT Press.

Greenspan, P. (1988) *Emotions and Reasons*, New York: Routledge.

Griffiths, P.E. (1997) *What Emotions Really Are: The Problem of Psychological Categories*, Chicago: The University of Chicago Press.

Lane, R.D., L. Nadel, J.J.B. Allen and A.W. Kaszniak (2000) "The study of emotion from the perspective of cognitive neuroscience", in: *Cognitive Neuroscience of Emotion*, R.D. Lane, L. Nadel, G.L. Ahern, J.J.B. Allen, A.W. Kaszniak, S.Z. Rapcsak and G.E. Schwartz, eds, New York: Oxford University Press, pp. 3-11.

LeDoux, J. (1996) *The Emotional Brain: The Mysterious Underpinnings of Emotional Life*, New York: Simon and Schuster.

Schwartz, N. and G.L. Clore (1996) "Feelings and phenomenal experiences", in: *Social Psychology: Handbook of Basic Principles*, E.T. Higgins and A. Kruglanski, eds, New York: Guilford, pp. 433-465.

Solomon, R.C. (2000) "The philosophy of emotions," in: *Handbook of Emotions* (Second Edition), M. Lewis and J.M. Haviland-Jones, eds, New York: Guilford, pp. 3-15.

Stocker, M. (1996) *Valuing Emotions*, Cambridge, England: Cambridge University Press.

EMOTION AND THE PROBLEM OF PSYCHOLOGICAL CATEGORIES

PAUL E. GRIFFITHS
Unit for History and Philosophy of Science, University of Sydney, Sydney NSW 2006, Australia

ABSTRACT

Emotion theory is beset by category disputes. Examining the nature and function of scientific classification can make some of these more tractable. The aim of classification is to group particulars into «natural» classes - classes whose members share a rich cluster of properties in addition to those used to place them in the class. Classification is inextricably linked to theories of the causal processes that explain why certain particulars resemble one another and so are usefully regarded as «of the same kind». The need to base categories on underlying causal processes explains why mere careful definition (including operational definition) need not produce categories that are productive objects of scientific study. Because different causal processes produce different patterns of similarity there is unlikely to be a single classification that is optimal for addressing all scientific questions. Cultural categories should not be contrasted to natural categories, but should be treated as *natural* classes generated by underlying *social* processes. Our capacity to introduce epistemically optimal categories is often restricted because categories play a role in social and political, as well as epistemic, projects. This account of classification has many implications for emotion theory.

1. The Search for Natural Categories

Emotion theory is beset by category disputes. The category of «emotion» itself has contested boundaries, as can be seen in attempts to distinguish between mood and emotion (Ekman & Davidson, 1994) and in discussions of whether the startle response is an emotion (Ekman *et al.*, 1985). A still better example is the debate over «basic emotions» (Ekman, 1992; Ortony & Turner, 1990). The questions at issue here include: (a) whether «the same» emotion can be identified despite individual or cultural differences; (b) the relation between a set of labels for emotion and emotions themselves; (c) the direct question of how a taxonomy of emotions is to be justified and how rival taxonomies are to be compared. All these questions can be illuminated by looking at the nature and function of scientific classification.

It is a commonplace that the aims of science are prediction and explanation. The classical analysis of these two activities is the «hypothetico-deductive» (H-D) model (Hempel, 1966). In the H-D model, science inductively confirms a hypothesis by successfully making some of the observations that are logically deducible from the hypothesis. Once confirmed, the hypothesis becomes a law or theory that licenses further, inductive predictions. According to the H-D model, explanation is simply the inverse of prediction or induction. If an observation *could have been* predicted from a theory, then that theory explains the observation.

The H-D model of how scientific theories are established has few adherents today and the literature on explanation is a series of more or less radical reactions against the H-D model (for a survey, see Salmon *et al.*, 1992). Nevertheless, one core element of the H-D understanding of induction and explanation still seems valid. Induction and explanation presume that some of the correlations between properties which we observe are «projectable» (Goodman, 1954). That is, having observed these correlations in a certain number of instances we can «project» them to new instances and expect to find them in force there too. Scientific classifications of particulars into categories embody our current understanding of where such projectable clusters of properties are to be found. The species category, for instance, classifies particular organisms into classes that represent reliable clusters of morphological, physiological and behavioral properties. Hence the properties of the species as a whole can be discovered by studying a few members of the species. I have never seen Socks the White House cat, but no doubt he has at least partially retractable claws: cats are like that. Conversely, we can explain the fact that an individual has certain properties by citing its species. Gina has retractable claws because she is a cat: cats are like that. In Nelson Goodman's original presentation the projectability of categories was judged on the basis of past success with those categories or with other categories that are part of the same taxonomic system. Success in using a species category inclines us to rely on that category and also to rely on other species categories generated using the same principles. However, this does not do justice to some of the more indirect ways in which we come to rely on categories. For example, one taxonomy of organisms might be preferred to another because it excludes data from DNA regions which current theory suggests are likely to have undergone a great deal of recent change. Anything in our theories and background knowledge that suggests that certain correlations will hold up in new instances can serve to support a particular taxonomic scheme.

Philosophers have traditionally called the categories on which we can rely for induction and explanation «natural kinds». The concept dates back perhaps as far as Plato's famous injunction to «carve nature at its joints». But in current academic English the phrase «natural kind» is unfortunately ambiguous. In some of the social sciences, «natural kinds» are categories that are used in many different cultures, such as bird or tree. The philosophical meaning is quite different: here the term «natural» implies not that it is natural for people to use the category, but that the category itself is part of the structure of nature. The chemical elements are perhaps the least controversial example of «natural» categories in the philosopher's sense. The periodic table carves nature at its joints, whereas a table that groups chemicals by their colors in pure form at room temperature does not. The idea that some categories are part of the structure of nature can be made somewhat less metaphorical if we say that natural categories are those which are the subjects of «laws of nature». The laws of chemistry describe how the elements behave when they interact with one another. There are no such laws about how substances defined by their colors interact chemically (although there are laws about how colors themselves interact). Similarly, it is a sensible project to seek laws concerning the effect of psychoactive drugs on persons with a particular mental disorder. Seeking to determine the special effect of a drug on people I don't like or on people with a particular star sign would be merely frivolous.

However, the traditional idea of a law of nature is too restrictive. Laws of nature are supposed to be universal and deterministic. They are exceptionless generalizations, which apply throughout the universe. Hence if natural categories were the subjects of laws of nature, there would be very few natural categories in the biological and social sciences where generalizations are often exception-ridden and/or not applicable at every time and place. Fortunately, it is easy to generalize the idea of a «law of nature» to include all statements that are to some degree «law-like». Laws of nature have what is called «counterfactual force»: it is not only true that all As are Bs, but that anything else which *were* an A *would* also be a B. If I *were* made of copper then I *would* be highly conductive. Counterfactual force is what makes laws different from mere widespread coincidences. It is easy to see that counterfactual force comes in degrees. If I *were* a smoker then there *would be* a certain probability of my getting lung cancer. A true, general statement is «law-like» if it has some degree of counterfactual force. The generalizations of the biological and social sciences are law-like to various degrees. Any generalization that is a better predictor of phenomena than a suitably designed null hypothesis has some counterfactual force and hence is at least minimally law-like. Take, for example, the generalization that increasing the money supply at or near full employment is inflationary. It is not perfectly reliable here and now, and given a new economic order it may become even less reliable, but as things are it gives us a grip on the structure of nature that is not to be despised. This broader conception of a law-like generalization leads easily to a broader definition of a natural category. A category is (minimally) natural if it is possible to make better than chance predictions about the properties of its instances. This, of course, is a very weak condition. Very many ways of classifying the world are minimally natural. The aim is to find categories that are a great deal more than minimally natural. Stronger forms of naturalness grow naturally out of this minimal definition, as I show in Section Four.

2. Pattern and Process

The ideas outlined in the previous section suggest that classification is inextricably linked to theories of the causal processes that give rise to or influence the objects in the domain of study. It is these theories which make it rational to extrapolate from past observations of a category and which confer force on explanations stated using a category. Both procedures only make sense if the category is a projectable one and, as Richard Boyd has argued, we judge a category to be projectable or natural when we have theoretical grounds for supposing that there is a causal explanation for the property correlations we have observed (Boyd, 1991; Boyd, in press). Boyd calls this postulated explanation a «causal homeostatic mechanism», because it causally explains the maintenance (homeostasis) of the same property correlations throughout the set of instances of the category.

Older theories of scientific classification emphasized one particular type of causal homeostatic mechanism - the existence of a shared microstructure in each instance of a natural category. The influential philosophical literature on «natural kinds» in the 1 970s argued that natural categories consisted of particulars which shared a microstructural essence and that the role of science was to uncover this essence (Kripke, 1980; Putnam, 1975). In the paradigm example of chemical elements, the causal homeostatic mechanism is indeed such a shared microstructure. It is because of their sub-atomic composition that the instances of a chemical element share their chemical properties. However, nothing in the idea of a causal homeostatic mechanism requires the mechanism to take the form of a set of intrinsic properties possessed by every member of the category and synchronically causally producing the properties characteristic of the category. This is fortunate, because the consensus in biology since the 1 940s has been that species, the other traditional paradigm of natural categories, do not have a genetic essence shared by all members of the species and differentiating them from other species (Hull, 1965; Mayr, 1976/1959; Sober, 1980). Instead, species are clusters of genetic variation. Organisms resemble one another not because of something inside each of them, but because of something outside each of them: the genealogical and ecological factors that make these organisms a population or a group of related populations. It is these factors that act as a causal homeostatic mechanism making species into useful categories for biological science (Griffiths, in press; Wilson, in press).

Despite the longstanding consensus in biology against «Aristotelian essentialism», there is a hankering for natural categories defined by shared microstructure. Some biologists have argued that biology cannot really be a science until it possesses a classification of organisms and their parts in terms of microstructure rather than a Darwinian definition in terms of common ancestry (Goodwin & Saunders, 1989; Goodwin & Webster, 1996). I have criticized this view elsewhere (Griffiths, 1996; Griffiths, in press). It draws much of its plausibility from the view that sciences without exceptionless, universal laws are inadequate and awaiting replacement by or reduction to some science that more closely approaches the traditional ideal. I believe it is deeply mistaken to cling to this view when well-confirmed theories about the domains of the «special sciences» (those whose domain of investigation is more limited than physics or chemistry) explain both why such exceptionless laws are not forthcoming and why various generalizations which are exception-ridden or of limited application are robust and reliable.

The same hankering after micro-structural essences may have an influence in emotion theory. Jaak Panksepp, for example, argues that our emotion categories will not be well-founded until they are based on the underlying mechanisms in the brain or on the genes involved in producing those mechanisms (Panksepp, 1992; Panksepp, 1994). But there is good reason to think that the relationship between genes and neural mechanisms and the relationship between neural mechanisms and behavior are not such as to facilitate such definitions. The best understood experimental organism with a nervous system is the nematode worm *C. elegans.* In a recent review of this organism Kenneth C. Schaffner suggests that the relationships between genes and neural structure and between neural structure and behavior are typically many-many and dependent on other, environmental parameters (Schaffner, in press). This is not a denial of the evolved, species-typical or otherwise «biological» nature of the behavior of *C. elegans.* The point, rather, is that the behavior of the organism is an object of study in its own right. Several developmental biologists and psychologists have suggested a framework for such studies and have gone some way towards implementing this framework (Elman *et al.*, 1996; Thelen, 1995; Thelen & Smith, 1994). The organism's genes are conceived as the parameter settings of a complex system, along with many other parameters representing all the causal inputs required for normal development. The stable outcomes of development are attractors of various sorts that emerge during a typical «run» of this complex system. The species-typical cognition and behavior of the adult organism are conceived in the same way, as attractors for the organism-environment system. These ideas have obvious application to emotion. Neil MacNaughton, for example, cites the finding that separation distress in rats pups, an obvious natural category of rat behavior, is only a reliable category because loss of the mother simultaneously deprives the offspring of milk and of warmth, setting in motion two separate physiological response systems (Hofer, 1972). MacNaughton argues that this provides a model for emotions, which may also reflect the typical interaction of a species with features of its environment (MacNaughton, 1989). The idea that species-typical behaviors can depend on the situated activity of the organism in its natural environment goes back to Konrad Lorenz and has led to a revival of interest in his work amongst students of «situated robotics» (Hendriks-Jansen, 1996). There are also resonances between these ideas and Klaus R. Scherer's concept of «modal emotions» - collections of universal emotional response elements which frequently occur together to give rise to recognizable overall emotional response types (Scherer, 1994).

If any of these ideas, originating in several independent traditions of research, are correct, then the natural categories of emotion and emotional behavior may be very far from having an internal micro-structural essence. Instead, they may be emergent properties of an organism and its operating environment. These categories would be none the less scientifically productive because of this.

3. Comparing Classifications

The idea that classification is a search for inductive and explanatory power and the consequent need to base categories on underlying causal processes explains why careful definition (including operational definition) need not produce categories that are productive objects of scientific study. Only natural categories are of the same kind» in respects other than those used to define them. Operational definition without reference to a theory of what causes similarity in the domain of study can only ever be the first step of a «bootstrapping» approach to locating natural categories. For example, a biological systematist might make use of easily measurable morphological or molecular properties to construct a cladogram of a group of species about which she knows nothing. The sub-set of these characters which fit together to give a coherent model of the evolutionary relationships between the species will be retained and the others discarded. The coherence between some of the initially arbitrary characters suggests that they may represent meaningful ways to atomize phenotypes in that group of species. Meaningless characters, such as the ratio of leg length to number of retinal receptors, would probably not be distributed across a group of species in the same way as meaningful developmental units (such as leg length or number of retinal receptors).

But agreeing that classification is a search for natural categories does not get us very far in itself. The criteria of minimal naturalness defined in section one is a very low hurdle and easily cleared. Almost all category disputes in science involve choice between competing natural classifications. These competing classifications must be compared along various dimensions relevant to the underlying goal of increasing inductive and explanatory power. The value of a law-like generalization can vary along two independent dimensions, which I call «scope» and «force». Force is a measure of the reliability of projections made using that generalization. Chemical laws have more force than economic generalizations. Scope is a measure of the size of the domain over which the generalization is applicable. Laws about isotopes have less scope than laws about elements. A theoretical category about which there are generalizations of considerable scope and force is, all other things being equal, more natural than one about which generalizations tend to have more restricted scope and lesser force. There will not always be a clear winner when we compare two sets of theoretical categories on the basis of scope and force. Scope and force may trade off against one another. Also, the scope of generalizations made with one set of categories may overlap rather than include the scope of generalizations made with the other taxonomy, so that neither taxonomy can be discarded without loss of understanding. Theoretical categories, as opposed to generalizations about a category, can also differ in «richness», the range of properties of the particulars in a category which are projectable in that category. Chemical isotopes are richer than chemical elements, and species richer than genera. As these examples make clear, the richness of a category will typically trade off against the scope of the generalizations we can make using the category. Although you can project more of the properties of species than of genera, claims about genera are applicable to a wider domain. Finally, theoretical categories are tied up in wider research programs whose relative prospects may cause us to prefer a set of categories despite deficiencies of scope, force and richness. The promise of «numerical taxonomy» in the 1 960s was not that it would outperform traditional, intuitive taxonomy, at least not in the short term, but that it would produce quantitative, objectively comparable results and hence turn taxonomy from an art into a science (Hull, 1988).

Although different classifications can be compared in these various respects, there is unlikely to be a single classification that is optimal for investigating all sorts of properties. The current received view in philosophy of science is that the dynamics of physical systems can only be adequately captured using a hierarchy of theoretical vocabularies, each irreducible to those below. Irreducibility is guaranteed by the fact that descriptions in one vocabulary can be made true by indefinitely many arrangements of the structures described in lower level vocabularies (Fodor, 1974; Jackson & Pettit, 1988; Lycan, 1990; Wimsatt, 1976a; Wimsatt, 1976b). Hence, in order to investigate the biological or economic properties of humans we need to classify them either as members of *Homo sapiens* or as members of various socio-economic groups. Characterizing them in quantum mechanical terms will not help, despite the impressive scope and force of quantum mechanical generalizations.

Comparing classifications is clearly a difficult and inconclusive business. Even with respect to a single, well-defined scientific project there are several desiderata for a set of categories and these are frequently incommensurable. But these potential difficulties are not always actual difficulties. In *What Emotions Really Are* (Griffiths, 1997), I made a case for the superiority of a classification of evolved emotional states in terms of cladistic homology to a classification in terms of adaptive function. A cladistic homologue contains all and only the traits that are copies by descent of a single ancestral trait. For example, the vertebrate limb is a homology. The legs of cats or of crocodiles, the legs and arms of humans and the legs and wings of birds or bats are all copies of the limbs of a common tetrapod ancestor. Fear in vertebrates is probably also homologous. A classification in terms of adaptive function is quite different. It groups together all and only those traits that are adaptations for a single ecological role. Thus, the wings of birds are classified with those of insects, and not with human arms. Fear in squid is classified with fear in vertebrates, despite their having evolved separately (since the last common ancestor had no nervous system). I argued that the scientific project of emotion theory is centrally concerned with understanding the mechanisms that produce emotional responses (where mechanism is understood so as to encompass the interaction of the organism with its natural environment described above, as well as internal, neural mechanisms). Hence it is with respect to these properties that we seek inductive and explanatory power. With that goal in mind I pointed out that that classifications in terms of adaptive function put together all the items produced by convergent evolution, and that it is a biological truism that resemblances due to convergence are «shallow». That is, traits that resemble one another by convergent evolution resemble one another only in their surface performance, not in the mechanism by which that performance is produced. Conversely, homologous traits resemble one another in their underlying mechanisms even when they have been used in more recent evolution for two different adaptive functions (e.g., the skeletal structure of the human arm and the bird's wing). Hence, homologies are richer than traits defined by adaptive function with respect to the properties we want to investigate. Since there are no opposing considerations to do with the scope and force of generalizations (rather the opposite) or the future promise of the different research programs (again, rather the opposite) we do not have to confront the problem of weighing up incommensurable considerations.

4. Natural Categories and Human Kinds

Natural categories are those where similarities between instances of the category are neither coincidental nor guaranteed by definition but supported (we think) by some causal process which produces similarity both in the respects we have already noticed and in unanticipated respects which we hope to discover. The essential property that makes something an instance of that category is its relation to that causal process - the causal homeostatic mechanism of the category. One exciting implication of this approach is that it breaks down the traditional distinction between natural categories and «socially constructed» categories - those generated by human agency. Categories of tool or ceremony can be the subject of law-like generalizations because the sociological processes that produce them can function as causal homeostatic mechanisms. These sociological processes guarantee with some degree of reliability in some suitably delimited domain that instances of the category will share a cluster of properties. Money, for example, has no microstructural essence, although it is a key node in many economic theories. The law-like generalizations about money, such as those connecting money supply to inflation or to interest rates, hold true in an economy because of a social convention treating some class of objects as a means of exchange and because agents in that economy try to maximize their utility. Neither of these circumstances is linked to any intrinsic property of the currency units.

Ian Hacking has argued against treating what he calls «human kinds» in a manner akin to natural categories because they have some highly distinctive properties (Hacking, 1991a; Hacking, 1991b). In particular, the existence of these categories can depend upon the practice of categorizing things in this way (Hacking calls this «the looping effect of human kinds»). The fact that people think certain things form a kind can function as the causal mechanism that causes instances of that kind to resemble one another. Hacking's best developed example of the looping effect is his analysis of contemporary multiple personality disorder (Hacking, 1995). He argues that both the contemporary disorder and the nineteenth century, mainly French, disorder of double personality are distinctive categories of mental disorder, but ones whose distinctive character is only guaranteed by the diagnostic scheme adopted by medical practitioners in certain countries and periods. The human beings classified by this diagnostic scheme are also shaped to fit it. I would add to Hacking's analysis by pointing out that because of the complexity of the systems involved, victims of these disorders will have many similarities that are not anticipated in the theory whose acceptance creates and sustains the disorder. The disorders can therefore be investigated and new knowledge about them generated. Hence the new category, created by our classificatory practices, is minimally natural. The distinctive «looping effect» of human kinds is likely to exist for some emotion categories. Culturally distinctive forms of emotion may well depend upon the exposure of developing individuals to cultural models of emotion. The mechanisms that sustain the distinctive Japanese experience of amae or the hypercognition of love in western cultures (Heelas, 1984/1986) may involve a «looping effect» of the practices of emotion classification in those cultures. The Duc de la Rochefoucauld was only exaggerating a little when he remarked that «No-one would love who had not read of love» (La Rochefoucauld, 1666/1959).

I do not agree with Hacking that the distinctive nature of human kinds is a reason to clearly distinguish them from other natural categories. First, I do not see that they are more distinctive than, say the Darwinian categories of contemporary biology. Second, I think it very plausible that there is a continuum of cases from characteristics that look like traditional «biological» traits to those that are paradigms of «social construction». Culture is a central feature of human biology, and the attempt to strip away culture to reveal the «biological» aspects of humans is simply incoherent. It is like seeking to investigate the true nature of an ant by removing the distorting influence of the nest! There was never a «naked ape» because humans have had a culture since before they were human (Griffiths & Stotz, in press; Ingold, 1995). Human beings and their cultures have co-evolved as surely as ants and hives or dogs and packs. Our «biological» nature is the product of a developmental matrix, which includes a great deal of cultural «scaffolding» that shapes how we grow. Thus I expect that emotional development in humans will depend upon a wide range of cultural resources. In some cases the emotional traits that develop will be pan-cultural and apparently «biological», despite their dependence on «cultural» developmental resources, just as normal psychosexual development in the rhesus macaque depends upon social interactions in infancy. In other cases emotional traits will differ between human lineages because of the stable inheritance of different cultural developmental resources, just as colony structure differs between different lineages in cases of «cultural inheritance» in ants (Keller & Ross, 1993). As the examples I have chosen indicate, I do not think that there is an interesting distinction between «biological» and «cultural» traits, or between biological and cultural aspects of human (or animal) psychology. The real distinctions concern the many different causal process that give rise to natural categories of developmental outcome, not some dichotomy between nature and culture.

Hacking's work on the distinctive nature of human kinds does raise one very real difficulty for the account of scientific classification I have given here. As well as arguing for out the «looping effect» of human classificatory practices, Hacking has pointed out that categories, including scientific categories, serve other human goals besides maximizing explanatory and inductive power. Categories, as represented by our concepts of things, also serve social and political ends. They are used to condemn, to promote attention to one aspect of a situation rather than another, and to induce conformity with certain norms of behavior. The use of different concepts promotes different agendas. Concepts may be contested on political grounds, not because of different views about the causal basis of the categories they represent. This seems particularly likely to occur in the case of emotion categories, which are frequently used to embody our ideals for human life. Introducing the concept of love may have served to create more humane relations between the sexes in medieval society (Hunt, 1959), but it need not have promoted self-understanding by the medieval mind. In fact, it may have achieved its purpose precisely by propagating to medieval people a fiction about how the mind works.

5. Implications for Emotion Theory

Almost every aspect of the account of classification given here has implications for emotion theory. I have argued that a scientifically productive scheme of classification should group particulars on the basis of postulated causal processes: The processes which current theory suggests cause the similarities we see between these particulars and which should cause other similarities of which we are currently unaware. This implies that a taxonomy of emotion is not something to be decided at the beginning of the science of emotion, but must develop in tandem with theories of emotion. Adopting a scheme of classification will influence the kind of research we do, but the outcomes of this research may cause us to revise or even abandon the scheme of classification.

I have argued that in the biological sciences natural categories need not be based on shared genetics or other intrinsic «essences. There are two distinctively Darwinian classification schemes that rely on extrinsic mechanisms to produce similarities in natural categories. These are classification by common descent (homology) and classification by shared adaptive function (analogy). Both of these represent possible approaches to classifying emotion. So the validity of an emotion category need not depend on any simple relationship between the emotion category and categories in genetics. In the case of classifications by homology, we have known since the 1940s that the apparent phenotypic uniformity in natural populations masks extensive genetic diversity. Important developmental outcomes are buffered against genetic variation as much as they are based on genetics. In the case of classifications by adaptive function, traits from quite diverse genetic lineages may resemble one another through selection for the same functional role. The validity of an emotion category defined using one of these Darwinian classification schemes need not depend on any simple relationship with neurobiology either. First, as George V. Lauder has been at pains to point out, function can be preserved in evolution while structure changes (Lauder, 1986; 1990; 1995). Second, as argued above, species typical patterns in behavior can emerge from organism environment interaction patterns, rather than being specified in any substantial way in the organisms neural structure. Such species typical results of interaction with the environment are legitimate objects of study in their own right.

Finally, I have briefly sketched some ways in which existing «folk» classifications of psychological traits might interact with an emerging scientific taxonomy of emotions. I have described how a folk classification scheme may «make itself come true» by shaping human being to fit their own cultural models of emotion. However, this simple outcome is only one possibility. Cultural models of emotion can play a critical role in shaping the developing psychological phenotype while not producing a result that they themselves accurately describe. Traditional patriarchal models of gender roles no doubt had a major influence on how human beings developed, but not one so simple as to create embodiments of these ideals of masculinity and femininity!

References

Boyd, R. (1991) "Realism, anti-foundationalism and the enthusiasm for natural kinds", *Philosophical Studies* 61:127-148.

Boyd, R. (In Press) "Homeostasis, species and higher taxa", in: *Species: New Interdisciplinary Essays*, R.A. Wilson, ed., Cambridge, M.A: MIT Press

Ekman, P. (1992) "Are there basic emotions?", *Psychological Review* 99:550-553.

Ekman, P. and R.J. Davidson, eds. (1994) *The Nature of Emotion: Fundamental Questions*, New York, Oxford: Oxford University Press.

Ekman, P., W.V. Friesen and R.C. Simons (1985) "Is the startle reaction an emotion?", *Journal of Personality and Social Psychology* 49:1416-1426.

Ekman, J.L., B.A. Bates, M.A. Johnson, A. Karmiloff-Smith, D. Parisi and K. Plunkett (1996) *Rethinking Innateness: A Connectionist Perspective on Development*, Cambridge, M.A: MIT Press.

Fodor, J.A. (1974) "Special sciences", *Synthese* 28:77-115.

Goodman, N. (1954) *Fact, Fiction and Forecast*, London: Athlone Press, University of London.

Goodwin, B.C. and P. Saunders (1989) *Theoretical Biology: Epigenetic and Evolutionary Order from Complex Systems*, Edinburgh: Edinburgh University Press.

Goodwin, B.C. and G. Webster (1996) *Form and Transformation: Generative and Relational Principles in Biology*, Cambridge: Cambridge University Press.

Griffiths, P.B. (1996) "Darwinism, Process Structuralism and Natural Kinds", *Philosophy of Science* 63:S1-S9.

Griffiths, P.B. (1997) *What Emotions Really Are: The Problem of Psychological Categories*, Chicago: University of Chicago Press.

Griffiths, P.B. (In Press) "Squaring the circle: natural kinds with historical essences", in: *Species: New Interdisciplinary Essays*, R.A. Wilson, ed., Cambridge, MA: MIT Press.

Griffiths, P. B. and K. Stotz, (In Press) "How the Mind Grows: A Developmental Perspective on the Biology of Cognition", *Synthese*.

Hacking, I. (1991a) "A tradition of natural kinds", *Philosophical Studies* 61:109-126.

Hacking, I. (1991b) "On Boyd", *Philosophical Studies* 6:149-154.

Hacking, I. (1995) *Rewriting the Soul: Multiple Personality and the Sciences of Memory*, Princeton, N.J.: Princeton University Press.

Heelas, P. (1984/1986) "Emotions across cultures", in: *The Social Construction of Emotion*, R. Harré, ed., Oxford: Oxford University Press, pp. 21-42.

Hempel, C.G. (1966) *Philosophy of Natural Science*, Englewood Cliffs, N.J: Prentice-Hall.

Hendriks-Jansen, H. (1996) *Catching Ourselves in the Act: Situated Activity, Interactive Emergence, Evolution and Human Thought*, Cambridge, Mass: MIT Press.

Hofer, M.A. (1972) "Physiological and behavioral processes in early maternal deprivation", in: *Physiology, Emotion and Psychosomatic Illnesses: CIBA Symposium 8*, R. Porter and J. Knight, eds, Amsterdam: Elsevier, pp.199-219.

Hull, D. (1988) *Science as a Process*, Chicago: University of Chicago Press.

Hull, D.L. (1965) "The effect of essentialism on taxonomy: two thousand years of stasis. Parts I & II.", *British Journal for Philosophy of Science* 15 & 16:314-326, 1-18.

Hunt, M. (1959) *The Natural History of Love*, NY: Alfred A. Knopf.

Ingold, T. (1995) "People like us: The concept of the anatomically modern human", *Cultural Dynamics* 7:187-214.

Jackson, F.C. and P. Pettit (1988) "Functionalism & broad content", *Mind* 97:381-400.

Keller, L. and K.G. Ross (1993) "Phenotypic plasticity and «cultural transmission» of alternative social organizations in the fire ant *Solenopsis invicta*", *Behavioral Ecology and Sociobiology* 33:121-9.

Kripke, S. (1980) *Naming & Necessity*, Cambridge, Mass.: Harvard University Press.

La Rochefoucauld, F., Duc de (1666/1959) *Maxims*, Harmondsworth: Penguin.

Lauder, G.V. (1986) "Homology, analogy, and the evolution of behavior", in: *Evolution of Animal Behavior: Paleontological and Field Approaches*, M.H. Nitecki and J. A. Kitchell, eds, New York: Oxford University Press, pp.9-40.

Lauder, G.V. (1990) "Functional morphology: Studying functional patterns in an historical context", *Annual Review of Ecology and Systematics* 21:317-340.

Lauder, G.V. (1995) "On the inference of function from structure", in: *Functional Morphology in Vertebrate Paleontology*, J.J. Thomason, ed., Cambridge: Cambridge University Press, pp.1-18.

Lycan, W.G. (1990) "The continuity of levels of nature", in: *Mind & Cognition: A Reader*, W.G. Lycan, ed., Oxford: Blackwells, pp.77-97.

MacNaughton, N. (1989) *Biology & Emotion*, Cambridge: Cambridge University Press.

Mayr, B. (1976/1959) "Typological versus populational thinking", in: *Evolution and the Diversity of Life*, Cambridge, Mass.: Harvard University Press, pp.26-29.

Ortony, A. and T.J. Turner (1990) "What's basic about basic emotions?", *Psychological Review* 97:315-331.

Panksepp, J. (1992) "A critical role for "affective neuroscience» in resolving what is basic about basic emotions", *Psychological Review* 99:554-560.

Panksepp, J. (1994) "The basics of basic emotion", in: *The Nature of Emotions: Fundamental Questions*, P. Ekman and R.J. Davidson, eds, New York: Oxford University Press, pp. 20-24.

Putnam, H. (1975) "The meaning of «meaning", in: *Mind, Language & Reality. Philosophical Papers, Vol. 2*, Cambridge: Cambridge University Press, pp. 251-271.

Salmon, M.H., J. Earman, C. Glymour, J.G. Lennox, P. Machamer, J.B. McGuire, J.D. Norton, W.C. Salmon and K.F. Schaffner (1992) *Introduction to the Philosophy of Science*, Englewood Cliffs, NJ: Prentice Hall.

Schaffner, K. (In Press) "Genes, Behavior and Developmental Emergentism: One Process, Indivisible?", *Philosophy of Science*.

Scherer, K.R (1994) "Toward a Concept of 'Modal Emotions'", in: *The Nature of Emotions: Fundamental Questions*, P. Ekman and R.J. Davidson, eds, New York, Oxford: Oxford University Press, pp.25-31.

Sober, B. (1980) "Evolution, population thinking and essentialism", *Philosophy of Science* 47:350-83.

Thelen, B. (1995) "Time-scale dynamics and the development of an embodied cognition", in: *Mind as Motion: Explorations in the Dynamics of Cognition*, R. F. Port, and T. van Gelder, eds, Cambridge, M.A: MIT Press, pp. 68-100.

Thelen, B. and L. Smith (1994) *A Dynamic Systems Approach to the Development of Cognition and Action*, Cambridge, M.A: MIT Press.

Wilson, R.A. (In Press) "Realism, essence and kind: resuscitating species essentialism" in: *Species: New Interdisciplinary Essays*, R.A. Wilson, ed., Cambridge, MA.: MIT Press.

Wimsatt, W. C. (1976a) "Complexity and organization", in: *Boston Studies in Philosophy of Science Vol. XXVII Topics in Philosophy of Biology*, M. Grene and G. Mendelsohn, eds, Reidel: Dordrecht, pp. 175-189.

Wimsatt, W.C. (1976b) "Reductionism, levels of organization and the mind/body problem", in: *Consciousness and the Brain*, G. Globus, G. Maxwell and I. Savodnik, eds, New York: Plenum Press, pp. 205-267.

THE NATURE OF TYPICAL EMOTIONS

AARON BEN-ZE'EV
Center for Inter-Disciplinary Research on Emotions, University of Haifa, Haifa 31905, Israel

ABSTRACT
Emotions are highly complex and subtle phenomena whose explanation requires careful and systematic analysis of their multiple characteristics and components. I suggest that the typical cause of emotions is a perceived significant change in our situation, the typical emotional concern is a comparative concern, and the typical emotional object is a human being. Typical emotions are considered to have a few basic characteristics - instability, great intensity, a partial perspective, and relative brevity - and basic components: cognition, evaluation, motivation, and feeling. These characteristics provide an initial answer to the classical question of "What is an Emotion?"

1. The Typical Cause, Focus and Object of Emotions

1.1 The Typical Cause of Emotion: A Perceived Significant Change

Emotions typically occur when *we perceive highly significant changes in our personal situation* - or in that of those related to us. A significant change is that which significantly interrupts or improves a smoothly flowing situation relevant to our concerns. Like burglar alarms going off when an intruder appears, emotions signal that something needs attention. When no attention is needed, the signaling system can be switched off. We respond to the unusual by paying attention to it. Emotions are generated when we deviate from the level of stimulation we have experienced for long enough to get accustomed to it (Frijda, 1988; Oatley, 1992; Ben-Ze'ev, 1996, 2000)

The importance of personal changes in generating emotions is evident from many everyday phenomena as well as scientific findings. People are very excited when facing changes in their lives: birth of a child; marriage; divorce; entering school for the first time; going to an interview which can significantly alter the course of one's life, and so on.

Spinoza most strongly emphasizes the importance of changes in our situation for the generation of emotions. He claims each individual strives to maintain its existence. When we undergo great change, we pass to a greater or lesser perfection, and these changes are expressed in emotions. As we change for the better we are happy and for the worse unhappy (Spinoza, 1677/1985).

The *evolutionary rational* for the important role that changes play in emotions is similar: for survival purposes it is crucial that the organism will pay special attention to significant changes which may increase or decrease survival chances. Being emotional, which is the opposite of being indifferent, forces the organism to pay such special attention. Responding primarily to changes is a highly economical and efficient way of using limited resources. From an evolutionary point of view, it is advantageous for us to focus our attention on changes rather than on stationary stimuli. Changes indicate that our situation is unstable, and awareness of this is important for survival. When we are already familiar with certain items, their mere repetition yields no new information and we can ignore them. Indeed, information theory measures the amount of information content by the extent of change which is brought about by a given operation. A change includes more information than repetition and as such is more exciting. Indeed, repetition reduces excitement and may have a relaxing function. In the case of mere repetition, no new activity is required, thereby resulting in an absence of consciousness. This is what people mean when they refer to a state of being "on automatic pilot".

Not only emotions, but consciousness in general, is strongly activated when the organism is confronted with changes. This is true, for example, of sensory sensitivity. Thus, if we were to suffer all our life from a toothache in a way that no change in our environment could alter the ache, then we would be unaware of it so that, in effect, we would have no pain. Without enough variety, the pleasure system tends to become satiated and our awareness decreases accordingly. We get bored when doing the same thing over and over, even if that activity was initially pleasant. Perceptual awareness is also connected with changes. Under normal conditions, we are unaware of air pressure even though it affects us constantly. We only perceive it when the level of air pressure changes, as when we take off or land in an airplane (Ben-Ze'ev, 1993; Lewis, 1929; Nussbaum, in press).

The importance of changes to consciousness in general and emotions in particular may be connected to our learning system, which must have a protective schema to prevent it from becoming trapped into endlessly repeating the same activity.

An important difference between the changes associated with consciousness in general and those associated with emotions is that emotional changes are of *highly personal significance*. Our attention may be directed to any type of change, but in order for the change to generate emotions, it must be perceived as having significant implications for us or those related to us.

The change relevant to the generation of emotions is a *perceived change whose significance is determined by us.* A significant emotional change may involve perception of changes that have actually taken place or imagined changes.

In addition to the specific changes which generate everyday emotions, our affective reactions are related to a more profound type of change connected with our contingent existence. Our possible death is always in the background of our existence: it reminds us of our profound vulnerability. This type of change expresses *our profound vulnerability* and dependence on external factors which we do not control.[4] Certain affective disorders, in particular anxiety and depression, are often related to such existential issues. Emotions themselves are typically concerned with more specific issues; the profound existential issues function as an important background framework influencing our specific emotional reactions. These differences are expressed, for instance, in the difference between the emotion of fear and the more general affective attitude of anguish.

Emotions may be viewed not merely as an expression of our profound vulnerability, but also as *a way to cope* with it. By attaching significance to specific, local changes in our current situation, we ignore in a way, the more profound type of change underlying our vulnerability; this is a type of self-deception. A certain measure of such self-deception is highly advantageous from an evolutionary point of view, as it enables us to protect our positive self-image and mobilize the required resources for facing daily changes. We deal with such changes as if our profound vulnerability is insignificant. This may seemingly reduce our vulnerability, but it does not significantly change it. The ninety-year-old-woman, who is enthusiastically studying for her graduate degree in history, is enriching her life in a way that seems to reduce the vulnerability of her age, but her basic vulnerability, expressed in her near death, remains unchanged. She is studying as if her near death is a factor which should hardly be considered. Indeed, the fact that in the long-run all of us will die does not imply that we should attach in the short-run no significance to specific changes.

1.2 The Typical Emotional Concern: A Comparative Personal Concern

Emotions occur when a change is appraised as relevant to our personal concerns. Concerns are our short- or long-term dispositions to prefer particular states of the world or of the self. Emotions serve to monitor and safeguard our personal concerns; they give the eliciting event its significance. Emotional meaning is mainly comparative.

Significance, or meaning, is by nature relational; it presupposes *order and relations*. To have meaningful information about something is to apprehend some relations in which it can be found. Meaning is closely analogous to a point in space: the meaning of a point is constituted by its relation to other points; its very essence is relational. The set of relations in which something stands constitutes its meaning. Attributing meaning is the setting of bounds and establishing of connections; what does not affect relation has no handle by which the mind can take hold of it (Lewis, 1929; Ben-Ze'ev, 1993).

The relational nature of meaning implies its *comparative nature* as well. Being in a certain relation means not being in a different relation. Understanding something implies grasping, to a certain degree, its alternative. We can understand what love is only if love can be compared to different states. Being in the water all the time, makes it hard for the fish to grasp the meaning of water.

Various philosophers have indicated the connection between reasoning and comparison. I would like to argue that not only reasoning, but emotions as well, are comparative by nature. The emotional environment contains not only what is, and what will be, experienced but also all that could be, or that one desires to be, experienced; for the emotional system, all such possibilities are posited as simultaneously there and are compared to each other.

The importance of the comparative concern in emotions is also connected with the central role of changes in generating emotions. An event can be perceived as a significant change only when *compared to a certain background framework.* If emotions occur when we confront a significant change in our situation, our concern is mainly comparative, referring to a situation different from the novel one.

The background framework against which emotional events are compared may be described as a personal baseline. The personal baseline, which actually expresses our values and attitudes, depends on many biological, social, personal, and contextual features; it is not a rigid entity, but a flexible framework enabling us to match it with our experiences. Such flexibility, however, is limited since our ability to change our values and attitudes is limited. The possibility of varying baselines is one reason why the same event occurring at different times may be associated with different emotional reactions.

The personal baseline determines the way in which we perceive our current, previous, ideal, and "ought" states, as well as these states in other people. We can compare our current, novel situation to a different one of ours or to that of significant others, such as parents, siblings, spouse, friends, or outstanding figures. The different state of ours can be a previous actual state, an ideal state in which we desire to be, or a state in which others think we ought to be. The different state of others can also be an actual state, an ideal state, or a state in which we think they ought to be. Emotions emerge whenever a significant discrepancy between our current personal state, or that of significant others, has changed (Festinger, 1954; Higgins, 1987).

The importance of the comparative concern is illustrated by the story of the man who was upset because he had no shoes - until he met a man with no feet. Satisfaction and happiness depends on comparative measures related, among other things, to our expectations and the fortune of relevant others. Thus, someone who receives a 5% raise might be happier than someone who receives an 8% increase if the former expected less than the latter. From an emotional viewpoint, comparative evaluations often override evaluations concerning our absolute position.

The comparative nature of emotional meaning implies that emotions go beyond the information given; hence, it involves an imaginary aspect. A perceived change may be *actual or imaginary.* Both types include a certain comparison and are present in typical emotional states. The actual and imaginary types of change may be in conflict, such as when we are satisfied with having won a small prize, but unsatisfied since we perceive ourselves as having just missed a much larger prize. Similarly, we would feel extremely lucky to escape from a car crash with only minor injuries, as it is easy to imagine far worse outcomes.

Whereas emotions in animals involve mainly, though not merely, the actual type of change, complex emotions of the imaginative type are more typical of human beings. Humans do not live exclusively in the immediate present. Through our mental capacities, we imagine what is likely to happen, what already happened, or what might happen.

The comparison underlying emotional significance encompasses the mental construction of *an alternative situation.* The more available the alternative, namely, the closer the imagined alternative is to reality, the more intense the emotion. A crucial element in emotions is, indeed, the imagined condition of "it could have been otherwise."

The notion of the availability of alternatives may explain many seemingly puzzling situations such as people who remain in unfulfilling marriages or jobs. Although their satisfaction is low, these people perceive other available alternatives to be even worse. There is much evidence indicating the tendency of people to react more strongly to those events for which it is easy to imagine a different outcome occurring. Therefore, the fate of someone who dies in an airplane crash after switching flights is perceived to be more tragic than that of a fellow traveler who was booked on the flight all along. Considering the importance that the availability of the alternative thus attains, «almost situations» or «near misses» come to have intense emotional effects (Heider, 1958; Kahneman & Miller 1986; Ortony, Clore, & Collings, 1988).

The comparative concern in emotions is mainly social. The social world is a principal theater of emotions since other people are most important for our well-being. Social comparison is not exercised indiscriminately; it typically refers to people and domains currently perceived to be relevant to our well-being or predominant in our concerns. We neither compare ourselves with everyone nor do we compare every aspect of ourselves.

In light of the importance of the social comparative concern in emotions, group membership is one of the most powerful factors in our emotional lives: the mere act of assigning people to different groups tends to accentuate the perceived cognitive and evaluative differences between them. Two particularly significant types of groups are: social group, which consists of those we have social relations with, and reference group, which consists of those who are important for determining our self-esteem.

1.3 The Typical Emotional Object: A Human Being

The typical emotional object is either the person experiencing the emotion or another person. People are more interesting to people than anything else. The things that people do and say, including the things that we ourselves do and say, are the things that affect us most. Emotions are typically directed toward agents who are capable of enjoyment and suffering. We can identify ourselves with other agents who are enjoying or suffering and this induces emotions. In light of the great similarity of other human beings to us, we can most easily identify ourselves with them and therefore their enjoyment and suffering have great impact upon us.

Although emotions are typically directed at a particular agent, they may sometimes be generalized and appear to be directed at a whole group of agents. Thus, we may hate people belonging to a particular ethnic group. Emotions may also be directed at living creatures such as dogs, cats, and birds. The more similar the creature is to human beings, the greater is the emotional intensity toward it. Emotions may also be directed at objects that are actually not agents but have some properties resembling agents or at least are construed to have such properties. Thus, we may feel anger toward our car or have compassion for an old house due for demolition.

2. Typical Characteristics: Instability, Intensity, Partiality, and Brevity

I suggest that instability, great intensity, a partial perspective, and relative brevity be considered as the basic characteristics of typical emotions. This characterization refers to «hot emotions,» which are the typical intense emotions. The more moderate emotions lack some of the characteristics associated with typical emotions. Hot emotions, or, simply, emotions, should be distinguished from other affective experiences such as moods, affective disorders, and sentiments (see Ben-Ze'ev, 1997b).

2.1 Instability

In light of the crucial role that changes play in generating emotions, instability of the mental (as well as the physiological) system is a basic characteristic of emotions. Emotions indicate a transition in which the preceding context has changed, but no new context has yet stabilized. Emotions are like a storm: as unstable states which signify some agitation, they are intense, occasional, and of limited duration. Another popular metaphor compares emotions to a fire.

The instability associated with intense emotions is revealed by their interference with activities requiring a high degree of coordination or control. One cannot easily thread a needle while trembling with fear or seething with anger. When we are in the grip of a strong emotion, our rational faculties no longer function normally, with the result that we «lose our heads» and act in ways which differ from our norm.

Emotional instability is applicable not only to the personal domain, but also to the *sociological* arena: emotions are more intense in unstable societies where, for example, the regime can rapidly change or people's personal status is subject to fluctuations. In stable, or static, societies, the availability of alternatives hardly exists, and hence emotional intensity is reduced. Envy, for instance, is less intense in such a society. The greater availability of alternatives in unstable societies also indicates greater individual insecurity, thereby intensifying most emotions.

2.2 Intensity

One of the typical characteristics of emotions is their relative great intensity. Emotions are intense reactions. In emotions the mental system has not yet adapted to the given change, and due to its significance, the change requires the mobilization of many resources. No wonder that emotions are associated with urgency and heat. One basic evolutionary function of emotions is indeed that of immediate mobilization. This function enables us to regulate the timing and locus of investment in the sense of allocating resources away from situations where they would be wasted, and toward those where investment will yield a significant payoff (Lazarus, 1991; Oatley & Jenkins, 1996).

Low intensity of the feeling dimension, as well as of other mental components, usually expresses neutral or indifferent states of the mental system. Emotions are the opposite of such states. Accordingly, it is preferable to consider low intensity states as nonemotional or non-typical. Although it is impossible to delineate the precise borderlines of emotional intensity, we can say that typical emotions have such an intensity which influences our normal functioning but not in a way that disables us completely - as is the case in affective disorders.

Typical emotions, characterized as possessing relatively great intensity, should be distinguished from extreme manifestations of affective disorders such as severe anxiety or depression which are the focus of a great deal of psychological research on pathology.

2.3 Partiality

Emotions are partial in two basic senses: they are focused on a *narrow* target as on one person or a very few people; and they express a *personal* and interested perspective. Emotions direct and color our attention by selecting what attracts and holds our attention; they make us preoccupied with some things and oblivious to others. Emotions are not detached theoretical states; they address a practical concern from a personal perspective. This perspective may also include considerations of those related to us. These people are like extensions of our egos, even though their emotional weight is typically of a lesser degree than the weight of personal considerations having direct bearing upon our own lives.

Not everyone and not everything is of emotional significance to us. We cannot assume an emotional state toward everyone or those with whom we have no relation whatsoever. The intensity of emotions is achieved by their focus upon a limited group of objects. Thus, one cannot love everyone; our romantic love is directed at a few people.

This limitation in the number of possible emotional objects forces us to focus upon those who are close to us. When we hear of the death of thousands of people in an earthquake occurring in a remote (that is, from our vantage point) part of the world, our emotional response comes nowhere near the intensity of our grief at the death of someone close to us. It does not even approach the level of feeling we experience in watching the suffering of a single victim of that same earthquake on television (thereby establishing some affinity with that particular victim).

The partiality of emotions is demonstrated by their cognitive, evaluative, and motivational components. The *cognitive* field of emotions does not offer varied and broad perspectives of our surroundings; it narrows and fragments our perspective. Selective abstraction, in which the focus of attention is on specific aspects, and over-generalization, which is the construing of a single event as representative of the whole situation, are frequently associated with emotions. Thus, sexual desire and envy considerably limit our focus of attention. The *evaluative* perspective of emotions is partial due to its highly polarized nature and its concern with very few objects. In comparison with other people, a typical emotional object is evaluated as being either highly positive or highly negative; it is also evaluated to be highly relevant to our well-being. Highly emotional people overestimate the degree to which events are related to them and are excessively absorbed in the event's personal meaning. The *motivational* field is narrow in the sense that the desired activity is often clearly preferred to any alternative. Even in emotions such as love, in which the range of activities concerning the beloved is wide, these are clearly preferred to other activities unrelated to the beloved; the latter are hardly considered at all.

In light of the partial nature of emotions, we may reduce emotional intensity by broadening our scope, and increase the intensity by further limiting it. Counting ten before venting our anger enables us to adopt a broader perspective that may reduce anger. A broader perspective is typical of people who can calmly consider multiple aspects of a situation; it is obviously not typical of people who experience an intense emotional reaction to the situation.

The partiality typical of emotions is less dominant in other mental capacities, such as perception, memory, and thinking: these capacities are usually directed at more objects and they typically include a less personal perspective. For example, although perception is limited in its scope to events and objects currently confronting us, we are able to perceive many things simultaneously. Similarly, memory may be limited to things we have experienced or learned in the past, but in a brief period we can remember quite a few people. Contrary to the partial nature of emotions, intellectual reasoning is not partial: it is focused on a broad, rather than narrow, target, and it is not done from a personal and interested perspective. Intellectual reasoning is a detached state: it looks at all implications of a current state; it takes us far beyond the current situation.

Laughter is similar to emotions in having a strong element of incongruity or change. Both emotions and humor combine two perspectives - the expected and the unexpected. However, whereas in emotions the simultaneous presence of incongruent perspectives is problematic, and hence the need for immediate practical actions, in humor the incongruity is enjoyable and requires no action. The ability to entertain several different perspectives is typical of humor and moderate positions, and is contrary to the partial nature of emotions. A sense of humor is thus often incompatible with an extreme emotional state.

2.4 Brevity

Typical emotions are essentially transient states. An emotional event may be compared to a large rock being thrown into a pool of still water: for a short time, emotional chaos reigns before calm gradually returns. The mobilization of all resources to focus on one event cannot last forever. A system cannot be unstable for a long period and still function normally; it may explode due to continuous increase in emotional intensity. A change cannot persist a very long time; after a while, the system construes the change as a normal and stable situation. The association of emotional intensity with change causes the intensity to decrease steadily due to the transient nature of changes. This association is a natural mechanism enabling the system to return within a relatively short period to normal functioning - which may be somewhat different from the previous normal functioning. If emotions were to endure for a long time regardless of what was occurring in our environment, then they would not have an adaptive value.

The transient nature of emotions does not imply that emotions must last no more than a few seconds: sometimes the transition from one stabilized state to another takes longer. Such a transition is not just a switch from one state to another; it involves profound changes in our plans and concerns and, as such, it may occupy us for some time. The typical temporal structure of an emotional response involves a swift rise-time, taking less than half a minute in most cases, followed by a relatively slow decay. After an emotional response reaches its peak, it can take hours, or even days to get back to the stable, normal state again (Gilboa & Revelle, 1994; Oatley, 1992). Consequently, the dispute concerning the duration of emotions can be settled by claiming that all typical and diagnostic features of emotions are indeed present for a very short time - typically a matter of seconds. The longer an emotion lasts, the more such features drop away.

3. Basic Components

In addition to the typical characteristics, we discern four basic components, namely, cognition, evaluation, motivation, and feeling. The difference between typical characteristics and basic components is that characteristics are properties of the whole emotional experience, whereas components express a conceptual division of the elements of this experience. It is arguable that one could perhaps find a few relevant characteristics other than those I have discussed; however, the conceptual division of emotions into four components is more comprehensive and is supposed to cover all possible components.

3.1 Intentionality and Feeling

I consider intentionality and feeling to be the two basic mental dimensions (Ben-Ze'ev, 1993). Intentionality refers to a subject-object relation, whereas feeling expresses the subject's own state of mind. When a person is in love with someone, the feeling dimension surfaces in a particular feeling, say a thrill, that is experienced when they are together; the intentional dimension is expressed in the person's knowledge of her beloved, her evaluation of his attributes, and her desires toward him.

Intentionality is the relation of "being about something." It involves our cognitive ability to separate ourselves from the surrounding stimuli in order to create a meaningful subject-object relation. The intentional object is something about which the person has some information. This object does not have to be a person or a certain thing; it can be a general situation or even an abstract concept. The intentional dimension includes several references to objects, such as those involved in perception, memory, thought, dreams, imagination, desires, and emotions.

The feeling dimension is a primitive mode of consciousness associated with our own state. It is the lowest level of consciousness; unlike higher levels of awareness, such as those found in perception, memory, and thinking, the feeling dimension has no meaningful cognitive content. It expresses our own state, but is not in itself directed at this state or at any other object. Since this dimension is a mode of consciousness, one cannot be unconscious of it; there are no unfelt feelings. In the intentional domain we play a more active role; feelings, on the other hand, just seem to surface, and can overcome us when they are intense.

The intentional dimension in emotions can be divided into three components: cognitive, evaluative, and motivational. The cognitive component consists of information about the given circumstances; the evaluative component assesses the personal significance of this information; the motivational component addresses our desires, or readiness to act, in these circumstances. Neither these three intentional components nor the feeling dimension are separate entities or states. Emotions do not entail the separate performance of four varieties of activity:

knowing, evaluating, desiring, and feeling. All four are distinct aspects of a typical emotional experience.

Typical mental states in human beings consist of both intentional and feeling dimensions. The relationships between the two dimensions vary in type and degree for different mental states. Whereas in emotions both dimensions are central, in most mental states only one of these is dominant. For example, the feeling dimension is dominant in painful experiences, thirst or hunger, and in affective disorders. The intentional dimension dominates the cognitive capacities of perception, memory and thinking. To a greater extent than other mental states, emotions include diverse components within their scope, ranging from intense and primitive feelings to complex, rational evaluations. The more fruitful approach to emotions, therefore, is to treat them as unique combinations of the entire range of mental components, rather than to account for them by referring merely to a single basic component.

3.2 The Cognitive Component

The cognitive component supplies the required information about a given situation. No emotional attitude toward something can emerge without some information about it - whether veridical or distorted. Whereas the cognitive component describes the object, the evaluative component addresses a certain assessment of the same.

The cognitive aspect in emotions is often *distorted.* This is due to several related features typical of emotions: (a) partiality, (b) closeness, and (c) an intense feeling dimension. Emotions may have some cognitive advantages that are due to our intimate acquaintance with the object.

It is commonly assumed that there are considerable cognitive differences between the emotional and intellectual systems. While both are comparative in nature, the intellect is concerned with the general and the stable whereas emotions with the particular and the volatile. The aim of the intellect is to see a specific event as a specific case of general regularities; the foundations of intellectual reasoning are features common to individual cases. Emotions prevail as long as a specific event can be seen as mutable and unique. Accordingly, the intellect has difficulties in understanding change and movement, whereas emotions have difficulties in prevailing under stable and universal conditions.

These differences have generated different evaluative attitudes toward these systems. A prevailing tradition has seen these differences as an indication of the shortcomings of the emotional system and hence drawn the conclusion that the intellectual system is the true essence of the mental realm. Plato, Descartes and Kant are prominent representatives of this tradition, which considers thinking to be the essence of the mental realm. In a modern formulation of this view, the mind is an intellectual processor of knowledge which sorts out information in a relatively unbiased manner and emerges with carefully drawn conclusions and well-considered decisions. From this perspective, the mind is envisaged as a sober little creature seeking the most intellectual answers. This attitude is still common in philosophy and psychology. It is clearly expressed in the computational approach to the mind which constitutes the prevailing view in the fields of the philosophy of mind and cognitive psychology.

The opposite view, represented by Bergson, considers the emotional system to be of greater cognitive value. Bergson's (1907) view is in clear opposition to the intellectualist tradition which assumes that rational thinking is the best, and in many cases, the only means to know reality. He considers the ultimate cognitive tool to be the instinct, which in many respects is similar to emotions. His criticism of human intellect is directed precisely at its need to work with stable.

Most people consider Spinoza to belong to the intellectualist tradition; I believe that Spinoza actually presents a different view from the two outlined above. He believes that the ultimate cognitive tool combines both the emotion and the intellect. Spinoza distinguishes between three different levels of knowledge. Knowledge stemming from singular (or unique) things, and which is based upon the senses and imagination, is considered to be confused and false; knowledge which is based upon common and universal notions is considered as necessarily true. However, the highest form of knowledge is an intuitive knowledge which combines elements from the other two types: it proceeds from singular things but expresses universal knowledge concerning the essence of things. For Spinoza this kind of knowledge is related to an emotional attitude: the intellectual love of God (e.g., see Spinoza, 1677/1985, II, p. 40s1,2; p. 47; V5, p. 33)

Like other notions referring to the unconscious realm, the notion of «unconscious emotions» is also problematic since it is unclear to what mental experience it refers. Take, for example, a case in which we hate someone without being aware of our hate. One may interpret this situation as referring to an unconscious emotion having all four basic components of which we are not aware. In that case, we should assume the existence of unconscious feeling, in other words, of «unfelt feeling.» This is an obvious absurdity. A more plausible explanation is that in the so-called «unconscious emotions,» not every component is unconscious. What is unconscious, or rather unknown, not realized, or mistakenly identified, is the nature of the emotional state, that is, our basic evaluative stand expressing our focus of concern. An unconscious emotion, then, is an emotion whose nature is unclear, while many of its components are known. Even Freud (1915, p. 78) claimed that, strictly speaking, «unconscious affect» is a contradiction in terms.

3.3 The Evaluative Component

The evaluative component is extremely important in emotions. Every emotion entails a certain evaluation. The evaluative component appraises the "cold" information presented by the cognitive component, in terms of its implications for personal well-being.

The *evaluative component* intrinsic to a certain emotional state should not be confused with *moral evaluation of the entire state*. Thus, pleasure-in-others-misfortune involves a positive evaluation of the misfortune of others, and hence its feeling component is agreeable. However, this emotion is often evaluated as negative from a moral viewpoint.

In accordance with the distinction between deliberate and schematic cognitive responses, we can distinguish between two major types of evaluations:

deliberate and schematic. Deliberate evaluations are present, for example, when we ruminate about a certain event and as result begin to feel angry. An example of a schematic evaluation is love at first sight. *Deliberate evaluations* typically involve slow and conscious processes, which are largely under voluntary control. Such processes usually function on verbally accessible, semantic information and they operate in a largely linear, serial mode. *Schematic evaluations* involve spontaneous responses depending on a more tacit and elementary evaluative system. Schematic activity is typically fast, automatic, and with little awareness. It is based upon ready-made structures or schemes of appraisal which have already been set during evolution and personal development; in this sense, history is embodied in these structures. Since the evaluative patterns are part of our psychological constitution, we do not need time to create them; we just need the right circumstances to activate them. Schematic activity largely occurs outside of focal awareness, can occur using minimal attentional resources, and is not wholly dependent on verbal information (Ekman, 1992; Lazarus, 1991; LeDoux, 1996; Leventhal & Scherer, 1987; Lyons, 1980; see also Ben-Ze'ev, 1993, chapter. 4).

The two types of evaluations may clash. Thus, we may persist in being afraid even when our conscious and deliberate judgment reveals that we are no longer in any peril. We can explain such cases by assuming that certain schematic evaluations become constitutive to a degree where no intellectual deliberation can change them.

The schematic nature of typical emotional evaluations enables us to consider emotions not as an isolated result of a cognitive inference, but as part of ongoing interaction. Deliberate evaluation is a preparatory process that precedes and is separate from its product. A schema is an active principle of organization which is constitutive in nature; it is not separate from the organized state, but part of it.

3.4 The Motivational Component

The motivational component refers to the desire or readiness to maintain or change present, past, or future circumstances. In the case of "passionate" emotions, such as anger and sexual desire, the desire is typically manifested in overt behavior; in "dispassionate" emotions, such as envy and hope, the behavioral element is less in evidence and appears merely as a desire. Emotions are not theoretical states; they involve a practical concern, associated with a *readiness to act.* Since emotions are evaluative attitudes, involving a positive or a negative stance toward the object, they also entail either taking action or being disposed to act in a manner which is compatible with the evaluation. Emotions typically express our most profound norms and beliefs; hence, they often express not merely superficial involvement, but deep commitment.

Different types of connection between the *motivational component* in emotions and *actual behavior* can be discerned: (a) a full-fledged desire which is expressed in actual behavior; (b) a desire or want which is not expressed in actual behavior because of external constraints; (c) a mere wish which is not intended to be translated into actual behavior.

3.5 The Feeling Component

The term "feeling" has several meanings: awareness of tactile qualities, bodily sensations, emotions, moods, awareness in general, and so forth. In this discussion, the term is confined to modes of awareness which express our own state and are not directed at a certain object.

The homogeneous and basic nature of feelings makes it difficult, though perhaps not impossible, to describe them. Indeed, there are few words for feelings, and we often have to resort to metaphors and other figures of speech in referring to them. No doubt feelings have intensity, duration, and some have location as well; but what about other qualities? The qualities of being painful or pleasurable are obvious. Some level of pleasantness or unpleasantness, albeit often of low intensity, is experienced by most people most of the time. In addition to pleasure and displeasure, the continuum of arousal may be a common aspect of the feeling dimension.

Despite the importance of feelings in emotions, equating the two is incorrect since emotions have an intentional component in addition to the feeling component.

3.6 Comparing the Different Components

The emphasis upon the evaluative component suggested here is not a new explanatory direction. It can be found in the writings of ancient (e.g., Aristotle, the Stoics, and Spinoza) and contemporary philosophers (De Sousa, 1987; Greenspan, 1988; Lyons, 1980; Nussbaum, in press; Solomon, 1976) and psychologists (e.g., Arnold, 1960; Lazarus, 1991; Ortony et al., 1988; Parkinson, 1995; Scherer, 1982). Indeed, evaluative theories are the foremost current approach to emotions in philosophy and psychology. The general assumption underlying these theories is that evaluations (appraisals) are the most crucial factor in emotions. This assumption may imply at least two different claims:

(A) Evaluative patterns distinguish one emotion from another;

(B) Evaluative patterns distinguish emotions from nonemotions;

These claims, which are not clearly distinguished by appraisal theorists, are not necessarily related. Accepting one of them does not necessarily imply acceptance of the other. I believe that whereas a simplistic formulation of (B) is false, (A) is basically true (see Ben-Ze'ev, 1997a)

4. Some Conceptual Aspects

So far I have described various characteristics and components of emotions; these can give us a general picture of a typical emotion. However, this picture does not answer more general conceptual questions:

(A) Is an emotion a mental capacity, as are perception, memory, imagination, and thought?

(B) Is an emotion a mental mode of reference, such as cognition, evaluation, and motivation?

(C) Is an emotion action or passion?

Let me summarize my position on these issues.

(A) Traditional description of mental phenomena suggest the presence of a few mental capacities (faculties) - for example, sensation (or feeling), perception, memory, imagination, thought, and the will. It is doubtful whether each of these capacities can be described as a single, unitary capacity. Thus, it has been suggested that memory is not a single capacity, but that what we call memory actually consists of various learning systems. Without entering into this debate, it seems that an emotion is not a single, unitary capacity. While experiencing an emotion, some of the above mental capacities, and often of all of them, are activated. Nevertheless, an emotion is not on the same conceptual level as each of them: an emotion involves, for example, perception and imagination, but it is not a type of perception and imagination; an emotion is a general term referring to a certain combination of such capacities.

(B) In addition to the above mental capacities, we may discern a few mental modes of reference: cognition, evaluation, and motivation. Not all mental capacities involve these modes. Sensation, which is the most primitive mental capacity, lacks any of these modes of intentional reference. The more complex mental capacities, such as perception and memory, have the cognitive mode of reference. The will utilizes the motivational mode, while imagination and thought may include all modes. These types of intentional references are essential components of emotions, but an emotion is not identical to any of them. Again, they belong to a different conceptual level to that of an emotion.

(C) The description of the mind into actions and passions is based on the issue of choice: actions, but not passions, are subject to our free choice. Thinking, remembering, and imagining are action while feeling is passion. In light of this division, an emotion is passion as we cannot choose our emotions: we do not willingly create the emotional state, but find ourselves in it. However, an emotion is a dynamic state including many actions.

We can see that traditional descriptions of mental phenomena are not suitable for describing the emotions because of their greater complexity. An emotion is then neither a mental capacity nor a particular mode of reference. An emotion is a complex system consisting of various mental capacities, modes of reference, attitudes, activities, and states. Accordingly, it is preferable to replace the substantial notion of emotion with a functional concept. For the purpose of an initial explanation we may consider an emotion to be an entity, but when a more scientific explanation is required, a functional explanation is in order.[1]

References

Arnold, M. (1960) *Emotion and Personality*, New York: Academic Press.

Ben-Ze'ev, A. (1993) *The perceptual system: A philosophical and psychological perspective*, New York: Peter Lang.

Ben-Ze'ev, A. (1996) "Typical emotions", in: *Philosophy of Psychology*, W. O'Donohue and R. Kitchener, eds, London: Sage, 1996, pp. 228-243.

Ben-Ze'ev, A. (1997a) "Appraisal theories of emotions", *Journal of Philosophical Research* 22:129-143.

Ben-Ze'ev, A. (1997b) "The affective realm", *New Ideas in Psychology* 15:247-259.

Ben-Ze'ev, A. (2000) *The Subtletv of Emotions*, Cambridge, MA: MIT Press.

Bergson, H. (1907) *Creative Evolution*, New York: Holt.

Cassirer, E. (1923/1953) *Substance and Function*, New York: Dover.

De Sousa, R. (1987) *The Rationality of Emotions*, Cambridge, MA: MIT Press.

Ekman, P. (1992) "An argument for basic emotions", *Cognition and Emotion* 6:169-200.

Festinger, L. (1954) "A theory of social comparison processes", *Human Relations* 7:117-140.

Freud, 5. (1915) "The unconscious", in: *The Standard Edition of the Complete Psychological Works*, vol.14, London: Hogarth Press.

Frijda, N. H. (1988) "The laws of emotion", *American Psychologist* 43:349-358.

Gilboa, E. and W. Revelle (1994) "Personality and the structure of affective responses", in: *Emotions: Essays on Emotion Theory*, S. H. M. van Goozen, N. E. van de Poll and J. A. Sergeant, eds, Hillsdale, NJ: Erlbaum.

Greenspan, P. (1988) *Emotions and Reasons*, New York: Routledge.

Griffiths, P. E. (1997) *What Emotions Really Are: The Problem of Psychological Categories*, Chicago, IL: The University of Chicago Press.

---

[1] I am grateful to Nico Frijda for helpful remarks in this regard. For the difference between substantial and functional explanations, see, e.g., Cassirer (1923); see also Ben-ze'ev (1993); Griffiths (1997).

Heider, F. (1958) *The Psychology of Interpersonal Relations*, New York: Wiley.

Higgins, E. T. (1987) "Self-discrepancy: A theory relating self and affect", *Psychological Review* 94:319-340.

Kalmeman, D. and D. T. Miller (1986) "Norm theory: Comparing reality to its alternatives", *Psychological Review* 93:136-153.

Lazarus, R. S. (1991) *Emotion and Adaptation*, New York: Oxford University Press.

LeDoux, J. (1996) *The Emotional Brain*, New York: Simon & Schuster.

Leventhal, H. and K. R. Scherer (1987) "The relationship of emotion to cognition: A functional approach to a semantic controversy", *Cognition and Emotion* 1:3-28.

Lewis, C. I. (1929/1956) *Mind and the World Order*, New York: Dover.

Lyons, W. (1980) *Emotion*, Cambridge: Cambridge University Press.

Nussbaum, M. C. (in press) *Upheavals of Thought: A Theory of the Emotions*, Cambridge: Cambridge University Press.

Oatley, K. (1992) *Best Laid Schemes: The Psychology of Emotions*, Cambridge: Cambridge University Press.

Oatley, K. and J. M. Jenkins (1996) *Understanding Emotions*, Cambridge, MA: Blackwell.

Ortony, A., G. L. Clore and A. Collings (1988) *The Cognitive Structure of Emotions*, Cambridge: Cambridge University Press.

Parkinson, B. (1995) *Ideas and Realities of Emotion*, London: Routledge.

Scherer, K. R. (1982) "Emotion as process: Function, origin and regulation", *Social Science Information* 21:555-570.

Solomon, R. C. (1976) *The Passions*, New York: Doubleday.

Spinoza, B. (1677/1985) "Ethics", in: *The Collected Works of Spinoza*, E. Curley, ed., Princeton: Princeton University Press (1985).

DETERMINANTS OF EMOTIONAL INTENSITY

AARON BEN-ZE'EV
Center for Inter-Disciplinary Research on Emotions, University of Haifa
Haifa 31905, Israel

ABSTRACT

People often talk about the intensity of their emotions: they tell us that their anger is overwhelming, that they feel extremely sad, or that they are madly in love. Despite the common usage of terms, which measure emotional intensity, the notion of «emotional intensity» is far from clear. In this paper I first clarify this complex notion and then discuss the circumstances in which emotions become intensified.

1. The Complexity of Emotional Intensity

The concept of "emotional intensity" is complex; it applies to different phenomena, not all of which are correlated. When Tom says to Ruth that he loves her now more than he has ever loved any other woman, what does he mean by this? He may mean several different things, such as: (a) his feeling toward her is the strongest he has ever experienced, (b) his love toward her has lasted longer than any other love of his; (c) he keeps thinking about her all the time; (d) he believes she is the most wonderful person in the world; (e) he is ready to do more for her than he has ever been ready to do for any other woman. Analyzing emotional intensity should take into account all such diverse features (see Ben-Ze'ev, 1996; 2000).

The diverse features of emotional intensity are expressed in two basic aspects: magnitude (*peak intensity*) and temporal structure (mainly, *duration*). If Tom's feeling component is very strong at this moment, but it lasts only a few minutes, we may say that his love is weaker than love of a similar magnitude lasting for a few hours. Similarly, if thinking about her occupies him most of the time, this is indeed intense love; when this preoccupation lasts several weeks, the love is more intense than if it lasts several days. Duration can vary dramatically with comparable levels of peak intensity. In one study, participants rated the positive emotion associated with having "someone you find attractive suggest you meet for coffee" as almost as high as the emotion experienced after "saving your neighbor's child from a car accident." However, the average estimated duration associated with the former was 20 minutes, whereas for the latter it was more than 5 hours. Similarly, respondents estimated that they would stop ruminating about the coffee suggestion after about two hours, whereas the experience of the car accident would lead to rumination for about a week (Gilboa & Revelle, 1994).

The two basic aspects of emotional intensity, namely, peak intensity and duration, are expressed in each of the four basic emotional components: feeling, motivation, evaluation, and cognition. We may speak about the peak intensity and duration of a certain feeling, urge to act, extremity of evaluation, and cognitive preoccupation. Highly significant emotional events are expressed by the two aspects of all four components.

Various difficulties in *determining the intensity* of different emotions exist. The central ones concern the relative weight of the various aspects and components of emotional intensity. Another set of difficulties concern the *accuracy of the measurements* of emotional intensity.

Despite the enormous difficulties in measuring emotional intensity, ordinary people can and do measure such intensity. In speaking of emotional intensity, we all resort to quantitative language. We speak of "more" or "less" emotional intensity, and quite often correctly estimate the intensity of the emotions of others and ourselves in our everyday behavior. Accordingly, psychologists have developed a variety of means and scales for measuring emotional intensity in general as well as the intensity of particular emotions. In addition to such psychological measures, emotional intensity is often estimated in scientific experiments by measuring the underlying physiological components, which indeed renders reliable results.

The concept of "emotional intensity" denotes a complex construct whose components seem to be incommensurable; nevertheless, the intensity of the whole emotional state can be estimated by comparison with similar states. Our ability to compare various emotional intensities is based on finding a *certain feature* whose changes are typically correlated with intensity changes of the whole state. Instability may be such a feature, as it is a basic characteristic of emotions, and it is easy to make comparative estimate of its value. Greater instability manifests itself in many obvious physiological and psychological aspects. Another factor like this may be overall felt intensity (See Frijda, Ortony, Sonnemans, & Clore, 1992; Green, 1992, pp. 136-138; Sonnemans & Frijda, 1994.

Although the concept of emotional intensity is complex, emotional intensity is often measured and compared in everyday life as well as in scientific experiments. This enables us to proceed with our discussion of the circumstances that determine emotional intensity even if the concept itself may not be entirely clear from a theoretical point of view.

2. Intensity Variables

Emotional intensity depends on the way in which we evaluate the *significance of events*. Although emotions arise from an immediate eliciting event, their intensity depends on broader sets of circumstances that circumscribe our sensitivity to such an event.

The various intensity variables may be divided into two major groups, one referring to the perceived *impact of the event* eliciting the emotional state and the other to *background circumstances of the agents* involved in the emotional state. The major variables constituting the event's impact are the strength, reality, and relevance of the event; the major variables constituting the background circumstances are accountability, readiness, and deservingness.[1] The suggested classification is not arbitrary; it expresses two major aspects of the emotional situation: the impact of the eliciting event, and the subjective background circumstances preceding it. The first group is crucial for determining our current situation; the importance of the second group is in realizing whether the situation could have been prevented and whether we deserve to be in such a situation.

2.1 Strength

The event's strength is a major factor in determining the intensity of the emotional encounter. It refers, for example, to the extent of the misfortune in pity, the extent of our inferiority in envy, the level of damage we suffer in anger, or the extent of beauty of the beloved. A positive correlation usually exists between the strength of the event as we perceive it and emotional intensity: the stronger the event is, the more intense is the emotion. Though positive, the correlation is not always linear: a stronger event may result in a more intense emotion, but the increase in intensity is not always proportional to the increase in the event's strength. In very strong events, an additional increase in their strength will hardly increase emotional intensity which is anyway quite high and almost at its peak. This kind of correlation is also typical of other variables. The typical curve of emotional intensity rises up to a point with increases in the given variable; from this point on, emotional intensity hardly changes with an increase in the given variable.

In some emotions, the event's strength can be specified between lower and upper limits. Thus, pleasure-in-others'-misfortune presupposes a certain degree of the other's misfortune; when this misfortune becomes very severe, it may exceed the upper limit typical of this emotion, and our emotion then turns into pity. Similarly, another person's improved fortune can make us happy up to the point where this person's fortune is so good that our emotion of happy-for turns into envy. There is also an upper limit of strength in events that cause embarrassment; beyond this limit, embarrassment may turn into shame. The positive correlation between the event's strength and emotional intensity is typically kept within specified limits.

2.2 Reality

The second major variable constituting the event's impact is its degree of reality: the more we believe the situation to be real, the more intense the emotion.

---

[1] Various psychologists have suggested lists of intensity variables, or basic appraisals, which have some similarity to my list. The most similar list to my own is discussed by Ortony *et al.* (1988); See also Frijda (1987); Frijda *et al.*(1989); Lazarus (1991); McCornack and Levine (1990); Roseman (1991); Scherer (1988); Smith and Ellsworth (1985).

This variable is particularly important in forward-looking emotions, such as hope and fear; it is, however, significant in other emotions as well. The importance of the degree of reality in inducing powerful emotions is illustrated by the fact that a very strong event, which may be quite relevance to our well-being, may not provoke excitement if we succeed in considering it as fantasy: the emotional intensity decreases accordingly. Thus, despite the horrifying impact of a potential nuclear holocaust, many people do not allow this to upset them, since they do not consider the event to be a real possibility (Frijda, 1986, p. 206; 1988, p. 352).

In analyzing the notion of "emotional reality" two major senses should be discerned: (a) *ontological,* and (b) *epistemological.* The first sense refers to whether the event actually exists or is merely imaginary. The second sense is concerned with relationships of the event to other events. The first sense expresses the "correspondence criterion" of truth where a claim is seen as true if its content corresponds to an existing event in the world. The second sense is related to the "coherence criterion" of truth in which truth is determined in light of whether the given claim is coherent with other claims we hold.

The ontological sense is often understood to imply physical existence; indeed, in modern discussions "real" is often identified with "physical" (or "material") and "unreal" with "mental" (or "spiritual"). Such identification is unwarranted even in the ontological sense of reality: we should assume the actual existence of mental states. When I say that a certain person is jealous or in love, this claim can be true in the first sense of reality, although it refers to mental states.

The epistemological sense of reality allows for a greater variety of real entities. In this sense, "real" and "unreal" are context-dependent attributes: something may be real in one context and unreal in another. Something is real in a certain context if it has relations to other things in that context. In the context of physical reality, moral values and feeling pleasure over the misfortune of others are not real. However, they are real at the psychological level of describing human experience since they directly influence such an experience.

In analyzing the perceived reality associated with our emotional experiences, the two senses of reality are relevant as well. The ontological sense is expressed in the actual existence of the emotional object, and the epistemological sense is typically expressed in its vividness. The degree of reality is highest *when the object is real in both senses.* Interesting cases are those with a conflict between the two senses, for example, when a fictional character is more vivid than a person we have just met. Both persons are real for us, and it is not obvious as to who may induce greater emotional intensity.

In light of the *crucial role imagination plays in emotions,* the importance of the degree of reality may be questioned. Thus, although works of art are understood to describe imaginary characters, they easily induce intense emotions. Art may in fact quite often induce more intense emotions than those we have toward real people, e.g., starving people in a (from our point of view) remote place in the world, about whose fate we read in the newspaper or hear on the radio. Responding to this difficulty requires taking account of both senses of reality. Works of art are obviously real in the epistemological sense of being vivid. They provide us with more vivid information than that reported about actual existing events. The degree of vividness is clearly different when reading a newspaper and watching a movie are compared. The detailed and concrete description we have of the life of a fictional character in a movie makes this character more vivid and closer to us than an actual existing person reported in a newspaper.

We are moved by a book or movie despite and not because of its being imaginary: its higher degree of reality in the sense of its being vivid generates intense emotions despite its low degree of reality in the sense of its actual existence. Indeed, various studies have suggested that the influence of television is greater when the characters are perceived to be more real. Such an influence is particularly significant in the case of young viewers who cannot easily make distinctions between reality and fantasy (Van Evra, 1990, pp. 85-88).

2.3 Relevance

The third major variable constituting the event's impact is its relevance: the more relevant the event is, the greater the emotional significance and hence intensity. Relevance is of utmost importance in determining the significance of an emotional encounter. What is irrelevant to us cannot be emotionally significant for us.

Emotional relevance typically refers either (1) to the achievement of our goals or (2) to our self-esteem. Goal relevance measures the extent to which a given change promotes or hinders our performance or the attainment of specific significant goals. Changes that promote our goals are associated with positive emotions, and those that hinder these goals with negative emotions. An enjoyable event may be negatively evaluated because it impedes the attainment of a certain goal. In light of the social nature of emotions, our self-esteem is an important emotional issue. We do not envy trees for their height or lions for their strength, since these are irrelevant to our personal self-esteem. The relevance component restricts the emotional impact to areas that are particularly significant to us.

The two related aspects of relevance are associated with all emotions, but to varying degrees. The aspect regarding goal achievement is more dominant in fear, hope, regret, and hate, whereas the aspect concerning our self-esteem is usually more evident in emotions such as envy, jealousy, shame, and pridefulness. Sometimes greater relevance changes the nature of a given emotion. If someone is better than we in an area that is of little relevance to our self-evaluation are, then our attitude toward this person may often be admiration. However, in a case of high relevance, other things being equal, the attitude is more likely to be envy.

Emotional relevance is closely related to *emotional closeness.* Events close to us in time, space, or effect are usually emotionally relevant and significant.

Closeness can be a crucial element in determining emotional relevance. Greater closeness typically implies greater significance and greater emotional intensity. Closeness sets the conditions for meaningful relationships and comparisons. When someone is too detached from us, we are unlikely to have to have any emotional attitude toward her. Closeness may be broken down into two factors: (1) similarity in background, for example, biological background, place of birth, education, significant experiences, and opportunities; and (2) proximity in current situation, for example, proximity in time, space, age, status, salary, or possession of a certain object.[2]

The correlation between relevance and emotional intensity is positive:

greater relevance leads to greater emotional intensity. Things become more complex if we discuss the constituents of emotional relevance, namely, goal relevance, relevance to self-image, similarity in background, and proximity in current situation. This becomes particularly complex when these constituents are dependent on each other; for instance, when an increase in one constituent causes a decrease in another. Thus, if two siblings, having a high degree of background similarity, want to remain emotionally close, they must reduce the relevance of each other's deeds to their own self-image; otherwise, envy will prevail and their relationship will be damaged (Elster, 1999; Tesser, Millar, & Moore, 1988.

As in other variables, relevance influences not merely the intensity of a given emotional state, but its nature as well. Thus, when our fortune is worse than that of another person our emotional attitude can be that of envy or happy-for. Relevance is an important factor determining which of these attitudes we may have: in case of high relevancy envy is more likely to emerge and in case of low relevancy happy-for is more likely to be our emotional attitude.

2.4 Accountability

Accountability refers to the nature of the agency generating the emotional encounter. Generally, the more responsible we are for the given change (for example, by having some control over the situation or by investing effort to bring it about), the more available is the alternative and hence the more intense the emotion. The major issues relevant in this regard are: (1) degree of controllability, (2) invested effort, and (3) intent.

Controllability may be divided into two major groups: (a) personal controllability, and (b) external controllability. Each group may be further divided

---

[2] Many philosophers and psychologists emphasize the importance of the closeness variable in emotions. Spinoza (1677/1985) claims that "men are by nature envious or are glad of their equals' weakness and saddened by their equals' virtue" (III, p55). Hume (1739-1740) says that "the great disproportion cuts off the relation, and either keeps us from comparing ourselves with what is remote from us, or diminishes the effects of the comparison" (pp. 377-378). Adam Smith (1790) argues that "we should be but little interested ... in the fortune of those whom we can neither serve nor hurt, and who are in every respect so remote from us" (p. 140). In the same vein, Festinger (1954) claims that "the tendency to compare oneself with some other specific person decreases as the difference between his opinion or ability and one's own increases" (p. 120).

into two subgroups. In the first group we can distinguish between events due to (a1) our deliberate behavior, (a2) behavior stemming from our character and habits, and (a3) our nondeliberate behavior. The second group may be divided into events due to (b1) others' deliberate behavior, (b2) others' nondeliberate behavior, and (b3) impersonal circumstances. The order of controllability is as follows: (a1), (a2), (a3), (b1), (b2) and (b3). It can be noticed that the division of personal controllability is more specified than that referring to the controllability of others: personal controllability refers to deliberate behavior, behavior stemming from our character and habits and nondeliberate behavior; others' controllability refers merely to deliberate and nondeliberate behavior. This difference stems from the fact that the varieties of our own accountability are of greater emotional concern than that of others. In any case, although classifications can be useful, we should not attach too much importance to these categories.

The order of emotional intensity is similar: events due to our deliberate behavior have the greatest emotional impact and those due to impersonal circumstances the least. The dependence of emotional intensity on the variable of controllability can be demonstrated by many everyday phenomena and empirical studies. People feel more entitled to (or frustrated by) an outcome they have helped to bring about than to (or by) an outcome resulting from the whim of fate or other powerful agents. Envy increases if our inferior position is due to our own failure, and frustration intensifies if the failure is attributed to us (Folger, 1984; Frijda, 1986; Smith & Ellsworth, 1985; Thibaut & Kelley, 1959).

In the movie *Sophie's Choice,* a Nazi officer demands that a Jewish mother choose which of her two small children will be sent to the gas chambers and which child will be allowed to live. The mother begs the Nazi to choose the child himself and thus to eliminate her control over the choice. The Nazi's cruelty is expressed, among other things, in his refusal to do so. By forcing the mother to choose which child will die, he creates one of the cruelest events for any parent: the death of her child, which is perceived to be due to the parent's behavior.

In accordance with the suggested positive correlation between emotional intensity and controllability, it has been found that we tend to overestimate our degree of control over positive outcomes and underestimate our control over negative outcomes. We also underestimate the degree of control of others over positive outcomes and overestimate their control over negative outcomes. Similarly, we attribute others' negative emotions equally to situational factors and to their personal dispositions, whereas our own negative emotions are attributed to the situation more than to personal dispositions. This helps us to maintain a positive self-image and prevent unflattering comparisons (Karasawa, 1995; Teigen, 1995; Whitley & Frieze, 1986). These considerations can explain why errors typically cause emotions. Errors do not merely involve an unexpected change, but also a change that was to a certain extent under our control.

In some cases people tend to take more personal responsibility for negative events. Thus, victims of wrongdoing often search for ways in which they are responsible for the wrong done to them. By doing this they avoid admitting that someone else has greater control over their lives. They trade the status of "victim" for that of "guilty agent" in order to retain a positive self-image, which entails having control over one's life.

There are various phenomena that *seem to contradict* the suggested positive correlation between controllability and emotional intensity. I believe that in all such cases, the correlation is absent because other variables besides controllability have also changed and these are responsible for the apparent exception to the general correlation.

Consider, for example, the strong tendency of both men and women to claim responsibility for initiating a breakup, regardless of who actually initiated it. In this case, controllability of the eliciting event decreases the intensity of the sadness and shame associated with a breakup. It is easier to accept and cope with the breakup if one views it as a controllable, desired outcome than as imposed against one's wish (Hill, Ruben, & Peplau, 1976). Although this case may appear to be an exception to the general positive correlation between controllability and emotional intensity, its explanation is actually different: it relates to a different emotional variable, namely, relevance. Claiming greater responsibility for the breakup reduces the relevance to our self-image and hence the emotional impact of the breakup decreases. A related situation is that in which, for example, someone wants to end a romantic relationship but she is worried that her partner will be hurt; in order to reduce his hurt, she leads him to believe that the breakup is actually his own decision, made for his own well-being. In such a case, the woman's concern for her partner's self-image causes her to exaggerate his accountability and hence to decrease the event's relevance to his self-image. Both situations do not express exceptions to the general positive correlation between accountability and emotional intensity.

The variable of controllability, referring to our past control over the circumstances that generated the given emotion, should not be confused with our present control of the emotional circumstances or our own behavior. Whereas a positive correlation exists between emotional intensity and past control, the correlation between emotional intensity and present control is negative. When an event is perceived to be under our personal control, it does not produce as much stress as one perceived to be uncontrollable (Taylor, 1989, pp. 75-76).

Effort is an additional factor constituting the variable of accountability. Like controllability, effort describes the extent of our involvement in the generation of emotions. Effort should be understood as including physical and mental effort as well as investment of all types of resources. Generally, the more effort we invest in something, the more significant it becomes and the more intense is the emotion surrounding it. As the saying goes: the more you pay, the more it is worth. The opposite is, of course, also true: when the stakes are greater, we invest more effort (Ortony, Clore, & Collings, 1988, pp. 71-73).

Effort is closely related to the variables which signify the impact of the event, especially that of relevance. Thus, we invest more effort in something that is relevant and hence significant for us; conversely, something we invest more effort in becomes more relevant and significant. The saying "easy comes, easy goes" expresses situations in which something we have gained without much invested effort is of lesser significance to us and hence we may lose it quite equably.

Intent is another factor constituting our accountability. If we intended to do something, then our involvement and responsibility are typically greater than when the event happened without our prior intention. Accordingly, the emotional intensity is typically greater. Thus, our anger is more intense if we believe that the other person intended to hurt us, and our shame is more intense if we intended to act the way we did.

Intent and controllability generally have a high co-variance: people intend to do what is controllable, and can control what is intended. But there are instances where intent and control do not coincide. For example, an over-achiever might intend to take some time off from work, but cannot control her working habits. The differentiation between intent and control lies at the heart of the distinction between murder and manslaughter: both involve control, but only murder is associated with intent as well (Weiner, 1985).

2.5 Readiness

The variable of readiness measures the cognitive change in our mind; major factors in this variable are unexpectedness (or anticipation) and uncertainty.

Unexpectedness, which may be measured by how surprised one is by the situation, is widely recognized as central in emotions. Since emotions are generated at the time of sudden change, unexpectedness is typical of emotions and is usually positively correlated with their intensity, at least up to a certain point. We are angrier if we happen to be expecting a contrary result, just as the quite unexpected fulfillment of our wishes is especially sweet. Unexpectedness may be characterized as expressing the gap between the actual situation and the imagined alternative expected by us. When the actual situation is better, pleasant surprise occurs; when it is worse, disappointment or remorse occurs.

In light of the importance of unexpectedness in determining emotional intensity, one way to decrease negative emotional impact is to lower our expectations. In doing so, we will be less frustrated by negative, and more surprised by positive, events. People who expect nothing will never be disappointed. However, their positive emotions will be limited as well, since no event will be perceived as a significant change to perceive an event as a significant change implies expectations of the normal situation from which the given event significantly deviates.

A factor related to, but not identical with, unexpectedness is *uncertainty*. We can expect some event to happen but may not be certain of its actual likelihood. Uncertainty is positively correlated with emotional intensity. The more we are certain that the eliciting event will occur, the less we are surprised at its actual occurrence and the lesser the emotional intensity accompanying it. In situations of certainty, the alternative to the situation is perceived as less available and hence emotions are less intense. Spinoza (1677/1985, IV, p. 50,s) emphasizes this variable, arguing that the wise man «who rightly knows that all things follow from the necessity of the divine nature, and happen according to the eternal laws and rules of nature, will surely find nothing worthy of hate, mockery or disdain, nor anyone whom he will pity.»

2.6 Deservingness

The perceived deservingness (equity, fairness) of our situation or that of others is of great importance in determining the nature and intensity of emotions. No one wants to be unjustly treated, or to receive what is contrary to one's wish. Accordingly, the feeling of injustice is hard to bear - sometimes even more so than actual hardship caused. When we perceive ourselves to be treated unjustly, or when the world in general is perceived to be unjust, this is perceived as a deviation and generates emotional reactions. The more exceptional the situation, namely, the more the situation deviates from our baseline, the more we consider the negative situation to be unfair or the positive situation to be good luck. In such circumstances, the issue of deservingness is crucial and emotions are intense. In some emotions, such as pity and envy, the variable of deservingness is very important; in others, such as fear, its role is less significant.

The characterization of deservingness is complex due to its similarity to, yet difference from, *moral entitlement.* Claims of desert, such as "I deserve to win the lottery," are based on our sense of the value of our attributes and actions. Claims based on moral right, such as "she is entitled to receive a raise in her salary," often refer to obligations constitutive of the relationships with other agents. Claims of desert are not necessarily grounded in anyone's obligations, but rather in the value persons perceive themselves to deserve. A major reason for the private nature of claims of desert is that they are often based on personal desires. Claims based on moral right refer to some mode of *treatment* by other persons, whereas claims of desert also refer to the *fairness* of the situation. When we perceive our situation to be undeserved, we do not necessarily accuse someone else of criminal or immoral behavior; we assume, however, that for us to be in such a situation is in some sense unfair. Similar considerations apply to circumstances in which we perceive our situation to be deserved.

In typical claims based on moral right the agent is a person with some responsibility, whereas in claims of desert an impersonal cause can also be an agent. Heavy rain may be the cause of undeserved but not of immoral circumstances. Similarly, whereas claims based on moral right are directed at humans and sometimes at other living creatures, claims of desert may also be directed at inanimate objects. One can say that «Cleveland deserves better publicity, since it is an interesting city.» (Sher, 1987)

Claims of desert are based on *perceived* undeserved or deserved situations that are not necessarily undeserved or deserved in a more objective sense. Perceived undeserved situations may be due to impersonal, arbitrary circumstances. Being unlucky may not involve any criminal or immoral deeds or attitudes of a particular agent, but the unlucky person may still be right to regard it as undeserved. It is not immoral for a rich person to win a big prize in the lottery or to marry another rich person; nevertheless, many poor people may consider it to be undeserved. Being born with a handicap may be considered unfair in the sense that no one deserves such a misfortune, but it does not entail a criminal or immoral deed.

Claims of desert are different from claims based on right, even when both refer to a mode of treatment other persons. Entitlement requires *eligibility* and satisfying some *general* rules, whereas deservingness requires satisfying certain conditions of *personal worthiness* which are not written down in any legal or official regulation. Sometimes claims based on entitlement are also claims of desert, for example, when the winning presidential candidate is the best-qualified person. The two types of claims may conflict if a person is entitled to something she does not deserve, or deserves something for which she is not entitled. An informer who betrays his brother is entitled to the advertised reward, but he does not deserve it; conversely, a defeated presidential candidate, who is the best-qualified person, deserves to be the president, but is not entitled to it. It is obvious that claims based on right do not exhaust the normative terrain of fairness.

Although emotions may sometimes involve claims based on moral right, claims of desert are more typical. Claims of desert typical of emotions are personal and are only rarely directly relevant to moral actions. Such claims are not considered as serious moral claims; dismissing many of them as morally irrelevant may be considered as an appropriate moral response. Moreover, in most cases there is no one to whom to address claims of desert, and anyhow these claims cannot be fulfilled in light of practical considerations. When David envies Adam's beauty, his envy involves a desert claim but not a serious moral claim. Telling David that he should be satisfied with his own good fortune in other domains is, in this case, a proper moral response. There is no one to blame for Adam's beauty, and it is impractical to try and change the situation by doing plastic surgery on David and all other people who are not as beautiful as Adam.

The relationship between *emotional intensity and deservingness* is quite complex due to the personal nature of deservingness. In describing this relationship, we should distinguish between the subject's and the object's deservingness as well as between good and bad situations. The general positive correlation typical of the relationship between emotional intensity and other variables is also present between emotional intensity and undeservingness of the subject's bad situations. Thus, envy, jealousy, anger, and hate are stronger the more we consider ourselves as undeserving of our current bad situation. The same holds for the object's bad situations, which we evaluate negatively. For example, pity is stronger, the less the object is considered as deserving the misfortune. In these situations, the subject's and the object's evaluations of the situation are similar, and therefore the correlation between the subject's bad situation and emotional intensity prevails here as well. When the subject's and the object's evaluations of the situations are different, the correlation is determined by the subject's evaluation. For instance, pleasure-in-others'-misfortune is stronger the more we believe that the object deserves her misfortune - even if the object believes otherwise.

Generally, undeserved situations are perceived to be less *normal;* hence the availability of an alternative is stronger and consequently emotional intensity is also stronger. Therefore, in many popular television series (such as *Murder She Wrote),* in which each episode involves the commitment of a murder, the person who is murdered is usually one who behaved criminally and in a certain sense deserves to be punished. Such deservingness decreases the intensity of the negative emotions we have while seeing the murder; it also allows us actively to enjoy the positive emotions we have when the murder is being committed, without feeling guilt.

In good situations, a positive correlation between the deservingness of the person (the subject or the object) enjoying the good fortune and emotional intensity prevails. It is as if "the good guys win." Thus, we are usually more proud about truly deserved praises. Similarly, we are happier with the success of a deserving friend than that of an undeserving one. It can also happen that we enjoy our undeserved good fortune more since we are more surprised at receiving it. In such circumstances, the variable of readiness is different and therefore the general correlation is not maintained.

3. Personal Make-Up

In assessing the significance of an emotional change, our personal make-up should be taken into consideration. Factors such as personality traits, world views, cultural background, and current personal situation are crucial for determining the emotional significance of given events. Differences in personal make-up may result in assigning different *significance* to given events, but they *do not undermine general regularities* concerning a certain intensity variable and emotional intensity. For example, different people may evaluate differently the reality of a given event: some consider the event to pose a real threat to their self-image, while others consider it to be imaginary. Thus, trivial social conversations between married women and other men may be perceived differently by their husbands depending on their personalities and cultural backgrounds. One man may perceive the situation as posing a real threat for him, while another will consider it as posing no real threat at all. The differences in attached significance will result in differences in the intensity of jealousy. However, in both cases the general correlation between the degree of reality and emotional intensity is maintained; the more real the event is perceived to be, the greater the emotional intensity it provokes.

In the same vein, the influence of culture is mainly in the perception and interpretation of the significance of events and not in shaping general appraisal regularities. Thus, cultural differences may determine differently the relevance to our well-being of a certain event, or the degree of accountability others have for their behavior, but these differences do not affect the general positive correlation between emotional intensity and the event's relevance or between emotional intensity and accountability. Although the emotional events that elicit emotions and the significance of emotions may differ appreciably from one culture to another, the dimensions of appraisal, the major patterns of appraisal, and the regularities of intensity variables are highly similar (Frijda & Mesquita, 1994; see also Ellsworth, 1994).

When we describe someone as emotional we refer, among other things, to the great sensitivity of the person: emotional reactions are easily invoked in the person. Highly emotional individuals perceive the events of their daily lives as being more significant than do those with less emotional sensitivity. The world of highly emotional people is a place where many events assume great significance. These people do not go out seeking emotionally charged situations, but react more strongly to everyday situations that are perceived by them as more significant. Highly emotional men would be more easily drawn to attractive women and more easily repelled by unattractive women than would men with a lower level of arousal. People of low emotional sensitivity have to look for unique events, or even create unique events (for example, a mountain climbing expedition), in order to be confronted with such significant events.

Differences in personal make-up, then, are not variables of emotional intensity, but factors determining the significance of given events. Such differences alter the significance of each variable in terms of the function describing the relationship between a certain emotional variable and emotional intensity, but basically they do not change the shape of the function. Since my main concern here is in understanding general relationships constituting emotional intensity, personal and cultural differences are not the focus of my discussion.

Personal make-up can be divided into two parts: (a) personality, and (b) personal current situation. Variables of the first group are relatively stable and include, for example, personality type (e.g., nervous or calm), sensitivity to other people, fundamental beliefs (e.g., moral and religious beliefs), gender, age, and cultural background. Variables constituting our current situation are more transient and include, for instance, our moods, attitudes, and personal resources.

4. Conclusions

The notion of "emotional intensity" has been found to be an extremely complex notion. Nevertheless, emotional intensity can be described and predicted. This paper has described six basic variables of emotional intensity: the strength of the event generating the emotion, the reality of the event, the relevance of the event, our accountability of the event, our readiness to the event and our deservingness to be in our present situation.

I have proposed a clear correlation between each variable and emotional intensity. A significant issue in describing the correlation between each variable and emotional intensity is that of the seeming exceptions to the general correlation. It may be argued that dealing with such exceptions requires assigning a negative sign to the variable in the exceptional situations. I have suggested leaving the value of the specific variable intact, but changing the values of other variables. The so-called "exceptions" are exceptions only from a local and partial perspective; from a general perspective, they tally with the overall function.

One way of examining the validity of my proposed framework is to construct a computer simulation of the suggested list of variables and their relationships. Although this may be a very complex task, it is possible to accomplish, at least in principle (see, e.g., Elliott, 1992).

The intensity variables are global in the sense that they are related to all emotions. This does not mean that they are necessarily prominent in every emotional situation. For instance, the issue of readiness may not be significant in sexual desire, but it is not entirely irrelevant to this emotion either - in some cases, readiness greatly influences the intensity of sexual desire. Generally, the more complex the emotion, the more that variables of emotional intensity are likely to be associated with its emergence. In addition to global variables associated with all emotions, there may be also local variables that derive from the particular nature of the given emotion.

Determining the influence of a certain variable should be limited to comparisons within a given emotion. Thus, it is misleading to say that anger, which is typically characterized as having a low degree of controllability, since it is primarily caused by others, would always be a less intense emotion than shame, which is characterized as having a high degree of controllability. The positive correlation between controllability and emotional intensity is maintained in both anger and shame: a greater degree of controllability will result in more intense anger and in more intense shame. Although the correlation between each variable and emotional intensity is positive in all emotions, the specific curve depicting the details of this correlation may vary from one emotion to another.

An important task for future research is that of determining the adequacy of the suggested correlations in specific emotions. There may be emotions in which the general curve is somewhat modified or may not apply at all. I do believe that the general correlations attributed to the relationships between emotional intensity and each variable are valid for all types of emotions, but this belief should be examined in more detailed empirical research.

The proposed framework for characterizing emotional intensity has important implications for understanding the emotional process and for emotional management. Regulating emotional experiences should refer, among other things, to the intensity variables.

References

Ben-Ze'ev, A. (1996) "Emotional intensity", *Theory and Psychology* 6:509-532.

Ben-Ze'ev, A. (2000) *The Subtlety of Emotions,* Cambridge, MA: MIT Press.

Elliott, C. (1992) *The Affective Reasoner: A process Model of Emotions in a Multi-Agent System*, Doctoral dissertation, Northwestern University.

Elster, J. (1999) *Alchemies of the Mind: Rationality and the Emotions*, Cambridge: Cambridge University Press.

Ellsworth, P. C. (1994) "Sense, culture, and sensibility",, in: *Emotion and Culture*, S. Kitayama and H. R. Markus, eds, Washington, DC: American Psychological Association.

Festinger, L. (1954) "A theory of social comparison processes", *Human Relations* 7:117-140.

Folger, R. (1984) "Perceived injustice, referent cognitions, and the concept of comparison level", *Representative Research in Social Psychology* 14:88-108.

Frijda, N. H. (1986) *The Emotions*, Cambridge: Cambridge University Press.

Frijda, N. H. (1987) "Emotion, cognitive structure, and action tendency", *Cognition and Emotion* 1:115-143.

Frijda, N. H. (1988) "The laws of emotion", *American Psychologist* 43:349-358.

Frijda, N. H., P. Kuipers and E. ter Schure (1989) "Relations among emotion, appraisal, and emotional action readiness", *Journal of Personality and Social Psychology* 57:212-228.

Frijda, N. H. and B. Mesquita (1994) "The social roles and functions of emotions", in: *Emotion and Culture*, S. Kitayama and H. R. Markus, eds, Washington, DC: American Psychological Association:

Frijda, N. H., A. Ortony, J. Sonnemans and G. Clore (1992) "The complexity of intensity: Issues concerning the structure of emotion intensity", *Review of Personality and Social Psychology* 13:60-89.

Gilboa, E. and W. Revelle (1994) "Personality and the structure of affective responses", in: *Emotions: Essays on Emotion Theory*, S. H. M. van Goozen, N. van de Poll and J. A. Sergeant, eds., Hillsdale, NJ: Erlbaum.

Green, 0. H. (1992) *The Emotions*, Dordrecht: Kluwer.

Hill, C.T., Z. Rubin and L.A. Peplau (1976) "Breakups before marriage: The end of 103 affairs", *Journal of Social Issues* 32:147-168.

Hume, D. (1739-1740/1978) *A treatise of Human Nature*, Oxford: Clarendon.

Karasawa, K. (1995) "An attributional analysis of reactions to negative emotions", *Personality and Social Psychology Bulletin* 21:456-467.

Lazarus, R. S. (1991) *Emotion and Adaptation*, New York: Oxford University Press.

McCornack, S. A. and T. R. Levine (1990) "When lies are uncovered: Emotional and relational outcomes of discovered deception", *Communication Monographs* 57:119-138.

Ortony, A., G. L. Clore and A. Collings (1988) *The Cognitive Structure of Emotions*, Cambridge: Cambridge University Press.

Roseman, I. J. (1991) "Appraisal determinants of discrete emotions", *Cognition and Emotion* 5:161-200.

Scherer, K. R. (1988) "Criteria for emotion-antecedent appraisal: A review", in: *Cognitive Perspectives on Emotion and Motivation*, V. Hamilton, G. H. Bower and N. H. Frijda, eds, Norwell: Kluwer.

Sher, G. (1987) *Desert*, Princeton, NJ: Princeton University Press.

Sonnemans, J. and N. H. Frijda (1994) "The structure of subjective emotional intensity", *Cognition and Emotion* 8:329-350.

Smith, A. (1790/1982) *The Theory of Moral Sentiments*, Indianapolis: Liberty Classics.

Smith, C. A. and P. C. Ellsworth (1985) "Patterns of cognitive appraisal in emotion", *Journal of Personality and Social Psychology* 48:813-838.

Spinoza, B. (1677/1985) "Ethics", in: *The Collected Works of Spinoza*, F. Curley, ed., Princeton: Princeton University Press.

Taylor, S. F. (1989) *Positive Illusions: Creative Self-Deception and the Healthy Mind*, New York: Basic Books.

Teigen, K. H. (1995) "How good is good luck? The role of counterfactual thinking in the perception of lucky and unlucky events", *European Journal of Social Psychology* 25:281-302.

Thibaut, J. W. and H. H. Kelley (1959) *The Social Psychology of Groups*, New York: Wiley.

Tesser, A., M. Millar and J. Moore (1988) "Some affective consequences of social comparison and reflection processes: The pain and pleasure of being close", *Journal of Personality and Social Psychology* 54:49-61.

Weiner, B. (1985) "An attributional theory of achievement motivation and emotion", *Psychological Review* 92:548-573.

Whitley, B. F. and I. H. Frieze (1986) "Measuring causal attributions for success and failure: A meta-analysis of the effects of question-wording style", *Basic and Applied Social Psychology* 7:35-51.

Van Evra, J. (1990) *Television and Child Development*, Hillsdale, NJ: Erlbaum.

EMOTIONAL QUALIA

CLOTILDE CALABI
Dipartimento di Filosofia, Università de gli Studi di Milano, Milan, Italy

ABSTRACT

The functionalist claims that mental states are functional states. He therefore identifies kinds of mental states with causal roles. One objection against functionalism is that qualitative states are mental states, but cannot be identified with any causal role. An argument in support of this objection is the qualia inversion argument, which runs as follows: if inverted qualia are possible in functionally equivalent systems, then qualia are not definable in functional terms, and functionalism cannot be an account of all psychological states and properties. A parallel argument concerns absent qualia. This chapter focuses on the inverted qualia argument and discusses it with reference to emotional qualia, a case of inversion which is more difficult to pin down than the case of color inversion. One reason for this difficulty is that we can describe both the space and the spectrum of colors and indicate laws governing them, whereas emotions are more resistant to be arranged in a spectrum. Why is this so? Because there seem to be no laws connecting emotional qualia to one another and to qualia from other domains (like the visual, the tactile or the acoustic domain), whereas for phenomenal colors the opposite appears to be true. Although we sometimes describe anger as red and envy as green, there are no conceptual connections between these emotions and the corresponding colors.[1] In fact, as it turns out, most of the a priori laws concerning emotions are about their cognitive bases.[2]

1. Introduction

In this chapter, the following issues are considered. First, is there a spectrum of emotions? Second, would inversion be possible without being detectable or would it rather undermine some conceptual truths concerning our emotions, therefore making itself detectable, as is the case with color inversion?

---

[1] There may be some contingent connections between emotional qualia and phenomenal colours, in that red is perceived as a warm hue, whereas blue and green are perceived as cold (cf. on this Hardin, 1988, p. 129). But even in this case, why should envy be considered as cold? Yet there are contrary opinions on this issue. For example, Shweder (1991, p. 246), claims that "The association of red with anger and black (and white) with bereavement is no historical accident" (quoted by Flanagan, 1992, p. 71).

[2] For example, regret is based on the following law: if I regret that p, then I necessarily believe that p has occurred in the past, that is regret is necessarily based on a belief concerning a past event. To regret about something that we know has not (yet) occurred is to violate a grammatical rule governing our concept of regret, and not simply to contradict some contingent facts.

2. Color Inversion

It has been argued that if color inversion occurs, then it is detectable. Various arguments have been presented in support of this thesis. I will confine myself to present only one of them very briefly since my primary goal is to see if we can construct a parallel argument for emotional qualia.

Color inversion for complementary colors has proved to be functionally detectable for various reasons, including the following. Philosophers working on color, in presenting the simplest case of inversion, invite us to imagine an individual for whom any color is reversed to its complementary: where we normally experience red he experiences green, where we experience blue he experiences yellow (see Harrison, 1973; Casati, 1990). To each hue of the spectrum corresponds a specific degree of brightness, for example, yellow is necessarily brighter than blue and conversely, blue is necessarily darker than yellow. Suppose now that colors (in this case analysis is confined to pure colors) are arranged in the shape of a solid, for example a sphere. It has been argued that in this representation, the positions of black and white must also be reversed (white is necessarily brighter than black). Now, black and white inversion is detectable. If we see in a succession a white patch on a black background and a black patch of the same size on a white background, the white patch appears bigger than the black one. Now, a person who undergoes this kind of inversion would see smaller what I see as bigger and vice-versa, and it would be easy to detect this difference between her and me. Hence, the ultimate reason which allows us to detect inversion of complementary colors in this case is the holistic character of visual perception: in this particular case, color perception is tied to perception of size.

Now, can we transpose this experiment to emotional qualia? More precisely, can we argue also in this case that if inversion occurs, then it would be detectable? Can we further argue for the stronger thesis that inversion would not even be possible because it would undermine some conceptual truths regarding emotions which necessarily hold given our emotional lexicon? The first step of the argument is to establish whether there is a spectrum of emotions and if it is characterized by specific asymmetries as is that of colors with respect to some perceptual concepts.

3. Is There a Spectrum of Emotions?

In order to answer the above question, we need to know if we can construct a spectrum of emotional qualia. Now, both problems of spectrum and inversion depend on the problem of classification. Notice that the second problem depends on the first one: in order to present a case of global inversion (i.e., not confined just to one couple of emotional qualia) we need to know how emotional qualia can be arranged. So, the first question is how can emotional qualia be classified. To start with, which mental phenomena should we consider emotions?

According to a widespread view, emotions are complex events with mental and physical components, related to each other by causal and internal relations. The physical component is a physiological change of the organism (for example, a change in blood pressure and galvanic skin response), which can be consciously felt but is not necessarily so. This change is a response of the organism to a stimulus and determines a bodily reaction (for example flight). The mental component is a cognition: emotions depend on the cognition[3] of a state of affairs or object, which is cognized as important, urgent and affecting the subject: this cognition is both a cause and a reason of the corresponding emotional state, that is, it is both in a causal and an internal relation with it.[4] Stimuli eliciting emotions, when consciously perceived, are intentional objects or states of affairs provided with natural and (sometimes) axiological properties.

Emotions are also constituted by a specific mode of action readiness, they are characterized by a specific valence, either positive or negative, and they involve focalization. In other words, emotions have as their components a feeling either of dominant pleasure or of a dominant pain (for example, in disgust pain is generally dominant over pleasure, whereas in hope pleasure is generally dominant) and, to some extent, an exclusive consciousness. Attention is identical to this exclusive consciousness: emotions capture attention and direct it in specific directions.[5]

Last but not least, an emotional experience is characterized by a certain feeling or felt quality, of which the felt pleasure or pain is just one part. This feeling has no content of its own and therefore lacks intentionality. It is what I have labeled an "emotional quale". All the elements I have indicated, excluding the quale are usually taken into consideration when one gives a functional account of emotion types. However, the qualia friend says that these other elements may be necessary, but they are not sufficient: one also needs qualia in order to account for emotions and according to the antifunctionalist, we cannot account for them functionally.

What characterizes emotional qualia, besides a certain amount of pleasure or pain? Among their features there are a certain degree of intensity and

[3] The cognition is not necessarily conceptual: some emotion types are necessarily based on beliefs, others are not.

[4] In this context, I do not consider the possibility that the cognitive basis of the emotion is different from the cognitive state which has caused it, that is, I do not consider the possibility that causes and reasons do not coincide.

[5] Exclusive consciousness can be functionally analyzed, if we consider the functional role of attention as that of an *excluder*. This means that attention is causally connected to omissions rather than to actions of particular kinds. And the omissions themselves are of a peculiar kind: if I read something with extreme interest, I may be ready to notice possible mistakes and not be prepared to realize, for example, that someone is ringing the door bell. My doing something attentively is compatible with doing it in a world in which many things could happen that I do not notice, that is of events with which I would not be in a relation of acquaintance: I would omit being acquainted with them.

localization.[6] But these features are not sufficient to classify and hence distinguish emotional qualia from one another. In fact, distinctions are often made on the account of the cognitive bases. Now, of course there is a difference between what it is like to be disgusted, what it is like to be hurt, or full of hatred, but this difference is not simply a difference in the respective cognitive properties, modes of action readiness and kinds of appraisal characterizing pain, disgust and hatred: there is also a different feeling associated with each of them. One way to indicate this difference is to say that the pain of disgust feels different from the way pain of hatred feels, that is, hurtfulness of disgust is different from the hurtfulness of hatred. But once again there is more to the feeling of disgust than a particular kind of hurtfulness, as there is more to the feeling of pride than a particular kind of pleasurableness

Now, does this complexity give us criteria for distinguishing emotional types? Can we make a distinction between the hurtfulness of mourning and the hurtfulness of disgust? Unfortunately we cannot: we can make a distinction between the two only by reference to their respective cognitive and evaluative bases (but not in terms of the corresponding qualia). So, the only parts we could detect in some way in the emotional quale are the dominant pleasure or pain. We shall now see if these components will allow us to arrange emotions in a spectrum.

With colors we have objective ways to indicate inversion: for example, the color solid can be rotated 180°. But what about emotional qualia? If we want to apply inversion to emotional qualia by using the model of color inversion, we should consider emotions as arranged along a line from more positive to more negative emotions, while considering their being positive or negative depending on their having a certain amount of pleasure or pain as dominant.[7] Along this line, we can see that emotions which are polarly opposed to one another in terms of their appraisal dimension, occupy opposite positions. Then we systematically invert their dominant pleasure and pain, without changing anything in the cognitive basis and in the other elements which form the emotional whole and see if it is possible to detect this inversion.

But there is a problem in this procedure. Although all emotions have valence (either positive or negative), we cannot establish which one characterizes a particular emotion token just by its belonging to one type: a dominant pleasure (or a dominant pain) does not necessarily belong to each emotion type in the manner that a particular degree of brightness belongs to each phenomenal color. For example, it may be possible that a token of disgust has pleasure as dominant, without changing any of the cognitive, evaluative and behavioral properties which

[6] Moods, as opposed to emotions, are not localized.

[7] The underlying presupposition if the argument is that emotions are necessarily either positive or negative: if they are positive they are internally related to pleasure, whereas if they are negative they are internally related to pain. Examples of opposite emotions are love and hatred, pride and humility, hope and despair, etc.

characterize it as token of that emotional type. This entails that emotional qualia types cannot even be characterized in terms of their specific pleasurableness or hurtfulness.

For this reason, there cannot be a spectrum of emotional qualia as there is a spectrum of colours: emotional qualia cannot be arranged, not even in a one-dimensional spacer.[8] Hence, we must stick to inversion of the pure feelings of pleasure and pain and see how it works, without extending the strategy to more complex emotional qualia: we cannot consider pleasure and pain as embedded in larger contexts. Let me return now to the inverter antifunctionalist.

4. Emotional Inversion

Can he prove that there may be a case of qualia inversion in this specific domain that does not make a functional difference? If he does, then he will have a strong argument against functionalism. If he does not, the functionalist can account for pleasure and pain, but not for more complex emotional wholes. Of course, the last resource open to the functionalist would be the possibility to "quine" off these more complex qualia, and simply confine his analyses to the emotions' cognitive bases. In this case, even if these qualia existed, they would be irrelevant to functional explanations in all cases but the cases of pure pleasure and pain.

In order to spell out the antifunctionalist strategy, let me consider the following example. Suppose that I have a toothache, I think that I should go to the dentist, pain does not allow me to sleep or to relax, makes me groan and does not allow me to think of anything but my toothache. Besides being in pain, I am also afraid of the pain the dentist will cause me. Now consider another individual (Twin l) who is in my same functional states, that is, she gets ready to do the very same things, she cannot sleep, she groans, has a fear of the dentist, etc., yet she does not feel pain, but instead intense pleasure.

A second individual (I call her Twin 2) in these same conditions feels an itch instead of pain: the state she is in occupies the causal role of pain, but she does not feel pain. She feels what we ordinarily feel when we are in a state that for most of us occupies the causal role of an itch. Some philosophers try to explain this situation in terms of cross-wired brains. For the moment I leave aside any concern for brains and their states and confine my analysis to the relation between functional states (no matter how they may be realized) on the one hand, and qualitative states on the other.

---

[8] Mulligan (1995) argues for the existence of a spectrum of emotions, but in fact, he mainly characterizes it in terms of the cognitive and evaluative bases. Polar opposition between admiration and contempt differs from polar opposition between like and dislike and between pleasure and pain. If my account is correct, the first opposition concerns the bases of admiration and contempt, and not the emotional qualia..

Are these two situations possible? The functionalist should claim that they are not. However, as I construe his position, he may try to work out some kind of response to the first case, but not to the second. As we will see at the end of the discussion, this kind of inversion appears unmanageable from a strictly functionalist point of view.

Concerning the first case one can reasonably claim that the *concept* of pain necessarily entails adversion against that which provokes pain, that it presupposes the validity of the law that the feeling of pleasure is better than the feeling of pain (necessarily, we prefer the feeling of pleasure to the feeling of pain). Adversion, on its part, is a functional state, that can manifest itself in various ways and has internal and causal relations with other emotional and cognitive states. If I prefer a to be, I necessarily think that a is (either all things considered or prima facie) better than b and my preferring a to b causes me to have some positive actions towards a rather than towards b. Now, is it possible that all these complex relations hold, that the same preferences hold, but that qualitative experience be different? Is it possible that the same functional states are accompanied by a feeling of pleasure instead of pain?

Let me suppose that this happens. Then, it is necessary that this person prefer the feeling of pleasure to the feeling of pain. Preferences are functional states that in principle are realizable in some particular behavior and are in functional relations to other mental states (they cause other types of mental states, for example, evaluations). However, by hypothesis, my twin will have my same adversions and adverse behavior: she will say, think and evaluate as I do, and do the same things I do. In this case these relations concerning adversion will be in contradiction with the functional relations which are at the basis of the preference of the feeling of pleasure to the feeling of pain (there will be, for example, opposed evaluations). Therefore, either of these two possibilities: a) the functional states of Twin 1 and my own cannot be identical; b) we cannot say that, while feeling pleasure, Twin 1 *prefers* the feeling of pleasure instead of the feeling of pain. By hypothesis, a) cannot be the case. Yet, concerning b, this preference is part of the definition of the concept of pleasure: a particular state cannot be judged as falling under the concept of pleasure (i.e. cannot be described as pleasure), if the one who is in this state does not prefer it to the opposite state. As a consequence, Twin 1's state will not be subject to an important functional law and if this functional law is essential to the corresponding state, we cannot state that Twin 1 is in a state of pleasure instead of being in a state of pain.

The one sketched above can be considered as a tentative a priori analysis of pure pleasure as a qualitative state. According to this analysis, pleasure is such that it is a priori preferable to the opposite state. Since to prefer is a functional state, according to the above analysis one could claim that we can give a functional analysis of the states of pleasure and pain: these states are necessarily such that to feel pleasure is better (more preferable) than to feel pain.

Of course, we may have cold preferences that determine us to act in certain ways and we may have thoughts concerning these preferences which entail other thoughts: for example, if I think that a is preferable to b and b is preferable to c, I may draw the further conclusion that a is preferable to c. There may also be a mismatch between these preferences and one's immediate likes and dislikes. We could describe this mismatch as one between one's non-cognitive and cognitive preferences. Contradiction may occur between what one likes prima facie and what one likes all things considered. But both states are functional states.

The inverter may try to argue that it is conceptually possible that Twin 1 prefers a to b and expresses her preferences through a particular adverse behavior and that at the same time she dislikes a and likes b more, but does not manifest her dislikes in any particular way nor connects them to any thought: they are causally and cognitively isolated. More generally, where the functionalist says there is a conceptual relation between functional preferences and raw feelings of pleasure and pain, the inverter objects that the relation between them is only a contingent one.

The functionalist, however, will counter that the relation is necessary: likes and dislikes are in an internal relation to adversion and desire since we cannot conceive a dislike that is not internally related to adversion. This does not entail that adversions necessarily manifest themselves in an adverse behavior. But we should not confuse this claim with the other claim. The objection misinterprets the functionalist's point of view as a strictly behaviorist point of view, which it is not.

Once this conclusion has been reached, the functionalist cannot extend it further. That is, although he has proved that pure pleasure and pain do not survive transposition, he cannot prove that complex emotional states do not survive it: in fact we have evidence that they do. Contrary to what Hume thought, it is not true that pride is intrinsically pleasurable and humility is intrinsically painful: we could in fact easily imagine someone for whom humility is a pleasurable state and who has the same cognitive, axiological and behavioral asset of the one for whom humility is a painful state. If what I said above is correct, inversion in this case is not detectable as it is in the case of pure pleasure and pain. The most we can say concerning emotional qualia is that they contain either a dominant pleasure or a dominant pain, but we cannot argue on a priori grounds which contains which: once again, one cannot assign to each emotion type a corresponding specific pleasure or pain, in the manner that a specific brightness corresponds to each hue. My first conclusion is that if the argument given above is correct, inversion cannot hold for pure states of pleasure and pain, but it may be possible in larger emotional contexts. In these larger contexts, this kind of inversion would not entail any failure of conceptual laws concerning our emotional lexicon. Essentialism holds only for pure pleasure and pain.[9]

---

[9] One may ask which are the essential properties of pure pleasure and in which contexts would pleasure occur. In fact, it may be argued that pure pleasure is more difficult to detect than pure

What about the itch-pain inversion? I do not think it would be detectable: itch has no internal relations with any emotional state, it has no cognitive content, hence no structure and *per se* involves no preferences, unless it is considered as intrinsically painful. But if it is intrinsically painful it would just be an example of pure pain.

Several further conclusions can be drawn at this point. Emotional qualia do not have much structure. Although they are complex, we cannot establish on a priori grounds which are their parts and, a fortiori, we cannot indicate how these parts are related to one another.[10] Interestingly, emotions are often considered a good example of holism characterizing mental life: they are contagious in that they influence our beliefs, direct our attention, and more generally they are at the basis of our cognitive make-up. But what exactly is in charge of this influence?

Unfortunately, although holism is an undeniable phenomenological truth, we cannot say much about it. Yet, despite Wittgensteinian-like arguments we cannot deny that emotional qualia exist, and we all know what it is like to be proud, surprised, hopeful or disgusted and there is much more to each of the corresponding episodes than hurtfulness (or for what matters pleasantness). Furthermore, it is plausible to think that this extra modifies in a substantial way a particular feeling of hurtfulness, when this latter occurs. However, it is not only that we cannot say what makes the difference between hurtfulness of an episode of disgust and hurtfulness of an episode of fear or of mourning, if not by reference to their cognitive basis: we cannot even say that disgust necessarily hurts. Hence, we have no criteria of identity for emotional qualia (with the exception of pure pleasure and pain) and they cannot be considered as functional states. Is there any other way we could account for them? Type identity theories may do the work, but their acceptability does not come without a significant price.

References

Casati, R. (1990) "What is wrong in inverting spectra", *Theoria,* 1:183-186.
Flanagan, O. (1992) *Consciousness Reconsidered,* Cambridge, MA: MIT Press.
Hardin, C.L. (1988) *Color for Philosophers: Unweaving the Rainbow,* Indianapolis, IN: Hackett,.
Harrison, B. (1973) *Form and Content,* Oxford: Basil Blackwell.

---

pain. I think that pure pleasures are of an Aristotelian variety, that is, they are pleasures taken in the performance of some activity. Aristotle's general thesis is that pleasure adds perfection to the activity itself. By this he means that pleasure taken in an activity modifies it in a substantial way, by making it more lively and more involving. Now, this species of pleasure can be identified with attention as I analyzed it above. Since we can give a functional account of attention (attention is an excluder, as I suggested above), we can also give a functional account of pure pleasure.

[10] By this I am not denying however that qualitative experience might also yield to neural decomposition. In fact holism is compatible with neural decomposition. See Flanagan (1992, p. 64) on this issue.

Horgan, T. (1984) “Functionalism, qualia and the inverted spectrum”, *Philosophy and Phenomenological Research,* 44:453-469.

Jackson, F. (1982) “Epiphenomenal qualia”, *Philosophical Quarterly,* 127-136. Reprinted in:, *Mind and Cognition. A Reader* (1990), W. Lycan, ed, London: Basil Blackwell.

Lewis, D. (1980) “Mad pain and Martian pain”, in: *Readings in the Philosophy of* Psychology, N. Block, ed, London: Methuen, , vol. I, pp. 216-222.

Mulligan, K. (1995) "Le spectre de l’affect inversé et l’espace des émotions", *Raisons Pratiques,* 6:65-82.

Shoemaker, S. (1982) “The inverted spectrum”, *The Journal of Philosophy,* 79:357-381.

KARL JASPERS' PHENOMENOLOGICAL APPROACH TO EMOTION IN HIS *GENERAL PSYCHOPATHOLOGY*

ANDREW L. GLUCK
392 Central Park West #8C, New York, New York 10025, USA

ABSTRACT

Karl Jaspers was a pioneer in the study of consciousness but a much neglected one. His interest in the subject resulted from his practical experience as a psychiatrist as well as his philosophical interests in human identity, freedom and world-views. In a sense he bridges the gap between the phenomenological school and the hard sciences that he respected so much. His interest in emotion stemmed from his studies of consciousness but unlike some treatments of consciousness, the study of emotion must take into consideration biological factors.

1. Introduction

Karl Jaspers is best known as an existentialist philosopher but that label is a somewhat misleading one. Unlike many of the other existentialists he always took a great interest in science and, indeed, began his career as a psychiatrist. His *General Psychopathology*, upon which this paper is based, was at one time a standard text in many European medical schools and has been translated into many languages.

Jaspers' phenomenological treatment of emotion is part and parcel of his concept of *Verstehen* in general and his psychology of meaningful connections in particular. He believed that no theory of mind that does not capture some of the richness of human experience could be adequate. But unlike the detached manner in which perception was often studied, the study of affective states can't be separated from either biological processes or world-views. While the former are constituent factors of human emotion, there is more to emotions than the physiology. And there is far more to conscious world-views than beliefs. There is a quality of comprehensibility underlying most psychic events. Human thought and perception also have an emotional quality though much of it is not noticed in ordinary life.

2. Verstehen

There are a number of terms that have been used somewhat interchangeably in the philosophy of the social sciences to depict a method which is distinct from the methodology of natural science which seeks out regularities under the categories of laws or causal connections. And just as there is some confusion between the *laws* of nature, *theories* (as human creations) and *causes* (as something existing in nature), similar confusion exists regarding this alternative methodology. A glance at the literature will reveal such terms as *Verstehen*, subjective meaning, interpretive understanding. *"Verstehen"* and "interpretive understanding" are roughly equivalent technical terms denoting methods employed by thinkers like Dilthey, Weber and Jaspers.

2.1 Max Weber

In our everyday thinking we explain human activities as the outgrowth of intentions and beliefs. But that kind of explanation, often called "folk psychology," seems to conflict with the scientific ideal of causal explanation. One of the most intractable problems in the philosophy of the social sciences has been the relationship between meaningful connections and causal connections. Weber was careful to distinguish them from one another conceptually while insisting that they are both necessary for an adequate social science. For example, in order to explain the influence of wage payment on worker performance it might be thought that one could simply correlate the payment of wages with certain objectively defined dependent variables but a difficulty arises. When a person receives a wage, he or she does not simply receive a piece of paper or metal but must also have an expectation regarding how others will react to that paper or metal. Description of wage payment has perforce an interpretive aspect. If we were to cleanse it of that, we could only say "he was given a piece of metal" but such descriptions would be inadequate; such social acts can only be understood in terms of beliefs and expectations. *Verstehen*, for Weber, refers to particular and unique phenomena.

> The type of social science in which we are interested is an empirical science of concrete reality (*Wirklichkeitswissenschaft*). Our aim is the understanding of the characteristic uniqueness of the reality in which we move. We wish to understand on the one hand the relationships and the cultural significance of individual events in their contemporary manifestations and on the other the causes of their being historically so and not otherwise. (Weber, 1949, p. 72)

Weber insisted on verifiable outcomes in social science. Causal / statistical analysis must coexist with the understanding of concrete individuals. Each is dependent upon the other.

> Statistical uniformities constitute understandable types of action, and thus constitute sociological generalizations, only when they can be regarded as manifestations of the understandable subjective meaning of a course of social action. Conversely, formulations of a rational

> course of subjectively understandable action constitute sociological types of empirical process only when they can be empirically observed with a significant degree of approximation.(Weber, 1978, p. 12)

2.2 Karl Jaspers' Psychology of Meaningful Connections

Many of Jaspers' arguments for a *verstehende* social science overlap with the better-known work of Weber and I will, therefore, concentrate on his psychology of meaningful connections. Jaspers views this psychology as hanging precariously between empirical psychology and philosophy. It is based on an immediate intuitive insight that is analogous to the intuition that we have about causal connections. Unlike causal analysis, however, it is not generalizable, tells us nothing about whether an event will occur again but only about the *meaning* of that particular event. Jaspers adapted the *Verstehen* approach of Weber to the study of psychiatry but in order to do that he also needed to adopt some descriptive phenomenological methods from the philosopher Edmund Husserl. For Jaspers, the overall state of consciousness is a self-evident feature of psychic life.

> When speaking of individual phenomenological data, we have temporarily pre-supposed that the total state of the psyche within which these data occur remains the same... But in actuality the total state of psychic life is extremely variable and the phenomenological elements are by no means always the same...Traditionally this fundamental fact has been emphasized by distinguishing the *content* of consciousness...from the *activity of consciousness itself*. (Jaspers, 1963, p. 137)

Phenomenology allows him to adapt the *Verstehen* method to the study of individuals.

> Psychic events "emerge" out of each other in a way which we understand. Attacked people become angry and spring to the defense, cheated persons grow suspicious. The way in which such an emergence takes place is understood by us, our understanding is genetic... The evidence for genetic understanding is something ultimate. When Nietzsche shows how an awareness of one's weakness, wretchedness and suffering gives rise to moral demands and religions of redemption, because in this roundabout way the psyche can gratify its will to power in spite of its weakness, we experience the force of his argument and are convinced... We might deny the object of psychological understanding altogether and

> maintain that phenomena, psychic contents, expression, extra-conscious mechanisms are all subjects for empirical research alone, while the possibilities of Existence itself are purely a matter for philosophy... But this meaningful psychology is always in balance between these two realms and we can never speak of it in isolation. It is related to them both and if there is to be a complete presentation they cannot be separated. (Jaspers, 1963, pp. 302-312)

Weber had described subjectively meaningful interpretations as hypotheses that had not been proven until statistical regularities could be found. But Jaspers, despite his obvious interest in causal connections, seems to go beyond that by saying that meaningful connections can stand on their own (to a point). Conceptually, the grasping of a meaningful connection requires no additional corroboration; but this does not render it truly scientific. As he points out, opposite interpretations are often equally meaningful and "understanding is inconclusive" because "That which is meaningful is itself inconclusive because it borders on the un-understandable, on what is given, on human existence and on the freedom of Existence itself." (Jaspers, 1963, p. 357). This should not be interpreted as an anti-scientific attitude. Jaspers states explicitly that there is no limit to the empirical and causal study of human phenomena. Every such phenomenon can be subjected to this kind of investigation *ad infinitum*. But humanity is always more than we can know and is never simply an object in the world.

3. Karl Jaspers' Treatment of Emotions

3.1 General Modes of Classification

1. Phenomenologically (the modes in which they appear): feelings may be
    (a) Aspects of conscious personality vs. lending color to object awareness
    (b) Grouped in opposites
    (c) Without an object (contentless) or directed upon an object
2. According to their object. The contrast is between "fantasy" feelings that are directed on to suppositions and reality feelings that are directed upon real objects.
3. According to the source. A distinction is made between different levels of psychic life: localized feeling-sensations, feelings regarding the entire body, psychic feelings like sadness and joy, spiritual feelings like the state of grace.
4. According to the biological purpose. Pleasurable feelings express advancement of those purposes and unpleasurable ones express their frustration.
5. Particular feelings are distinguished from all-inclusive ones. The latter is a temporary whole called a "feeling-state." Examples are irritable feeling-states, increased (decreased) excitability.

6. Feeling, Affect and Mood. Feelings are unique commotions of the psyche. Affects are momentary, complex emotional processes of great intensity and visible bodily changes. Moods are states of feeling that accompany prolonged emotion and color the psychic life.
7. Feelings and sensations. Feelings are states of the self. Sensations are elements of perception of the environment or one's body. There are feeling-sensations that are both, i.e. hunger, thirst, fatigue and sexual excitement.

3.2 Abnormal Feeling States

These can be classified broadly into two categories. The first emerge in "understandable fashion from some experience" but are exaggerated. The second consist of affective states that "defeat understanding" and must be explained in terms of factors beyond consciousness. One can distinguish between normal homesickness, pathological but understandable homesickness and depression that is interpreted by the patient as homesickness. This is a further categorization:
(a) Changes in bodily feeling. Bodily feelings are always associated with emotion but in these pathological conditions the connection become strange and difficult to comprehend.
(b) Changes in feeling of capacity. We ordinarily have a feeling of confidence in our abilities without even recognizing it. In these cases the patient becomes distressed from feelings of insufficiency.
(c) Apathy. The patient is completely conscious but with flat affect. In extreme cases there is no emotion whatsoever and he/she would die if not for external intervention.
(d) The feeling of having lost feeling. This is somewhat different from apathy because the patient has the unpleasant feeling of not having any feeling whatsoever.
(e) Changes in the feeling-tone of perception. Ordinary experiences take on a peculiarly unpleasant feeling.
(f) Unattached feelings (free-floating feelings) Some examples are as follows. 1) Free floating anxiety, 2) Anxiety linked with feelings of restlessness, 3) abnormal feelings of happiness, mystical ecstasy, etc.
(g) The growth of private worlds from unattached feelings. Unfamiliar feelings act as an incentive to create a private world. This also occurs as a result of epileptic auras or drug-induced experiences and often takes on a religious or spiritual character.

In a section entitled "Attention and Fluctuations in Consciousness" under the heading "Clouding of consciousness" Jaspers describes some pathological effects of emotion on consciousness.

> When there are *violent affects*, as in anxiety states and deep

> melancholia as well as in manic states, it becomes much more difficult to concentrate on anything external, contemplate anything, reach a judgment, or even think of anything... For this reason... the contents of delusion-like ideas go unscrutinised by the patient and there is no reality-judgment concerning possible sense-deceptions. Consciousness is completely filled by the affect, and judgment and attitude become very disturbed in an understandable way. This is even more the case in depressive states when primary inhibition of function is added. All the above states, however, deserve the name of abnormal consciousness, which may become a persisting emptiness of consciousness in the last named instance. (Jaspers, 1963, p. 14)

4. Concluding Summary

Jaspers' treatment of emotion links those states with physiology on one side and with world-views on the other. Emotional states condition one's world-view and one's world-view conditions one's emotional states. The subject of world-views is distinguished from the more discrete study of cognition and sensation as it contains volitional elements that color the person's entire psychic life. It is the individual analog of culture and has not been given the credit it deserves as to its effect on human experience. Hopefully, this deficiency will be addressed by the emerging field of consciousness studies.

References

Jaspers, K. (1963) *General Psychopathology*, Chicago: University of Chicago Press.

Weber, M. (1949) "'Objectivity' in social science," *The Methodology of the Social Sciences*, New York: The Free Press.

Weber, M. (1978) *Economy and Society*, Berkeley, CA: University of California Press.

EMOTIONS ASSOCIATED TO COGNITIVE REVISION AS A BASIS FOR VALUES

PIERRE LIVET
Department of Philosophy, University of Aix-Marseille I, CEPERC, 29 Avenue R.Schuman, 13621 Aix-en-Provence, CEDEX 1, France

ABSTRACT

Emotion (in contrast to feelings and sentiment) is an affective reaction to a situation (peculiar or exceptional) differing from our expectations about the way things are normally going on. As such, it can trigger a cognitive revision, carried out in such a way as to minimise changes of our high priority expectations. Accommodation of emotion (a long-term habituation) is achieved when revision is. But when an emotion links local expectations and more fundamental ones, revision can be stuck because of the conflict between the minimisation of revision and the emotion questioning our high priority expectations. As our preferences are revealed by our choices, our "real" values are revealed by the fact that emotion resists accommodation, and our "strong" values by the fact that this resistance is the case even when cognitive revision is achieved.

1. Introduction

I propose to focus not directly on value judgements, but on the relationship between emotion, values, and revision. Revision is the process by which we cancel the premises (beliefs and preferences) that lead to our expectations (beliefs and plans of action) when a discrepancy appears between them and new facts (for cognitive revision) and their consequences (for revision of preferences leading to self-defeating consequences). As we are interested in the dynamical aspects of cognition, premises are supposed here to be expectations, that is, revisable rules of inference indicating as their conclusions the way things are normally going on. These expectations are mostly implicit and even unconscious. They can have perceptual, motor, affective and conceptual content.

2. Emotion and Revision

Emotion can then be defined as the affective and physiological experience caused by the occurrence and the perception of a situation which does not match the conclusions of such expectations about the way things are normally going on. Emotion is positive or negative; depending on weather the discrepancy is in favor of our preferences and plans of action or conflicting with them. Emotion differs from feelings because feelings are not related to cognitive revision, and from sentiment because of its rather acute dynamical profile (sentiments are more stable). Emotion decreases in two ways. It cannot be kept continuously at the same level of intensity, if repeated at a short interval, so habituation occurs. And as the same situation is repeated again and again, but at a longer interval, most emotions, but not all, are subject to a kind of long-term habituation that can be called accommodation.

When the event, which is the cause of emotion, is in conflict with our expectations, it triggers a cognitive revision. The simpler kind of revision consists in categorizing the event as an exceptional one, so that we just cancel out the conclusion of our expectation, but keep it as a rule for further times. In addition, emotion changes the expected action and triggers another reaction. But when the situation and the emotion are repeated from time to time, we have to question our expectations, and to cancel out some of them. Many different sets of expectations and several sets of the same amount could be cancelled out to restore consistency. So we choose the ones to be cancelled out according to some order of priority (an entrenchment order, says Gärdenfors). This revision obeys a principle of minimization: revise, but minimize your revision. This is achieved by canceling the set of expectations that have the lowest rank of priority. But there is no other way to know this order of priority that to observe the way we carry out these revisions. As preferences are revealed by our choices, the order of priorities that guide our changes of preferences is revealed by our revisions. The emotional effect of the completion of revision is accommodation: emotion does not reappear, or only in a weaker way. So if an emotion resists accommodation, this reveals that one of our fundamental priority is at stake. The two kinds of revision, the revision of the conclusions of expectations and the revision of the expectations themselves, can be related to the two neurophysiological paths of emotion, the short and limbic one and the long and cortical one - grafted in a "shunt" onto the first one.

3. Imagined Emotion

An objection may be raised: we can imagine a situation as an exceptional one, as differing from the way things are normally going on, and expect the realization of this imagined situation. So we experience an emotion but the imagined situation does not differ from our expectation. But in order to experience an emotion when simply imagining an exceptional situation, we have to possess and to activate the appropriate implicit or explicit expectations about the ways things are normally going on, so we experience also the difference of our imagined expectation with our normal expectations.

Is imagined emotion related to revision, as is non-imagined emotion? The answer appears to be no, not in the same way. The repetition of an emotional situation in imagination cannot put into question our expectations, as we imagine precisely the situation as an exceptional one, outside of the scope of our expectations about how things normally go on. But when we repeatedly evoke this situation, we focus on it more than on another goal or desire and this leads us to an indirect revision of our preferences and priorities. This focussing is more efficient when the imagination is accompanied by action (as it is the case for artists, but we ordinary people are also used to double our action with its imaginary qualification, as in Sartre's example of the waiter playing to act as a waiter).

4. Values, Emotion and Revision

This imaginative process is one way for activating values. But our “real” values are revealed by their associated emotion, when this emotion resists accommodation. Our real but maybe momentary preferences are revealed by our choices, our real order of priorities, the one that guide our changes of preferences is revealed by our revisions, our real values, values resisting most of our changes of preferences, are revealed by the fact that our emotions resist accommodation. Real values reveal themselves in two ways: either our expectations are frequently satisfied by reality, but nevertheless this satisfaction produces positive emotion, or they are rarely or even never satisfied, and nevertheless dissatisfaction produces negative emotion. So imagining the satisfying event as an exceptional one (even if it is in fact a rather frequent one) is a way to give it a positive value.

"Strong” values can be distinguish from other ones (even the real ones) by the way they behave regarding the resistance of their associated emotions against completion of cognitive revision. If the behavior of a person reveals that the cognitive revision (of the belief basis of her repeatedly unsatisfied expectations) has been done, but her emotional behavior (expressions and the like) reveal that the emotion is reoccurring with a similar intensity, we can then infer that this reveals one of her strong values.

Of course, we can make value judgements without experiencing co-occurring emotions. Emotions are first order reaction to differences with our basic expectations. Values are second order attitudes, for they imply an expectation about the satisfaction of our expectation, and, when the reality is persistently not in accordance with our values, the expectation that the situation satisfying our expectation is exceptional. Imagined emotions also imply these second order expectations. But if these imagined emotions are supported by actual desires, which have actually priorities for us, we cannot distinguish this kind of experience from the one of values (letting aside the distinction between affective and conceptual values). So we can make value judgements on the basis of second order expectations and our priorities, without experiencing real emotions. But we could not distinguish “real” and “strong” values from pretended or merely claimed values if we were not able to experience emotions, as emotions reveal to other people and even to ourselves what are our values, when we experience emotions resisting habituation.

5. Revision Can Be Stuck by Emotion

If the cause of a repeated emotion is also a candidate for triggering a revision, is emotion helping or hindering revision? In some cases, the two things are possible. Suppose that the cause of an emotion is the link between some unsatisfying situation and some of our fundamental and high priority expectations. For example, I fail to find the solution of some easy problem of mathematics, and being good in mathematics is a high priority expectation for me. The reason of my emotion is the creation of a link between these two data. Emotion produces as usually the first kind of revision, the cancellation of the ordinary expected conclusion (I fail to find the solution, contrary to what I have expected). My high priority expectation could be immune to the second kind of revision, the revision of my high priority expectation, if I could consider the situation as exceptional and not repeatable. But here we have to take into account the cognitive dynamics involved in emotions. When I am trying to improve my status in some domain, I first expect this improvement not to be easily obtained, and as emotion is a reaction to a difference of the present situation with our expectations, I am more emotionally sensitive to the positive clues of this improvement. But when I become a bit accustomed with this better status, my expectations are now positive, so I become more emotionally sensitive to negative clues. My failure on this easy problem questions my capacity to be a good mathematician. So the revision of this higher expectation is triggered. In this second kind of revision, I have to minimize. I cannot cancel but the expectations of lower priority (this revision is mostly an unconscious process, as consciousness implies some stability of representations, and revision is destabilization). But each time I try to revise only low priority expectations (being good in this specific and local kind of problem), remembering my previous failure raises the emotion, which questions the more general and higher priority expectation. I am stuck in a conflict between revising and minimizing revision. Revision cannot be obtained, as neither the minimal nor the maximal one can be done, the maximal one because a very entrenched priority cannot be revised when a minor change could done the job, and the minimal one because emotion links this local problem with my fundamental priority. Emotion and revision create two links between higher and lower priorities working in reciprocal and opposite directions. When such a loop is the case, cognitive revision cannot be achieved (I will have tremendous difficulties remembering the right solution of the problem) and emotion will reappear again and again as many times as this problem is evoked by association. So revision can be stuck by emotion, but it cannot disappear as long as revision has not been achieved.

Two points in conclusion. (1) Imagining bad consequences of the preference for an option in the near future over a more remote one can help us revising this bias as this repeated focussing changes our priorities. But if focussing is a remedy for previous biases, it is also the cause of some new ones. (2) As resistance to accommodation even after the completion of cognitive revision, our criteria for "strong" values, can be the effect of this imaginative focusing, our emotional experiences give us no absolute guarantee that our "strong" values are the “true” ones, that is, the ones which will resist any justified revision.

References

de Sousa, R. (1987) *The Rationality of Emotions*, Cambridge, MA: MIT Press.

Gärdenfors, P. (1988) *Knowledge in Flux*, Cambridge, MA: MIT Press.

Frijda, N. (1986) *The Emotions*, Cambridge: Cambridge University Press.

Ledoux, J. (1996) *The Emotional Brain*, New York: Simon and Schuster.

Mulligan, K.(1998) “From appropriate emotions to values”, *The Monist*.

Rimé, B., C. Fingenauer, O. Luminet, E. Zech and P. Philippot (1998) “Social Sharing of Emotion: New Evidence and New Questions”, in: *European Review of Social Psychology*, vol 9. W. Stoebe and M. Hewstone, eds, New York: John Wiley and Sons, pp. 145-189.

Tappolet, C. (1996) *Les Valeurs et Leur Epistémologie*, Geneva, PhD, dissertation.

EMOTION AND INTERSUBJECTIVE PERCEPTION: A SPECULATIVE ACCOUNT

SHAUN GALLAGHER
Department of Philosophy and Cognitive Science, Canisius College, Buffalo, New York 14208, USA

ABSTRACT

This paper suggests an account of the intersubjective perception of emotion that is consistent across the disciplines of philosophy, developmental psychology, and neuroscience. It explores the relationship between phenomenological accounts of how we re-cognize others to be persons like ourselves, the absence of this kind of recognition in autism, the possibility of imitation in neonates, and the recent discovery of mirror neurons in the premotor cortex. Phenomenologists suggest that the recognition of another mind depends upon the empathetic perception of the other person's body or face. Several problems with this approach are identified, including the secondary position granted to empathy. It is well known that autistic individuals have problems with precisely this kind of recognition of other people, and some theorists propose that this problem is best explained in terms of a lack of ability to perceive emotion. Studies of neonate imitation can help to clarify these issues. These studies suggest that there is an innate, intermodal mechanism that helps to explain the perception of emotion in others. The recent discovery of mirror neurons in the premotor cortex suggests a neuronal explanation of this mechanism. One can speculate that problems with both the recognition and imitation of others in autism may be in part due to the malfunction of these neurons.

1. Introduction

A variety of philosophers and scientists have argued that an understanding of another person's feelings or emotions is based on an empathetic perception of the other person's body, especially the face (Husserl, 1929; Merleau-Ponty, 1964; Wittgenstein, 1980; and more recently Hobson, 1993; 1998; Cole, 1997). I think this is right. Two questions remain unanswered, however. First, what is the precise nature of the em-pathetic perception? Second, how is it possible? To address these questions I propose an account that is consistent across the disciplines of philosophy, psychology, and neuroscience.

2. Husserl's Phenomenological Account

One theory of intersubjectivity, proposed by Husserl (1929), claims that in recognizing another person *as another person* we make a conscious analogy ("analogizing apprehension") based on the "similarity" between the appearance of the other person's body and our own embodied feeling (p. 111). This involves what Husserl calls an "appresentation," that is, a move beyond mere bodily appearance towards the other person's interior life. There are several problems with Husserl's account that need to be sorted out. First, the nature of the proposed analogy remains unclear. Husserl contends that the analogizing apprehension is not an act of inference, but is a conscious act which "transfers" already instituted meaning gained in previous experience. He traces this possibility of transferal back to what he calls a "primal instituting" in early childhood,[1] but he does not develop it any further except to say that this primal instituting is, in this case, a specific instance of a more general form of perceptual association called "pairing" (p. 112). Thus, Husserl argues that the analogizing apprehension which allows us to perceive the body of another *as* the body of *another person,* comes, through experience, to be built into perception in an implicit way. In this case the analogizing apprehension seems to be part of the experiential structure of mature consciousness (part of the way our consciousness functions), but it is not something we are explicitly conscious of.

This answer, which depends on the notion of a certain kind of prior experience involving associative pairing, leads on to a second problem. Husserl explains "primal instituting," and "pairing," by claiming that my own living body is always present for me, and that I perceive a *similarity* between it and other bodies. It has been pointed out, however, that since I experience my own body in a way that is not similar to my perception of the other's body (Schutz, 1932; 1966),[2] the analogizing apprehension must in some way transcend the non-similarity involved. Related to this issue, the phenomenological account is not clear on precisely what aspect of the other person's body we perceive in such instances - whether it is the *appearance* of the other's body, or the *movement, action,* or *behavior* of the other.[3] So either we need to explain precisely what it is that we experience and how we can transcend that experience toward an under-standing of the other as other, or we need to rethink Husserl's account.

A third problem concerns the role of affect. Husserl considers emotion and empathy as "further consequences" that involve "higher psychic spheres," but that

---

[1] Husserl writes: "The child who already sees physical things understands, let us say, for the first time the final sense of scissors; and from now on he sees scissors at the first glance as scissors - but naturally not in an explicit reproducing, comparing, and inferring" (1929, p. 111).

[2] I experience my body in a way that is very dissimilar to the way in which I perceive the other person's body. In contrast to the visual perception of the other's body, I visually experience my body with characteristic perspectival distortions - for example, I do not visually perceive my own face without the aid of mirrors.

[3] Husserl does realize that continued confirmation of the initial apprehension comes by way of the other person's "harmonious behavior" (1929, p. 114) or perceived conduct (p. 119) - something that gives evidence that the animate organism is indeed what it appears to be. But Husserl does not indicate behavior or conduct to be the original basis for the perceived similarity. On this point Merleau-Ponty makes some progress. One does not perceive the outward appearance of the other, but the action of the other. "Thus it is in [the other's] conduct, in the manner in which the other deals with the world, that I will be able to discover his consciousness." The other presents me with "themes of possible activity for my own body" (1964, p. 117).

follow a similar course as perception by involving a reflexive reference to our own somatic experience, analogizing apprehension, and so forth (1929, p. 120). On this account, empathy is not seen as an original mode of access to other persons, but as derived from an original perceptual pairing.

3. The Lack of Emotional Recognition in Autism

There are other theorists who give affect/emotion a more prominent place in the explanation of how we come to understand other persons. The importance of the perception of the face has been emphasized by Cole (1997; 1998), for example. Through the face, he argues, "we can reveal, or attempt to conceal, our emotions. It also, perhaps most importantly, allows shared emotions and relationships between people to a level and refinement not observed in other species" (1997, p. 467). This can be seen, he suggests, in an examination of various cases involving facial problems, including autism.

In autism, for example, subjects have difficulty recognizing the emotional states of others. In this respect, Asperger (1944) points out that "autistic children have a paucity of facial and gestural expression. In ordinary two-way interaction they are unable to act as a proper counterpart to their opposite number, and hence they have no use for facial expression as a contact-creating device" (cited in Cole 1997, p. 476). In counterpoint to theory-of-mind approaches to autism -- which push emotion to second place, and which are easily translated into Husserl's terms (that is, autism is a complete failing of analogizing apprehension) - Hobson (1986; 1993; Hobson, Ouston & Lee, 1988) has argued that it is best to understand autism as a disturbance of affect -primarily the inability to recognize the emotional states of other persons. Moore, Hobson, and Lee (1997) conducted experiments that show autistic individuals, in contrast to normal and mentally retarded individuals, are unable to read emotion from the bodily movements of others. Hobson (1998) has also shown that autistic subjects are unable to properly imitate actions performed by others. From the study of autism, which involves the complete failure of the emotional recognition of the other person as another person, we find evidence that the inability to perceive emotion in facial and bodily gesture is associated with the in-ability to imitate.

4. Neonate Imitation

I think that studies of neonate imitation can help to clarify these issues, and indeed, resolve some of the problems in Husserl's account of intersubjectivity. These studies suggest that there is an innate, intermodal (visual-proprioceptive and sensory-motor) mechanism that helps to explain the perception of others as other persons, and that this perception is conditioned by emotional content from the beginning of life (Gallagher & Meltzoff, 1996; Meltzoff, 1990; Meltzoff & Gopnik, 1993; Meltzoff & Moore, 1992). Here I can only summarize some of the details very quickly:

- Neonates under the age of 1 hour are capable of imitating facial gestures (mouth openings, tongue protrusions) presented by a model (Meltzoff and Moore, 1977). Gestures that can be imitated include affective gestures such as smile, frown, etc. (Field *et al.*, 1982).
- This kind of imitation involves a proprioceptive sense of the neonate's own face, as well as the visual perception of the other person's face.
- The neonate must be capable of translating between its proprioceptive sense and its visual sense in order to perform the motor action required for the imitation.
- Neonates will not attempt to imitate non-human models. (Legerstee, 1992).

Specifically, these imitation studies lead to the conclusion that it is not the perception of the objective appearance of the other person's body, but the perception of the other person's gesture, *behavior, or* action that forms the basis of intersubjective experience. Emotional communication depends more on perceived motility and physical expression than on static appearance, shape or image of the other's body. The infant does not perceive the other as an object, but as another entity like itself; the infant recognizes at the behavioral level that the perceived movement or expression is one that the infant itself can make.

5. Intermodal Mechanisms, Mirror Neurons, and the Communication of Emotion

One unresolved issue in the developmental literature concerns the nature of the intermodal mechanism which allows for intrasubjective communication between visual and proprioceptive perception, or more generally between perceptual and motor systems, and on that basis, the intersubjective communication of emotion. Meltzoff and Moore (1997), two of the leading researchers in this field, go no further than to propose a psychological-cognitive model, a set of theoretical black boxes representing "comparison function," "act equivalence," "recognition of my own capability," etc. I propose that in part this model can be filled in on the neurological level by the functioning of mirror neurons (Gallese, 1998; Gallese *et al.*, 1996; Rizzolatti *et al.*, 1996).

Mirror neurons located in the premotor cortex of the macaque monkey and (there is good evidence) in premotor cortex and Broca's area in the human, respond *BOTH* when a particular motor action is performed by the subject *AND* when the same action performed by another individual is observed. Mirror neurons thus constitute an intermodal link between the visual perception of action or dynamic expression, and the *intrasubjective, proprioceptive* sense of one's own capabilities. Even if, at this time, no direct link between mirror neurons and emotional response can be posited (Rizzolatti *et al.*, 1996), it is not inconsistent with the neurological account that the emotional meaning of another person's action or gesture is mediated by just such a neuronal mechanism capable of registering perceived actions or gestural expressions of others on the same neuronal machinery that generates a matching motor action in oneself.

This neurological speculation (a speculation in more than one sense) not only fulfills the quest for an intermodal mechanism in developmental psychology, capable of explaining the innate capacity for imitation, but resolves a number of the problems encountered in the philosophical account of the analogical apprehension of another person's emotional state. Analogical apprehension is an innate (not fully experience-based) ability. We can find its basis built into certain kinds of neuronal activity that are already "on line" from at least the time of birth. The fact that this apprehension is inter-modal resolves the problem involving the differences between the way that I experience my body and the way I experience the other's body. It also makes it clear that the important factor for intersubjective recognition is not the visual appearance of the other's body, but how that body moves, gestures, and acts. Movement, gesture, and action, as Cole, Hobson, and others have shown, is laden with emotion. Finally, if autism is considered a developmental disorder involving a deficit in the perception of affect (Hobson, 1993), it seems quite reasonable to speculate that the failure of a mediating sensory-motor, intermodal mechanism may be responsible. Thus, across the disciplines of philosophy, developmental psychology, and neuroscience, there is the possibility of a consistent answer to the problem of the perception of emotional states in others.

Acknowledgments

Part of the research for this paper was completed at the National Endowment for the Humanities Summer Institute on Mind, Self, and Psychopathology, directed by Louis Sass and Jennifer Whiting, July-August 1998, at Cornell University.

References

Cole, J. (1998) *About Face,* Cambridge, MA: MIT Press.

Cole, J. (1997) "On being faceless: Selfhood and facial embodiment", *Journal of Consciousness Studies* 4:467-84.

Field, T. M., R. Woodson, R. Greenburg and D. Cohen (1982) "Discrimination and imitation of facial expression by neonates", *Science* 218:179-181.

Gallagher, S. and A. Meltzoff (1996) "The earliest sense of self and others: Merleau-Ponty and recent developmental studies", *Philosophical Psychology* 9:211-233.

Gallese, V. (1998) "Mirror neurons: from grasping to language", Paper read at *Tucson III Conference: Towards a Science of Consciousness* (Tucson 1998).

Gallese, V., L. Fadiga, L. Fogassi and G. Rizzolatti (1996) "Action recognition in the premotor cortex", *Brain* 119:593-609.

Hobson, P. (1998, July) "Imitation in Autism", *Seminar at NEH Institute on Mind, Self, and Psychopathology*, Cornell University.

Hobson, P. (1993) "The emotional origins of social understanding", *Philosophical Psychology* 6:227-49

Hobson, P. (1986) "The autistic child's appraisal of expressions of emotion", *Journal of Child Psychology and Psychiatry* 27:321-342.

Hobson, P., J. Ouston and A. Lee (1988) "What's in a face? The case of autism", *British Journal of Psychology* 79:441-453.

Husserl, E. (1929/1970) *Cartesian Meditations*, The Hague: Nijhoff.

Legerstee, M. (1991) "The role of person and object in eliciting early imitation", *Journal of Experimental Child Psychology* 51:423-433.

Meltzoff, A. (1990) "Foundations for developing a concept of self: The role of imitation in relating self to other and the value of social mirroring, social modeling, and self practice in infancy", in: *The Self in Transition: Infancy to Childhood*, D. Cicchetti and M. Beeghly, eds, Chicago: University of Chicago Press, pp. 139-164.

Meltzoff, A. and A. Gopnik (1993) "The role of imitation in understanding persons and developing a theory of mind", in: *Understanding Other Minds: Perspectives from Autism*, S. Baron-Cohen, H. Tager-Flusberg and D. Cohen, eds, Oxford, New York: Oxford University Press, pp. 335-366.

Meltzoff, A. and M. K. Moore (1997) "Explaining facial imitation: A theoretical model", *Early Development and Parenting* 6:179-192.

Meltzoff, A. and M. K. Moore (1992) "Early imitation within a functional framework: The importance of person identity, movement, and development", *Infant Behavior and Development* 15:479-505.

Meltzoff, A. and M. K. Moore (1977) "Imitation of facial and manual gestures by human neonates", *Science* 198: 75-78.

Merleau-Ponty, M. (1964) *The Primacy of Perception*, Evanston: Northwestern University Press.

Moore, D. G., R. P. Hobson and A. Lee (1997) "Components of Person Perception: An investigation with autistic, non-autistic retarded and typically developing children and adolescents", *British Journal of Developmental Psychology* 15:401-423.

Rizzolatti, G., L. Fadiga, V. Gallese and L. Fogassi (1996) "Premotor cortex and the recognition of motor actions", *Cognitive Brain Research* 3: 131-141.

Schutz, A. (1966) *Collected Papers, Vol. 3*, The Hague: Nijhoff.

Schutz, A. (1932/1974) *Der sinnhafte Aufbau der sozialen Welt*, Frankfurt: Suhrkamp.

Wittgenstein, L. (1980) *Remarks on the Philosophy of Psychology*, Chicago: University of Chicago Press.

BIOLOGICAL PERSPECTIVES

INTRODUCTION: BIOLOGICAL PERSPECTIVES

ALFRED W. KASZNIAK
Center for Consciousness Studies, Departments of Psychology, Neurology & Psychiatry, University of Arizona, 1503 E. University, Tucson, Arizona 85721, U.S.A.

Biologic changes have long been known to be important characteristics of emotion. The second century Greek physician, Galen, clearly recognized that alterations in the pulse accompanied emotions such as fear (Reiss, Sushinsky, & Kaszniak, 1977). By 1806, Sir Charles Bell, the noted physiologist, was writing about the anatomy and physiology of emotional expression (cited in Darwin, 1872/1965), a topic which Darwin's (1872/1965) treatise on emotional expression in humans and other animals brought to broad scholarly and popular attention. Darwin's influential work was soon followed by William James' (1884) article entitled "What is an Emotion?" James' formulation of emotional experience as due to feedback from bodily responses drew increased attention to questions about biological processes in emotion. This interest was further enhanced by Walter Cannon's (1929, 1932/1963) detailed studies of the physiology of motivational and emotional states in rabbits and cats, which challenged James' hypothesis and proposed his own neural theory of emotion. Cannon's theory was based, to a large extent, upon research carried out in his laboratory by Philip Bard (1929), who performed lesion studies to determine areas of the brain necessary for rage expression in animals. Based on these experiments, both Bard and Cannon concluded that the hypothalamus was central to emotion.

The next major contribution to scientific thinking about the biology of emotion came when James Papez (1937) synthesized available anatomic, physiologic, and evolutionary research. Papez proposed that the "stream of feeling" (i.e., emotional experience) was mediated by evolutionarily old aspects of the medial cortex (the cingulate gyrus), the hypothalamic mammillary bodies, and the anterior thalamus. Subsequently, Paul MacLean (1949, 1952) expanded on Papez' theory, and proposed that the "visceral brain" (including the cingulate cortex, the hypothalamus, and the hippocampus) formed the anatomical core of a "limbic system" subserving emotion. Different anatomic components of MacLean's limbic system (see Mega, Cummings, Salloway, & Malloy, 1997) are now known to play roles in both emotion and memory (as well as other functions). Further, the validity of the general concept of a limbic system has been questioned (see LeDoux, 1996). However, MacLean's insight that brain evolution is important to understanding emotion has remained. This insight has motivated much of modern animal research on the neurobiology of emotion (e.g., LeDoux, 1996, 2000; Panksepp, 1998; Rolls, 1999).

The authors of papers within this second section of the present volume collectively provide a comprehensive survey of current research on the biology of emotion. Among the many questions addressed by these authors are: How might evolutionary theory best be applied to an understanding of emotion? Can positive emotions, such as joy, be scientifically studied in rats? What can be learned about the neurobiology of specific emotions, such as fear, from systematic animal studies employing multiple experimental methods? How is the amygdala involved in the processing of emotional information? How do subcortical brain circuits and different neuropeptides generate the variety of different affective states? What can the neurobiological study of pain teach us about emotional qualia? How have human psychophysiological studies informed our understanding of emotional perception and emotion-memory relationships? What brain systems subserve emotional imagery and emotion appraisal? How do the two cerebral hemispheres differ in their roles in human emotion and autonomic physiologic response? How might research and theory on brain systems involved in emotional arousal be used to motivate psychotherapeutic approaches to treating emotional disorder? What role does autonomic balance play in emotional experience? What do the nonlinear dynamics of cardiovascular emotion response tell us about functional aspects of emotion in relation to behavioral performance? Does emotion play a fundamental role in the genesis of all conscious experience?

The research examined in the following papers illustrates both the present yield and future promise of biological approaches to understanding emotion. This research also makes it clear that emotion, cognition, and consciousness itself are interrelated at the level of basic biological processes. Unraveling the specific neurobiological details of these relationships is high on the agenda for future research in this area.

References

Bard, P. (1929) "The central representation of the sympathetic system: As indicated by certain physiological observations", *Archives of Neurology and Psychiatry* 22:230-246.

Cannon, W.B. (1929) *Bodily Changes in Pain, Hunger, Fear, and Rage, Volume 2*, New York: Appleton.

Cannon, W.B. (1932/1963) *The Wisdom of the Body*, New York: W.W. Norton.

Darwin, C. (1872/1965) *The Expression of the Emotions in Man and Animals*, Chicago: University of Chicago Press.

James, W. (1884) "What is an emotion?", *Mind* 9:188-205.

LeDoux, J. (1996) *The Emotional Brain: The Mysterious Underpinnings of Emotional Life*, New York: Simon and Schuster.

LeDoux, J. (2000) "Cognitive-emotional interactions: Listen to the brain", in: *Cognitive Neuroscience of Emotion*, R.D. Lane, L. Nadel, G.L. Ahern, J.J.B. Allen, A.W. Kaszniak, S.Z. Rapcsak and G.E. Schwartz eds, New York: Oxford University Press, pp. 129-155.

MacLean, P.D. (1949) "Psychosomatic disease and the "visceral brain": Recent developments bearing on the Papez theory of emotion", *Psychosomatic Medicine* 11:338-353.

MacLean, P.D. (1952) "Some psychiatric implications of physiological studies on frontotemporal portion of limbic system (visceral brain)", *Electroencephalography and Clinical Neurophysiology* 4:407-418.

Mega, M.S., J.L. Cummings, S. Salloway and P. Malloy (1997) "The limbic system: An anatomic, phylogenetic, and clinical perspective", *The Journal of Neuropsychiatry and Clinical Neurosciences* 9:315-330.

Panksepp, J. (1998) *Affective Neuroscience*, New York: Oxford University Press.

Papez, J.W. (1937) "A proposed mechanism of emotion", *Archives of Neurology and Psychiatry* 79:217-224.

Reiss, S., L.W. Sushinsky and A.W. Kaszniak (1977) "Psychophysiologic disorders and sexual dysfunctions", in: *Abnormality: Experimental and Clinical Approaches*, S. Reiss, R.A. Peterson, L.D. Eron and M.M. Reiss, eds, New York: Macmillan, pp. 311-338.

Rolls, E.T. (1999) *The Brain and Emotion*, New York: Oxford University Press.

EVOLUTIONARY PERSPECTIVES ON EMOTION

PAUL E. GRIFFITHS
Unit for History and Philosophy of Science, University of Sydney, Sydney NSW 2006, Australia

ABSTRACT

Evolutionary Psychology links the methodology for cognitive science associated with the late David Marr to evolutionary theory. The mind is conceived as a bundle of modules which can be described at three theoretical levels. Each module represents an adaptation to some specific ecological problem. Evolutionary psychologists try to derive the highest level of description using a heuristic method called 'adaptive thinking'. This paper questions the value of the official EP methodology and reasserts the value of the earlier methodology associated with classical ethology, in which the structural and comparative analysis of the products of evolution precedes the investigation of their origin by natural selection. The adaptive heuristic is shown to be unhelpful. It tends to restrict psychology to work designed to confirm our preconceptions. It is argued that evolutionary psychology must be retrodictive and explanatory, rather than predictive, and that a powerful method for testing postulated claims about the adaptive origins of traits is available in modern versions of the comparative method. Some arguments for the inapplicabilitv of the comparative method to human psychoevolution are examined and shown to be specious. These various points are exemplified in evolutionary work on emotions.

1. Evolutionary Psychology

Ever since Charles Darwin's *Descent of Man...* (Darwin, 1871), evolutionary biology has served as an Archimedean point on which groups of psychologists have stood in an attempt to move the rest of their discipline to a new theoretical orientation. Psychology was strongly influenced by the recapitulationist biology of Ernst Haeckel and his contemporaries (Gould, 1977). From the 1950s the methods of classical ethology were extended to the study of human behavior. Human behavior was studied in its natural ecological setting and in comparison to the behavior of related species, and suitable behaviors were interpreted as the outcome of human evolution (Eibl-Eibesfeldt, 1979). This approach met with considerable success in the case of the emotions. A slightly different emphasis was introduced by human sociobiology in the 1970's, with game-theoretic models of adaptation coming to occupy a prominent position (Trivers, 1985; Wilson, 1975). Sociobiology attempted to show that many human behaviors were 'evolutionarily stable strategies': strategies that cannot be outcompeted by any other strategy once they have evolved. Continuity with the ethological tradition was evident, however in the 'Darwinian anthropology' of the same period (Chagnon & Irons, 1979).

Sociobiology recognized the emotions as a legitimate area of interest (Weinrich, 1980), but until the work of Robert Frank there was no systematic attempt to treat specific emotional behaviors as solutions to adaptive problems characterized in game-theoretic terms (Frank, 1988).

Evolutionary Psychology (EP) represents the latest attempt to transform psychology. It has been embraced quite widely in both psychology and anthropology and, like classical ethology and sociobiology before it, been the subject of some of the best-selling popular science of its day. EP differs from the sociobiology of the 1970s and 1980s in two main respects (Symons, 1992; Tooby & Cosmides, 1992). First, it does not attempt to show that current human behaviors are solutions to problems in evolutionary game theory. Human beings no longer live in the environments in which they evolved. Hence EP does not expect human behavior to be currently adaptive, nor does it expect that the behavior we exhibit today will closely resemble the behavior exhibited in our evolutionary past. This leads naturally to the second main difference between EP and sociobiology. EP tries to explain underlying mental mechanisms rather than the behavior they produce. Behavioral research is used only to reveal these mechanisms.

EP takes from cognitive science a particular model of the mental mechanisms it expects to find. The mind will be a collection of 'modules' - specialized devices designed to solve particular problems. EP also accepts the model of psychological explanation associated with the modularity thesis and famously outlined by David Marr in his book *Vision* (Marr, 1982). Marr proposed that psychology must make use of three levels of description. The highest level is the 'task description' - what the psychological system accomplishes for the organism. The visual system takes patterns of stimulation on the retina and produces an interpretation of the world in terms of moving, three-dimensional objects with color. The second level describes the computational methods by which the system accomplishes its task. The lowest level describes how these computational processes are implemented in the brain. Evolutionary Psychology adds to this vision the idea that the task description of a module is a description of an adaptive, evolutionary problem. The modular structure of the mind reflects the range of separate adaptive problems faced by our ancestors. EP proposes that evolutionary theory can assist psychology by specifying these task descriptions:

> «Knowledge of the adaptive problems and ancestral conditions that human hunter-gatherers faced can lead to new hypotheses about the design of psychological mechanisms that evolved to solve them. Such heuristic analyses can supply crucial guidance in the design of experiments to discover previously unknown psychological mechanisms - investigations that researchers who neglect functional analysis would not have thought to conduct.» (Cosmides *et al.*, 1992, p.10).

In the best known example of this method of 'adaptive thinking', Leda Cosmides and John Tooby have provided a novel task description for human reasoning which is based on game theoretic models of the evolution of cooperation (Cosmides & Tooby, 1992). They argue that there is a specific module for reasoning about social exchange, and that this module is designed to detect cheating in contexts described by the evolutionary game of 'prisoners dilemma'.

2. Adaptive Thinking

EP proposes to determine the modular structure of the mind and the task descriptions of the various modules by 'adaptive thinking'. Ideally, adaptive thinking would take the form of building optimality models or game-theoretic models of the relationship between organism and environment. In the case of human psychoevolution this is often impractical and adaptive scenarios must be stated in informal, narrative terms. The idea that these models or scenarios will allow us to predict unanticipated features of an organism's phenotype presupposes that selective problems are very strongly associated with particular solutions to those problems. I have argued elsewhere that problems and solutions do not correspond in this straightforward fashion (Griffiths, 1994; 1995; 1996; 1997; Sterelny & Griffiths, in press). The view that they do is one important element of 'adaptationism', a term that refers to several related controversies over the proper conduct of evolutionary explanation. Evolutionary psychologists are aware of the most egregious error of adaptationism. They accept that the solution cannot simply be inferred from the problem. A hypothesis about mental structure cannot be *established* merely by producing an adaptive scenario in which that mental structure would be advantageous. An empirical demonstration that the mind is actually structured in that way is also required. As Donald Symons notes 'Although selectional thinking is an important source of inspiration for the evolutionary psychologist, nature always gets the last word' (1992, pp.143-144). This is a step forward, but it does not solve the problem. Evolutionary psychologists think that the fact that a particular feature 'makes evolutionary sense' should increase our expectation that this feature actually exists. That is what it means to say that adaptive thinking is an 'heuristic'. The leading evolutionary psychologists Cosmides, Tooby and Barkow claim that an hypothesis backed by a plausible adaptive scenario is more likely to be true than the alternatives and hence more worth investigating (see the quotation above). They do not accept one of the central claim of recent anti-adaptationist writing, which is that adaptive scenarios are very easy to generate. Given any presupposition about how an organism is structured we can devise a scenario in which that structure 'makes evolutionary sense'. Those of us who take this latter point of view see 'adaptive thinking' not as a way of discovering things we would otherwise miss, but as a recipe for only doing experiments designed to confirm your existing beliefs.

Consider, for example, Cosmides and Tooby's view that adaptive thinking can reveal the nature of human reasoning about social exchange. They argue that the dominant mode of thought will be 'cheater detection' - determining whether you have obtained appropriate reciprocation when you have performed your side of a bargain. This is the most important thing to determine if; as Cosmides and Tooby believe, the evolutionary stable strategy in the prisoners dilemma, the game they use to model social exchange, is 'tit-for-tat'. My objection is not to Cosmides and Tooby's conclusion, for which they offer some direct empirical evidence in just the way that Symons recommends, but to the idea that 'adaptive thinking' gave them solid grounds for supposing that this was the right hypothesis to test. Cosmides and Tooby certainly think so. They describe the situation as follows:

> «The iterated Prisoner's Dilemma is an abstract description of the problem of altruism between nonrelatives. By studying it, one can derive a set of general constraints that the cognitive problems *[sic]* of virtually any species must satisfy to be selected for under these circumstances.» (Cosmides & Tooby, 1992, p.178).

Compare this to a typical statement from a game-theoretic discussion of the evolutionary prisoners dilemma:

> «If players can make probabilistic choices, taking into account their co-player's previous actions, a strategy known as 'generous tit-for-tat' dominates the long-term behavior of such a population. If they can also take into account their own previous action, a strategy of 'win stay, lose shift' dominates instead. These models assumed that participants make their decisions in synchrony, which seems improbable in many biological situations. If individuals make their decisions at different times, neither of the above strategies survives given the usual payoffs.» (Frean, 1994, p.75)

For reasons that we will discuss below, the predictions of models of adaptation are usually sensitive to historical assumptions about the actual problem faced by organisms at the time the trait in question evolved. Rather than yielding general constraints on any possible solution, they describe a range of solutions that will be favored under different ecological conditions. Genuine 'adaptive thinking' involves reconstructing evolutionary history.

The great danger of 'adaptive thinking' is that it can induce complacency in the face of unreliable or ambiguous data, because that data or interpretation of the data makes adaptive sense. Parent/offspring conflict is a good example. Evolutionary Psychologists often cite Robert L. Trivers' demonstration that the long term interests of parents need not be identical with those of their offspring (Trivers, 1974). The parent wants to conserve its resources for future offspring, whereas the offspring wants as much as it can get. This model was taken to explain observations of squabbling between parents and offspring around the time of weaning in primates, including humans. It also created the expectation that offspring would deceive their parents about their needs in an attempt to get more resources. Parent/offspring conflict is 'inherent to the human condition' (Pinker & Bloom, 1992, p. 483) and its 'inevitability' is cited in support of the existence of psychodynamic mechanisms of deceit and self-deceit. Yet the empirical evidence for parent/offspring squabbling over weaning is ambiguous. Patrick Bateson summarizes various studies that failed to find aggressive interactions at weaning in a wide range of species, studies which found offspring weaning themselves, and studies which found both parties engaging in reliable signaling in order to co-ordinate weaning (Bateson, 1994). Rather than expand our sense of the possible, as Cosmides and Tooby suggest, adaptive thinking has narrowed it and given us more faith in views we were inclined to accept anyway.

The evolutionary psychology of emotion contains a striking example of the ability of adaptive thinking to actually drive out empirical findings in favor of views without no support other than 'making adaptive sense'. Tooby and Cosmides have predicted that emotions will turn out to be behavioral programs that are deployed in response to frequently recurring ecological situations:

> «Each emotion state - fear of predators, guilt, sexual jealousy, rage, grief and so on -will correspond to an integrated mode of operation that functions as a solution designed to take advantage of the particular structure of the recurrent situation these emotions respond to.» (Tooby & Cosmides, 1990, p. 410)

The existence of stereotyped, pan-cultural emotional responses has been suspected since Darwin's own investigation of the subject (Darwin, 1872/1965). The existence of such responses was substantially confirmed by the work of Paul Ekman and his collaborators a century later (Ekman, 1972). These 'affect programs' are complex responses involving several physiological systems and over which we have little conscious control. It is certainly not unreasonable to interpret them as the output of mental 'modules' of the sort favored by EP. I myself argued for this interpretation in an article published in the same year as Cosmides and Tooby's claims about emotion, although I no longer find these arguments convincing (Griffiths, 1990). Hence Tooby and Cosmides cite work related to the affect program theory as if it confirmed the predictions of their 'adaptive thinking' about emotion. But in many important respects what we know about the affect programs contradicts their predictions.

Tooby and Cosmides suggest that emotions are solutions to very specific ecological problems. They suggest that the emotion module will be programmed to respond to cues such as 'looming approach of a large fanged animal' (for fear) and 'seeing your mate have sex with another' (for a postulated emotion of sexual jealousy). But although the *output* of an affect program emotion is stereotyped and pan-cultural, the *input* to the affect program - the emotional stimulus - is very flexible and varies between cultures and individuals. The 'stimulus appraisal mechanism' controlling the affect programs is not programmed to respond to specific stimulus situations like those Tooby and Cosmides describe. Newborn babies respond to loud sounds and loss of balance with fear, to prolonged restraint with rage, and to gentle forms of skin stimulation with pleasure. They are also sensitive to human facial expressions (Izard, 1978; Meltzoff & Moore, 1977; Trevarthen, 1984). In later life fear seems to be produced by any stimuli an individual has come to associate with danger, sadness by stimuli associated with loss, and so forth. Contrary to the predictions of the evolutionary psychologists, affect programs are designed to cope with quite general evolutionary problems, and the affect program system is designed to redefine those problems as the environment changes.

This does not mean that the 'input side' of emotion cannot be understood in evolutionary terms. Learning itself is a complex ability with an evolutionary history. Fear of classic phobic stimuli such as snakes must indeed be acquired, but associations between these stimuli and fear may be easier to acquire and harder to lose than associations between fear and other stimuli like flowers and colored shapes (Öhman *et al.*, 1976). They may be 'prepared associations' that the organism makes easily and discards with difficulty (Seligman & Hager, 1972). The sources of information from which emotion stimuli are learnt are also interesting from an evolutionary perspective. Associations are acquired through the child's own experience, but they are also acquired through observations of adult emotion (Klinnert *et al.*, 1983). Children need not be hurt in the dark to learn to fear darkness. They need only witness fear of the dark in adults.

The failure of adaptive thinking in this case reflects many of the dangers of this method. First, it is much harder than its proponents seem to suppose. The affect program system may represent a subtle compromise between the need for flexibility in a changing world and the need to learn without many expensive trials. This particular compromise could not be predicted in advance by imaginatively reconstructing the evolutionary process. Defenders of adaptationism are fond of telling us that evolution is more subtle than we suppose. We should not take our inability to guess the adaptive function of a trait to indicate that it has none because 'evolution is cleverer than you are' (Dennett, 1995, p.74). But they fail to recognize the flip-side of this maxim - 'evolution is more complex than you think' (Griffiths, 1996, p.517). If we try to evaluate the selective pressures on a trait by just thinking about the evolutionary environment, we will typically get them wrong.

A second problem for adaptive thinking exemplified by Cosmides and Tooby's work on emotion is what Kim Sterelny and I have called the 'grain problem' for evolutionary psychology (Sterelny & Griffiths, in press). Tooby and Cosmides used an overly fine-grained analysis of the adaptive problem, treating 'fear of predators' as a separate problem in itself. Evolution treated it as a part of a single larger problem - 'fear in general'. Where one adaptive problem ends and another begins depends on how the organism can adjust its phenotype, since this determines whether various 'problems' can be disentangled one from another. Such matters of the developmental structure of the organism are, of course, central to the literature on anti-adaptationism and they are a closed book to 'adaptive thinking'. Sterelny and I argue that this is a particularly hard problem for evolutionary *psychology*, since it aims to delineate the modular structure of the mind through 'adaptive thinking' (Sterelny & Griffiths, in press). But it is precisely the structure of the mind that makes one adaptive problem in psychoevolution separate from another. Hence we need to know the modular structure of the mind before we can describe the adaptive problems, not the other way around. Imagine, for example, an organism that detects its prey using two sensory pathways, each independently driving the same behavioral response, such as striking at the prey. This organism faces two well-defined problems in signal detection theory, each defined by the information content of the signals in the relevant sensory pathway. These problems dictate the selection pressures on the organism's response system and also the pressures for improvement of the two sensory systems. Contrast this case to an organism that integrates the signals from the same two sensory pathways into a single representation that drives its behavioral response. This organism faces a single, complex problem in signal detection theory and the selection pressures on the response system and the two sensory systems will be quite different. For example, the first organism may use each sensory pathway to detect an optimal signal in that pathway, given its noise-signal ratio. The second organism may use one sensory pathway for sensitivity and the other for discrimination, perhaps exploiting the natural advantage of each pathway in one of these respects.

Both these problems reflect the inherent difficulty of describing the selective environment of the evolving organism. Authors such as Richard Lewontin have long emphasized the distinction between the physical surroundings of an organism and its selective environment (Lewontin, 1982). Lewontin argues that the selective environment must be conceived in relation to the organism that occupies it. An organism's niche is defined by the variables that affect that organism's ability to reproduce itself. Two organisms occupying the same physical environment, an elephant and a tree for example, may occupy areas of niche space which have very few common dimensions. This suggests that the idea of first characterizing the niche and then using this to predict what sort of organism will occupy it may be fundamentally misguided. There are indefinitely many different niches in a particular physical environment, and our ability to distinguish a particular niche and the selection pressures it creates depends to a great extent on our prior understanding of the organism that occupies that niche and its ecology (See also Brandon, 1990). Kim Sterelny and I have suggested that insofar as we are able to characterize a niche prior to understanding its occupant this will be because we understand other, similar organisms (Sterelny & Griffiths, in press). 'Adaptive thinking' thus requires an understanding of *comparative biology* - an idea that will be explored further below.

3. Adaptation and the Environment

Models of adaptation describe how the fitness of an organism is determined by the design it adopts and the adaptive environment it faces. It is easy to make the mistake of supposing that optimality and game-theoretic models do not involve any particular assumptions about evolutionary history. They seem to involve only general principles about which traits are most efficient. In the prisoner's dilemma a 'defector' will always gain more fitness that an 'unconditional cooperator' when the two meet. However, these functional considerations typically will not make specific predictions about evolution unless we specify the particular historical conditions under which the trait evolved. In her adaptive explanation of human pregnancy sickness, Margie Profet postulates that, in the period in which it evolved, human mothers had a diet rich enough in vitamins to reduce the cost of lost nutrition due to pregnancy sickness below the threshold required for it to evolve in her model (Profet, 1992). In her explanation of human menstruation she postulates that in the period in which menstruation evolved females abstained from sex for several months after birth, so that menstruation correlates better with promiscuous copulation, the driving force in her explanation (Profet, 1993). History also creeps in when choosing the range of competing designs that are to be evaluated. Without knowing what sort of ancestors an organism had it is impossible to say which alternatives competed to produce the form we see today. History has a third role because evolution is a stochastic process. Conventional evolutionary theory says that many important innovations occur when organisms are isolated in small 'allopatric' populations. Evolutionary 'drift' can be very important in these populations. Taking all these factors into account, the role of particular historical facts in evolution is very large. An adaptive model must make many assumptions of historical fact.

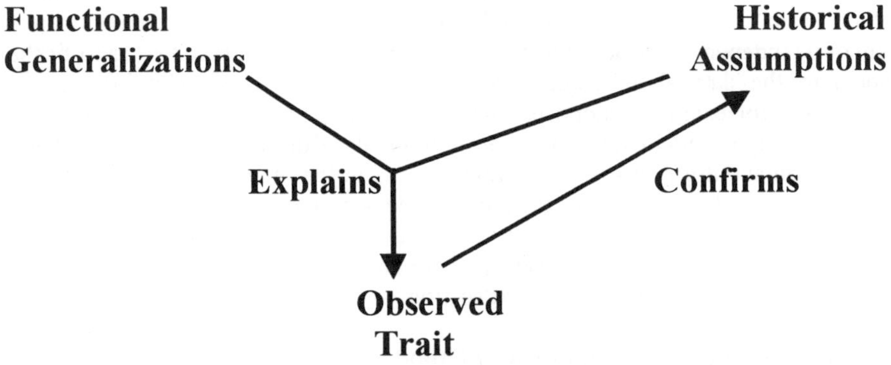

Figure 1. The 'adaptationist abduction'. This argument to the best explanation is supposed to avoid the need to independently test the historical assumptions of postulated adaptive scenarios. The fit between the model and the observed data provides an argument confirming the historical assumptions that the model requires.

Adaptationists have tried to avoid the problem of determining the accuracy of their historical assumptions by conceiving an adaptive explanation as a simultaneous 'abductive' argument for the truth of the assumptions which it requires (Figure 1). Abduction, or 'argument to the best explanation' is an important form of scientific reasoning. If one theory explains the data better than any other then it is reasonable to accept that theory. The adaptationist argues that if they make certain historical assumptions they can neatly explain the actual trait. Therefore, by argument to the best explanation, we have grounds for accepting these historical assumptions.

But for many of the adaptationist hypotheses central to contemporary evolutionary theory, arguments to the best explanation are too blunt an instrument. Optimality modeling, evolutionary game theory and the like are powerful engines for generating possible explanations. So there are often a number of potentially adequate explanations. This is particularly true when the ecological situations are themselves theoretical scenarios in the distant past. The rapid expansion of brain size in our primate ancestors has been explained as the effect of an upright stance and the consequent freeing of the hands for complex manual work. According to the engaging aquatic ape hypothesis, it is the effect of a period when our ancestors were supposedly surviving and foraging in shallow coastal waters. The less popular but very well developed 'radiator theory' suggests that brain size expansion was the effect of removing a developmental constraint on the thermoregulation of the brain (Falk, 1990). Perhaps the most popular current view is the 'Machiavellian Intelligence' hypothesis. Brain expansion was caused by the social structure of hominid societies. In these social groups, as in chimpanzees today, the ability to form and manipulate personal relationships was the key to success. A person who could form a more complex system of alliances and remembered favors would do well (Byrne & Whiten, 1988; Whiten & Byrne, 1997). All these models make assumptions about the historical conditions under which brain size expansion occurred. The value of many of these parameters cannot be independently determined, leaving them free to be tuned so as to fit the model to the data about actual brain size expansion. The social intelligence hypothesis, for example, supposes that human groups became larger, giving rise to more complex interactions between individuals. But we have no independent information about group size. So in this case, as in the others, the fit between the model and the data does not really constitute a test of the model. It might be significant if no other model could be 'tuned' to fit the data, but this is transparently not the case. Fortunately, there are other ways to test these models, methods that are described in the next section.

4. Adaptation and the Comparative Method

A powerful method for directly testing postulated claims about the adaptive origins of traits is available in the form of the 'comparative method'. This term refers to the range of techniques that infer how one organism evolved by comparing what evolution produced in that case with what it produced in other cases. The comparative method is one of biology's main windows on the past. The simplest comparative tests check the actual sequence of evolutionary changes to see if it is the one presumed by the adaptive hypothesis. Some models of sexual selection presume that exaggerated male traits develop in response to an arbitrary female preference. Others presume that the male trait is correlated with fitness in some way (perhaps advertising the male's ability to survive with a 'handicap') and that females are selected for their ability to respond to the trait. Alessandro Basolo discriminated between two such hypotheses in the case of the exaggerated 'sword' tails of fish of the genus *Xiphophorus.* She showed that females of a related species in which males do not have a sword would prefer males with a sword if they were available. Thus, female preference seems to have evolved in a more distant common ancestor than the exaggerated male tail (Basolo, 1990). Mary McKitrick provides another simple example (McKitrick, 1993). It has been suggested that the low birthweight characteristic of the genus Ursa - the bears - is the result of an adaptive trade-off. It is the price they pay for altering their physiology in order to allow hibernation. But a reconstruction of bear phylogeny shows that this cannot be the case. Low birth weight emerges before hibernation, and exists on branches on which hibernation never originated. Tests of this sort have wide application. The 'aquatic ape' hypothesis claims as a particular strength its ability to explain a wide range of human characters: for example, upright posture, bipedalism, hair loss, our layer of subcutaneous fat, our diving reflex and many more. All these are said to have evolved together as an adaptive complex when our ancestors made a return to a semi sea-going life. Since the theory suggests that these characters emerged together in a single phase of hominid evolution, we can test it by determining when they appeared on the tree for hominids and their relatives. If the traits appeared at different times, they will be inherited through to different chunks of the hominid family tree. If the characters emerge at various different points in the tree if they do not, in fact, evolve together then however neatly the theory explains them, it cannot be the correct one.

A second important role for the comparative method is to directly test the idea that adapted traits are responses to particular features of an organism's environment. Adaptationist hypotheses can be supported by finding a covariation between certain traits and habitat factors. These correlations suggest that the habitat factor has something to do with the evolution of the trait. Suppose we are interested in a group of seabird species in which some species nest in burrows, have plain white eggs, and do not remove the egg shells after hatching. Other species nest on ledges, have patterned shells (camouflage, we suspect), and remove their shells after hatching. We reconstruct the phylogeny of the group and discover (a) the ancestor species nested in a burrow, (b) it had plain white eggs, (c) and it did not remove eggshells after hatching. We find that on several occasions when a descendant species changed its nesting habits from burrow to rock face, its shell pattern and behavior changes too. Here the inference of an adaptation to the new nesting condition would be enormously powerful.

It is often alleged that human psychoevolution is too complex and idiosyncratic to be studied using these classic biological techniques, but consider the comparative test performed by Beverley Strassman on Margie Profet's theory of the evolution of menstruation (Strassman, 1996). Profet's 'promiscuous primate' theory suggests that menstruation is designed to flush out sperm-born parasites. Some primates menstruate, whilst others display complete or partial endometrial re-absorption. Profet's theory predicts that menstruation should be preferred over re-absorption when females have many partners and hence are at more risk of sexually transmitted disease. By mapping promiscuity and menstruation onto an independently established primate phylogeny, Strassman was able to show that there is no association between the evolution of the two traits. Similar tests must surely be possible of Profet's equally well known contribution to evolutionary psychology - her adaptive theory of pregnancy sickness.

5. 'Good Old Darwinian Procedures'

Like Evolutionary Psychology, classical ethology had an official methodology. Konrad Lorenz described this methodology colorfully as 'these good old Darwinian procedures' (Lorenz, 1966). He claimed that these methods were first used by Darwin in The Expression of the Emotions in Man and Animals (Darwin, 1872/1965). According to Lorenz, ethology begins with the structural analysis of behavior, proceeds to comparative analysis and finally to adaptive analysis. Ethology begins by observing organisms in their natural environments with the aim of identifying species-typical behavioral sequences. These behaviors are then given a comparative interpretation. Central to this is the identification of homologies, features in different species which are copies of the same ancestral trait. Thus, for example, Darwin was concerned to show that the same muscle movements were utilized to produce the very different superficial expressions of emotion in humans and other primates. Finally, these homologies are given evolutionary explanations in terms of natural selection. I believe that we can still learn a good deal from these 'good old Darwinian procedures', and that paradoxically, we can make progress in evolutionary psychology by going 'back to the future'.

The first stage of the classical ethological method is likely to raise the hackles of many contemporary evolutionary psychologists. The movement regards its shift of emphasis from behavior to psychological mechanisms as one of its main achievements. Rejecting the classical ethological methodology on these grounds would be mistaken for two reasons. First, the idea that we should start with descriptive natural history rather than evolutionary theory can be applied to a theory that deals with underlying mechanisms as well as to a theory dealing with behavior. After all, one of Lorenz's inspirations was the tradition of comparative *anatomy* that provided so much of the evidence for Darwin's theory. Many of the criticisms of EP that I have made in this paper have had the import that we are better off seeking evolutionary explanations for psychological phenomena than seeking psychological phenomena to fit evolutionary explanations. It may be disappointing that evolutionary theory is a retrodictive and explanatory discipline rather than a predictive one, but disappointment is not an argument.

The second reason not to reject Lorenz's methods as 'behavioristic' is that contemporary cognitive science is rediscovering behavior as an independent level of analysis. Some of the most exciting contemporary work in artificial intelligence is research into 'situated robotics' and 'embodied cognition'. The focus of this research is the ability of relatively simple cognitive systems to accomplish complex tasks by exploiting the regularities in their environments (For an introduction see Clark, 1997). Embodied cognition research suggests that many cognitive tasks are solved in ways that are not 'visible' in the internal organization of the cognitive system, but represent order that emerges in the organism-environment interaction. Some situated robotics research actually proceeds by observing unanticipated regularities produced by a very simple system in its intending operating environment and then building control structures that exploit these regularities. When the system is removed from its environment its internal structure becomes meaningless. Hence researchers in this tradition have shown a great deal of interest in Lorenz's 'behavioristic' method of field-observation as a way to reveal how the human mind works (Hendriks-Jansen, 1996).

The second stage of the classical ethological method is to interpret behaviors in a comparative perspective. I believe that this is critical in dealing with either mental mechanisms or with behavior. I showed in the last section how it is a critical step to making adaptive explanations of behavior testable. In my previous work on emotion I have also argued that homology - common descent from an ancestral form - provides a powerful way to organize psychological traits into categories that are in a deep sense 'of the same kind'. In particular, homology is our best guide to which non-human animals models are most relevant to understanding human emotions (Griffiths, 1997). Another valuable contribution of comparative analysis is that it gives order to our observations. EP argues that adaptive thinking provide a rational structure into which to fit the many apparently meaningless 'effects' discovered by empirical psychology (Tooby & Cosmides 1992). Comparative biology provides another such structure, and one that I have argued is more tractable.

Finally, the classical ethological method deals with adaptation and natural selection. The data to be explained are evolutionary homologies: traits which arise at some specific point in an evolutionary tree from some postulated earlier state and which exist in the descendants of that evolutionary event. Various traits of a species develop at various points in the tree. These traits are modified in the course of evolution, giving rise to more detailed homologies between more closely related species. The current state of the human mind would thus not be seen as a suite of adaptations fitting humans to their pleistocene 'environment of evolutionary adaptedness'. Instead it is a set of traits acquired, retained and modified as the lineage leading to *Homo sapiens sapiens* passed through many speciation nodes and many different selective environments. The origin of these traits will be reflected in their distribution in the nested hierarchy of primate taxa, hominid taxa and human populations. Elsewhere I have called such attempts to reconstruct evolutionary history 'adaptive-historical explanations' (Griffiths, 1996). Adaptive-historical explanation contrasts to adaptationist explanation, which tries to show that a trait is optimized to some ecological parameter. From an adaptive-historical viewpoint traits are only optimized in the sense that they are the best a lineage can do given the resources it brings to the problem and, furthermore, the nature of the evolutionary problem is partly a function of the existing state of the organism. Different lineages face different problems in the same physical environment.

An adaptive-historical perspective is likely to be particular valuable in the case of emotion. Darwin established a characteristic pattern of explanation for emotional expressions, which is clearly adaptive-historical in form. Many of the communicative elements of emotional responses are arbitrary. Their form is not intrinsically suited to their function. There is no reason why a human being should use one facial expression to convey a particular signal, rather than another. Darwin's explanation of such traits can be disassociated from his Lamarckian theory of inherited habits to yield an explanation that holds up very well today. He argued that behaviors that originally served some more utilitarian function in the situation associated with the emotion, such as tooth-baring in primate anger, were recruited as signals of emotional state. They were retained, perhaps in a modified form, when their original functions ceased to be important. Such expressions can only be understood as the result of successive modifications of a single structure through a number of different adaptive phases. The response of the lineage to each phase was not predictable unless the particular resources it brought to the problem were taken into consideration. Tooth-baring and pilo-erection would not have their current communicative function if it had not been for the earlier, agonistic functions of which they are vestiges:

> «...with mankind some expressions, such as the bristling of the hair under the influence of extreme terror, or the uncovering of the teeth under that of furious rage, can hardly be understood, except in the belief that man once existed in a much lower and animal-like condition. The community of certain expressions in distinct though allied species as in the movements of the same facial muscles during laughter by man and by various monkeys, is rendered somewhat more intelligible, if we believe in their descent from a common progenitor.» (Darwin, 1872/1965, p12)

I have argued in *What Emotions Really Are...* (Griffiths, 1997) that the adaptive-historical nature of Darwin's project accounts for the great productivity of the research traditions derived from his work as compared to those inspired by a more adaptationist conceptions of evolution.

6. The Comparative Method and Human Evolution

It is commonly thought that claims about human cognitive adaptations will be more difficult to subject to rigorous comparative testing than many other aspects of evolution. Human cognitive adaptations are supposed to have evolved after the separation of hominids from other primates. *Homo sapiens* is the only living representative of the lineage involved. It is therefore not possible to test hypotheses about these adaptations by looking at the distribution of homologous traits in related species or higher taxa. This is a legitimate worry, but it should not be overstated. Natural history can creep up on the period in which our lineage became distinctively human from both sides. Many emotion theorists have crept up from behind on what EP calls 'the pleistocene'. Darwin and his modern successors found extensive homologies between human emotions and those of other primates (Chevalier-Skolnikoff, 1973). New findings about the psychological and neurological bases of emotion can only enrich this collection of homologies. Similar results should be possible in many other areas of psychology. The discovery of primate homologies does more than illuminate the history of the homologous structures themselves. Those structures and their history provide a framework with which evolutionary scenarios for uniquely hominid features must be consistent. This consistency requirement creates extensive opportunities for testing those scenarios, as described above.

The possibility of creeping up on 'the pleistocene' from in front has been created by developments in molecular biology and historical linguistics. There is a discernible phylogenetic structure within the species *homo sapiens,* as there is in most widespread species. Modern molecular biology makes possible extensive investigations of the history and biogeography of human populations (Cavalli-Sforza *et al.*, 1994). Further access to this structure comes from historical linguistics. Phylogenetic trees for human populations can be constructed on the basis of their languages. The congruence between the structures discovered by these two means is very considerable (Penny *et al.*, 1993). These discoveries make it possible to discern homologies within *Homo sapiens* and to infer the 'ancestral' condition of many traits (the condition which existed before the separation of current human populations). It is these traits which are the best candidates for explanation by descent from an ancient hunter-gatherer lifestyle. The access to history provided by these methods is limited by the fact that human populations only began to separate 150,000-200,000 years ago. Nevertheless, determining whether a trait was ancestral in humanity as a whole should be an essential precursor to any attempt to explain that trait by conditions in 'the Pleistocene'. Hypotheses about the evolution of aesthetic preferences for landscape or of sexual attraction might well be disconfirmed by such a test. This test should also be applied to many of the putative emotional adaptations, such as the 'sense of fairness', postulated by Robert Frank (Frank, 1988).

The use of the comparative method in this intra-species role should replace the antiquated procedure of arguing that a trait is 'culturally universal' before attempting to give an evolutionary explanation. Traits that are specific to one or more cultures may nevertheless have an evolutionary history. In one of the most original and progressive elements of EP, Tooby and Cosmides argue that some culturally specific traits may represent facultative adaptations, adaptations which only develop under certain ecological conditions (Tooby & Cosmides, 1992). I myself would argue for the still more radical possibility that evolved features of the human psychological phenotype can be sustained by epigenetic inheritance and so may be lost by cultural change in some populations despite these populations being genetically very similar to others. After all, this phenomena is now widely accepted for other organisms (Jablonka & Lamb, 1995; Jablonka & Szathmáry, 1995). The idea that the biological nature of humans is something that we discover by a deprivation experiment that removes or manipulates culture is as bizarre as supposing we should investigate the biological nature of the ant by removing the distorting influence of the hive (Griffiths & Stotz, in press). Human beings have had a culture since before they were human and human development presumes and makes use of culture just as surely as ant development presumes and makes use of the nest and its pheremonal culture. In both these new theoretical frameworks it makes sense to use the comparative method on a range of human populations to infer the ancestral state of a trait which varies across human populations as a result of differences in the cultural environment. It then makes sense to ask if this trait is a candidate for evolutionary explanation. This is surely an exciting prospect for the evolutionary psychology of culturally variable traits like human emotion.

Acknowledgements

In preparing this paper I have drawn extensively on my work with Kim Sterelny for our forthcoming book *Sex and Death: An Introduction to the Philosophy of Biology*.

References

Basolo, A. L. (1990) "Female preference predates the evolution of the sword in sword-tailed fish", *Science* 250:808-810.

Bateson, P. (1994) "The dynamics of parent-offspring relationships in mammals", *TREE* 9:399-402.

Brandon, R. (1990) *Adaptation and Environment*, Princeton: Princeton University Press.

Byrne, R. W. and A. E. Whiten (1988) *Machiavellian Intelligence*, Oxford, New York: Oxford University Press.

Cavalli-Sforza, L. L., P. Menozzi and A. Piazza, (1994) *The History and Geography of Human Genes*, Princeton: Princeton University Press.

Chagnon, N. and W. Irons, eds, (1979) *Evolutionary Biology and Human Social Behavior: An Anthropological Perspective*, North Scituate: Duxbury Press.

Chevalier-Skolnikoff, S. (1973) "Facial expression of emotion in non-human primates", in: *Darwin and Facial Expression: A Century of Research in Review*, P. Ekman, ed, New York and London: Academic Press, pp.11-89.

Clark, A. (1997) *Being There: Putting Brain, Body and World Together Again*, Cambridge, MA.: MIT Press.

Cosmides, L. and J. Tooby (1992) "Cognitive adaptations for social exchange", in: *The Adapted Mind: Evolutionary Psychology and the Generation of Culture*, J. H. Barkow, L. Cosmides and J. Tooby, eds, Oxford, New York: Oxford University Press, pp.163-228.

Cosmides, L., J. Tooby and J. H. Barkow (1992) "Introduction: Evolutionary Psychology and Conceptual Integration", in: *The Adapted Mind: Evolutionary Psychology and the Generation of Culture*, J. H. Barkow, L. Cosmides and J. Tooby, eds, Oxford, New York: Oxford University Press, pp.3-15.

Darwin, C. (1871) *The Descent of Man and Selection in Relation to Sex*, London: John Murray.

Darwin, C. (1872/1965) *The Expression of the Emotions in Man & Animals*, Chicago: University of Chicago Press.

Dennett, D. C. (1995) *Darwin's Dangerous Idea*, New York: Simon and Schuster.

Eibl-Eibesfeldt, I. (1979) "Human ethology: concepts and implications for the sciences of man", *The Behavioral and Brain Sciences* 2:1-57.

Ekman, P. (1972) *Emotions in the Human Face*, New York: Pergamon Press.

Falk, D. (1990) "Brain evolution in *Homo*: the 'radiator' theory", *Behavioral and Brain Sciences* 13:333-381.

Frank, R. H. (1988) *Passions Within Reason: The Strategic Role of the Emotions*, New York: Norton.

Frean, M. (1994) "The prisoners dilemma without synchrony", *Proceedings of the Royal Society of London B* 257:75-79.

Gould, S. J. (1977) *Ontogeny & Phylogeny*, Cambridge, MA: Belknap/Harvard University Press.

Griffiths, P. E. (1990) "Modularity and the psychoevolutionary theory of emotion", *Biology & Philosophy* 5:175-196.

Griffiths, P. E. (1994) "Cladistic classification and functional explanation", *Philosophy of Science* 61(2):206-227.

Griffiths, P. E. (1995) "The Cronin controversy", *British Journal for the Philosophy of Science* 46:122-138.

Griffiths, P. E. (1996) "The historical turn in the study of adaptation", *British Journal for the Philosophy of Science* 47:511-532.

Griffiths, P. E. (1997) *What Emotions Really Are: The Problem of Psychological Categories*, Chicago: University of Chicago Press.

Griffiths, P. E. and K. Stotz (in press) "How the mind grows: A developmental perspective on the biology of cognition", *Synthese*.

Hendriks-Jansen, H. (1996) *Catching Ourselves in the Act: Situated Activity, Interactive Emergence, Evolution and Human Thought*, Cambridge, MA: MIT Press.

Izard, C. (1978) "On the development of emotions and emotion-cognition relationship in infancy", in: *The Development of Affect,* M. L. Lewis and L. Rosenblum, eds, New York: Plenum Press, pp.389-413.

Jablonka, E. and M. J. Lamb (1995) *Epigenetic Inheritance and Evolution*, Oxford, New York, Tokyo: Oxford University Press.

Jablonka, E. and E. Szathmary (1995) "The evolution of information storage and heredity", *TREE* 10:206-211.

Klinnert, M.D., J.J. Campos, J. F. Sorce, R. N. Emde and M. Svejda (1983) "Emotions as behavior regulators: social referencing in infants", in: *Emotion: Theory Research and Experience*, R. Plutchik and H. Kellerman, eds, New York: Academic Press, pp.57-86.

Lewontin, R. C. (1982) "Organism & environment", in: *Learning, Development, Culture*, H. Plotkin, ed, New York: John Wiley, pp.151-170.

Lorenz, K. (1966) "Evolution of ritualization in the biological and cultural spheres", *Philosophical Transactions of the Royal Society of London* 251:273-284.

Marr, D. (1982) *Vision*, New York: W.H. Freeman.

McKitrick, M. (1993) "Phylogenetic constraint in evolutionary theory: has it any explanatory power?", *Annual Review of Ecology and Systematics* 24:307-330.

Meltzoft, A. N. and M. K. Moore (1977) "Imitation of facial and manual gestures by neonates", *Science* 198:75-78.

Öhman, A., M. Fredrikson and K. Hugdahl (1976) "Premise of equipotentiality in human classical conditioning", *Journal of Experimental Psychology* 105:313-337.

Penny, D., E. E. Watson and M. A. Steel (1993) "Trees from languages and genes are very similar", *Systematic Zoology* 42:382-385.

Pinker, S. and P. Bloom, (1992) "Natural language and natural selection", in: *The Adapted Mind: Evolutionary Psychology and the Generation of Culture*, J. H. Barkow, L. Cosmides and J. Tooby, eds, Oxford, New York: Oxford University Press, pp.451-494.

Profet, M. (1992) "Pregnancy sickness as adaptation: A deterrent to maternal ingestion of teratogens", in: *The Adapted Mind: Evolutionary Psychology and the Generation of Culture*, J. H. Barkow, L. Cosmides and J. Tooby, eds, New York, Oxford: Oxford University Press, pp.327-366.

Profet, M. (1993) "Menstruation as a defense against pathogens transported by sperm", *Quarterly Review of Biology* 68:335-386.

Seligman, M. E. P. and J. L. Hager, eds. (1972) *Biological Boundaries of Learning*, New York: Appleton, Century, Crofts.

Sterelny, K. and P. E. Griffiths (in press) *Sex and Death: An Introduction to the Philosophy of Biology*, Chicago: University of Chicago Press.

Strassman, B. I. (1996) "The evolution of endometrial cycles and menstruation", *Quarterly Review of Biology* 71:181-220.

Symons, D. (1992) "On the use and misuse of Darwinism in the study of human behavior", in: *The Adapted Mind: Evolutionary Psychology and the*

Generation of Culture, J. H. Barkow, L. Cosmides and J. Tooby, eds., Oxford: Oxford University Press, pp.137-159.

Tooby, J. and L. Cosmides (1990) "The past explains the present: Emotional adaptations and the structure of ancestral environments", *Ethology and Sociobiology* 11:375-424.

Tooby, J. and L. Cosmides (1992) "The psychological foundations of culture", in: *The Adapted Mind: Evolutionary Psychology and the Generation of Culture*, J. H. Barkow, L. Cosmides and J. Tooby, eds, Oxford and New York: Oxford University Press, pp.19-136.

Trevarthen, C. (1984) "Emotions in infancy: regulators of contact & relationship with persons", in: *Approaches to Emotion*, K. Scherer, and P. Ekman, eds, Hillsdale, New Jersey: Erlbaum, pp.129-162.

Trivers, R. L. (1974) "Parent-offspring conflict", *American Zoologist* 14:249-264.

Trivers, R. L. (1985) *Social Evolution*, Menlo Park, CA: Benjamin-Cummings.

Weinrich, J. D. (1980) "Towards a sociobiological theory of emotions", in: *Emotion: Theory, Research and Experience*, R. Plutchik and H. Kellerman, eds, New York: Academic Press, Volume 1, pp.113-140.

Whiten, A. and R. W. Byrne, eds, (1997) *Machiavellian Intelligence II. Extensions and Evaluations*, Cambridge: Cambridge University Press.

Wilson, E. 0. (1975) *Sociobiology: The New Synthesis*, Cambridge, MA: Harvard University Press.

TOWARDS A GENETICS OF JOY: BREEDING RATS FOR "LAUGHTER"

JAAK PANKSEPP, JEFF BURGDORF and NAKIA GORDON
Department of Psychology, Bowling Green State University, Bowling Green, OH 43403, USA

ABSTRACT

Laughter is a simple and robust indicator of joyful social affect. All too commonly it has been considered to be a unique emotional capacity of humans and perhaps a few other higher primates. If more primitive mammals also exhibit such emotional responses, it would suggest that joyful affect emerged much earlier within mammalian brain evolution than is generally believed. Evidence for the recent discovery of laughter in lower animals is summarized. We have discovered that one can evoke vigorous 50 kHz chirping in young rodents during tickling, and evidence that elevation and reduction of this response tendency can be transmitted genetically is now provided. A variety of lines of evidence suggest that a study of this response may help us decipher the neural basis of joy and positive emotional consciousness within the mammalian brain.

1. Affective Processes Exist as Genetic Birthrights in All Mammalian Brains

It is now generally agreed that on the average, half the temperamental variability of humans arises directly from psychoneural mechanisms that are provided as genetically delivered abilities/dispositions. The other half is from learning mechanisms that allow individuals to developmentally navigate the environmental and psychological terrain in which they find themselves (Bouchard, 1994; Tellegen *et al.*, 1988; Plomin, 1990). In this sense, we are no different than other mammals, although what we can learn is obviously more subtle and sophisticated than what most other creatures care about. This singularly important fact has qualified the widespread assumption of 20th century psychology, that all the important human behaviors are learned. Still, it has been quite difficult for psychologists and other mind scientists to specify and agree upon the types of aboriginal abilities the genes provide. This is because genes do not directly control behavior, but typically genetic potentials provide subtle predispositions to behave in certain ways. Also, it is difficult to specify what is genetically provided since the epigenetic terrain of real lives, arising from the interaction of genetic potentials and real-life environmental complexities, can be remarkably complex (Loehlin, Willerman, & Horn, 1988).

The genes surely control how the sensory-perceptual apparatus of each species operate, and how the motor apparatus is prone to promote species-typically ways of behaving. However, the deeper intervening mechanisms are much more difficult to see. Still, we must suspect that the inherited potentials penetrate into the subtle organizational structures of the "great intermediate net" of the brain/mind that intervenes between inputs and outputs. Those brain functions would include not only various learning mechanisms, but also motivational and emotional skills and dispositions. Within the *terra incognita* of the brain, we are bound to find various neural operating systems--integrative systems that allow animals to affectively and cognitively respond to the world in many intrinsically adaptive ways. The natural order of these ancestral memories has remained a great mystery for the mind-sciences ranging from anthropology to philosophy. A few of the most obvious emotional systems that exist in all mammalian brains have recently been revealed (see Panksepp, 1998 for summary). However, the genetic analysis of these systems remains essentially nonexistent, with some recent work indicating that tendencies to exhibit separation distress responses can be genetically selected (Brunelli *et al.*, 1997).

2. A Synopsis of Evidence for Laughter and Other Affective Vocalizations in Rats

In this paper we will summarize the beginning of a research program on the genetics of positive emotions in rats. We are presently breeding rats with differential proclivities to "laugh" (i.e., chirp) in response to tickling (Panksepp & Burgdorf, 1998), in the hope of eventually determining some of the genetic underpinnings of joy in all mammals. This "laughing" response consists of cascades of 50-kHz vocalizations that young rats emit when they are tickled. This vocalization pattern was first observed in rodents during their normal rough-and-tumble (R & T) play activities (Knutson, Panksepp & Burgdorf, 1997), and further work affirmed that the response may represent an ancient form of "laughter." The vocalizations are clearly reflective of a positive affective state as indicated by the attractiveness or behaviorally reinforcing properties of the tickling (i.e., as indicated by acquisition of instrumental runway responses and conditioned place preference). Animals run for such stimulation as well as to associated stimuli (Panksepp & Burgdorf, 1999, and unpublished observations). This response can be classically conditioned to predictive cues that lead to either play or tickling, and it is dramatically reduced by all kinds of emotionally negative situations such as exposure to bright lights and the odor of potential predators (e.g., cats). We believe this vocal response can be used a measure of affective self-report in animals.

In general, an underlying assumption of this work is that affective states in animals can be best achieved through a study of their emotional vocalizations (Hauser, 1997; Panksepp, Newman & Insel, 1992). Rats exhibit three distinct types of ultrasonic vocalizations across their life span which appear to communicate different affective states to conspecifics. Infant rats exhibit a 45-kHz "distress" call during such noxious events as exposure to social isolation and cold stress (Blumberg *et al.*, 1992; Hofer & Shair, 1991). Adult rats emit a similar "distress" call at 22-kHz, which is heard during foot shock, social defeat, as well as during both morphine and cocaine withdrawal (Cuomo *et al.*, 1988; Miczek *et al.*, 1995; Mutschler & Miczek, 1998; Sales & Pye, 1974; Thomas *et al.*, 1983). In contrast to these long distress vocalizations, rats exhibit a third vocalization called the 50-kHz call which typically are heard in short and rapid pulses which have some outward temporal resemblance to human laughter. These vocalizations are heard during such affectively positive events as R & T play, the proceptive phases of sexual behavior, and as we have now found, during manual tickling (Barfield *et al.*, 1979; Knutson *et al.*, 1997). Since this type of vocalization, when evoked manual tickling by humans, is strongly related to rat playfulness (Panksepp & Burgdorf 1999), we decided to use it as the endpoint measure to initiate a breeding program for the positive social temperament of fun-loving playfulness.

Initially, We have simply been interested in determining how well genetic selection for such an affective response would proceed, and also, whether positive selection for this behavioral endpoint would also lead to heightened playfulness. Since past work on our population of rats had indicated the two are related (Knutson *et al.*, 1997; Panksepp & Burgdorf, 1999), we predicted that rats that "laugh" most in response to tickling would also play the most among each other. The overarching premise of this work is that other animals do have various emotional experiences (Panksepp, 1998), and that their behaviors, especially their vocal behaviors, can be used to index such central psychological states. The validity of this premise would be based on predictions that could be made from animal research to human subjective experiences of affect (e.g., Kramer *et al.*, 1998; Panksepp, Lensing, Leboyer & Bouvard, 1991). To provide a broader context for the present work, let us first briefly summarize what is known about the heritability of emotional characteristics in animals and humans.

3. The Behavioral Genetics of Temperament

Although our scientific understanding of emotional temperament has not advanced all that much from the recognition of the four classical types--choleric, melancholic, phlegmatic and sanguine--we do now have abundant knowledge how certain traits are genetically controlled (Plomin, 1997). Recent work indicates how gender preference (Hamer & Copeland, 1994) and various aspects of mood (Barondes, 1998) can be inherited. Using modern linkage analysis, one can even specify which parts of which chromosomes, if not yet which specific genes, are critical for such traits (Leboyer, & Gorwood, 1995; Young, 1995).

As an alternative strategy, there are now procedures to generate single-gene deletions (e.g., knock-out mice) and if these animals survive, we can have some idea what contribution single-genes make to the behavioral competence of animals (Gerlai, 1996; Wehner, Bowers, & Paylor, 1996). A large number of animals have been generated, which exhibit striking learning changes (Overstreet, 1993; Tonegawa *et al.*, 1995) and motivational changes ranging from extreme aggressiveness to hypersexuality (Nelson *et al.*, 1995; Saudou *et al.*, 1994). In other words, the consequences of single gene products can have extreme consequences for the behavioral typology of an animal. These effects are not surprising for the power of heritability has long been recognized.

Under the broad rubric of behavior genetics, it has been known that we can select for emotional traits ranging from aggression (Maxson *et al.*, 1982) to tendencies to respond for rewarding electrical stimulation of the brain (Ganchrow, Lieblich, & Cohen, 1981). Unfortunately, our ability to specify exactly what has been selected for in such breeding experiment remains primitive. Most emotional traits are probably not controlled by single-genes but rather through the interaction of many genes. Although there is very little data concerning the genetics of positive emotions such as joy and playfulness (Barondes, 1998), there are some preliminary neurochemical findings as to what may control social responsivity of primates (Higley *et al.*, 1993).

The aim of the following research was to determine the patterns of vocal amplification and reduction that could be achieved by breeding rats which had high and low tendencies to laugh (i.e., exhibit 50 kHz chirps) to tickling, and to relate those changes to changes in playfulness. In this report, we summarize the results derived from the first four generations of our breeding program for this response.

4. Testing and Genetic Selection Procedures for Rodent Laugher

The breeding program started with the thorough analysis of tickle induced chirping and juvenile playfulness in 40 litters of Long-Evans rats that we have been utilizing for play research for the past 20 years. Our original breeding stock has been sustained by random matings with occasional replenishment from commercially available stock. The standard procedure was to wean and individually house the animals at 21-22 days of age, and starting 24 hrs after isolation, to evaluate first tickling responses and then play responses for 4 successive test cycles. In other words, animals were tested for tickling at about 24, 28, 32 and 36 days of age, and for play at 26, 30, 34 and 38 days of age. The specific testing procedures were as follows:

Animals were placed into 45 x 35 x 20 cm high test arena covered with corn cob bedding, and tested for 2 minutes, starting with 15 secs of no stimulation followed by 15 sec of full body tickling, a cycle which was repeated three additional times. 50 kHz chirping was recorded on-line with a "bat-detector," with an observer counting number of chirps on line for each successive 15 sec period with the aid of a manual digital counters.

On the intervening test days, pairs of animals matched for source litter, gender, and body weight were given access to play for two-minute periods. Testing was done under red illumination. For each animal separately, the number of play-solicitive dorsal contacts (sustained contact of one animal with the dorsal surface of the other animal) and pins during active wrestling-type of engagements (one animal on its back with the other on top) were monitored with manual joysticks connected to a special-purpose computer. The total number of chirps made by each play pair was also counted.

The last generation at the point of this preliminary report was also evaluated for several additional behaviors in subsets of animals: (1) the rates of separation-induced 40-kHz isolation-type distress vocalizations (DVs) in 10 day old pups during two minute periods of separation in a 4 inch diameter jar (i.e., individual pups being removed directly from home cage and litter-mates); (2) The amount of motor activity and 50-kHz chirping exhibited by young adults (60 days of age) during a 5 minute test in a novel 50 x 50 x 30 cm high open-field, ruled off into 9 equal size squares for monitoring of line-crossings.

From the initial group of animals, six lines were established. These lines consisted of: (1) the two male-female pairs which exhibited the highest overall levels of chirping during the 4 tickling sessions (high line); (2) the two male-female pairs which exhibited the lowest overall levels of chirping during these sessions (low line); and (3) two pairs of arbitrarily selected male-female pairings from the remaining animals (the random line). During all subsequent generations, matings were brother-sister, with the highest pairs being mated in the "high line" litters, the two lowest pairs in the "low line" litters, and 3) two arbitrarily selected pairs from the so-called "random line" litters.

5. Selection Results for Rodent Laugher

5.1 Overall Results

The mean results for frequency of tickle-induced chirping, tabulated separately for the 15 sec. no-tickle periods (top, baseline) and 15 sec. tickle periods (bottom), averaged across test days for the four generations are summarized in Figure 1. By the third generation, there was a separation of the high lines from the random lines, and the low lines were separated from the others from the very outset, except for the inexplicable elevation of responding in the third generation, which we would attribute to error variance.

5.2 Fourth Generation Results for Tickling

The remaining results are summaries for the 4th generation, with data provided for the successive test days. As is evident in Figure 2, animals in all groups chirped much more during the tickle than the non-tickle sessions, with a good separation of lines in all comparisons. This separation only emerged gradually during the baseline sessions, since the gradual elevation of chirping during these sessions reflects an acquisition curve for contextual conditioning. In any case, the overall analysis indicated a reliable differentiation among lines during the baseline ($F(2,45) = 14.03$, $p < .0001$), as well as during the tickle sessions ($F(2,45) = 36.42$, $p < .0001$). The individual comparisons among lines were reliably different from each other in all possible contrasts (p's $<.001$) except for low vs. random comparisons for baseline ($p <.03$).

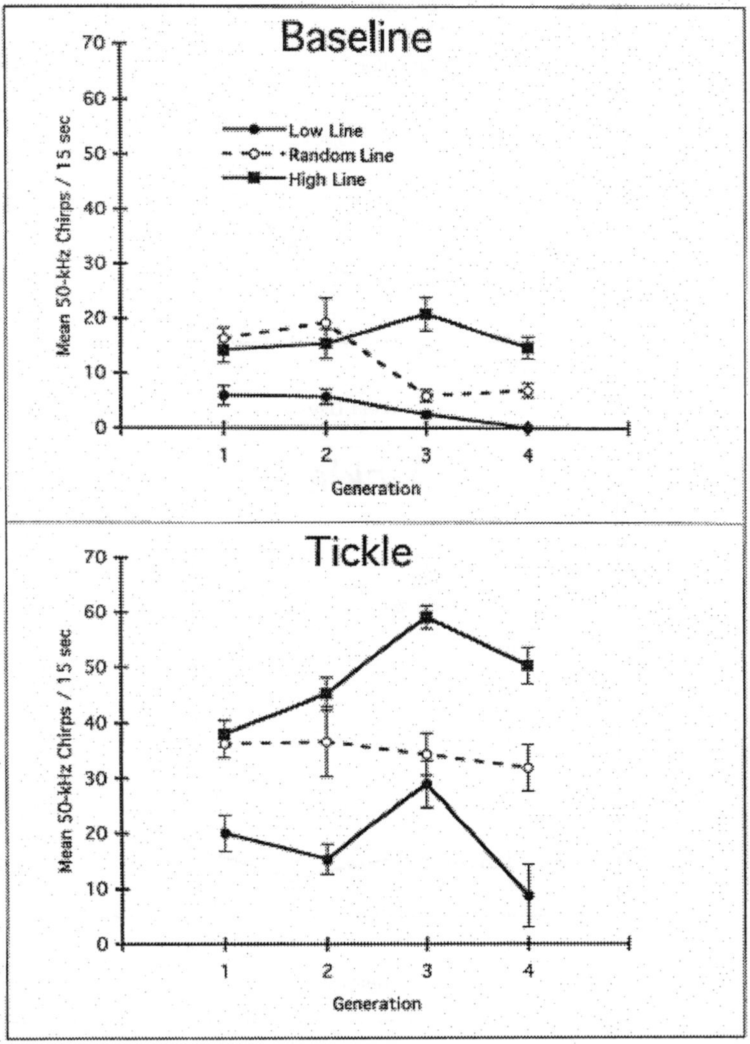

Figure 1. Mean (±SEM) levels of 50-kHz chirping in 4 generations of rats during standardized 2 minute tickling sessions, with successive periods of 15 secs of no stimulation (i.e., baseline) and 15 secs of full body tickling stimulation (i.e., tickle).

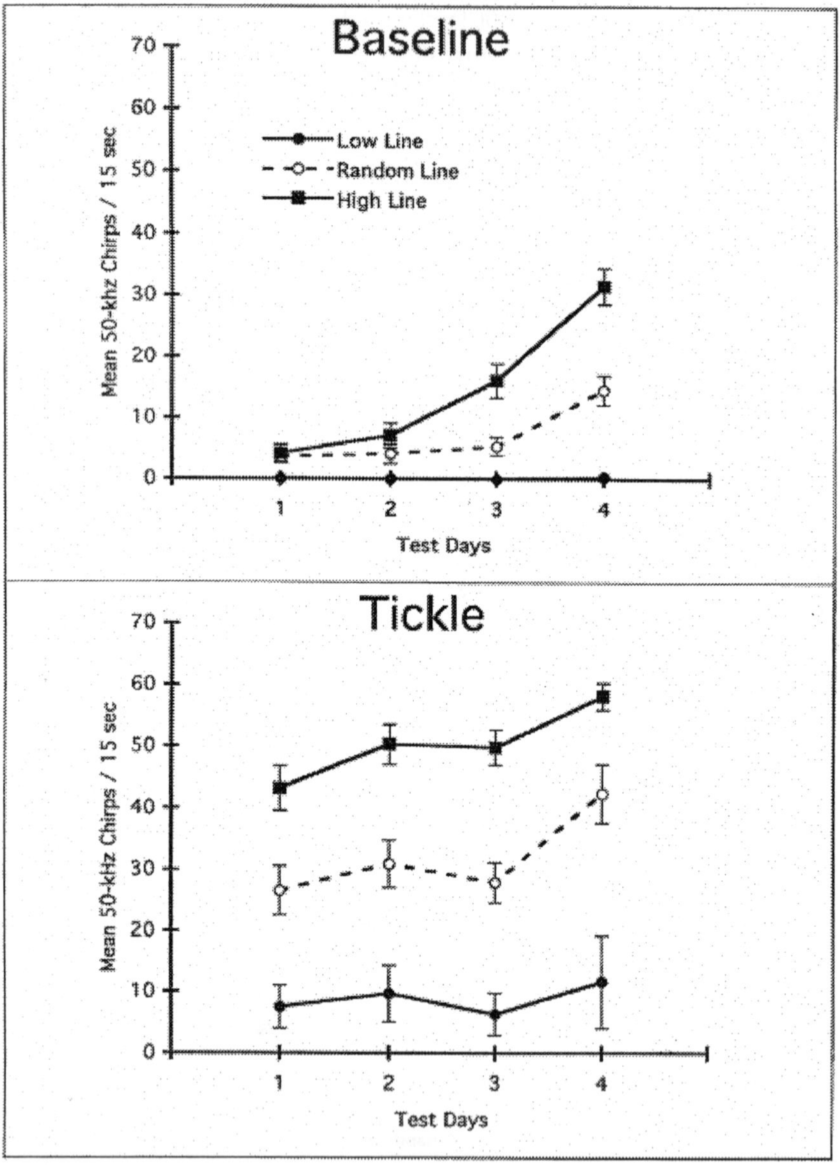

Figure 2. Mean (±SEM) levels of 50-kHz chirping in the 4th generation of animals during four successive standardized 2 minute tickling sessions, with successive periods of 15 secs of no stimulation (i.e., baseline) and 15 secs of full body tickling stimulation (i.e., tickle).

5.3 Fourth Generation Results for Rough-and-Tumble Play

As summarized in Figure 3, the high line exhibited reliably more pins than either the random or low lines (p's < .001), with no difference among the random and low lines. A comparable differentiation was evident in vocalizations, with the high line exhibiting more chirping than both of the other lines, but the difference between low and random lines was only marginally (p <.05) different.

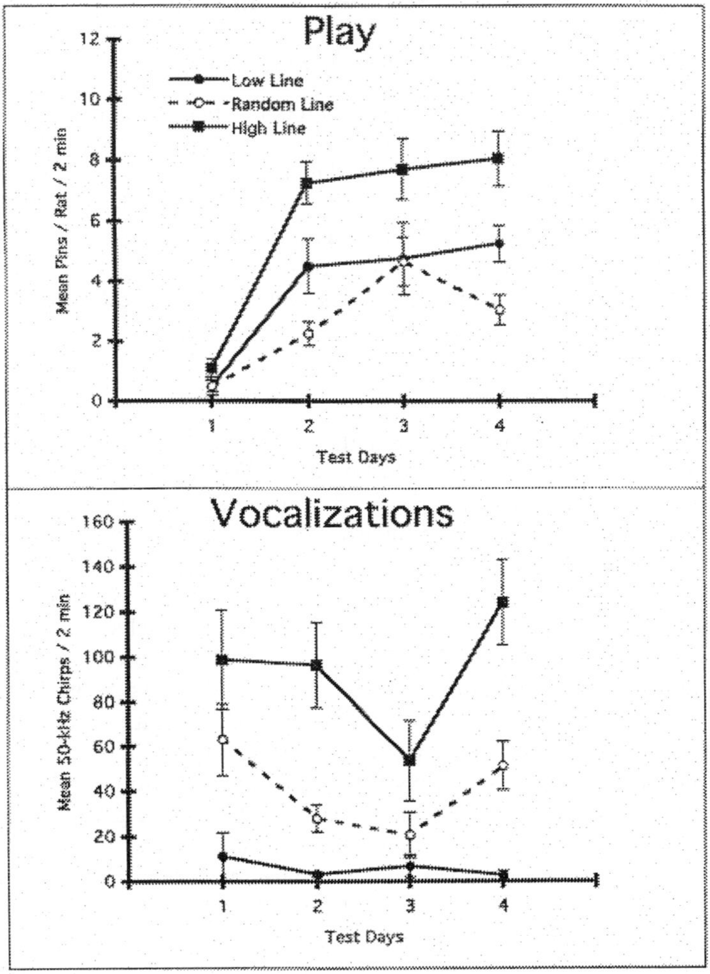

Figure 3. Mean (±SEM) levels of play (as measured by pinning during standard 2-min dyadic play encounters (top), and concurrent levels of vocalizations by pairs of animals in the respective groups (bottom).

5.4 Fourth Generation Results for Separation Distress

The various selection lines exhibited different levels of DVs during brief periods of social isolation even though there was a great deal of variability among animals (low mean ±SEM: 20 (±8.5); random: 65 (±11.4); high: 92 (±18.7)). Only the comparisons between high and low lines was statistically different ($p < .03$), with the comparison between low and random lines being $p = .10$, with no clear tendencies for high and random lines to exhibit differential DVs.

5.5 Fourth Generation Results for Open-Field Activity

There were no difference among the lines in open-field line crossings (High Line: 52.1 ± 9.1; Random Line: 53.9 ± 6.1; Low Line: 53.3 ± 5.5). However, there were substantial differences in the amounts of 50 kHz vocalizations exhibited by the different lines ($F(2, 21) = 7.52$, $p < .005$). The high line exhibited many more vocalizations (188.0 ± 37.7) than the low line (14.9 ± 4.2) ($p < .001$) and also modestly more than the random line (94.4 ± 39.4) ($p < .05$). The apparent difference between random and low lines was not statistically significant ($p < .09$).

6. Implications

Clearly, the breeding program was quite successful, with quite similar trends being evident in each of the high and low lines. We would like to conclude from this that we were able to increase the level of joy-responses in the high-tickle line, while reducing those tendencies in the low-lines. Of course, this conclusion would require additional data regarding the motivational levels of the animals. We did conduct a final experiment with animals from each of the lines of the 4th generation of the breeding program. This test was conducted starting at 61 days of age, immediately following the open field activity test described above. At the start of this test, individual animals were placed under a 16.5 cm diameter, 11.5 cm. high glass half-sphere with a 6 cm diameter exit hole at the top. The animals were given four test trials each day for four successive days, where they were given access to tickling immediately upon leaving the "safe haven" of their glass spheres. After exiting (with a 3 minute cut-off), they were given an opportunity for up to two minutes of tickling (using the 15 second tickle, 15 second no-tickle testing procedure used during the standard baseline testing). During this period, all animals could terminate the tickling simply by returning to their safe havens. By the 4th test day; the high line animals were all departing from their safe havens in less than 6 seconds, while the low and random animals were taking an average of 70 and 84. seconds respectively. Each of the high line animals spent only an average of 7 seconds back in the "safe haven" (typically exiting promptly as soon as they had entered the "safe" area), while the low and random animals returned back to the "safe haven" an average of 47 and 66 seconds respectively. This pattern suggests to us that the high line was especially eager for the tickling experience.

It is also important to emphasize that it was the high line that exhibited the highest levels of play of the three groups, and they chirped much more during the play than the other animals. We take this data to strongly indicate that the high-line animals had a much higher motivation for the tickling and joyful-play experiences than the other two lines. This suggests to us that we had indeed successfully selected for some type of positive emotional aspect of social engagement. The lower degree of differentiation between the low and random lines suggests that we were not as successful in selecting for reduced levels of social-joy motivation, as compared to the randomly bred lines of animals.

Our high lines also exhibited some additional differences in the outcome measures that were harvested: They did vocalize more during social isolation, but this tendency did not differentiate them from the random lines, only from the low lines. Also, although their overall motor activity in open-field tests was no different than other groups, they did exhibit more 50-kHz vocalizations than either the low or random lines during this non-social test (which may reflect a generalized conditioned response). This would tend to suggest that their overall background rates of ultrasonic chirping were higher than normal, which may indicate that their trait-"happiness," in the absence of any explicit social stimulation, was higher than normal. However, since this test required experimenter handling in placing the animal in the open-field, it is possible that it simply reflected conditioned solicitive vocalizations reflecting the recognition on the animal's part that a friendly human hands may have been nearby. Obviously, more research is needed to discriminate between these interesting possibilities.

Of course, we would like to know what precisely had been selected for in our high-laughter lines, but we have no knowledge at that level. Indeed, it would be very hard to track down the genes that had been differentially assorted. Thus, at the present time, the only thing this piece of research demonstrates is that it is rather easy to select for what appears, on the face of it, to be a credible index of animal joyfulness. Presumably there are some distinct nervous system differences, probably in the vigor of key neurochemical systems, which distinguish these animals from their more dour companions. We assume that if we could identify those chemistries, we would have a much clearer understanding of the neural basis of joyful affect/consciousness in all mammals. Such knowledge could also have important implication for generating new ways to treat psychiatric disorders (e.g., melancholia/depression and mania), which appear to be conditions where the potential for experiencing joy has changed markedly within the nervous system.

References

Barfield, R.J., P. Auerbach, L.A. Geyer and T.K. McIntosh (1979) «Ultrasonic vocalizations in rat sexual behavior», *American Zoologist* 19:469-480.

Barondes, S. (1998) *Mood genes: Hunting for Origins of Mania and Depression*, New York: W.H. Freeman and Company.

Blumberg, M.S., I.V. Efimova and J.R. Alberts (1992) «Ultrasonic vocalizations by rat pups: The primary importance of ambient temperature and thermal significance of contact comfort», *Developmental Psychobiology* 25: 229-250.

Bouchard, T.J. (1994) «Genes, environment, and personality», *Science* 264:1700-1701.

Brunelli, S.A., D.D. Vinocur, D. Soo-Hoo and M.A. Hofer (1997) «Five generations of selective breeding for ultrasonic vocalization (USV) responses in N:NIH rats», *Developmental Psychobiology* 31:255-265.

Cuomo, V., K. Cagiano, M.A. De Salvia, M.A. Maselli, G. Renna and G. Racagni (1988) «Ultrasonic vocalization in response to unavoidable aversive stimuli in rats: effects of benzodiazepines», *Life Sciences* 43:485-491.

Ganchrow, JR., I. Lieblich and F. Cohen (1981) «Consummatory responses to taste stimuli in rats selected for high and low rates of self-stimulation», *Physiology and Behavior* 27: 971-976.

Gerlai, K. (1996) «Gene-targeting studies of mammalian behavior: is it the mutation or the background genotype?», *Trends in Neursciences* 19:177-181.

Hamer, D. and P. Copeland (1994) *The Science of Desire: The Search for the Gay Gene and the Biology of Behavior*, New York: Simon and Schuster.

Hauser, M.D. (1997) *The Evolution of Communication*, Cambridge, MA: MIT Press.

Higley, J.D., W.W. Thompson, M. Champoux, D. Goldman, M.F. Hasert, G.W. Kraemer, J.M. Scanalan, S.J. Suomi and M. Linnoila (1993) «Paternal and maternal genetic and environmental contributions to cerebrospinal fluid monoamine metabolites in rhesus monkeys (Macaca Mulatta)», *Archives of General Psychiatry* 50:615-523.

Hofer, M.A. and H.N. Shair (1991) «Independence of ultrasonic vocalization and thermogenic responses in infant rats», *Behavioral Neuroscience* 105:4-48.

Knutson, B., J. Panksepp and J. Burgdorf (1998) «Anticipation of play elicits high-frequency ultrasonic vocalizations in young rats», *Journal of Comparative Psychology* 112:65-73.

Kramer, M.S. *et al.* (1998) «Distinct mechanisms for antidepressant activity by blockade of central substance P receptors», *Science* 281:1640-1645.

Leboyer, M. and P. Gorwood (1995) «Genetics of affective disorders and schizophrenia», in: *Advances in Biological Psychiatry, Volume 1*, J. Panksepp, ed., Greenwich, CT: JAI Press, pp.27-66.

Loehlin, J.C., L. Willerman and J.M. Horn (1988) «Human behavior genetics», *Annual Review of Psychology* 39:101-133.

Maxson, S.C., P. Shrenker and L.C. Vigue (1983) «Genetics, hormones, and aggression», in: *Hormones and Aggressive Behavior,* B.B. Svare, ed., New York: Plenum Press, pp.179-196.

Miczek, K.A., B.M. Weerts, J.A. Vivian, and H.M. Barros (1995) «Aggression, anxiety and vocalizations in animals: GABA and 5-HT anxiolytics», *Psychopharmacology* 121:38-56.

Mutschler, N.H. and KA. Miczek (1998) «Withdrawal from i.v. cocaine "binges" in rats: Ultrasonic distress calls and startle», *Psychopharmacology* 135:161-168.

Nelson, R.J., G.E. Demas, P.L. Hunag, M.C. Fishman, V.L. Dawson, T.M. Dawson and S.H. Snyder (1995) «Behavioural abnormalities in male mice lacking neuronal nitric oxide», *Nature* 378:383-386.

Overstreet, D.H. (1993) «The Flinders Sensitive Line rats: A genetic animal model of depression», *Neuroscience and Biobehavioral Reviews* 17:51-68.

Panksepp, J. (1998a) *Affective Neuroscience: The Foundations of Human and Animal Emotions,* New York: Oxford University Press.

Panksepp, J. and J. Burgdorf (1998) «Laughing Rats? Playful tickling arouses 50KHz ultrasonic chirping in rats», *Society for Neuroscience Abstracts,* 24:691.

Panksepp, J. and J. Burgdorf (1999) «Laughing rats? Playful tickling arouses high frequency ultrasonic chirping in young rodents», in: *Toward a Science of Consciousness - III,* S. Hameroff, D. Chalmers and A. Kaszniak, eds, Cambridge, MA: MIT Press, pp. 231-244.

Panksepp, J., P. Lensing, M. Leboyer and M.P. Bouvard (1991) «Naltrexone and other potential new pharmacological treatments of autism», *Brain Dysfunction* 4:281-300.

Panksepp, J., J.D. Newman and T.K. Insel (1992) «Critical conceptual issues in the analysis of separation distress systems of the brain», in: *International Review of Studies on Emotion, Volume 2,* K.T. Strongman, ed., Chichester, UK: John Wiley & Sons, pp.51-72.

Plomin, K. (1990) «The role of inheritance in behavior», *Science* 248:183-188.

Plomin, K., J.C. DeFries, G.E. McClearn and M. Rutter (1990) *Behavioral Genetics* (3rd ed.), New York: W.H. Freeman.

Sales, G. and D. Pye (1974) *Ultrasonic Communication by Animals,* London: Chapman & Hall.

Saudou, F., D.A. Amara, A. Dierich, M. LeMeur, S. Ramboz, L. Segu, M.C. Buhot and K. Hen (1994) «Enhanced aggressive behavior in mice lacking 5-HT1B receptors», *Science* 265:1875-1878.

Tellegan, A., D.T. Lykken, T.J. Bouchard, K. Wilcox, N. Segal and S. Rich (1988) «Personality similarity in twins reared apart and together», *Journal of Social and Personality Psychology* 54:1031-1039.

Thomas, D.A., L.K. Takahashi and R.J. Barfield (1983) «Analysis of ultrasonic vocalizations emitted by intruders during aggressive encounters among rats»,

Journal of Comparative and Physiological Psychology 93:201-206.

Tonegawa, S., Y. Li, R.S. Erzurumlu, S. Jhaveri, C. Chen, Y. Goda, K. Paylor, A.J. Silva, J.J. Kim and J.M Wehner (1995) «The gene knockout technology for the analysis of learning and memory, and neural development», *Progress in Brain Research* 105:3-14.

Wehner, J.M., B.J. Bowers and K. Paylor (1996) «The use of null mutant mice to study complex learning and memory processes», *Behavioral Genetics,* 26:301-312.

THE NEUROSCIENCE OF FEAR: PERSPECTIVES FROM ANIMAL RESEARCH

JOSEPH E. LEDOUX
Center for Neural Science, New York University, 4 Washington Place, New York, NY 10003, USA

ABSTRACT

Considerable progress has been made in elucidating the brain pathways involved in detecting and responding to threatening stimuli, and learning about novel threats. Much of this progress has come from studies of fear conditioning, in which a relatively neutral stimulus, like a tone, acquires aversive properties after being paired with a noxious event. The pathways involve transmission of information from sensory processing areas in the thalamus and cortex to the amygdala. The lateral nucleus of the amygdala receives and integrates sensory information and sends the outcomes of its processing to the central nucleus. The central nucleus, in turn, is the interface with motor systems controlling automatic or reflexive fear responses of various types (behavioral, autonomic, endocrine). Sites of plasticity within this circuitry, and some cellular mechanisms involved, have also been identified. In addition to developing fear responses to the specific stimulus that is paired with the noxious event, fear also conditions to the general context in which the noxious stimulus occurs. So-called context conditioning requires the hippocampus and inputs to the basal nucleus of the amygdala, which in turn projects to the central nucleus. An important issue is how we get rid of fear. It seems that the mesial frontal cortex and its connections to the amygdala are involved. Thus, lesions of the mesial cortex lead to an intensification of fear and an increased resistance to extinction. Given that stress can adversely affect the hippocampus and mesial frontal cortex, and that patients with psychiatric conditions often suffer from stress, it is possible that stress-induced changes in these areas contribute to the intense, therapeutically resistant, contextually free fears that psychiatric patients often have. We are thus beginning to uncover the neural mechanisms, from systems to cellular levels, underlying emotional processing, including emotional learning and memory, at least within the fear system. These findings are beginning to elucidate mechanisms that may be relevant to understanding emotional disorders.

1. Introduction

The focus of this paper is on the neuroscience of fear and, more generally, on how to "cheat" our way into the study of emotion. As noted by other contributors to this volume, emotion seems to be in the eye of the beholder. Many of the phenomena that we call emotions are only loosely related to one another. These phenomena are more related to each other than to other things that we don't call emotions, so we feel comfortable lumping them together. However, if you take three purported emotions, A and B may be related to each other in a particular way, but this may not be the same way that B and C are related to each other. At some point we must address the question of whether such phenomena do form some kind of natural category and the answer to this question is presently far from clear. It is with these kinds of concerns that I try to cheat my way into the study of emotion and I do that by actually performing four or five different cheats.

The first cheat is that I completely ignore, at least at the beginning, the question of where our feelings come from. So, this paper will not address that question at all. The second cheat is that the paper will focus on only the emotion of fear, so that I don't have to define the whole emotion space. Third, I will be concerned with only one way of measuring fear and will not discuss any other way of measuring this emotion. Fourth, I'm going to describe experiments that employ a very simple stimulus that is very unnatural in life. Using such a simple and highly repeatable stimulus means that there will be very minimal additional kinds of processing load on the organism. I'm also going to use a very simple response that is expressed the same way every time in every animal that I've studied, in this case rats. Finally, I will examine how he brain processes the stimulus on its first pass through the brain. Thus, I've created a very artificial situation that allows me to go deep into the neurobiology of the system but perhaps not very broad into the psychology of the system.

1.1 The Classical Fear Conditioning Paradigm

The experiments to be described employ the research paradigm of classical fear conditioning. In this paradigm, a tone or a light is paired with a foot shock. Typically, after only a single trial of such pairing, when the rat hears the tone or sees the flash of light it expresses a variety of different behaviors. As soon as the tone comes on, the rat (or other animal) freezes its movement, blood pressure goes up, pain reactivity is suppressed, reflexes are potentiated (e.g., the startle reflex), and stress hormones (e.g., corticosteroids), epinephrine, and norepinephrine are released (Blanchard & Blanchard, 1989; Bouton & Bolles, 1980; Mason et al., 1961; McAllister & McAllister, 1971). All these responses occur essentially upon presentation of the artificial stimulus (tone or light) that has been paired with the shock. However, the same pattern of responses is seen if a laboratory-born rat, never having been out in the wild or previously exposed to a cat, is put him in a room with the cat. So, the natural trigger, in this case the cat, is activating circuits in the brain that produce a pattern of responses that can also be elicited by this artificial alarm trigger. The learning in this case is not response learning because the rat comes into life having these responses built into its brain by evolution. Rather, the rat learns what particular stimuli out there in the world are going to turn this system on. In this case learning allows the brain to use evolutionarily perfected ways of responding to new dangers that are similarly associated with old dangers.

2. Fear Conditioning and the Amygdala

All of this is controlled by a small structure deep within the brain, the amygdala (Fig. 1). Whether the eliciting stimulus is a cat or a tone that has been paired with a shock, the amygdala produces this pattern that we call defense or fear responses. Nowhere in this chain of events is there the feeling of fear, at least from my point of view. My focus will be on how these responses are controlled and come to be controlled by external stimuli (for more extensive review, see LeDoux, 1996; 2000).

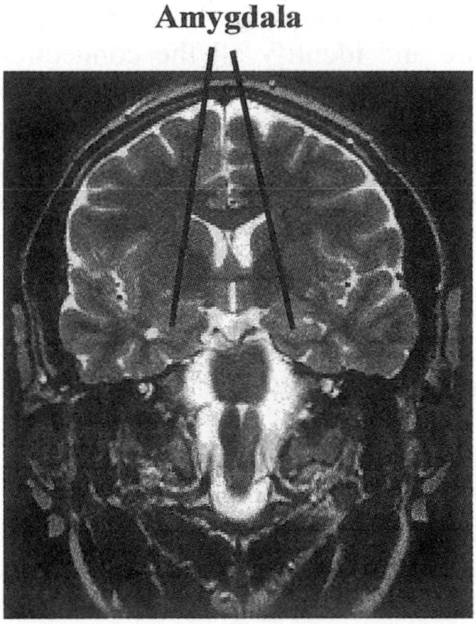

Figure 1. Human Brain Magnetic Resonance Image (MRI), showing location of the amygdala in each hemisphere.

How does the brain, with its billions of neurons and trillions of connections, do something like this? How can we actually get inside and figure this out? We have to follow some kind of strategy. In the fear conditioning paradigm we have a tone producing responses such as changes in blood pressure and freezing. The auditory neurons in the brain which process the tone are somehow connected with the motor neurons that control the blood pressure and freezing responses. If we can determine the links between the auditory system and the motor system, we will know something about what the circuits are that controls the responses. The assumption is that the plastic changes that underlie the learning are somewhere along this circuit that carries the stimulus from the outside world to the motor neurons. That is what my research has been looking for. Where is the circuit and what is the nature of its plasticity? How do we do that? How do get from the beginning to the end?

It is important to have a starting point, and that is why having a stimulus like the tone is key. There have been many different types of approaches to the study of fear. Some studies have used various forms of avoidance conditioning, where there is no explicit stimulus, but instead the animal is moving around in some environment, presumably using vague internal or external cues about where it is in space. The lack of ability to precisely specify the stimuli in such experiments makes it very difficult to trace the conditioning process through the brain. The key to discovering the important neural circuitry is to have a stimulus that you can start at one point and follow through the brain, and then also have a well-defined response. This is why fear conditioning is such a nice model. If there is a starting point, say the auditory system, a lesion can then be made somewhere in the auditory system that results in conditioning being blocked. We may already know that that the lesioned structure has 3, or 4, or 15 connections. If you put a tracer into that brain structure and identify all the connections, you can then make a lesion in each of those and observe whether only one has an effect on the response. We can then put a tracer into another structure in the system, find the connections, lesion these, and determine whether only one has an effect, and so on. By following this approach of lesion and tracing, lesion and tracing, we can get from A to Z. That is what I have been trying to do for a long time now (LeDoux, 1996, 2000), the results of which indicate the following: In the center of the whole system is the amygdala. The amygdala receives inputs from the sensory thalamus and sensory areas of the cortex, which allow sensory features and objects to activate the amygdala. The amygdala is also getting information from higher areas of the cortex that may deal with conceptual information, and from the hippocampus which is involved in spatial context and long term memory (see Cohen & Eichenbaum, 1993; Nadel & Willner, 1980; O'Keefe & Nadel, 1978; Sutherland & Rudy, 1989). Thus, the amygdala is like the hub in a wheel. It receives numerous inputs, and the particular input that it is receiving and using at the moment will drive the amygdala and then produce the defense or fear responses. For example, because of input from the hippocampus, a long term memory of some aversive situation could trigger the amygdala into producing the fear responses.

2.1 Amygdala Outputs and the Control of Fear Response Components

The outputs for these responses are all coming from one small area of the amygdala, called the central nucleus (Fig. 2). The central nucleus projects to a variety of brain stem areas, each of which controls a particular response. If one of these brain stem nuclei, the central gray is lesioned, the freezing component of the set of fear responses is blocked, but autonomic responses (e.g., changes in blood pressure) and endocrine responses (e.g., the release of steroids) are not altered (LeDoux, Iwata, Cicchetti, & Reis, 1988). In contrast, if the paraventricular hypothalamus (which received inputs from the central amygdala directly and by way of the bed nucleus of the stria terminalis) is lesioned, the endocrine response components are blocked, but not the freezing or endocrine responses (Gray et al., 1993). If the lateral hypothalamus (also receiving input from the amygdala central nucleus) is lesioned, autonomic, but not the freezing or endocrine, responses are blocked (LeDoux et al., 1988). Each of the areas receiving input from the central nucleus of the amygdala are thus controlling individual fear response modalities. If, however, the central nucleus of the amygdala itself is lesioned, all of these responses are blocked (for review see Davis, 1992; Kapp, Wilson, Pascoe, Supple, & Whalen, 1990; LeDoux, 1995). The central nucleus is the last point in the circuit that is controlling the *integrated* fear responses. Prior to the central nucleus, the amygdala is dealing with information that is independent of the response. In fear conditioning, the lateral nucleus of the amygdala receives information directly from both the thalamus and the sensory cortex. These two pathways will be the focus of much of the remainder of this paper.

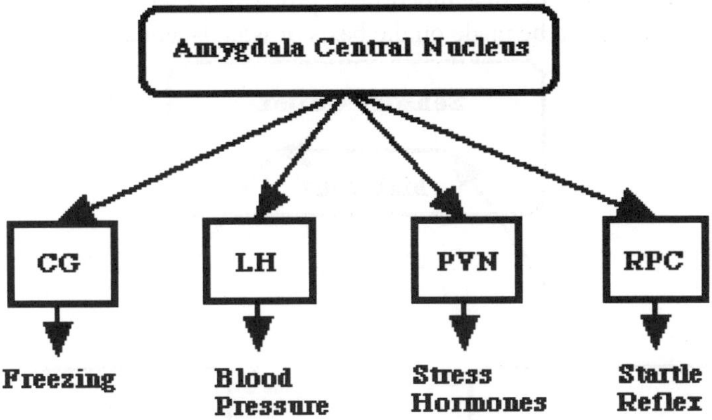

Figure 2. Outputs from the amygdala central nucleus, controlling different components of the conditioned fear response. CG=central grey; LH=lateral hypothalamus; PVN=paraventricular hypothalamus; RPC=reticulopontis caudalis. (Modified from LeDoux, 1996, p. 160)

2.2 Amygdala Inputs in the Conditioned Fear Response

Thalamic input to the amygdala provides low level, relatively crude information. The cortical input provides a much broader, richer representation, obviously because that's what the cortex does. One way of thinking about what the thalamus is doing in relationship to the amygdala is to think of a situation where you hear a loud noise, such as a bomb going off. The first time you hear the bomb you are startled. Your amygdala doesn't need to know what the nature of that noise is in order to produce the initial response. What the amygdala needs to know is that high intensity stimuli tend to be dangerous. Intensity is a good clue to distance, and distance is a good clue to danger. The amygdala detects this clue by a very simple computation. We know from the recording of amygdala neurons that they ignore stimuli at 40 decibels but respond very reliably to 80 decibel stimuli (see Bordi & LeDoux, 1992; LeDoux, 1996; Weinberger, 1995). The thalamus can provide the amygdala with information about the loudness of a stimulus without providing information about the specific nature of that stimulus (Bordi & LeDoux, 1994a, 1994b). In order to distinguish two different kinds of music, the auditory information has to be discriminated at the cortical level (the "high road"). But intensity itself can be handled through the "low road" of the thalamus (Fig. 3). There are many ways to think about the importance of this low road which activates the amygdala directly and quickly, preparing to defend against danger. For example, the crude visual information passed from the thalamus when a snake is seen on the ground can result in the amygdala initiating the integrated fear response. Although the stimulus could actually be a stick rather than a snake, the low road of the thalamic-amygdala circuit quickly initiates a response to possible danger. From an evolutionary point of view, the organism is better off treating the stick as a snake than treating the snake as a stick (LeDoux, 1994). The high road of thalamocortical processing, which takes longer, then allows for a decision to be made on the basis of what is actually out there.

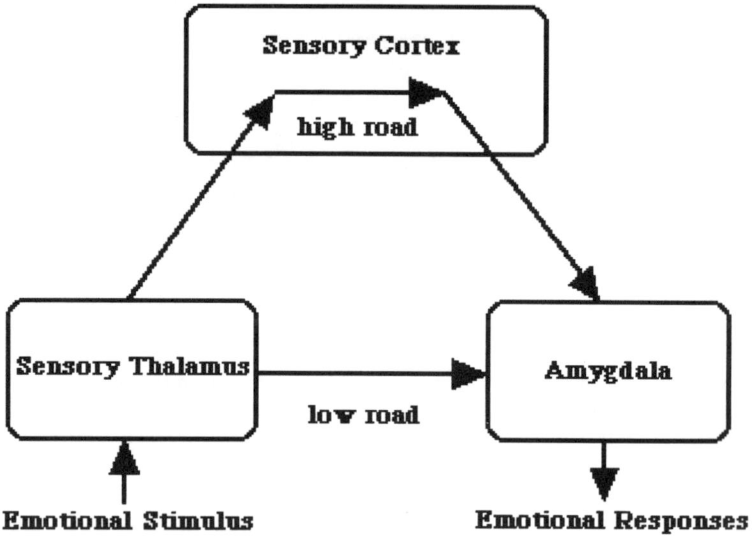

Figure 3. The low and the high roads to the amygdala. (Modified from LeDoux, 1996, p. 164)

One of the reasons that the system might be organized this way is that neurons can be made to fire more easily by an input if they are brought close to firing threshold or if they are already firing. Thus, if neurons that are not firing are given a stimulus, it is much harder to get them going than if they're already firing. Through the thalamic-amygdala low road, amygdala cells are stimulated to firing or are brought close to firing. When the stimulus then comes to the amygdala from the high road of cortical processing, the cells are much better able to respond than if they were just sitting there silently (Li, Stutzmann, & LeDoux, 1996).

Within the auditory system, information splits at the level of the thalamus so that the information that goes to the cortex allows it to perform higher-order discriminations (e.g., two kinds of music). This information goes through a different channel in the thalamus than the information that is going directly to the amygdala. This later auditory information also goes to the cortex but it is not carrying tonotopic information. The high road is the main channel for tonotopic representation. As noted above, the relatively, non-specific information within the low road can carry things like intensity or brightness but not quality. For quality, information must come to the amygdala via the cortex. The high and low road pathways converge on single cells in the lateral nucleus of the amygdala, which is the sensory gateway into the structure (Li et al, 1996). Within the auditory system, lesions in either the midbrain (where information from the ear initially arrives) or the medial geniculate nucleus of the thalamus (the auditory relay nucleus) disrupt fear conditioning (LeDoux, Sakaguchi, & Reis, 1984). However, if the auditory cortex is lesioned there is no effect. That must mean that the stimulus goes through the mid-brain to the thalamus and then goes someplace else besides the auditory cortex.

If a tracer (wheat germ agglutinin conjugated horseradish peroxidase; WGA-HRP) is injected into the medial geniculate, it is not surprising that anterograde transport of the tracer, revealed by reactive staining, shows projections into the auditory cortex. However, projections are also found into the striatum and down into the amygdala (LeDoux et al., 1984). That was our first clue that there might be some way of getting the information directly to the amygdala without going to the cortex. We then lesioned the different areas of projection from the auditory thalamus that had been revealed by the tracer technique (LeDoux, Sakaguchi, Iwata, & Reis, 1986). As noted above, lesioning of the auditory cortex had no effect on classical fear conditioning. Lesioning of the striatum also had no effect, but when the amygdala was lesioned, conditioning was blocked. When the freezing and blood pressure change data from the amygdala-lesioned rats are compared to that from animals that have received pseudo-conditioning (a control procedure where instead of giving the shock at the end of the tone, the tone and shock are randomly related), the responses look identical. Thus, the amygdala lesion reduces the animal to the state of being able to be sensitized to the shock so that any stimulus that comes along will produce these kinds of low level responses. However, there is no learning involved, since it is a non-associative response that remains.

2.3 Intra-Amygdala Processing in Fear Conditioning

The information from the thalamus and cortex both come into the lateral nucleus of the amygdala. Once this information is processed in the lateral amygdala, it is then distributed to the intra-amygdala nuclei. There, it is further processed (and integrated with other incoming inputs), and then sent to the central amygdala nucleus which serves as the main output station for the amygdala's control of emotional responses (Fig. 4). The question, however, is what is going on within the amygdala? The amygdala has about 13 different nuclei (Amaral, Price, Pitkänen, & Carmichael, 1992), so we need to know something about its internal structure.

Not all of these nuclei are relevant to fear conditioning. In order to understand which nuclei are relevant, it is necessary to digress briefly and describe two different ways of doing conditioning. One way is the kind that I have been discussing thus far, in which a tone and a shock are presented, following which the response to the tone itself is measured. However, if a rat has been in a particular box when the tone (followed by the shock) comes on, it will show fear responses not only to the tone, but also to the box itself, even in the absence of the tone (Kim & Fanselow, 1992; Phillips & LeDoux, 1992). This phenomenon is called contextual conditioning. Contextual conditioning is dependent upon the hippocampus because the hippocampus is involved in spatial representation, creating complex representations between stimuli and the spatial context. If the hippocampus is lesioned, when the rat is placed back into the box it does not act afraid until the tone comes on. However, if the hippocampus is intact, the rat acts afraid as soon as it is placed in the box (Phillips & LeDoux, 1992, 1994, 1995).

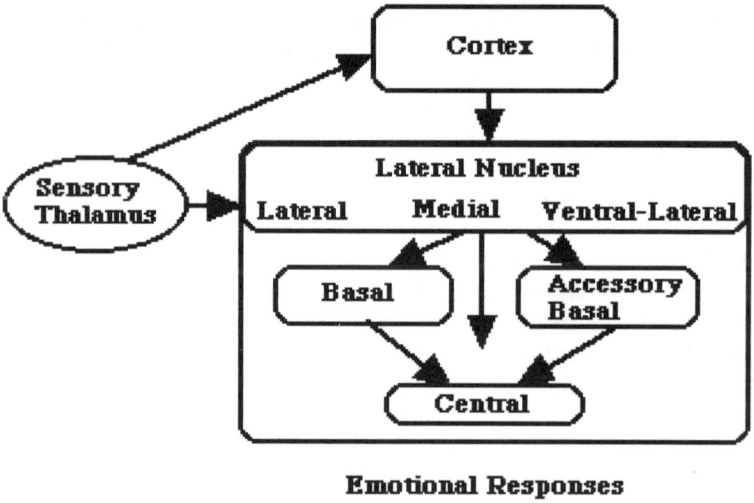

Figure 4. Intra-amygdala Nuclei and connections.

We can now return to our question about intra-amygdala processing. If auditory information, initiated by a tone in the classical fear conditioning paradigm, is coming in from the auditory thalamus and cortex, lesioning the lateral nucleus of the amygdala will block fear conditioning, as will lesioning of the central nucleus (for review see Pitkänen, Savander, & LeDoux, 1997). However, lesions of all the other nuclei do not block fear conditioning. For context conditioning, lateral amygdala nucleus lesions have no effect, but lesions of the basal and accessory basal amygdala nuclei do. These differential effects occur because the auditory system projects to the lateral nucleus and the hippocampus projects to the basal and accessory basal nuclei. The anatomy of the system is telling us what parts of the amygdala should be involved in fear conditioning, and when those structures are lesioned the appropriate kind of conditioning is blocked. The flow of information through the amygdala thus depends on the kind of stimulus that is driving it. The output, however, will always be through the central nucleus.

The next question concerns how the learning actually takes place. We know that cells in the lateral nucleus are responsive to tones and shocks, so it has the right inputs to be put together to result in conditioning. Information from the shock is going to come in through the spinothalamic tract and also through the parabrachial nucleus in the brain stem. The parabrachial nucleus sends information directly to the central amygdala. The spinothalamic tract goes to the medial division of the medial geniculate nucleus, which also receives information from the tone and sends both tone and shock information directly into the lateral nucleus (for a review of these pathways, see Kandel, Schwartz, & Jessell, 2000).

This means that the lateral nucleus is the first place in the amygdala where the shock and tone are coming together. This provides the opportunity for integrating the conditioned stimulus (CS) of the tone and the unconditioned stimulus (US) of the shock In the basal nucleus of the amygdala there is no US information so we would not necessarily expect basic plasticity there. In the accessory basal nucleus of the amygdala the CS context can be integrated with the US. The central amygdala nucleus is a place where the US arrives. However, it doesn't have the information from the CS itself, but instead has intra-amygdala information. Thus, the shock can be integrated with higher order information at this point. The lateral nucleus is integrating the tone and the shock information and that representation can then be interacting with the shock information itself again in the central nucleus. There is thus opportunity for several points of plasticity within the amygdala. However, I want to focus on the first step to answer some questions about the simplest form of plasticity there.

To make things simple, I will focus the lateral nucleus. The lateral nucleus has three parts; the dorsolateral, the medial, and the ventral lateral areas. The connections among these intra-amygdala areas have been demonstrated by injections of tracers into the various nuclei (Pitkänen et al., 1995; Savander et al., 1995, 1996a,b,c). The dorsolateral part of the amygdala receives sensory information, and this information doesn't come to the other areas. The dorsolateral area projects to the medial area, and the medial area tells the rest of the amygdala what the dorsolateral area of the lateral nucleus is doing. There is a systematic flow of information through this set of structures, and we know what the connections are and how the information is flowing. However, now we are at an impasse because of being beyond the resolution of lesion and tracer injection techniques to study the lateral nucleus itself. In one case it took us 112 lesions to get 6 that were confined to the lateral nucleus. To get 6 lesions confined just to the dorsal part of the lateral nucleus would take thousands of animals, and that would probably be immoral. Because of these limitations, we have turned to other techniques.

The main alternative technique is the use of single unit recordings where we can put very fine wires into the amygdala that will be localized to various nuclei. For example, a rat would have an electrode array implanted that has about 30 very fine wires that can all be placed in the small dorsolateral part of the lateral nucleus. There are a number of advantages to that technique. One, is that we can record many cells simultaneously. In addition, it allows us to watch the interaction between cells that are simultaneously being recorded. By computing cross-correlations and doing other forms of ensemble recording we can begin to see how the population of cells within this region is responding as a whole, as opposed to how each cell is responding alone.

Our work has focused on the lateral nucleus because it is the entry point of sensory processing in the amygdala (Quirk, Armony, & LeDoux, 1997; Quirk, Repa, & LeDoux, 1995). These studies have shown that the most prominent conditioning-induced increases in firing rates in the lateral nucleus occur at the earliest response latency (< 15 msec after the CS onset). This reflects changes in the efficacy of signal processing in the direct thalamo-amygdala pathway. In contrast, conditioned changes observed in other amygdala nuclei occur later (e.g., 30-50 msec after CS onset in the central nucleus; Pascoe & Kapp, 1985). These latency differences and the relationship observed among conditioned changes seen at each of the amygdala nucleus sites suggests that significant processing is occurring within amygdala circuits between input and output stages.

Latency differences in response can also be seen among the subnuclei of the amygdaloid lateral nucleus. If recordings are made from the dorsal part of the lateral nucleus, early (10 to 20 msec) conditioned changes are observed. If, however, the recording is made from below that area, in the ventral part of the lateral nucleus, response latencies are longer (Quirk et al., 1995). The lateral nucleus responds, and begins to stop responding, and the area that it projects to then begins to respond and passes information on to the next area. There is thus a flow of information from the top of the lateral nucleus to the bottom of the lateral nucleus and then out to the rest of the amygdala.

Response latencies also provide information concerning the source of amygdala inputs (Quirk et al., 1997). As noted above the earliest possible amygdala response begins at 12 msec. The medial geniculate nucleus of the thalamus responds in 7 msec to a tone, before any kind of conditioning. If the medial geniculate is stimulated and recording is made in the lateral amygdala, the latency difference between stimulation and response indicates that it takes 5 msec for information to go from the auditory thalamus to the lateral amygdala. Thus, the 12 msec response latency in the amygdala is accounted for by the 7 msec it takes auditory information to get to the medial-geniculate and 5 msec it takes to then get to the amygdala. If recording is made in the auditory cortex, it also takes 12 milliseconds to there. The fastest the amygdala can respond when auditory information is coming to it via the cortex is about 19 to 20 msec. Any observed response latency between 12 and 20 msec can thus be assumed to be coming from the thalamus.

What is the nature of the plasticity? Where does the conditioning occur? It is possible that the cortex learns over trials. Let us assume that the cortex learns in two trials and then teaches the amygdala something that would allow it to respond more quickly to the thalamic information. However, electrophysiological recording studies have shown that the lateral amygdala learns within the first 3 trials, whereas the auditory cortex does not learn until 6-9 trials (Quirk et al., 1997. The auditory cortex is both slower to respond than the amygdala within a trial and it takes longer for it to learn over trials. The information reflected in the early spike recording in the amygdala at the beginning of the stimulus is thus accounted for by entirely thalamic input and not by cortical input.

2.4 The Nature of Amygdala Plasticity

Because we wanted to ask questions about detailed mechanisms of plasticity that occur during fear conditioning in the amygdala, we turned to a preparation where we stimulate fibers from the auditory cortex all the way to the thalamus and record from neurons within the lateral nucleus of the amygdala (Clugnet & LeDoux, 1990; Rogan & LeDoux, 1995). These studies have used the technique of inducing long-term potentiation (LTP), where an electrical training stimulus in the thalamo-amygdala pathway can produce a long-lasting facilitation of synaptic activity. Typically, LTP is dependent upon calcium influx through N-methyl-D-aspartate (NMDA) receptors, which are a class of glutamate receptors important in plasticity (for review see Staubli, 1995). NMDA receptors normally allow calcium to enter the cell, which serves as the key signal that triggers the entire second messenger cascade that stabilizes the memory. In the case of LTP in the lateral amygdala, it is also dependent on calcium entry, not through the NMDA receptor but instead through voltage gated calcium channels. When the membrane of the cell is depolarized, calcium can enter the cell either through the NMDA receptor or through calcium channels directly. Amygdala plasticity in the lateral nucleus appears dependent upon these voltage gated calcium channels. We (Clugnet & LeDoux, 1990; Rogan & LeDoux, 1995) first demonstrated that LTP in the thalamo-amygdala pathway results in enhanced processing of auditory stimuli through the pathway. This was step one, showing that the brain could use artificial LTP to respond to a natural stimulus.

The next step was to substitute the artificial simulation completely and instead condition the animal. We paired the tone and shock, and then tested with the tone, measuring the field response which is typically used for measuring LTP (Rogan, Staubli, & LeDoux, 1997). This showed that fear conditioning and LTP induction produce the same kind of changes in the field response in the amygdala and that these responses correspond very nicely with the learning of the behavioral response. In the estimation of many investigators, this comes the closest demonstrating that LTP has anything to do with real-life memory and has encouraged continuing research on how the amygdala learns and on the mechanisms of plasticity.

One of the key questions concerns where the plasticity resides in the brain. There are a number of ways to approach this, but one way is to temporarily turn the amygdala off during the learning process. Permanent lesions, including those that occur in human brain disease, damage not only the amygdala but also the areas to which the amygdala projects. One way to avoid such complications is to create a temporary inactivation of the amygdala. A drug can be used that facilitates gamma-aninobutyric acid (GABA), which is an inhibitory neurotransmitter that shuts the amygdala down. When such a drug is injected into the amygdala during fear conditioning, no learning occurs. Thus, the amygdala has to be active during the conditioning process in order for learning to occur. The initial physiologic activity in amygdala cells is needed in order to trigger the entry of calcium which then sets off, among other things, the synthesis of new proteins. If protein synthesis is blocked in the amygdala, the formation of the long-term memory is also blocked. Similarly, if the second messenger system for cyclic AMP is interrupted, the formation of long-term memory is blocked, without affecting the short-term memory. This is true for both the context and tone memory. Thus, we are beginning to understand how the cells learn and what goes on within the cells as a result of the learning process.

These kinds of classically conditioned fear or defensive responses occur throughout the animal kingdom, from lizards to humans (see LeDoux, 1996). In every animal that has an amygdala, damage to this structure interferes with fear conditioning. Fear conditioning thus appears to be an evolutionarily old solution to the problem of how to detect and respond to new dangers. The amygdala provides the substrate for this solution, at least within the vertebrates. It is interesting to note that not all animals respond with exactly the same responses to danger. This raises the question of how different responses can be maintained throughlut vertebrate species across different kinds of bodily parts and ways of responding. It is likely the brain system that is conserved through evolution, rather than the response itself.

2.5 The Extinction of Conditioned Fear Response

Fear responses tend to be quite persistent, which has obvious survival advantages. Learned fear responses reflect a record of previously encountered threatening experiences and allow quick response to similar future situations. Nonetheless, it is also important to be able to learn that a stimulus no longer signals danger. Otherwise, unnecessary fear responses would be elicited by innocuous stimuli, and this could interfere with other important routine tasks. In the laboratory, learned fear responses can be reduced (extinguished) by repeatedly presenting the CS without the US. It is of interest to examine what happens during extinction, comparing the cortex to the amygdala. If the CS sound is presented repeatedly, the lateral amygdala stops responding (extinction). In contrast, the auditory cortex does not extinguish. This may have relevance for how we think about clinical phenomena, such as phobias. A patient with a phobia who is successfully treated can be compared to the rat for whom a conditioned fear response is being extinguished. Following treatment, the presence of the phobic stimulus (e.g., a snake) does not produce the phobic response. However, the patient still knows that he or she was afraid of snakes. We might think of the cortical representation as reflecting the declarative memory of having been afraid of snakes. The auditory cortex is connected to the medial temporal system involved in declarative memory. Because the cortical response does not extinguish, the patient remembers that he or she was afraid of snakes. However the patient no longer shows a fear response to snakes because the amygdala response has extinguished.

Despite the fact that fear can be extinguished in phobia, the phobic response can be brought back. For example, in a person with a phobia who has been successfully treated, the death of the person's mother can result in a return of the phobia (Jacobs & Nadel, 1985). This begs the question of how the phobia continues to live in the brain at a time when the phobic stimulus is not producing the fear response. How does the memory trace survive in the presence of extinction? An approach to addressing this question can take advantage of the technique of multiple simultaneous cell recording and cross-correlation. Applying this technique, we have found something that gives us the beginning of an explanation. Assume that we are recording from 2 cells, A and B, where A fires before conditioning but B does not. After conditioning B begins to fire. This is what Donald Hebb (1949) called a cell assembly. Conditioning is creating a linkage or assembly between the neurons and this is a memory. When the CS occurs, it activates the memory and produces the response. What extinction does is to weaken the ability of the input to get to the memory to produce the response. Thus, in the case of a phobia, therapy is operating not on the memory itself, which is extinction-resistant. Fear conditioning experiments suggest that neocortical areas, particularly areas of the prefrontal cortex, are involved in the extinction process. When the medial prefrontal cortex is lesioned, there is a potentiation of fear responses and a retardation of the extinction (Morgan & LeDoux, 1995; Morgan, Romanski, & LeDoux, 1993). Thus, the medial prefrontal cortex, possibly in conjunction with other neocortical regions (LeDoux, Romanski, & Xagoraris, 1989), may be involved in regulating amygdala responses to stimuli based on their current affective values. These findings suggest the possibility that fear disorders may be related to a dysfunction of the prefrontal cortex that makes it difficult for patients to extinguish acquired fears. It is of interest to note that studies have demonstrated that stress has the same fear exaggeration effects as lesions of the medial prefrontal cortex (Corodimas, LeDoux, Gold, & Schulkin, 1994). Stress can also adversely affect the hippocampus (see Sapolsky, 1992). Since patients with psychiatric conditions offen suffer from stress, it is possible that stress-induced changes in the mesial frontal cortex and hippocampus contribute to the intense, therapeutically resistant, contextually free fears that these patients often have. We are thus beginning to uncover the neural mechanisms, from systems to cellular levels, underlying emotional processing, including emotional learning and memory, at least within the fear system. These findings are beginning to elucidate mechanisms that may be relevant to understanding emotional disorders.

3. Summary and Conclusions

In summary, the amygdala is the key to the entire fear conditioning system. It doesn't matter whether the CS is a light, tone, or context. It doesn't matter whether it's rats, humans monkeys, or lizards, and it doesn't matter what kind of response is measured. The amygdala remains the center piece of the fear conditioning system. There is a great deal of enthusiasm about the amygdala in neuroscience, psychology, and the popular press today. We are almost in danger of replacing the old limbic system theory of emotion with an amygdala theory of emotion. I think that would be a big mistake. All of what I have described in this paper, and most all of what we know about the amygdala, has been learned from the very narrow procedure of fear conditioning. We must remain very careful about generalizing to other kinds of tasks and other kinds of emotional preparation.

References

Amaral, D.G., J.L. Price, A. Pitkanen, and S.T. Carmichael (1992) "Anatomical organization of the primate amygdaloid complex", in: *The Amygdala: Neurobiological Aspects of Emotion, Memory, and Mental Dysfunction*, J.P. Aggleton, ed, New York: Wiley-Liss, pp. 1-66.

Blanchard, D.C., and R.J. Blanchard (1989) "Experimental animal models of aggression: what do they say about human behavior?" in: *Human Aggression: Naturalistic Approaches*, J. Archer and K. Browne, eds, New York: Routledge, pp. 94-121.

Bordi, F., and J. LeDoux (1992) "Sensory tuning beyond the sensory system: An initial analysis of auditory properties of neurons in the lateral amygdaloid nucleus and overlying areas of the striatum", *Journal of Neuroscience* 12:2493-2503.

Bordi, F., and J.E. LeDoux (1994a) "Response properties of single units in areas of rat auditory thalamus that project to the amygdala. I: Acoustic discharge patterns and frequency receptive fields", *Experimental Brain Research* 98:261-274.

Bordi, F., and J.E. LeDoux (1994b) "Response properties of single units in areas of rat auditory thalamus that project to the amygdala. II: Cells receiving convergent auditory and somatosensory inputs and cells antidromically activated by amygdala stimulation", *Experimental Brain Research* 98:275-286.

Bouton, M.E., and R.C. Bolles (1980) "Conditioned fear assessed by freezing and by the suppression of three different baselines", *Animal Learning and Behavior* 8:429-434.

Clugnet, M.C., and J.E. LeDoux (1990) "Synaptic plasticity in fear conditioning dircuits: Induction of LTP in the lateral nucleus of the amygdala by stimulation of the medial geniculate body", *Journal of Neuroscience* 10:2818-2824.

Cohen, N.J., and H. Eichenbaum (1993) *Memory, Amnesia and the Hippocampal System*, Cambridge, Massachusetts: MIT Press.

Corodimas, K.P., J.E. LeDoux, P.W. Gold, and J. Schulkin (1994) "Corticosterone potentiation of learned fear", *Annals of the New York Academy of Sciences* 746:392-393.

Davis, M (1992) "The role of amygdala in conditioned fear", in *The Amygdala: Neurobiological Aspects of Emotion, Memory, and Mental Dysfunction*, J.P. Aggleton, ed, New York: Wiley-Liss, pp. 255-306.

Gray, T.S., R.A. Piechowski, J.M. Yracheta, P.A. Rittenhouse, C. L. Bertha, and L.D. van der Kar (1993) "Ibotenic acid lesions in the bed nucleus of the stria terminalis attenuate conditioned stress induced increases in prolactin, ACTH, and corticosterone", *Neuroendocrinology* 57:517-524.

Hebb, D.O. (1949) *The Organization of Behavior: A Neuropsychological Theory*, New York: John Wiley and Sons.

Jacobs, W.J., and L. Nadel (1985) "Stress-induced recovery of fears and phobias", *Psychological Review* 92:512-531.

Kandel, E.R., J.H. Schwartz, and T.M. Jessell, eds (2000) *Principles of Neural Science (Fourth Edition),* New York: McGraw-Hill.

Kapp, B.S., A. Wilson, J. Pascoe, W. Supple, and P.J. Whalen (1990) "A neuroanatomical systems analysis of conditioned bradycardia in the rabbit", in: *Learning and Computational Neuroscience: Foundations of Adaptive Networks*, M. Gabriel and J. Moore, eds, Cambridge, Massachusetts: MIT Press, pp. 53-90.

Kim, J.J., and M.S. Fanselow (1992) "Modality-specific retrograde amnesia of fear", *Science* 256:675-677.

LeDoux, J.E. (1994) "Emotion, memory and the brain", *Scientific American* 270:32-39.

LeDoux, J.E. (1995) "Emotion: Clues from the Brain", *Annual Review of Psychology*, 46:209-35.

LeDoux, J. (1996) *The Emotional Brain: The Mysterious Underpinnings of Emotional Life*, New York: Simon and Schuster.

LeDoux, J. (2000) "Cognitive-emotional interactions: Listen to the brain", in: *Cognitive Neuroscience of Emotion* , R.D. Lane, L. Nadel, G.L. Ahern, J.J.B. Allen, A.W. Kaszniak, S.Z. Rapcsak, and G.E. Schwartz eds, New York: Oxford University Press, pp. 129-155.

LeDoux, J.E., J. Iwata, P. Cicchetti, and D.J. Reis (1988) "Different projections of the central amygdaloid nucleus mediate autonomic and behavioral correlates of conditioned fear. *Journal of Neuroscience* 8:2517-2529.

Ledoux, J.E., L.M. Romanski, and A.E. Xagoraris (1989) "Indelibility of subcortical emotional memories:, *Journal of Cognitive Neuroscience* 1:238-243.

LeDoux, J.E., A. Sakaguchi, J. Iwata, and D.J. Reis (1986) "Interruption of projections from the medial geniculate body to an archi-neostriatal field disrupts the classical conditioning of emotional responses to acoustic stimuli in the rat", *Neuroscience* 17:615-627.

LeDoux, J.E., A. Sakaguchi, and D.J. Reis (1984) "Subcortical efferent projections of the medial geniculate nucleus mediate emotional responses conditioned by acoustic stimuli", *Journal of Neuroscience* 4:683-698.

Li, X.F, G.E. Stutzmann, and J.L. LeDoux (1996) "Convergent but temporally separated inputs to lateral amygdala neurons from the auditory thalamus and auditory cortex use different postsynaptic receptors: in vivo, intracellular and extracellular recordings in fear conditioning pathways", *Learning and Memory* 3:229-242.

Mason, J.W., G. Mangan, J.V., Brady, D. Conrad, and D.M. Rioch (1961) "Concurrent plasma epinephrine, norepinephrene and 17-hydroxycorticosteroid levels during contitioned emotional disturbances in monkeys", *Psychosomatic Medicine* 23:344-353.

McAllister, W.R., and D.E. McAllister (1971) "Behavioral measurement of conditioned fear", in *Aversive Conditioning and Learning*, F.R. Bush, ed, New York: Academic Press, pp. 105-179.

Morgan, M., and J.E. LeDoux (1995) "Differential contribution of dorsal and ventral medial prefrontal cortex to the acquisition and extinction of conditioned fear", *Behavioral Neuroscience* 109:681-688.

Morgan, M.A., L.M. Romanski, and J.E. LeDoux (1993) "Extinction of emotional learning: Contribution of medial prefrontal cortex", *Neuroscience Letters* 163:109-113.

Nadel, L., and J. Willner (1980) "Context and conditioning: a place for space", *Physiological Psychology* 8:218-228.

O'Keefe, J., and L. Nadel (1978) *The Hippocampus as a Cognitive Map*, Oxford: Clarendon Press.

Pascoe, J.P., and B.S. Kapp (1985) "Electrophysiological characteristics of amygdaloid central nucleus neurons during Pavlovian fear conditioning in the rabbit", *Behavioral Brain Research* 16:117-133.

Phillips, R.G., and J.E. LeDoux (1992) "Differential contribution of amygdala and hippocampus to cued and contextual fear conditioning", Behavioral *Neuroscience* 106:274-285.

Phillips, R.G., and J.E. LeDoux (1994) "Lesions of the dorsal hippocampal formation interfere with background but not foreground contextual fear conditioning", *Learning and Memory* 1:34-44.

Phillips, R.G., and J.E. LeDoux (1995) "Lesions of the fornix but not the entorhinal or perirhinal cortex interfere with contextual fear conditioning", *Journal of Neuroscience* 15:5308-5315.

Pitkänen, A., V. Savander, and J.L. Ledoux (1997) "Organization of intra-amygdaloid circuitries: An emerging framework for understanding functions of the amygdala", *Trends in Neuroscience* 20:517-523.

Pitkänen, A., L. Stefanacci, C.R. Farb, C.G. Go, J.E. LeDoux, and D.G. Amaral (1995) "Intrinsic connections of the rat amygdaloid complex: Projections originating in the lateral nucleus", *Journal of Comparative Neurology* 356:288-310.

Quirk, G.J., J.L. Armony, and J.E. LeDoux (1997) "Fear conditioning enhances different temporal components of tone-evoked spike trains in auditory cortex and lateral amygdala", *Neuron* 19:613-624.

Quirk, G.J., J.C. Repa, and J.E. LeDoux (1995) "Fear conditioning enhances short-latency auditory responses of lateral amygdala neurons: Parallel recordings in the freely behaving rat", *Neuron* 15:1029-1039.

Rogan, M.T., and J.E. LeDoux (1995) "LTP is accompanied by commensurate enhancement of auditory-evoked responses in a fear conditioning circuit", *Neuron* 15:127-136.

Rogan, M.T., U. Staubli, and J.E. LeDoux (1997) "AMPA-receptor facilitation accelerates fear learning without altering the level of conditioned fear aquired", *Journal of Neuroscience* 17:5928-5935.

Sapolsky, R.M. (1992) *Stress, the Aging Brain, and the Mechanisms of Neuron Death*, Cambridge, Massachusetts: MIT Press.

Savander, V., C.G. Go, J.E. LeDoux, and A. Pitkänen (1995) "Intrinsic connections of the rat amygdaloid complex: projections originating in the basal nucleus", *Journal of Comparative Neurology* 361:345-368.

Savander, V., C.G. Go, J.E. LeDoux, and A. Pitkänen (1996a) "Intrinisic connections of the rat amygdaloid complex: projections originating in the accessory basal nucleus", *Journal of Comparative Neurology* 374:291-313.

Savander, V., J.E. Ledoux, and A. Pitkänen (1996b) "Interamygdala projections of the basal and accessory basal nucleus of the rat amygdaloid complex", *Neuroscience* 76:725-735.

Savander, V., J.E. Ledoux, and A. Pitkänen (1996c) "Topographic projections from the periamygdaloid cortex to select subregions of the lateral nucleus of the amygdala in the rat", *Neuroscience Letters* 211:167-170.

Straubli, U.V. (1995) "Parallel properties of long-term potentiation and memory", in: *Brain and Memory: Modulation and Mediation of Neuroplasticity*, J.L. McGaugh, N.M. Weinberger, and G. Lynch, eds, New York: Oxford University Press, pp. 303-318.

Sutherland, R.J., and J.W. Rudy (1989) "Configural association theory: The role of the hippocampal formation in learning, memory, and amnesia", *Psychobiology* 17:129-144.

Weinberger, N.M. (1995) "Retuning the brain by fear conditioning", in: *The Cognitive Neurosciences*, M.S. Gazzaniga, ed, Cambridge, Massachusetts: MIT Press, pp. 1071-1090.

AMYGDALA AND PROCESSING OF INFORMATION WITH EMOTIONAL CONTENT

PASQUALE CALABRESE[1], ANNA NEUGEBAUER[2],
HANS J. MARKOWITSCH[3], HERBERT F. DURWEN[4],
ANDREAS FALK[4], ALBRECHT G. HARDERS[5],
KIRSTEN SCHMIEDER[5] and WALTER GEHLEN[1]
[1] *Department of Neurology, Knappschaftskrankenhaus, Faculty of Medicine, Ruhr-University, Bochum, Germany*
[2] *Istituto Internazionale di Genetica e Biofisica, CNR, Napoli, Italy*
[3] *Faculty of Psychology, University of Bielefeld, Germany*
[4] *Institute of Radiology, Knappschaftskrankenhaus, Faculty of Medicine, Ruhr-University, Bochum, Germany*
[5] *Department of Neurosurgery, Knappschaftskrankenhaus, Faculty of Medicine, Ruhr-University, Bochum, Germany*

ABSTRACT

Although lesion studies in rodents and primates have provided strong evidence of a crucial role of the amygdala in the processing of emotionally loaded information, its precise role in the human neuronal emotion-cognition network is far from being understood. We therefore studied patients with selective amygdalar damage and also normal subjects neuropsychologically (memory tests, face recognition tests, subjective emotional ratings) and neuroradiologically (fMRI) using stimuli either with high emotional value, or non-emotional content. By this procedure, we where able to show an additional amygdaloid activation in normal individuals. The functional neuroanatomical aspects connected to the neuropsychological ones are discussed in the frame of the hypotheses both of a "bottleneck" function of amygdala, and of disinhibition of CA3 hippocampal neurons.

1. Introduction

The limbic system (Mark *et al.*, 1995) has traditionally been regarded as the major processing unit for emotional information (Papez, 1937). Today, especially structures like the amygdala (Adolphs *et al.*, 1994; Cahill *et al.*, 1995; Markowitsch *et al.*, 1994) or the septal nuclei (Cramon *et al.*, 1985) are those limbic regions which are recognized as centrally implicated in emotional information processing. In the cortico-limbic system the sensory inputs are thought to converge with the reinforcing emotional ones (Neugebauer *et al.*, 1997). Based on these assumptions the question arises concerning a possible emotional "gating mechanism" that "filters" the sensory inputs.

2. Neuropsychological investigations on amygdala damaged patients

We will try to strengthen the hypothesis that only early amygdaloid damage or malfunctioning will result in an impaired processing ability for affective facial information. One of the tests used to study the role of the amygdala in emotion is recognizing facial emotion (Adolphs *et al.*, 1994; Young *et al.*, 1993; 1996). Recently, a controversy evolved as to the universality of the deficit in judging emotional facial expression after amygdala damage and Hamann *et al.* (1996) failed to detect impairments in recognizing pictures of faces with specific emotional expressions (happyness, sadness, surprise, etc.) in two patients with encephalitis based brain damage which included both amygdalae. Adolphs *et al.* (1994), on the other hand, had found a profound impairment in this ability in one patient with congenital Urbach-Wiethe disease (resulting in bilateral amygdala damage). Hamann *et al.* (1996) assumed that amygdala damaged patients may show this defect most likely "...if these lesions occur early in development, rather than in adulthood".

Here we report the case of a patient whose performance would fit this assumption. P.S., a 33-year old female patient, was tested neuropsychologically before surgery of a cavernoma which completely infiltrated her left amygdala (Fig 1).

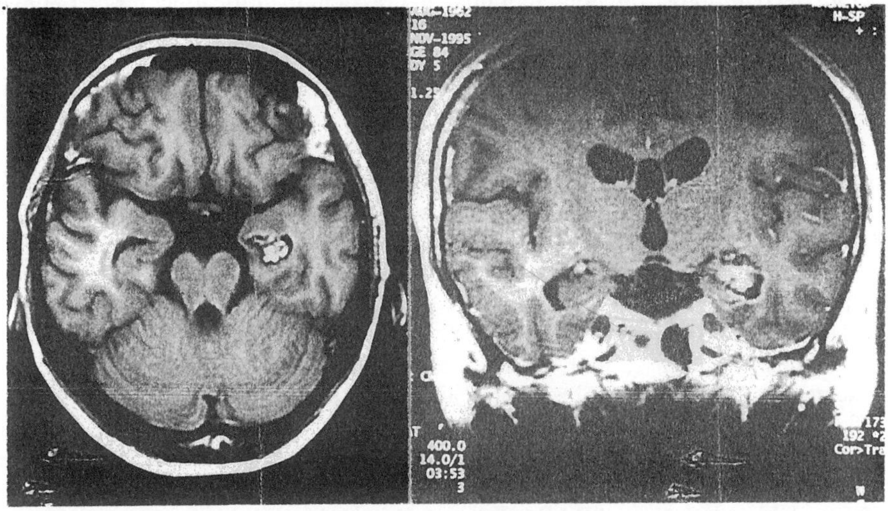

Figure 1. Amygdaloid lesion of patient P.S. (Left cavernoma). Left scan, horizontal plane. Right scan, coronal plane.

The patient was of average intelligence (IQ=96). She had some minor memory problems as measured by the Wechsler Memory Scale-revised (WMS-r) (Wechsler, 1987). WMS-r indexes were 90 for Attention-Concentration, 87 for Verbal Memory, 112 for Visual Memory, 93 for General Memory and 87 for Delayed Recall. She reported frequently perceiving human faces as deadlike, expressionless masks ("like zombies"). We gave her the "Emotional Faces Test" (Young *et al.*, 1993), which consists of 24 photographs of male and female faces, representing 6 different emotional expressions (anger, fear, disgust, happiness, sadness, surprise). While control subjects of her age perform at a mean of 22+1.97 (John P. Aggleton and Andrew W. Young, personal communication), she only gained 13 expressions correct by choicing "fear" as the most common response (10 from 13).

As an explanation for her choice of "fear" we suggest that she either selected what appeared likely to her, given that she was uncertain, or this response selection may be due to her left-hemispheric damage which more likely than right-hemispheric or bilateral brain damage may result in negative emotional feelings [cf. Goldstein's "catastrophic reaction" (Goldstein, 1939)]. In a picture recognition task with 20 neutral and 20 emotional pictures (*e.g.*, surgical intervention *vs* landscape), which later had to be re-identified out of the double number of items, she recognized 75% of the emotional and 65% of the neutral pictures correctly (in addition, she erroneously made 3 false positives). These values are similar to those obtained in a previous study on two patients with amygdaloid damage (Markowitsch *et al.*, 1994), but below the scores of control subjects (88.3% and 95% for neutral and emotional picture recognition). On non-emotional expression identification tasks (animate, inanimate objects) she performed normally.

3. Neuroradiological investigations on normal subjects

Previous results of fMRI activation studies related to the presentation of verbal stimuli showed activity in the amygdalo-hippocampal area and the fronto-temporal and prefrontal cortices in experiments using stimuli with high emotional value. It was also noticed the differential activation between fMRI activation patterns upon presentation of "emotional" vs "non-emotional" verbal stimuli (Neugebauer *et al.*, 1998).

Here we report the methods used in these previous experinents, valid also for the actual ones: On day 1, four normal healthy individuals (age 25-32 years) were presented 30 different neutral sentences auditorily (*e.g.*, "The hand was cold") and 30 different emotionally loaded phrases (*e.g.*, "The baby was killed"). The emotional and neutral sentences where presented intermixed. Each sentence was presented three times. On day 2 the same individuals had to listen to the same sentences which where presented according to an ABABAB-design (where A are the neutral sentences and B the emotionally loaded ones) under fMRI conditions. Later, on the same day 2, the sentences where presented only fragmentarily (*e.g.*, "The hand was...", or "The baby was..."), thus they had to be completed by the subjects without speaking aloud. The functional activation patterns were recorded with a SIEMENS VISION MRI scanner (1.5 T, TR 1.68 ms; TE 64 ms, FOV 220 mm; 114* 128 Matrix) using a circularily polarized head coil and Echo-Planar-Imaging technique (EPI).

The above findings where further substantiated by extending the subject sample where the additional amygdaloid activation was marked in the emotional condition. New additional data (Calabrese *et al.*, in preparation) demonstrated an activation in the left mesial temporal cortex (hippocampal area), that was even more pronounced and extended to amygdaloid region when the retrieval concerned the emotion loaded verbal stimuli in comparison to the neutral ones (Fig. 2). Furthermore, in an extension of the before mentioned study, by using only the stimuli with extremely emotional or neutral values, we where able to show an activation increase which depended on the intensity of the emotional stimuli (*e.g.*, shoe *vs* massacre) (Calabrese *et al.*, in preparation), and thus we corroborated our previous findings.

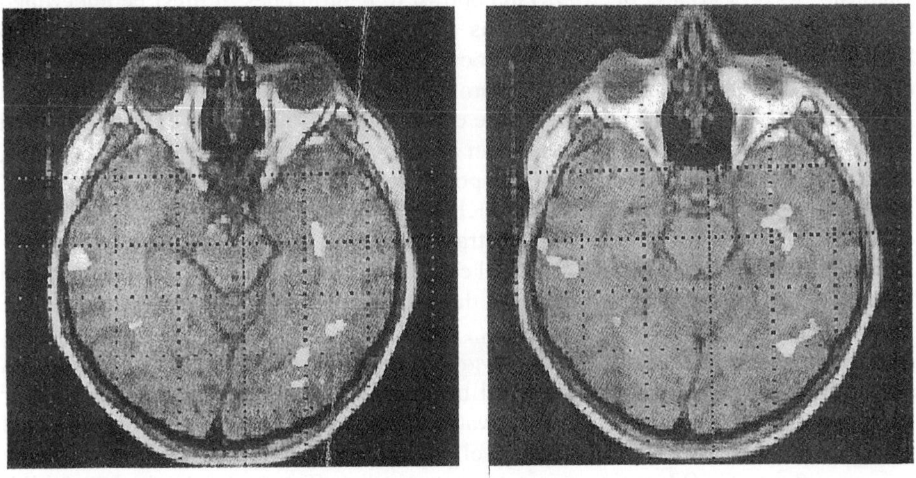

Figure 2. Transversal section through the human brain. Left scan, hippocampal activation with neutral words in the left mesial temporal lobe. Right scan, additional amygdaloidal activity in the "emotion loaden words"-condition.

4. Discussion

With respect to the data on amygdala damaged patients, as the cavernoma of P.S. most likely was of congenital origin, this case confirms the second line of Hamann *et al.*'s (1996) argumentation, namely the idea that the degeneration of the amygdalae and the periamygdaloid regions in cases with Urbach-Wiethe disease (Cahill *et al.*, 1995; Markowitsch *et al.*, 1994) impairs limbic functions. It seems that for more basic emotional functions, such as the detection of social signals, the brain largely relies on a pre-wired network of phylogenetically older structures (limbic system, in particular the amygdala) and is consequently unable to compensate for amygdalar processing failure. *Vice versa*, when the patterns for the interpretation of social signals for facial expression have been established, they are integrated in a wider net and combined with additional perceptual signals (Markowitsch, 1988), so that the failure of one component within this net (the amygdala) has less devastating consequences. Thus, we argue that the amygdala is one of those "bottleneck structures" which may be of special importance in the processing of socially and emotionally demanding context.

For what it concerns the data on the normal subjects stimulated with material of high emotional content, the fMRI activations included hippocampus, amygdala and the fronto-temporal and prefrontal cortices, thus embracing structures involved in the information processing and storage, while the extension of the activation to the amygdaloid region is linked to the presentation of emotional stimuli (cf also Damasio, 1994; LaBar *et al.*, 1998) in contrast to neutral ones. It must be underlined that there exist anatomical connections between amygdala and septum and also between the septum and the CA3 hippocampal area. Already before the finding of disinhibition of the CA3 pyramidal neurons of the hippocampus by septal afferents (Tóth *et al.*, 1997), the theta waves were compared to a "gate" that defines a limited period of time during which a neuron can be activated by an input (Fox *et al.*, 1983), or to a phase comparator of the two hippocampal input signals with participation of theta waves as a temporal filter (Klemm, 1976; Vinogradova, 1995). The disinhibition requires a synaptic substrate in which inhibitory afferents contact inhibitory interneurons, but not principal cells of a brain region. The activation of the inhibitory afferents may then suppress the inhibitory cell firing and so reduce the inhibition of principal cells (Ito *et al.,* 1968; Tóth *et al.*, 1997). The flow of emotional information from amygdala *via* septum to the CA3 hippocampal area can evoke the disinhibition of its pyramidal neurons. It can be hypothesized that under conditions without any emotional activation the sensory inputs from the external world may be blocked at the CA3 level by inhibitory interneurons. Some results suggest that GABA-ergic septo-hippocampal afferents selectively inhibit hippocampal inhibitory interneurons and so disinhibit pyramidal cells (Tóth *et al.*, 1997). We may expect that - if for a short time an emotional influx from amygdala *via* septum "opens the gate" disinhibiting pyramidal CA3 neurons - specific sensory information that impinge at this specific moment on the brain may freely flow through Schaffer collaterals up to the cerebral cortex for its final storage.

References

Adolphs, R., D. Tranel, H. Damasio and A. Damasio (1994) "Impaired recognition of emotion in facial expressions following bilateral damage to the human amygdala", *Nature* 372:669-672.

Cahill, L., R. Babinsky, H.J. Markowitsch and J.L. McGaugh (1995) "Involvement of the amygdaloid complex in emotional memory", *Nature* 377:295-296.

Cramon, D.Y., N. Hebel and U.A. Schuri (1985) "A contribution to the anatomical basis of thalamic amnesia", *Brain* 108:993-1008.

Calabrese, P., A. Neugebauer, H.F. Durwen, A. Falk, L. Heuser and L. Gehlen (1998) "fMRI values depend on emotional stimulus strenght", in preparation.

Damasio, A.R. (1994) *Descarte's error: Emotion, reason and human brain*, New York: Grosset-Putnam.

Fox, S.E., S. Wolfson and J.B. Ranck (1983) "Investigating the mechanisms of hippocampal theta rhythms: Approches and progress", in: *Neurobiology of*

Hippocampus, W. Seifert, ed., pp. 303-319, New York: Academic Press.

Goldstein, K. (1939) *The Organism: A Holistic Approach to Biology, Derived from Pathological Data in Man*, New York: American Book.

Hamann, S.B., L. Stefanacci, L.R. Squire, R. Adolphs, D. Tranel, H. Damasio and A. Damasio (1996) "Recognizing facial emotion", *Nature* 379:497.

Ito, M., N. Kawai, M. Udo and N. Sato (1968) "Cerebellar evoked inhibition of Deiters neurons", *Exp. Brain Res.* 2:723-725.

Klemm, W.R. (1976) "Hippocampal EEG and information processing. A special role for theta rhythm", *Progr. Neurobiol.* 7:197-214.

LaBar, K.S., J.Ch. Gatenby, J.C. Gore, J.E. LeDoux and E.A. Phelps (1998) "Human Amygdala activation during conditioned fear acquisition and extinction: a mixed-trial fMRI study", *Neuron* 20: 937-945.

Mark, L.P., D.L. Daniels, T.P. Naidiich and L.E. Hendrix (1995) "Limbic connections", *AJNR* 16:1303-1306.

Markowitsch, H.J. (1988) "Individual differences in memory performance and the brain", in: *Information processing by the brain*, H.J. Markowitsch, ed., pp. 125-148, Toronto: H. Huber.

Markowitsch, H.J., P. Calabrese, M. Würker, H.F. Durwen, J. Kessler, R. Babinsky, D. Brechtelsbauer, L. Heuser and W. Gehlen (1994) "The amygdala's contribution to memory - A PET-study on two patients with Urbach-Wiethe disease", *NeuroReport* 5:1349-1352.

Neugebauer, A., P. Calabrese, H.F. Durwen, A. Falk, L. Heuser and W. Gehlen (1998) "Possible selection mechanisms in declarative memory", in: *Neuronal bases and psychological aspects of consciousness*, C. Taddei-Ferretti and C. Musio, eds, World Scientific, Singapore, New Jersey, London, Hong Kong, in press.

Papez, J.W. (1937) "A proposed mechanism of emotion", *Archs Neurol. Psychiat.* 38:725743.

Tóth, K., T.F. Freund and R. Miles (1997) "Disinhibition of rat hippocampal pyramidal cells by GABAergic afferents from the septum", *J. Physiol.* 500:463-474.

Wechsler, D. (1987) *Wechsler Memory Scale-revised. Manual*, San Antonio, CA: Psychological Corporation.

Vinogradova, O.S. (1995) "Expression, control and probable functional significance of the neuronal theta-rhytm", *Progr. Neurobiol.* 45:523-583.

Young, A.W., J.P. Aggleton, D.I. Hellawell, D.J. Johnson, M. Brooks and J.R. Hanley (1996) "Facial expression processing after amygdalotomy", *Neuropsychologia* 34: 31-39.

Young, A.W., F. Newcombe, E.H.F. de Haan, E.H. Small and D.C. Hay (1993) "Face perception after brain injury: Selective impairments affecting identity and expression", *Brain* 116:941-959.

NEURO-AFFECTIVE PROCESSES AND THE BRAIN SUBSTRATES OF EMOTION: EMERGING PERSPECTIVES AND DILEMMAS

JAAK PANKSEPP
Dept. of Psychology, Bowling Green State University, Bowling Green, OH 43403

ABSTRACT

The neurobiological systems that mediate the basic emotions are beginning to be understood. They appear to be constituted of genetically coded, but experientially refined executive circuits situated in subcortical areas of the brain which can coordinate the behavioral, physiological and psychological processes that need to be recruited to cope with a variety of primal survival needs. The various emotional circuits are coordinated by different neuropeptides, and the arousal of each system may generate distinct affective/neurodynamic states. Although these central states cannot yet be empirically monitored using neurophysiological tools, viewpoints of how they are controlled in the brain and how they may modulate consciousness can now be advanced. The aim of the following essay is to discuss the underlying conceptual issues.

1. On The Nature of Emotions

Twenty five hundred years ago, Hippocrates wrote in his classic *The Sacred Disease* that "Men ought to know that from the brain and the brain alone arise our pleasures, joys, laughter and jests, as well as our sorrows, pains, grief and tears. Through it... we think, we see, we hear, and we distinguish the ugly from the beautiful, the bad from the good, the pleasant from the unpleasant . . . . By the same organ, we become mad or delirious, and fears and terrors assail us . . . and dreams and untimely wanderings." (2, p. 344 or as quoted in Eisenberg, 1995 p. 1564). In the intervening two and a half millennia, modern psychology and neuroscience, not to mention philosophy, have had a difficult time coming to terms with the deep biological nature of emotions, especially the affective feelings that the brain creates. This is because feelings reflect evolutionarily ancient, genetically prescribed types of neurodynamics that cannot yet be measured directly in humans or animals. They must be estimated from the types of linguistic feedback that can be provided by humans or inferred from the external bodily signs that we can also study in animals.

Because of the potential interpretive difficulties with such approaches, the study of emotional feelings (internal sensations replete with valence and arousal) has always taken a back seat to the study of the many other brain functions that are more evident to our senses. It has been much easier to agree upon the existence of cognitive structures that are created from our various sensations and perceptions (i.e., non-affective feelings or qualia) than the internal affective feelings (evolutionary qualia or *equalia*) bequeathed to us by our ancestral past.

If emotional feelings exist in brains as evolutionary birthrights - as the internal "voices" of primitive genetically constrained mechanisms of the brain - we may ultimately be unable to fathom the nature of the brain unless we begin to analyze the nature of feelings at the neurobiological level. Although our introspective access to our own minds clearly suggests that many of our cognitive responses to the world are modulated by emotional feelings, this alone has not opened up an empirical doorway to understanding the fundamental way that *equalia* are created in the brain. To achieve that, one must triangulate among three lines of evidence: 1) the study of emotional behaviors and other bodily changes, as well as 2) the brain substrates for these processes, especially in other mammals, and 3) the subjective self-reports of humans (Panksepp, 1991, 1998a)

Here I will summarize some of the dilemmas we will eventually face in our attempts to understand the mammalian brain if we do not begin to consider the existence of such ancient brain functions more widely. My aim for the past three decades has been to take a naturalistic, evolutionary approach to the study of the basic emotions, assuming that coordinated affective dynamics (behavioral, physiological and psychological) emerge from specific circuits of the brain that generate strikingly energized forms of action readiness. These forms of action readiness can be objectively analyzed to provide the essential measures we must use in decoding the nature of *equalia* in the brain. In this venture, the vocal expressions of experimental animals - from separation calls to animal laughter (see our other paper in these proceedings)--provide some of the most compelling evidence for the neural organization of affect. It is assumed that the vocal expressive apparatus is especially closely related to the internal dynamics of affective consciousness.

Based on current cladistic perspectives concerning evolutionary relations among species, it is reasonable to assume that the executive mechanisms for various basic emotions are homologous in all mammalian species. In other words, they are governed by essentially the same basic principles while varying in detail, depending on evolutionary branchings that have resulted from the ecological constraints confronting each species. For instance, guinea pigs are much better subjects than rats for studying separation distress, because they exhibit sustained patterns of social bonding, while rats are a more optimal species for fathoming the brain systems that mediate playfulness (Panksepp, Newman & Insel, 1992).

Our empirical aim has been to identify the necessary brain substrates for emotions rather than to claim that these substrates are sufficient to explain all aspects of emotionality. In other words, the strategy is a materialistic one that does not aspire to a comprehensive "nothing but" reductionism. In philosophic terms, the aim is to reveal supervenience relationships between certain key psychological and bodily attributes of emotions and the underlying brain substrates. In order to understand affective feelings, we first need to comprehend the epigenetically provided executive neural structures for the various emotional action systems of the brain. This strategy has now yielded enough fundamental and coherent neurological data (Panksepp, 1998a), that we can envision proceeding further into a study of affective neurodynamics and perhaps even the fundamental sources of primary-process or core consciousness - processes that lie at the interface of the unconscious - within the brain (Panksepp, 1998b). In this scenario, it is assumed that every moment of normal waking consciousness is built upon affective states - being especially strong and prominent in infancy and childhood but declining in salience as one matures.

There are good reasons to believe that affective neurodynamics lie at the very evolutionary roots of consciousness within relatively primitive parts of the brain. The basic emotional systems may help establish various global states of waking existence within the nervous systems. Such affective states can be envisioned to arise from neurochemical and neuroelectric tides that initially controlled the global shifts in behavioral and physiological dispositions, and with accruing brain evolution, shifts in the cognitive substrates as well (Panksepp, 1998a, Watt, 1998). The reason such global, organic brain states continue to be neglected is they are considerably more difficult to study objectively than the discrete information-transfer functions of the brain (LeDoux, 1996), but that can lead to various misconceptions. Since I have recently summarized those issues thoroughly (Panksepp, 1998a), my main aim here is to focus on some emerging conceptual dilemmas.

2. Affective Processes as Global State Variables of the Brain

Brain research has had a remarkably difficult time dealing with global state variables such as emotions. Central states that emerge from widespread shifts of brain activity are not easily objectified using the traditional micro-analytic techniques that are commonly used in modern brain research. At present, perhaps the most striking visual images we can generate of the widespread consequences of emotionality throughout the brain are the complex patterns of neurons that are activated during emotional episodes as highlighted with cFos *immunocytochemistry* or *in situ hybridization*. The distributions of neurons aroused by emotional episodes are remarkably widespread in the brain. Indeed, they are much more widespread than might have been expected from attempts to visualize emotions in the human brain using procedures such as PET and fMRI, which often highlight only a few pixels in the amygdala during emotional tasks (George *et al.*, 1996; Irwin *et al.*, 1996). Although other studies exhibit broader patterns of brain arousal during emotions, they rarely highlight diencephalic and midbrain areas which animal work indicates are essential for emotionality (Lane *et al.*, 1996; Paradisio *et al.*, 1996)

In contrast, with neuronal proto-oncogene imaging a large array of subcortical areas "light up" when animals are challenged emotionally. Most spectacular is the extremely widespread transcription of cFos in neurons of the cortex during such states (Beck & Fibiger, 1995; Kollack-Walker, Watson & Akil, 1997). However, the hyper-aroused cortex does not appear to be essential for generating the various basic forms of emotional arousal. Decorticate animals generally behave in the same emotional manner as normal animals - indeed they typically appear to be more emotionally aroused, suggesting that a major function of the cortex is to inhibit emotional processes (Panksepp, Normansell, Cox, & Siviy, 1994). This suggests that emotions markedly modulate the functions of the cortex, but that the fundamental urge to behave emotionally does not emerge from cortical functions. Still, the emotion regulatory effects of many cortical areas - of medial frontal, cingulate, insular and perirhinal cortices - is bound to be profound. It is there presumably were cognitive attributions interface with emotional urges. Indeed, with the cFos approach, we can analyze how powerfully previously experienced emotional states guide subsequent emotional responses (Bruijnzeel, Stain, Compaan, Croiset, Akkermans, Olivier & Wiegant, 1999).

Unfortunately, the issue of whether decorticated animals actually feel emotional states has never been empirically addressed. If one wished to do so, there are no options but to utilize behavioral indicators as potentially valid indices of internal states. However, few neuroscientifically-oriented investigators deem such inferences to be valid. Indeed, a prevailing opinion in behavioral neuroscience continues to be that even neurologically intact animals might, in reality, have no internal affective experiences. However, if this presupposition is incorrect (which seems highly likely), it will necessarily lead to a very impoverished view of how the living brain operates. Accordingly, I have chosen to advance the proposition that other animals do have certain basic emotional feelings (Panksepp, 1998a). Indeed, such feelings may reflect primitive ancestral memories - the communicative "voices" of the genes (Buck & Ginsberg, 1997)-that help regulate social and other behavioral processes. Affects may serve as set points and error signals in the long-term regulation of action tendencies. In other words, internally experienced affective states may allow organisms to coordinate their social and other activities in accordance with basic biological values that were designed to control behavioral tendencies essential for sustaining life. Indeed, it is hard to imagine how animals could coordinate their social affairs and other major survival concerns if they did not have internal value indicators that could refine behavior through the "law of effect" (i.e., that behavior is controlled by rewarding consequences). That was one vague affective principle accepted even by radical behaviorists. Now it, as well as many other affective processes of the brain need to be cashed out in terms of neural processes.

3. Paradigmatic Conflicts in Affective/Cognitive Neuroscience

Within modern neuroscience there has been little discussion concerning the issues mentioned above, and emotion research continues to languish, except for specific areas such as fear processes (Damasio, 1994; LeDoux, 1996) and several other vigorous pockets of activity (see Panksepp, 1998a; Watt, 1998). Indeed, many prominent behavioral/cognitive neuroscientists, continue to deny that affective feelings are critical cross-species aspects of emotional brain functions; if feelings exist, they are not commonly deemed to be causally important in the control of behavior. Instead of considering that emotions may be global state functions of the nervous system that establish long-term behavioral strategies, investigators of radically behaviorist persuasion are more likely to envision emotions as discrete, short-term, unconscious information transfer functions of the brain that are best studied through the analysis of rapid conditioned reflexes, their latencies and accompanying brain activities. At best, feelings are conceived as simple informational components, not much different from cognitive components, within the higher working-memory mechanisms of the brain (LeDoux, 1996).

Such viewpoints aspire to construct mind from relatively simple parts interacting with each other in fairly linear and traditional ways. Let us call this conservative viewpoint the Cognitive or Component Parts (COP) approach to understanding emotions. If this view is correct, then it should be possible for the two disconnected hemispheres in split-brain patients to experience two very distinct emotions, but good evidence for that proposition is scarce. Individuals with corpus callosum sections outwardly tend to exhibit singular and coherent emotional responses, even though the cognitions that precipitate emotions can certainly be restricted to one or the other hemisphere (Gazzaniga & LeDoux, 1978). Of course, there is now abundant evidence that the left hemisphere is more prone to promote negative feelings while the right hemisphere promotes positive ones (see Gainotti's overview in these proceedings), but this does not necessarily mean that the two hemispheres normally experience different moods. It only indicates that the two hemispheres tend to focus on different features of situations and to respond differentially. Of course, this is an open issue, as are the affective capacities of other species.

Various investigators, like myself, believe that all mammals do experience a variety of affective states, that these states help establish long-term behavioral strategies, and that the neurodynamics underlying such global neuropsychological states may be best monitored through the study of the population dynamics of ensembles of neurons. Such study can be achievable with EEG, in vivo neurochemical measures, perhaps cFos immunocytochemistry, and most clearly through the study of a diversity of complex instinctual behaviors which may reflect brain emotional states (Panksepp, 1998a; 1999). Although investigators who subscribe to this approach do accept that emotional systems serve important roles in guiding information-processing within working memory, emotional systems are conceptualized more broadly as giving rise to a variety of distinct global state functions of the brain that exist independently of, but work interactively with, working memory.

In this view, global neuronal broadcasting at great distances in the brain (via either traditional neuronal or more diffuse hormonal/paracrine effects) are critically important functions of emotional processes in the brain. Emotions do not arise computationally through the discrete digital-type interactions of local information nodes that appear to create many of the specific cognitive contents of the mind. Instead, emotions are widely embodied states of the brain whose fundamental character is analog and our understanding of such processes may have to be based on a much more broadly conceived systems analysis of brain functions than is traditional in behavioral neuroscience. In this view, action potentials are mere stitches in a widespread neurodynamic fabric that should be conceived in more global terms. Let us call this perspective the Central Affective Programs (CAP) approach - a view of emotions that was first envisioned in clinical psychology by investigators such as Tomkins (1962).

If this view is correct, then the two disconnected hemispheres in split-brain patients would be expected to commonly experience shared emotional feelings (albeit with differential cognitive nuances), and there is an abundance of anecdotal data to support such a conclusion (Gazzaniga & LeDoux, 1978). These effects may be elaborated by brain stem arousal systems that can access both sides of the brain concurrently. Indeed, the massive cortical innervation of key subcortical emotion-integration zones such as the periaqueductal gray (Shipley, Ennis, Rizvi & Behbehani, 1991) could help coordinate emotional states on both sides of the brain. This may also evoke a coherent sense of self, even within split-brain individuals (Panksepp, 1998b)

Thus, it may be instructive for us to consider that even though cognitions may not be able to pass from one hemisphere to another following collosotomy, affects may achieve this through various subcortical routes. Of course, perceptually mediated behavioral cueing could also mediate such interhemispheric affective coherences after collosotomy (Springer & Deutsch, 1998). In any event, there are good reasons to hypothesize that distinct types of emotional arousal do arise from global brain dynamics, which are quite unlike those, that mediate sensations or exteroceptively derived qualia. Affects may be ancient evolutionary memories-*equalia* that can be triggered by environmental events but which are not created from those events.

From the CAP perspective, even though affective and cognitive processes are highly interactive, the two could be profitably conceptualized as distinct types of brain entities (Panksepp, 1990). Thus, while the COP approach sees affect to be only one of the many possible cognitive contents of working memory (LeDoux, 1996), the CAP approach sees emotions to reflect global states of the brain that are controlled by distinct affect programs. The fact that these states can be evoked by cognitive attributions is not denied within such a view. The distinction between COP and CAP approaches boils down to the extent to which one allows emotional processes a coherent, independent integrity, rather than viewing them simply as routine types of information operating within cognitive workspaces of the brain.

Since emotions are fundamentally pre-propositional, it does seem a bit illogical to view them as cognitive (i.e., propositional) states. Indeed, neuro-evolutionary evidence strongly affirms that emotional processes, as generators of global states of the nervous system, evolved long before cognitive competence emerged from the cortical expansions of the brain. The primitive emotional systems seem to generate simple response strategies and to establish robust value structures within the brain that are quite distinct from those produced by affect-free perceptual processes. Of course, one could easily define "cognitive" so broadly as to cover everything organisms do, but that would be an arbitrary blurring of essential distinctions. In the present context, cognitions are defined largely as cortically mediated information functions that are created by the brain processing of the exteroceptive experiences of an organism. Affects, on the other hand, reflect certain global states of the nervous system that bias the whole brain and body apparatus to behave in certain characteristic ways.

Such distinct points of view, similar in a sense to the paradigms of pre- and post- Newtonian physics, are bound to characterize this research area for some time to come. Rapprochement and productive synthesis between these views is slower than desirable. Investigators who subscribe to the COP approach, typically pay little attention to the characteristic affective states postulated by CAP theorists. On the other hand, those who subscribe to CAP strategies, readily accept the existence and importance of the many cognitive components that contribute to emotional processing, but they tend to downplay the importance of those factors for understanding how affect is proximally created within the brain. CAP theorists are prone to assert that the supervenient essences of emotional feelings arise not from working-memory but from more primitive executive command circuits for the individual emotions. Since they have not sought to study those circuits as much as associated learning mechanisms, COP theorists are more likely to ignore the issue of emotional feelings or to simply view emotional feelings as causally irrelevant aspect of human working-memory abilities. CAP theorists see working memory simply as a mechanism to extend emotional feelings in space and time, which helps increase the subtlety of emotional live.

There is abundant room for compromise - a comprehensive understanding of emotions requires the cultivation of both views. Although I will not detail the main points of agreement, let me simply note that both types of theorists would probably agree that the apparent adaptive "rationality of emotions" must certainly be constructed from the interactions of affective and cognitive processes. The affects provide the "energy" to sustain certain courses of thought and action, and cognitions provide the ability to embed actions in socially meaningful contexts. The most obvious way to synthesize these seemingly antithetical views is to assume that the more cognitive aspects of emotion supervene on the more primitive affective command-system substrates. This would allow the more cognitive views to be grounded on evolutionarily integrated, embodied neural agencies, which appear to be essential for any neurologically coherent theory of mammalian emotionality. Gray (1995) has recently pursued this synthesis most faithfully.

The dialectic between these approaches has yet to weave its way toward a satisfactory empirical/theoretical resolution (because all too commonly they tend to ignore each other's arguments and databases), but a substantive middle ground could be constructed on the available evidence. After all, both camps subscribe to materialist views of how psychological functions and consciousness are created within the brain. Perhaps rapprochement has not yet been achieved or sought largely because of different axiomatic convictions concerning the existence and causal efficacy of internal emotional experiences in animals - issues that remain difficult to resolve definitively. In any event, two key issues that will continue to deserve the attention of future brain researchers are the possibility that other animals do have affective experiences, and if they do, how these experiences might be elaborated in their brains. At present, the COP and CAP views do lead to very different research priorities and predictions of how affect can be most productively studied.

Because of the widespread success of cognitive neuroscience (Gazzaniga, 1998), the COP approach has more richly penetrated the current intellectual *zeitgeist.* This seems to be inhibiting investigators from initiating studies of affective processes on their own terms as opposed to simply being conceptualized as cognitive flotsam. From my perspective, that is a pity. I believe that an understanding of the deep neural nature of animal emotional system has great potential interfacing with many urgent human issues (e.g., biological psychiatry and developmental issues) (e.g., Panksepp, 1998c), and can serve as a new foundation for psychoanalytic thought (Panksepp, 1999).

4. The Time Scale of Affective Processes in the Brain

Investigators who subscribe to COP approaches, often conceptualize emotions within remarkably short time frames, highlighting the fact that various emotional reactions can be precipitated unconsciously (often with latencies of less than a second). On the other hand, investigators of CAP persuasion are prone to believe that such studies only examine emotional information processing and the most dramatic eruptive emotional reflexes, without really focussing on affective feelings. Such feelings are held to operate along much longer time lines, typically lasting for many minutes or hours, and potentially extending for days. In other words, emotional feelings establish global "field" states within organisms that do not simply control behavior reflexively, but also in time frames which provide ample opportunity for the emergence of conscious affective processing from the underlying arousal of specific neural substrates. However, such global fields take longer to be established and much longer to decay than can be measured in typical rapid-fire classical conditioning experiments. For instance, specific emotional states may be promoted by neuropeptidergic neuromodulators, which can sustain certain types of brain activities for extended time periods. The tendency to utilize short-term and longer-term emotional paradigms by CAP and COP oriented research groups has probably also lead to differential emphasis on conscious and unconscious modes of information processing by such theorists.

5. Conscious and Unconscious Processing of Emotional Information vs. Feelings

The fact that the processing of exteroceptive affective information (e.g., facial expressions) can proceed at an unconscious level is certain (Öhman, 1993), but such data are not pertinent for the issue of whether affective feelings can be unconscious. They only indicate that emotional information can be subconsciously processed. Anyone who would use such evidence to argue that affective feelings can exist at an unconscious level, would be making a category mistake. A feeling is by definition an event that has entered consciousness, and that may require a minimum of many seconds rather than milliseconds of time. Indeed, short-term eruptive emotional responses may not be directly indicative of affective arousal; they may simply be neural preconditions that gradually precipitate distinct affective feelings that require much longer spans of time to unfold. In other words, affective feelings probably emerge from neural effects that are distinct from eruptive emotional responses.

That affective background feelings, not typically monitored by cognitively oriented investigators, may be operating in rapid-fire perceptual recognition tasks, is suggested by recent evidence (see Öhman's presentation, these proceedings). Subjects who do not consciously recognize very brief presentations of emotional faces are able to report that vague emotional feelings can be aroused by such stimuli.

Thus, many of the information-processing paradigms that are presently used in human emotion research are not optimally designed for studying the neural substrates of affective processes. Although exteroceptive emotional information can, most certainly, be processed unconsciously, that straightforward fact in no way indicates that emotional feelings can also exist unconsciously. This is not to deny that individuals whose emotional experiences are mild or over-intellectualized may experience difficulty admitting or identifying the mild types of affective arousal that assail them. Indeed, higher emotional regulatory strategies can dampen affective experience. Too much cortical activity may repress and obfuscate emotional experience altogether.

Thus, it may well be that alexithymic individuals do experience distinct forms of affective arousal but they do not semantically recognize them as being emotional, even though essentially the same types of arousal are typically reported to be consciously perceived feelings in most other individuals. This kind of disjunctive response may have various causes: Peripheral autonomic responses may only be correlates of emotions rather than causes, and in certain individuals the peripheral and essential central substrates may not be well coordinated, becoming dissociated for various reasons. For instance, it is possible that emotional arousal is always accompanied by felt affective experiences during early development, but with the maturation of higher cognitive abilities affective experiences may tend to fade from consciousness because of repression and other ego-defense mechanisms. An analogy here may be the fading from consciousness of motor acts as one learns to perform various skilled actions such as riding a bicycle. Indeed, it may be a general principle of the brain that all organisms confronted by new situations and challenges respond to situations emotionally initially, but that after they have learned the contingencies of the environment, they begin to respond more habitually with very little concurrent emotional arousal.

Indeed, in many adult circumstances where emotional arousal may only be a hindrance for generating the most adaptive behavior patterns, it may be wise to repress the disruptive and intrusive psychological effects that can emerge from primitive forms of affective arousal. From this perspective, young children and animals may actually have more intense affective experiences than adults. Parenthetically, it may be worth considering that scientists, as a population, may be much less emotional than most other adults. That may tend to skew and bias the types of emotion theories that they would be prone to create and the types of experiments they would be likely to undertake. The role of investigators' temperaments in the construction of psychological theories probably deserves more attention than it has received. Perhaps we should be more open to views that are most prevalent among ordinary people. Most folk-psychological views of emotions do emphasize the importance of affective feelings as centers of mental gravity in everyday life and decision making. I believe such naturalistic views are fundamentally correct.

6. Global Emotional Dynamics

My evaluation of the available literature is that there exists a cornucopia of evidence suggesting that similar internally felt emotions probably exist in humans and other animals (summarized in Panksepp, 1998a). If so, the many underlying neural homologies, especially in subcortical reaches of the brain, will tell us much about human emotions, and I believe credible empirical strategies can now be mapped out to monitor the various distinct affective states of the mammalian brain (Panksepp, 1999). I believe these internal neurodynamics will be reflected predictably in the outward dynamics of emotional behaviors in both animals and humans, especially as they unfold characteristically as a function of time and various environmental events (e.g., Schultz, 1998). One of the great goals of future emotion research should be to try to characterize and monitor these valenced arousal states (i.e., raw feels) as directly as possible from the brain. I do believe that will eventually be capable of being done both electrophysiologically and neurochemically (Panksepp, 1999).

It is already possible to detect the dynamic expressions of emotions in the flow and force of bodily movements (facial and otherwise) as well as the sound characteristics of the voice. Such dynamics are also evident in emotionally expressive music; even young children are adept at identifying the emotion in musical selections they have never heard (Terwogt & Van Grinsven, 1991). The case of music is especially important, suggesting that emotions can be aroused without the mediation or engagement of specific cognitive-attributional processes. It would be marvelous if the emotion-specific dynamics could be detected directly from surface EEG recordings from the human brain, but such attempts have generally met with marginal success (see Panksepp & Bekkedal, 1997).

To my knowledge, the only individual who has tried to measure such dynamics in the abstract - namely in the sense that emotions are characterized by relatively pure dynamic forms - is Manfred Clynes (1977). He has reported remarkable success in characterizing the time and force characteristics of emotional dynamics as voluntarily expressed via a single finger pressure exerted on a sensitive force transducer (Clynes, 1988). Such empirical approaches certainly deserve to be more widely utilized than they have in the past.

What we now need is a general purpose device that can tap into the emotional feelings in various provocative experimental settings and which may be implemented quantitatively to help adjudicate among various competing theoretical perspectives. What I suggest is not novel, and the following is based on a system that Marcel Zentner of the University of Geneva is presently developing for studying of emotion in the context of listening to music. Let me detail briefly: A computer "mouse" serves as the prototypic dependent measure output device which each subject could guide flexibly in an explicit two-dimensional topographic affective state-space while experiencing emotions in real time. The video screen could be programmed to provide the topography of a variety of emotional spaces based on one's theoretical preconceptions. From my preferred point of view, a good place to start would be with the generally accepted primes (especially those affirmed by brain research) laid out as pie wedges, with felt intensity going from the center (a neutral zone) outwardly as emotional intensity increased. One could easily navigate from one emotion to another during the presentation of any of a variety of externally presented or internally imaged stimuli. Special locations are provided for unique emotional experiences such as chills, tears, etc. It might require a bit of practice, but the rich data yield should be computationally manageable . One could imagine different topographies for different emotion theories. Perhaps such an approach could also adjudicate among theories: The map that led to the most consistent (least variable) patterns of navigation might be deemed a winner in the theoretical sweepstakes. In a sense, animals could be tested in similar but much simplified schemes, which would be variants of place - preference paradigms. Obviously, there is abundant room for methodological development to achieve better behavioral measures of affect in both animals and humans.

7. Executive Neural Systems for the Basic Emotions

I will now summarize my preferred approach that aims to identify the natural ordering of "affective kinds" within the nervous system. Since all emotional functions are represented at various hierarchical levels of the brain, this is a difficult enterprise that presently only permits us to seek necessary rather than sufficient solutions to the problem that affective experience poses for understanding the brain and emotions.

On the basis of a century of animal brain research, we can now be confident that a short list of distinct psychoneural processes constitutes the basic emotions that animals and humans share. For a long time practically all theorists have agreed that anger, fear, sadness and joy are fundamental potentials of the human and animal nervous systems, but there is abundant brain evidence that several other affective processes should also be deemed basic.

Converging evidence for functional circuit localization derived from studies using localized electrical and chemical stimulation of the brain, indicate that a variety of basic executive systems for basic emotional tendencies exist in the mammalian brain (e.g., SEEKING, RAGE, FEAR, LUST, CARE, PANIC and PLAY). However, most of the details, at neuroanatomical, neurophysiological and neurochemical levels, remain to be worked out. The capitalization of labels helps remind us that we are designating specific brain systems that are necessary substrates for certain emotions rather than being concepts that are subsuming all the attributes that constitute each type of emotional response.

In the study of emotional systems, we must distinguish very generalized neurochemical state control systems that modulate each and every psychological function (e.g., norepinephrine, serotonin and acetylcholine) from those that coordinate distinct types of emotional responses. The specific controls appear to be neuropeptidergic. However, the most prolific excitatory and inhibitory amino acid transmitters, glutamate and GABA operating in very specific areas of the brain, also provide detailed resolution to all emotional and cognitive responses; their specificity lies largely in the details of their circuit connectivities. In other words, excitatory amino acids appear to constitute the core signaling system in each emotional circuit. On the other hand, the neuropeptide systems bias and sustain which of the executive throughput circuits prevails at any one moment of time from among the organism's vast, evolutionarily prepared, emotional readiness repertoire.

In sum, many of the specific emotional systems of the brain appear to be organized around discrete neuropeptide circuits that can sustain arousal of an integrated emotional/mood/motivational response within the nervous system. I will now briefly outline the types of neural circuits that integrate the seven basic emotional responses described above. Details can be obtained from Panksepp (1998a).

1) The SEEKING/Expectancy system is a generalized motivational "module" that allows organisms to acquire a variety of rewards from their environments-from food to sex (Panksepp, 1981, 1986). This system is outlined by mesolimbic and mesocortical dopamine circuits arising from the ventral tegmental area of the midbrain, projecting to nucleus accumbens and frontal cortical regions, that can mediate a highly energized form of self-stimulation behavior. The dopamine neurons are especially responsive to novelty, but they sustain their firing in response to environmental events if those events are followed predictably by biologically important consequences (i.e., rewards or punishments) (Schultz, 1998). A variety of neuropeptide influences converge on this system (e.g., opioids, neurotensin and CCK), presumably mediating specific types of motivational controls. This system is essential for the arousal of seeking/wanting processes related to all the natural of rewards, as well as unnatural ones such as drugs of abuse that can activate receptors normally activated by the environmental incentives. Presumably the affective responses triggered by arousal of this systems are not ones of "pleasure" but feelings of curiosity, interest and appetitive urges at mild levels of arousal, and cravings and ecstatic feelings at higher levels, depending upon the associated contextual circumstances. Pleasure or liking induced by rewards is mediated by closely related opioid and benzodiazepine systems of the brain (Berridge & Robinson, 1998).

2) A RAGE/Anger system that promotes attack behaviors courses between corticomedial amygdaloid areas through the anterior lateral hypothalamus to the periaqueductal gray (PAG). Substance P is a key player in the amygdaloid to hypothalamic component of the overall circuit, and anger facilitatory influences descend from there to the PAG via excitatory amino acids such a glutamate. Opioids are powerful inhibitory influences in this emotional system (Siegel & Schubert, 1995). Presumably the cognitive processes that normally arouse anger are centered in frontal cortical areas specialized for the detection of reward presence and absence. The withdrawal of rewards automatically yields a frustrative-anger response. In other words this system is aroused when expected rewards are not forthcoming and when other irritations assail an organism.

3) A FEAR/Anxiety system courses along and interdigitates with the RAGE system (Panksepp, 1990, 1996; Panksepp *et al.*, 1991). This system emerges from the basolateral and central amygdaloid nuclei, descends via anterior and medial hypothalamic zones to the dorsal PAG. Learned fearful associations converge on the amygdaloid components of this emotional system from the surrounding temporal cortical areas, and they are solidified by facilitation of glutamatergic inputs into the system (LeDoux, 1996). Critical neuropeptide modulators include CRF, ACTH, MSH, and DBI. The differential components of the integrated

responses that arise from these influences remain to be worked out in any detail. Whether the feeling of fear is generated by one level of organization within this hierarchical system is not known. It is evident that activation of the lower components can unconditionally arouse affective states (Panksepp *et al.*, 1990; Panksepp, 1996), while the higher components are more essential for the acquisition of learned fears (LeDoux, 1996; Rosen & Schulkin, 1998). It is especially intriguing, that even a single experience with fear, sensitizes the nervous system to be hyper-responsive to quite different subsequent fear experiences (Bruijnzeel *et al.*, 1999). Whether there axe multiple fear or alarm systems in the brain remains an open issue.

4) The various LUST/sexuality systems of the brain provide a solid foundation for the organism's search for reproductive fitness (for summary see, Panksepp, 1998a). The distinct modes of male and female sexuality are organized respectively around vasopressinergic (VP) in and oxytocinergic (OXY) influences in the preoptic area and the ventromedial hypothalamus. Sexual consummatory urges are sensitized by widely distributed hormone receptors along the trajectory of these systems, with testosterone promoting arousability of VP and estrogen the relevant OXY neurons. The appetitive urges for sexual contact are mediated by the higher reaches of these systems (centered in corticomedial amygdala and bed nucleus of the stria terminalis (BNST)) connected to the above consummatory response zones, both of which converge on the PAG where the sexual urges may interact with many other emotional systems. The orgasmic/reward component of sex appears to have a strong oxytocinergic and opioid influences in both males and females. The jealous/aggressive aspects of sexuality, especially male sexuality, appear more strongly linked to vasopressingeric arousal in the brain (Window, et al., 1993). Many other neuropeptides including LH-RH and CCK are also important in mediating well-integrated sexual responses, but how each contributes to sexual urges is not well understood.

5) A CARE/nurturance system is based on prolactin and oxytocinergic systems of the brain - the same hormones that mediate milk synthesis and delivery in the periphery (Carter, 1998). The oxytocin systems arouse maternal intent partly by activating dopaminergic appetitive systems, and the pleasure of opioid release may provide feedback to the mother concerning the adequacy of her maternal behaviors. Social attachments appear to be mediated by all three of the neuropeptides implicated in maternal behavior - oxytocin, opioids, and prolactin (Nelson, & Panksepp, 1998; Numan, 1994)

6) A PANIC/separation response - most prominently characterized by a distinct distress vocalization - rapidly emerges in young animals that are isolated from caretakers with whom they have established social bonds (Panksepp, et al., 1985, 1988). This system may promote feelings of human sadness, and it is certainly distinct from the brain systems of FEAR. The trajectory of PANIC circuitry, as estimated with localized brain stimulation procedures, courses between anterior basal forebrain areas such as BNST, dorsal preoptic and ventral

septal areas, descending via the dorsomedial thalamic zones to the PAG. The most rostral aspects of the system are represented within anterior cingulate areas. The core of the system appears to be glutamatergic, and the neuropeptides that most clearly promotes arousal of this emotional response is Corticotropin Releasing Factor (CRF). Indeed, the anatomy of CRF circuits highlight brain areas from where isolation-type distress vocalizations (DVs) can be evoked with localized electrical stimulation of the brain. The neuropeptides that strongly suppress arousal of this system are the same as those that regulate nurturant CARE, namely oxytocin, opioids and prolactin (Nelson & Panksepp, 1998).

To highlight the type of robust data upon which these conclusions are based, I will briefly summarize two studies analyzing the distress vocalizations (DVs) of young domestic chicks. In the first (Figure 1) robust and sustained elevations of DVs evoked by 1 microgram of CRF microinjected into the 4th ventricle of the brainstem in 3-week-old chicks, whose natural DVs had declined substantially because of maturation. The elevations lasted for more than 4 hrs. The second (Figure 2) summarizes 3 hr isolation sessions in pairs of 1 week old birds following administration of the CRF-type neuropeptide, urocortin, which some have claimed does not evoke emotional responses (Spina *et al.*, 1996). Clearly this peptide is very effective in promoting indices of separation distress in birds.

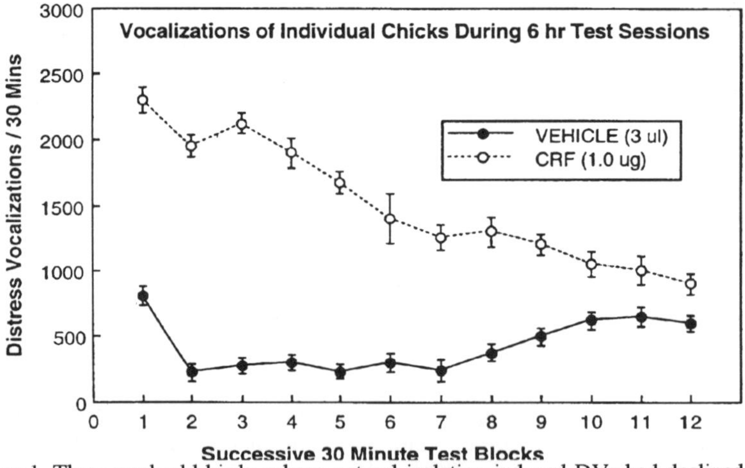

Figure 1. Three-week old birds, whose natural isolation-induced DVs had declined to low levels vocalized like new-born birds following administration of 1 ug of CRF into their ventricular systems. This same response could be evoked again at the end of 6 hrs indicating that the gradual decline was due to clearance of the CRF rather than fatigue, providing a possible mechanism by which emotional states are normally sustained in time. Values are means ±SEMs.

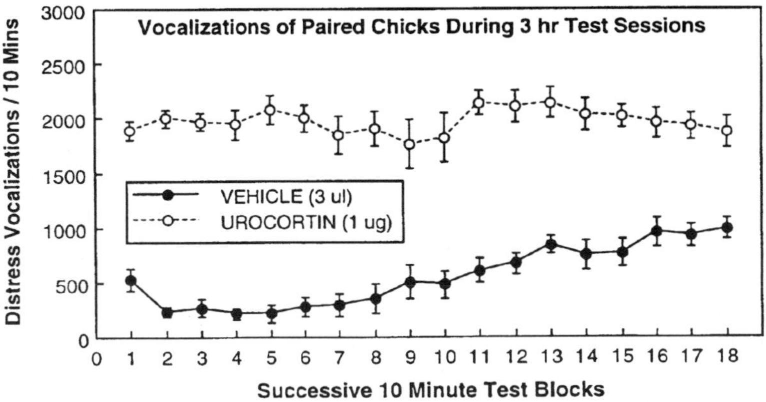

Figure 2. One-week old birds tested in pairs (to reduce baseline levels of DVs) exhibited sustained CRF-type elevations in DVs following injection of 1 ug of urocortin into the 4th ventricle region. Values are means ±SEMs.

7) A basic PLAY/Joy system for rough-and-tumble social engagement exists in the mammalian brain. However, little is known about its neuroanatomy, except for the fact that somatosensory inputs are especially important for the activation of play, and an epicenter for playful impulses exists in the parafascicular area of the thalamus that integrate nonspecific somatosensory information. Many neurochemical systems modulate the arousal of playfulness, but no neuropeptidergic "command" influence has yet been clearly identified (Panksepp, Siviy, & Normansell, 1984; Vanderschuren, Niesink, & Van Ree, 1997). A simplified approach to studying play circuitry in rats has recently been revealed by discovery of a "laughter" type of response (i.e., 50 kHz chirping) in rats that predicts playfulness (Panksepp & Burgdorf, 1999; and also see our other contribution in this volume).

At present, the study of these circuits provides the best strategy for understanding the brain organization of emotionality. This should not be considered an exclusive list. There may well be other key systems that will need to be distinguished, for instance, brain systems for power and social dominance (Kollack-Walker *et al.,* 1997). Of course, each of these systems is complex with widespread influences throughout the brain. Recently, we have visualized the activated neural substrates in rat brains during brief episodes of rough-and-tumble play using cFos immunocytochemistry. We are humbled by the number of brain areas (from the giant cells of the brainstem to practically all areas of the cortex) that are aroused by this type of emotional response. As already mentioned, this same type of broad arousal is evident for various other emotional responses (Beck & Fibiger, 1995; Bruijnzeel *et al.,* 1999). Obviously, an enormous number of brain areas are recruited by this and every other emotional state, and it will require a massive empirical effort to discriminate essential neural components from secondary influences.

8. The Construction of Emotional Feelings in the Brain

One of the key issues of emotion research is how the brain constructs emotional feelings. Three general possibilities have been proposed: (1) that feelings are created by "somatic markers" which reflect bodily changes that accompany emotions (a modern variant of the James-Lange perspective advocated by Damasio, 1994); (2) that feelings arise from various subcortical systems interacting with higher "working memory" systems (LeDoux, 1996); and (3) that feelings emerge from the intrinsic neurodynamics of emotional command systems interacting with a neurosymbolic "virtual body" depicted in the brain, which may constitute a primordial representation of "the self" (Panksepp, 1998b). An attractive aspect of the last view is that it permits emotional values to interact directly with the Extended Reticular-Thalamic Activating System (ERTAS), which helps govern conscious awareness of external events (Baars, 1996; Newman, 1997), providing a way in which the attentional searchlight can be optimally directed within the brain (for more on this see Watt's contribution to these proceedings).

In other words, emotional systems interacting with the other workspaces of consciousness may generate internally experienced states that permit organisms to face the world with various archetypal psychological and behavioral attitudes. These basic emotional states presumably allow organisms to sustain various intentional attitudes toward events and occurrences in the world. Of course, considering the complexity of underlying neural issues, there is abundant room for all three types of processes mentioned above to contribute to the integrated neurodynamic states that we recognize as the various emotional feelings.

There are probably significant supervenience relationships among all three of the postulated contributory brain processes and emotional feelings, and the critical question now is how much of the variance might be explained by each. My reading of the evidence is that the strongest relationship to affect is to be found among the various subcortical emotional operating systems and especially in key convergence zones such as the hypothalamus and PAG (Panksepp, 1998b). At this early stage of contrasting these views, no one has yet plotted a strategy whereby the degree of such supervenience relationships can be empirically contrasted among the various theoretical approaches.

In sum, we presently have an abundance of well-characterized neurobiological substrates upon which different affective states could be built. To the extent possible, future theories should seek to integrate these perspectives into a comprehensive structure and, wherever possible, to evaluate empirically the differential predictions of the various theoretical viewpoints.

References

Baars, B.J. (1996) "Understanding subjectivity: Global workspace theory and the resurrection of the observing self", *Journal of Consciousness Studies* 3:211-216.

Berridge, K.C. and T. Robinson (1998) "What is the role of dopamine in reward: hedonic impact, reward learning or incentive salience", *Brain Research Reviews* 22:347-354.

Beck, C.H.M. and H. Fibiger (1995) "Conditioned fear-induced changes in behavior and the expression of the immediate early gene c-fos: With and without diazepam pretreatment", *Journal of Neuroscience* 15:709-720.

Bruijnzeel, AW., R. Stam, J.C. Compaan, G. Croiset, L.M.A. Akkermans, B. Olivier and V.M. Wiegant (1999) "Long-term sensitization of Fos responsivity in the rat central nervous system after a single stressful experience", *Brain Research* 819:15-22.

Buck, R. and B. Ginsburg (1997) "Communicative genes and the evolution of empathy", in: *Empathic Accuracy*, W.W. Ickes, ed., New York: Guilford.

Carter, S.C. (1998) "Neuroendocrine perspectives on social attachment and love", *Psychoneuroendocrinology* 23:749-1014.

Clynes, M. (1977) *Sentics: The Touch of Emotions*, New York: Doubleday/Anchor.

Clynes, M. (1988) "Generalized emotion: How it may be produced, and sentic cycle therapy", in: *Emotions and Psychopathology*, M. Clynes and J. Panksepp, eds, New York: Plenum Press, pp. 107-170.

Eisenberg, L. (1995) "The social construction of the human brain", *American Journal of Psychiatry* 152:1563-1575.

Gazzaniga, M.S. and J. LeDoux (1978) *The Integrated Mind*, New York: Plenum Press.

George, M.S., T.A. Kettner, T.A. Kimbrell, J.M. Steedman and R.M. Post (1996) "What functional imaging has revealed about the brain basis of mood and emotion", in: *Advances in Biological Psychiatry, Vol. 2*, J. Panksepp, ed., Greenwich, CT: JAI Press, pp. 63-114.

Irwin, W., R.J. Davidson, M.J. Lowe, F.J. Mock, J.A. Sorenson and P.A. Turski (1996) "Human amygdala activation detected with echo-planar functional magnetic resonance imaging", *NeuroReports* 7:1765-1769.

Kollack-Walker, S., S.J. Watson and H. Akil (1997) "Social stress in hamsters: defeat activates specific neurocircuits within the brain", *Journal of Neuroscience* 17:8842-8855.

Lane, R.D., E.M. Reiman, M.M. Bradley, P.J. Lang, G.L. Ahern, R.J. Davidson and G.E. Schwartz (1997) "Neuroanatomical correlates of pleasant and unpleasant emotion", *Neuropsychologia* 35:1437-1444.

Nelson, E. and J. Panksepp (1998) "Brain substrates of infant-mother attachment: contributions of opioids, oxytocin, and norepinephrine", *Neuroscience and Biobehavioral Reviews* 22:437-452.

Numan, M. (1994) "A neural circuitry analysis of maternal behavior in the rat", *Acta Paediatrica* 397 (suppl.):19-28.

Newman, J. (1997) "Putting the puzzle together, Part 1: Towards a general theory of the neural correlates of consciousness", *Journal of Consciousness Studies* 4:47-66.

Öhman, A. (1993) "Fear and anxiety as emotional phenomena: Clinical phenomenology, evolutionary perspectives, and information-processing mechanisms", in: M. Lewis and J.M. Haviland, eds, *Handbook of Emotions*, *New* York: Guilford, pp. 511-536.

Panksepp, J. (1981) "Hypothalamic integration of behavior: Rewards, punishments, and related psychobiological process", in: *Handbook of the hypothalamus, Vol. 3, Part A. Behavioral studies of the hypothalamus*, P.J. Morgane and J. Panksepp, eds, New York: Marcel Dekker, pp. 289-487.

Panksepp, J. (1986) "The anatomy of emotions, in: *Emotion: Theory, Research and Experience Vol. III. Biological Foundations of Emotions*, R. Plutchik, ed., Orlando: Academic Press, pp. 91-124.

Panksepp, J. (1990) "The psychoneurology of fear: evolutionary perspectives and the role of animal models in understanding anxiety", in: *Handbook of Anxiety, Vol. 3: The Neurobiology of Anxiety*, G.D. Burrows, M. Roth and R. Noyes Jr., eds, Amsterdam: Elsevier, pp. 3-58.

Panksepp, J. (1991) "Affective Neuroscience: A conceptual framework for the neurobiological study of emotions", in: *International Reviews of Emotion Research*, K. Strongman, ed., Chichester, UK: Wiley, pp. 59-99.

Panksepp, J. (1996) "Modern approaches to understanding fear: From laboratory to clinical practice", in: *Advances in Biological Psychiatry. Vol. 2*, J. Panksepp, ed., Greenwich, CT: JAI Press, pp. 209-230.

Panksepp, J. (1998a) *Affective Neuroscience: The Foundations of Human and Animal Emotions*, New York: Oxford University Press.

Panksepp, J. (1998b) "The periconscious substrates of consciousness: Affective states and the evolutionary origins of the SELF", *Journal of Consciousness Studies*, 5:566-582

Panksepp, J. (1998c) "Attention deficit hyperactivity disorders, psychostimulants, and intolerance of childhood playfulness: A tragedy in the making?", *Current Directions in Psychological Sciences* 7:91-98.

Panksepp, J. (1999) "Emotions as viewed by psychoanalysis and neuroscience: An exercise in consilience", *Neuro-Psychoanalysis* 1:31-87.

Panksepp, J. (1999b) 'The neurodynamics of emotions: An evolutionary - neurodevelopmental view", in: *Emotion, Self-Organization, and Development*, M.D. Lewis and I. Granic, eds, New York: Cambridge University Press.

Panksepp, J. and M.Y.V. Bekkedal (1997) "The affective cerebral consequence of music: Happy vs. sad effects on the EEG and clinical implications", *International Journal of Arts Medicine* 5:18-27.

Panksepp, J. and J. Burgdorf (1998) "Laughing Rats? Playful tickling arouses 50KHz ultrasonic chirping in rats", *Society for Neuroscience Abstracts* 24:691.

Panksepp, J. and J. Burgdorf (1999) "Laughing rats? Playful tickling arouses high frequency ultrasonic chirping in young rodents", in: *Toward a Science of consciousness III*, S. Hameroff, D. Chalmers and A. Kaszniak, eds, Cambridge, MA: MIT Press, pp. 231-244.

Panksepp, J., J.D. Newman and T.R. Insel (1992) "Critical conceptual issues in the analysis of separation distress systems of the brain", in: *International Review of Studies on Emotion, Vol. 2*, K.T. Strongman, ed., Chichester, UK: John Wiley & Sons, pp. 51-72.

Panksepp, J., L. Normansell, J.F. Cox and S.M. Siviy (1994) "Effects of neonatal decortication on the social play of juvenile rats", *Physiology and Behavior* 56:429-443.

Panksepp, J., L.A. Normansell, B. Herman, P. Bishop and L. Crepeau (1988) "Neural and neurochemical control of the separation distress call", in: *The Physiological Control of Mammalian Vocalization*, J.D. Newman, ed., New York: Plenum Press, pp. 263-300.

Panksepp, J., D.S. Sacks, L.J. Crepeau and B.B. Abbott (1991) "The psycho- and neuro-biology of fear systems in the brain", in: *Aversive Events and Behavior*, M.R. Denny, ed., New York: Lawrence Erlbaum, pp. 7-59.

Panksepp, J., S. Siviy and L. Normansell (1984) "The psychobiology of play: Theoretical and methodological perspectives", *Neuroscience and Biobehavioral Reviews* 8:465-492.

Panksepp, J, S.M. Siviy and L.A. Normansell (1985) "Brain opioids and social emotions", in: *The Psychobiology of Attachment and Separation*, M. Reite and T. Fields, eds, New York, Academic Press, pp. 3-49.

Paradisio, S., R.G. Robinson, N.C. Andreasen, J.E. Downhill and R.J. Davidson (1997) "Emotional activation of limbic circuitry in elderly normal subjects in a PET study", *American Journal of Psychiatry* 154: 384-389.

Rosen, J.B. and J. Schulkin (1998) "From normal fear to pathological anxiety", *Psychological Review* 105:325-350.

Schultz, W. (1998) "Predictive reward signal of dopamine neurons", *Journal of Neurophysiology* 80:1-27.

Shipley, M.T., M. Ennis, T.A. Rizvi and M.M. Behbehani (1991) "Topographical specificity of forebrain inputs to the midbrain periaqueductal gray: Evidence for discrete longitudinally organized input columns", in: *The Midbrain Periaqueductal Gray Matter*, A. Depaulis and R. Bandler, eds, New York: Plenum Press, pp. 417-448.

Siegel, A. and K. Schubert (1995) "Neurotransmitters regulating feline aggressive behavior", *Reviews of Neuroscience* 6:47-61.

Spina, M., E. Merlo-Pich, R.K.W. Chan, A.M. Basso, J. Rivier, W. Vale and G.F. Koob (1996) "Appetite-suppressing effects of urocortin, a CRF-related neuropeptide", *Science* 273:1561-1565.

Springer, S.P. and G. Deutsch (1998) *Left Brain, Right Brain: Perspectives from Cognitive Neuroscience*, 5th ed., New York: W.H. Freeman.

Terwogt, M.M. and Van F. Grinsven, (1991) "Musical expression of mood states", *Psychology of Music* 19:99-109.

Tomkins, S.S. (1962) *Affect, Imagery, and Consciousness*, New York: Springer.

Vanderschuren, L.J., R.J. Niesink and J.M. Van Ree (1997) "The neurobiology of social play behavior in rats", *Neuroscience and Biobehavioral Reviews* 21:309-326.

Watt, D.F. (1998) "The Association for the Scientific Study of Consciousness", *Electronic Seminar on Emotion and Consciousness*, Sept. 21 - Oct 9, 1998, http://www.phil.vt.edu/assc/esem.html.

Winslow, H.J.T., N. Hastings, C.S. Carter, C.R. Harbaugh and T.R. Insel (1993) "A role for central vasopressin in pair bonding in monogamous prairie voles", *Nature* 365:544-548.

THE AFFECTIVE DIMENSION OF PAIN: MECHANISMS AND IMPLICATIONS

C. RICHARD CHAPMAN and YOSHIO NAKAMURA
Department of Anesthesiology, University of Washington, Seattle, Washington, 98195-6540 USA

ABSTRACT

Pain is an unpleasant sensory and emotional experience associated with actual or potential tissue damage, or described in terms of such damage. Neuroscience has characterized it as a predominantly or entirely sensory experience, but a review of basic mechanisms reveals that pain involves extensive limbic processes. Evidence from animal studies and human PET studies demonstrates that tissue trauma initiates a complex pattern of central processing that involves both the thalamocortical sensory pathways and the limbic brain. We suggest that pain is an emotion with sensory features rather than a sensory experience with emotional sequelae. The phenomenal experience of pain seems to involve at least two superimposed qualia: sensory and affective. Review of findings obtained from patients with damage to the insular cortex and pain asymbolia suggests that the affective quale of pain is critical in initiating adaptive/protective actions. We propose that the affective quale of pain represents the potential threat of an injurious event to the biological integrity of the individual. As such it contributes to defensive behavior and enables adaptive function.

1. Introduction

Emotion, consciousness and qualia are among the last remaining frontiers for science. Workers in these fields engage some of the most intriguing problems in the psychological and biological sciences today. One of the primary challenges is identifying and fostering a fruitful domain of inquiry. We contend that pain research provides an ideal domain to pursue frontier issues. In this chapter, we provide a basic tutorial on the mechanisms of pain with strong focus on its affective dimension. We then offer an updated review of brain imaging studies of experimental and clinical pain. In the last section, we discuss asymbolia for pain and its implications for understanding qualia. We conclude by reviewing recent speculations about what functions qualia might serve and suggest an approach to how functional theories/models of consciousness can begin to illuminate the problem of qualia.

1.1 What is Pain?

The International Association for the Study of Pain (IASP) provides the standard scientific definition: "Pain [is] an unpleasant sensory and *emotional* experience associated with actual or potential tissue damage, or described in terms of such damage" (Merskey 1979, p. 250, italics added). This definition clearly emphasizes the role of affect as an intrinsic component of pain. Nonetheless, a sensory neurophysiology framework has dominated pain research from its inception. The neurophysiological definition of pain holds that it is a sensory message of peripheral tissue trauma: specifically and accurately coded in peripheral nerves as well as in pathways of central neural transmission and in the brain.

Who or what interprets the signals that complete their journey from periphery to cortex is not at all clear. The neurophysiology model tacitly assumes that a conscious entity receives and interprets pain alarm signals, like a person attending to shouts of "Fire!" while watching a film in a motion picture theater.

Contemporary understanding of pain reflects the strong influence of Descartes, who in the 17th Century described bodily processes as clockwork mechanics. Descartes held that body and mind were separate entities. Pain was a specific modality - a straight-through sensory projection system that moved injury signals from damaged tissue to the brain where the mind could appreciate them. This perspective went unchallenged for two centuries, and it still exerts considerable subtle influence. Scientists and physicians alike assumed, until the 1960s, that tissue trauma activates specific receptors and that signals of tissue trauma follow specific pain pathways through the spinal cord to a pain center in the brain (Bonica, 1953). In classical neurophysiological thinking, pain is the sensory end product of an essentially passive information transmission process that operates as a biologically adaptive mechanism.

2. Basic Sensory Mechanisms

In this review of mechanisms, we use the language of sensory neurophysiology, constraining though it is, because it is the language of the knowledge base upon which we draw. Pain involves four processes: *transduction* (the conversion of the energy in an injurious stimulus to neural activity), *transmission* of the signals produced by transduction to the brain, *central representation* (this generally involves a Cartesian-like appreciation of the signals when they arrive at somatosensory cortex), and *modulation* (attenuation of transmission by descending inhibitory processes). Because receptor sensitization and damage to neural structures can affect transduction and transmission, we also discuss these processes.

2.1 Transduction

The transduction of tissue trauma into neural signals occurs via sensory end organs known as nociceptors (Besson & Chaouch, 1987; Heppelmann *et al.*, 1991; Willis & Westlund, 1997). The free nerve endings of thinly myelinated A∂ fibers function as thermal and/or mechanical nociceptors, conducting impulses at 4-44 m/s. In addition, certain unmyelinated C fibers that conduct slowly (roughly .5-1 m/s) act as polymodal nociceptors, responding to various high intensity mechanical, chemical and thermal stimuli. Both types of fibers distribute widely in skin and in deep tissue. Repetitive stimulation of these receptors produces pain. In addition, some primary afferents act as "silent nociceptors." Normally these end organs will not respond to harmless sensory stimuli, but noxious events or chemical changes can sensitize them so that they function thereafter as nociceptors (McMahon & Koltzenburg, 1990; Willis, 1993).

Nociceptors innervate skin, muscle, fascia, joints, tendons, blood vessels and visceral organs. From a sensory perspective, these tissues group into cutaneous, deep and visceral types. Nociception appears to serve somewhat different functions in the three types of tissues, and the quality of the pain that ensues from their activation varies across types. Most cutaneous pain is well localized, sharp, pricking or burning. A∂ fibers produce sharp, pricking pain sensations of short duration while C fibers typically generate burning sensations. Deep tissue pain usually seems diffuse and dull or aching in quality, although deep tissues can produce bright, sharp pains under certain conditions (e.g., muscle rupture). Visceral pain is very diffuse, often referred to the body surface, perseverating, and frequently associated with a queasy quality that patients describe as "sickening." Severe visceral pain typically produces an accompaniment of profuse sweating, nausea and vomiting.

The adequate stimuli for nociception differ across tissue types. Cutaneous receptors detect injurious stimuli from the surrounding environment, and so they respond to severe mechanical and thermal events such as cutting, burning or freezing. Nociceptors in deep tissue such as muscle detect overuse strain, deep mechanical injury like tearing and contusion, spasm or cramping, and ischemia. Their function resembles that of nociceptors in cutaneous tissue, but their responses may link more intimately to flexor reflexes than are those of their counterparts in skin. Muscle pain tends to foster muscle stiffness and splinting, which serves a protective function by bracing or supporting injured muscle. Visceral nociceptors do not respond to cutting or burning injury like their counterparts in cutaneous tissue and instead fire in response to pathological change. A hollow viscus needs to identify and transduce distention, stretch, and isometric contraction. A solid organ needs signal distention of the capsule that contains it and inflammation. Gebhart (1991) listed the following as naturally occurring visceral stimuli: distention of hollow organs, ischemia, inflammation, muscle spasm, and traction. The peripheral origins of pain vary markedly, depending on whether the nociceptors involved lie in superficial or deep tissues.

2.2 Sensitization of Nociceptors

Sensitization of nociceptors plays a major role clinical pain states (Alexander & Black, 1992). As nociceptors become sensitized, pain thresholds diminish (allodynia) and the painful qualities of subsequent noxious stimuli increase (hyperalgesia). Such alterations may reflect changes in the transduction process, central changes that facilitate the transmission of noxious messages, or both.

Sensitization of nociceptors can result from either repetitive stimulation of nociceptors or inflammation. Enhanced sensitivity is usually adaptive since it promotes recuperation and repair, minimizing further injury by discouraging all contact rather than just contact with noxious stimuli.

Once traumatized, tissue normally becomes inflamed. It is now clear that the process of inflammation sensitizes nociceptors and thereby increases their signal generating capability (Woolf, 1989). Chemical by-products of inflammation, such as the prostaglandins, alter the chemical environment of nociceptors, lowering their thresholds for firing and in some cases recruiting other fibers to function as nociceptors. Thus, injured peripheral tissues can become extraordinarily sensitive because of local chemical changes.

A key feature of sensitization is that it can awaken nociceptors that are otherwise silent -- so-called sleeping nociceptors (McMahon & Koltzenburg, 1990). Furthermore, it can recruit sensory endings that are normally not nociceptive to function, like volunteer firemen, as nociceptors. Sensitization drastically alters the process of transduction.

2.3 Neuropathic Mechanisms vs. Transduction

Some painful conditions, collectively termed neuropathic pain, arise from dysfunction of the peripheral or central nervous system. When pain originates in disturbed neural function, it is neurogenic in origin. Patients with neurogenic pain may experience ongoing or episodic electrical sensations or paresthesias, painful paroxysms, or a general hypersensitivity that makes harmless stimuli exquisitely painful (Davar & Maciewicz, 1989; Bowsher, 1991; Elliott, 1994; Galer, 1995). Injury to a peripheral nerve can produce pathophysiological changes in electrical excitability that generate abnormal ongoing and evoked discharge (Devor, 1991). Changes in the afferent impulse barrage can induce long term shifts of central synaptic excitability as well as changes in spinal cord cell excitability (Wall, 1991).

Chronic nerve root compression, e.g., a herniated disc, can generate pain by causing severe demyelination and fibrosis (Boulu & Benoist, 1996). Certain mono- and polyneuropathies associated with diabetes (Boulton, 1992), or alcoholism (Galer *et al.,* 1991) sometimes produce persisting pain. In addition, pain can arise from iatrogenic or adventitious injury to peripheral nerves or neural plexuses (Vecht, 1989) or to central structures such as the spinal cord (Siddall *et al.*, 1995). Severed nerves occasionally form neuromas that generate abnormal impulse discharge (Fried *et al.*, 1991). Bowsher (1991) estimated that such cases make up about a quarter of the patient population of most pain clinics; however, neuropathic pain is rare and afflicts at most about 1% of the general population. Not all neuropathy is painful; why some lesions produce pain and others do not is still an enigma.

Vecht (1989) described a classical iatrogenic neuropathic pain syndrome. Breast cancer patients who have undergone radical breast amputation with an axillary lymph node dissection sometimes develop electric-shock-like pain in the axilla, inner side of the upper arm and/or shoulder. This syndrome occurs when the surgical procedure produces a lesion of the intercostobrachial nerve. This is a difficult pain to control.

Damage to neural tissue may disturb a central regulating mechanism and thereby produce a condition in which the sympathetic nervous system plays a role in nociception (Roberts, 1986). Terms for this include Sympathetically Maintained Pain and Complex Regional Pain Syndrome (Stanton-Hicks *et al.*, 1995). Such conditions are rare, but excruciatingly painful, conditions in which altered function of the sympathetic nervous system contributes to a painful hypersensitivity in an affected area of the body. Abnormal skin color, temperature change, abnormal sudomotor activity, and edema accompany this type of pain. There are two types of Sympathetically Maintained Pain. The first occurs without a definable nerve injury, and the second, commonly called causalgia, occurs in response to a definable nerve lesion. Causalgia illustrates the complexity of this type of pain state.

Causalgia typically appears after a high velocity wound (a bullet, shrapnel or knife injury) that has damaged a major nerve in a limb (Bonica, 1990). Most patients experience surface pain of a burning quality immediately in the periphery of the injured extremity, and they develop shiny skin and edema in the affected area. The pain worsens and evolves into a constant hyperesthesia and allodynia (everything touching the area causes pain). With time, the pain spreads and eventually involves the entire limb. Temperature changes, light touch, friction from clothing, blowing air, movement of the limb, and any stimulus that affects the patients emotional state can exacerbate the pain. Minor events like the cry of a child, the rattling of a newspaper, or watching a television program can provoke intense pain. Any stimulus that activates the sympathetic nervous system, even social stimuli that are emotion-eliciting, can provoke severe pain. Consequently, patients suffer greatly, become reclusive and withdrawn, and become tragically incapacitated by the pain.

2.4 Transmission

The centripetal transmission of noxious signals takes place in the spinal cord. Nociceptive afferents enter the spinal cord primarily through the dorsal route, terminating principally in lamina I (the marginal zone) but also in laminae II (the substantia gelatinosa) and V of the dorsal horn (Craig, 1991). The spinal and medullary dorsal horns are much more than simple relay stations; these complex structures participate directly in sensory processing, performing local abstraction, integration, selection and appropriate dispersion of sensory impulses (Bonica, 1990; Perl, 1984; Willis, 1988; Jänig, 1987). Upon entry, nociceptive afferents form synaptic connections with projection neurons that convey information to higher centers, facilitory interneurons that relay input to projection neurons, and inhibitory interneurons that modulate the flow of nociceptive signals to higher centers (Jessell & Kelly, 1991). Similar neural processing occurs in the spinal cord and the medullary dorsal horn.

The spinal cord contains a complex network of interneurons. These networks not only relay signals to higher levels of the central nervous system; they also modulate signal transmission and initiate motor reflexes. Peripheral trauma can sensitize dorsal horn nociceptive neurons, making them sensitive to normal inputs and also excessively responsive to those inputs (Woolf & King, 1990; Willis & Westlund, 1997). The exaggerated response of transmission cells in the spinal cord is central sensitization. Enduring central sensitization could cause persisting pain.

There are two principal types of projection neurons in the spinal cord: nociceptive specific and multireceptive or wide dynamic range (WDR) neurons (Jänig, 1987). The former convey only tissue trauma signals; the latter respond to stimuli of increasing intensity. Ascending tracts include spinothalamic, spinoreticular, spinomesencephalic, spinocervical, and postsynaptic dorsal cord tracts. Willis & Westlund (1997) and Besson & Chaouch (1987) provide useful reviews of nociceptive transmission mechanisms. In classical thinking, the spinothalamic tract is clearly the most important. Lesions of the anterolateral quadrant of the spinal cord result in a loss of pain sensation below the segmental level of the lesion on the contralateral side of the body (Bonica, 1990).

2.5 Central Registration

The thalamus is a gateway and relay center for afferent input reaching the brain; therefore, it is the key structure in central registration. It consists of several functionally distinct nuclei that are reciprocally connected to many parts of the limbic system and the cortex (Willis & Westlund, 1997). Medial and ventrobasal thalamic nuclei relay noxious signals to the primary and secondary somatosensory cortices (SI, SII) where refined localization and discrimination occur. In classical thinking, the appreciation of pain occurs in these cortical areas.

Recent work acknowledges the existence of spinoreticular, spinomesencephalic and spinolimbic nociceptive pathways (Willis & Westlund, 1997), but to date neurophysiologists do not link them to appreciation of pain sensation. Chapman (1996) suggested that spinolimbic and spinoreticular pathways play a major role in the emotional component of pain, and that this determines the aversive quality of the pain experience.

2.6 Modulation

Pain is the end product of modulated transmission. The concept of modulation revolutionized biomedical thinking about pain. Historically, Gate Control Theory (Melzack & Wall, 1965) brought modulation to the forefront in pain research. What had been a rigid, bottom-up information transmission system incorporated a top-down influence when the gate control concept came onto the scene. Gate Control Theory postulated a gating mechanism at the dorsal horn of the spinal cord that could modulate the transmission of noxious signaling. The signal dampening action of the gate depends upon the relative amount of activity in large versus small diameter fibers in the periphery. It also suggested descending inhibitory influences from higher centers.

Numerous, and more elegant, models of modulation have emerged over the past three decades, and they have displaced the original gate control concepts many times over. Currently, the dominant model is the Diffuse Noxious Inhibitory Control (DNIC) concept that focuses on counter-irritation -- the phenomenon of one painful stimulus reducing the pain caused by another noxious stimulus applied concurrently to a distant part of the body (Talbot *et al.*, 1989). In humans, counter-irritation induces parallel decreases in the sensation of pain and the RIII nociceptive spinal flexion reflex simultaneously evoked by electrical stimulation of the sural nerve (Willer *et al.*, 1989). The mechanism of DNIC is at issue and apparently involves, but may not be limited to, inhibition of the activity of wide-dynamic-range (WDR) neurons in the dorsal horn.

Before moving on, we return to our movie viewer analogy in light of the modulation concept. We see that the movie viewer hearing alarming sounds now finds that the shouting varies in clarity as a function of the sound and activity level of the ongoing movie. Furthermore, the clarity of the shouted message may increase or diminish as a function of how much interest the viewer has in receiving alarming news messages.

2.7 Summary of the Classical Neurophysiology Model

In sum, the classical sensory neurophysiological model of pain holds that nociception, transmission of noxious signaling, modulation and sensory registration of pain are biologically predetermined processes. This is a predominantly bottom-up, unidirectional, sequential information-processing model, rooted in Cartesian dualist assumptions. Although mechanisms of modulation exist, pain is something that happens in the awareness of an injured or sick person, like a written telegram that arrives. This position also has major problems in explaining how a sensory experience can contribute so powerfully to suffering: why pain hurts is still unclear.

3. Mechanisms of the Affective Dimension of Pain

The principal ascending tracts are the spinothalamic and spinoreticular. We propose that sensory and affective processes subserving pain share common input from afferent sources, injury-sensitive A∂ and C primary afferents. Differentiation of sensory and affective processing begins at the dorsal horn of the spinal cord with sensory transmission following spinothalamic pathways and affective transmission taking place in spinoreticular pathways.

The spinothalamic tract delivers noxious signals to medial and lateral thalamus. These structures in turn activate areas in primary and secondary somatosensory cortex. Detailed reviews of spinothalamic processing appear in Willis (1985), Fields (1987), Peschanski and Weil-Fugacza (1987), Bonica (1990), and Craig (1991). The processes associated with these structures equip the individual with a capability for determining the nature of the traumatic event, its location, its duration and to some extent its severity. Spinothalamic processing plays an important role in the perception of injurious cutaneous events that one can escape, but it performs poorly for trauma in deep tissues and visceral structures.

The spinoreticular tract, which has received far less attention to date, also participates in nociceptive centripetal transmission (Villanueva *et al.*, 1989). Spinoreticular axons possess receptive fields that resemble those of spinothalamic tract neurons projecting to medial thalamus, and, like their spinothalamic counterparts, they transmit tissue injury information (Bonica, 1990; Fields, 1987; Villanueva *et al.*, 1990). Most spinoreticular neurons carry nociceptive information and many of them respond preferentially to noxious input (Bowsher, 1976; Willis, 1985; Bing *et al.*, 1990). We suspect that the spinoreticular tract conveys nociceptive signaling to higher central nervous structures that undertake affective (in contradistinction to sensory) processing of those signals. These higher structures are primarily noradrenergic. We emphasize them here because: 1) most of the literature on pain overlooks them; 2) they implicate limbic structures in pain perception; 3) the emotional aspect of pain plays a greater role in clinical pain problems than its sensory counterpart; and 4) these pathways link pain and neuroendocrine responses.

3.1 Nociception and Central Limbic Processing

Central sensory and affective pain processes share common sensory mechanisms in the periphery. A-delta and C fibers serve as tissue trauma transducers (nociceptors) for both, the chemical products of inflammation sensitize these nociceptors, and peripheral neuropathic mechanisms such as ectopic firing excite both processes. In some cases neuropathic mechanisms may substitute for transduction as we classically define it, producing afferent signal volleys that appear, to the central nervous system, like signals originating in nociceptors. Differentiation of sensory and affective processing begins at the dorsal horn of the spinal cord. Sensory transmission follows spinothalamic pathways and transmission destined for affective processing takes place in spinoreticular pathways. For more detail on the sensory processing of nociception see Willis and Westlund (1997).

Nociceptive centripetal transmission engages multiple pathways: spinoreticular, spinomesencephalic, spinolimbic, spinocervical and spinothalamic tracts (Villanueva *et al.* 1989; Willis & Westlund, 1997). The spinoreticular tract contains somatosensory and viscerosensory afferent pathways that arrive at different levels of the brain stem. Spinoreticular axons possess receptive fields that resemble those of spinothalamic tract neurons projecting to medial thalamus, and, like their spinothalamic counterparts, they transmit tissue injury information (Villanueva *et al.* 1990; Craig, 1992). Most spinoreticular neurons carry nociceptive signals and many of them respond preferentially to noxious activity (Bing *et al.* 1990; Bowsher 1976). The spinomesencephalic tract comprises several projections that terminate in multiple midbrain nuclei, including the periaqueductal gray, the red nucleus, nucleus cuniformis, and the Edinger-Westphal nucleus (Willis & Westlund, 1997). Spinolimbic tracts include the spinohypothalamic tract, which reaches both lateral and medial hypothalamus (Burstein *et al*; 1988; 1991) and the spinoamygdalar tract that extends to the central nucleus of the amygdala (Bernard & Besson, 1990). The spinocervical tract, like the spinothalamic tract, conveys signals to the thalamus. All of these tracts transmit tissue trauma signals rostrally.

Central processing of nociceptive signals to produce affect undoubtedly involves multiple neurotransmitter systems. Four extrathalamic afferent pathways project to neocortex: the dorsal noradrenergic bundle (DNB) originating in the locus coeruleus (LC); the serotonergic fibers that arise in the dorsal and median raphe nuclei; the dopaminergic pathways of the ventral tegmental tract that arise from substantia nigra; and the acetylcholinergic neurons that arise principally from the nucleus basalis of the substantia innominata (Foote & Morrison 1987). Of these, the noradrenergic and serotonergic pathways link most closely to negative emotional states (Gray 1982; Gray 1987; Bremner *et al.*, 1996). The set of structures receiving projections from this complex and extensive network corresponds to classic definition of the limbic brain (MacLean 1990; Papez 1937; Gray 1987; Isaacson 1982).

Although other processes governed predominantly by other neurotransmitters almost certainly play important roles in the complex experience of emotion during pain, we emphasize the role of central noradrenergic processing here. This limited perspective offers the advantage of simplicity, and the literature on the role of central noradrenergic pathways in anxiety, panic, stress, and post-traumatic stress disorder provides a strong basis (Bremner *et al.*, 1996; Charney & Deutch, 1996). This processing involves two central noradrenergic pathways: the dorsal and ventral noradrenergic bundles.

3.2 Locus Coeruleus and the Dorsal Noradrenergic Bundle

The pontine nucleus, locus coeruleus (LC), resides bilaterally near the wall of the fourth ventricle. It has three major projections: ascending, descending and cerebellar: The ascending projection, the dorsal noradrenergic bundle (DNB), is the most extensive and important (Fillenz, 1990). Figure 1 illustrates the DNB, along with noradrenergic projections to cerebellum. It reaches from the LC throughout limbic brain and to all of neocortex, accounting for about 70% of all brain norepinephrine (Svensson, 1987; Watson *et al.*, 1986). The LC gives rise to the majority of central noradrenergic fibers in spinal cord, hypothalamus,

thalamus, hippocampus (Aston-Jones *et al.*, 1985; Levitt & Moore, 1979) and its projections extend to limbic cortex and all of neocortex.

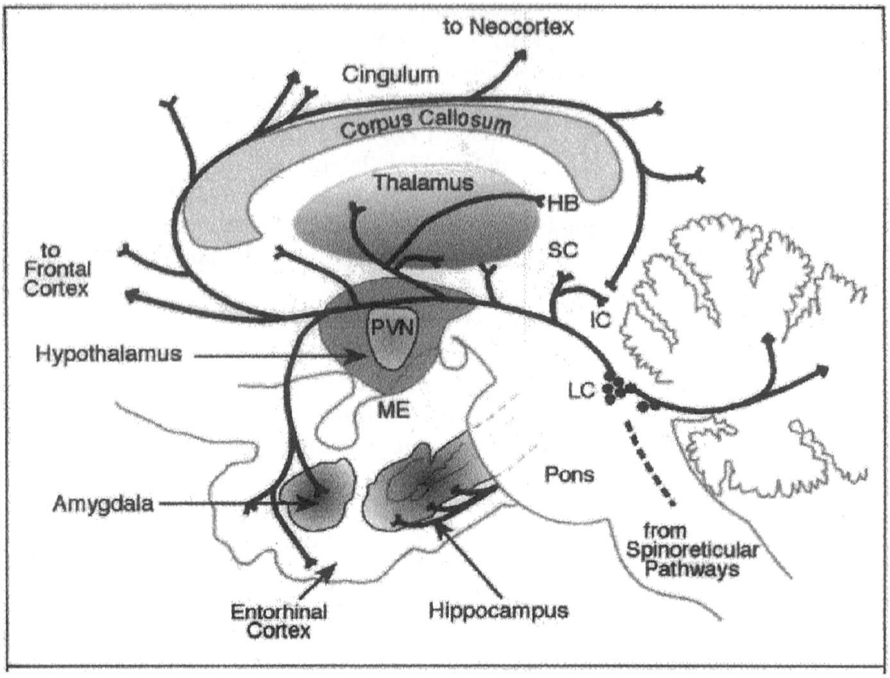

Figure 1. The Dorsal Noradrenergic Bundle (DNB) appears as a dark line. It originates at the Locus Coeruleus (LC). The cell bodies of the neurons that produce noradrenaline appear as circles. ME: median eminence; HB: habenula; IC: inferior colliculus; SC: superior colliculus; PVN: hypothalamic paraventricular nucleus.

The LC responds to sensory stimuli that potentially threaten, or signal injury to, biological integrity. Nociception inevitably and reliably increases activity in neurons of the LC, and LC excitation appears to be an inevitable response to nociception (Korf *et al.*, 1974; Stone, 1975; Svensson, 1987; Morilak *et al.*, 1987). This does not require cognitively mediated attentional control because it occurs in anesthetized animals. Foote, Bloom and Aston-Jones (1983) reported that slow, tonic spontaneous activity at LC in rats changed under anesthesia in response to noxious stimulation. Experimental electrical stimulation of the LC causes alarm and apparent fear in primates (Charney, 1990; Redmond & Huang, 1979), and lesions of the LC eliminate normal heart rate increases to threatening stimuli (Redmond, 1977).

How does this relate to tissue trauma? The LC responds consistently, although not exclusively, to tissue injury. However, increased LC activity also follows nonpainful threatening events such as strong cardiovascular stimulation

(Elam, *et al.*, 1985; Morilak *et al.*, 1987) and certain distressing visceral events such as distention of the bladder, stomach, colon or rectum (Elam, 1986b; Svensson, 1987). Thus, while it reacts to nociception, the LC is not a nociception-specific nucleus. It responds to events that represent biological threat, and tissue trauma is such an event.

Studies of negative emotion and vigilance behavior implicate the DNB as the largest and the most important LC projection for emotional processing of nociception. The DNB makes possible vigilance, and orientation to threatening and detection of novel stimuli can occur because of the DNB; it also regulates attentional processes and facilitates adaptive responses (Elam *et al.*, 1986a; Foote & Morrison, 1987; Gray, 1987; Svensson, 1987). Direct activation of the DNB and/or associated limbic structures produces sympathetic nervous system response and releases patterns of emotional behaviors in animals such as defensive threat, fright, enhanced startle, freezing and vocalization (McNaughton & Mason, 1980). In normal circumstances activity in this pathway increases alertness. Fundamentally, tonically enhanced LC and DNB discharge corresponds to hypervigilance and heightened emotion (Butler *et al.*, 1990; Foote *et al.*, 1983). The LC and DNB foster survival by making possible global vigilance for threatening and harmful stimuli.

We speculate that the affective dimension of pain shares central mechanisms with vigilance, a biologically important process. Vigilance intensified by tissue trauma signals, threats from the environment, or a combination of these can develop into hypervigilance and beyond it into panic. Extrapolated to subjective experience, the emotional aspect of pain corresponds to the emotional awareness of potential threat.

3.3 The Ventral Noradrenergic Bundle and the Hypothalamo-Pituitary-Adrenocortical Axis

The ventral noradrenergic bundle (VNB) enters the medial forebrain bundle and links neurons in the medullary reticular formation to the hypothalamus (Bonica, 1990; Sumal *et al.*, 1983). Sawchenko and Swanson (1982) identified two VNB-linked noradrenergic and adrenergic pathways to paraventricular hypothalamus in the rat and described them using the Dahlström and Fuxe (1964) designations: the A1 region of the ventral medulla (lateral reticular nucleus, LRN), and the A2 region of the dorsal vagal complex (the nucleus tractus solitarius, NTS) that receives visceral afferents. These medullary neuronal complexes supply 90% of catecholaminergic innervation to the paraventricular hypothalamus via the VNB (Assenmacher *et al.*, 1987a,b). Regions A5 and A7 make comparatively minor contributions to the VNB.

The VNB is important for emotion research because it innervates the hypothalamus. The noradrenergic axons in the VNB respond to noxious stimulation (Svensson, 1987) as does the hypothalamus (Kanosue *et al.*, 1984). Moreover, nociception-transmitting neurons at all segmental levels of the spinal cord project to medial and lateral hypothalamus and several telencephalic regions (Burstein *et al.*, 1988). These considerations suggest that threatening events, and particularly tissue trauma, excite the hypothalamoadrenocortical (HPA) axis via several routes. As the HPA axis controls the stress response, its reactions to tissue trauma are important corollaries of the pain state and may contribute to the emergence of pathological pain (Griep *et al.*, 1993).

The hypothalamic paraventricular nucleus (PVN) coordinates the HPA axis. Neurons of the PVN receive afferent information from several reticular areas including ventrolateral medulla, dorsal raphe nucleus, nucleus raphe magnus, LC, dorsomedial nucleus, and the nucleus tractus solitarius (Lopez *et al.*, 1991; Peschanski & Weil-Fugacza, 1987; Sawchenko & Swanson, 1982). Still other afferents project to the PVN from the hippocampus and amygdala. Nearly all hypothalamic and preoptic nuclei send projections to PVN.

The PVN responds to potentially injurious or tissue traumatizing stimuli by initiating a complex series of events that prepare the individual to cope powerfully with the threat at hand (Selye, 1978). Cannon (1929) described this "flight or fight" capability as an emergency reaction. Described another way, such responses constitute stress. The PVN must integrate these signals and coordinate a response.

The hypothalamus contributes to autonomic nervous system reactivity (Panksepp, 1986). Psychophysiologists have long considered diffuse sympathetic arousal to reflect, albeit imperfectly, negative emotional arousal (Lacey & Lacey, 1970). The PVN invokes autonomic arousal through neural as well as hormonal pathways. It sends direct projections to the sympathetic intermediolateral cell column in the thoracolumbar spinal cord and the parasympathetic vagal complex, sources of preganglionic autonomic outflow (Krukoff, 1990). In addition it prompts the release of epinephrine and norepinephrine from the adrenal medulla. These considerations implicate the HPA axis in the neuroendocrinologic and autonomic manifestations of affective changes during pain.

In addition to controlling neuroendocrine and autonomic nervous system reactivity, the HPA axis coordinates emotional arousal with behavior (Panksepp, 1986). Direct stimulation of hypothalamus can elicit well-organized patterns of behavior, including defensive threat behaviors, accompanied by autonomic manifestations (Jänig, 1985). The existence of demonstrable behavioral subroutines suggests that the hypothalamus plays a key role in matching behavioral reactions and bodily adjustments to challenging circumstances or threatening stimuli. The HPA system appears to coordinate behavioral readiness with physiological capability, awareness, and cognitive function. The acute stress response serves this purpose.

The stress response probably interacts with pain, and when pain is limited in duration, it may ameliorate it. Glucocorticoids released by the HPA axis during stress response diminish inflammation and block the sensitization of nociceptors in injured tissue. At the same time, HPA arousal releases ACTH and other proopiomelanocortin derived peptides including beta-endorphin into the blood stream. Moreover, stress hormones, especially glucocorticoids, may affect central emotional arousal, lowering startle thresholds and influencing cognition (Sapolsky, 1992). Saphier (1987) observed that cortisol altered the firing rate of neurons in limbic forebrain. When pain persists, however, the stress response may progress to a "burnout" of physiological coping resources, a condition Selye (1978) labeled "distress". Disturbed circadian rhythm (sleep, appetite) and fatigue can ensue. Griep *et al.* (1993) hypothesized a link between HPA axis dysfunction and the chronic pain of primary fibromyalgia.

4. Supporting Findings from Brain Imaging Studies

Our model predicts that noxious stimulation, whether of experimental or pathological origin, will generate massive, parallel distributed processing in the central nervous system. This processing must involve limbic structures as well as sensory pathways.

Studies involving positron emission tomography (PET) of regional cerebral blood flow (rCBF) in volunteers experiencing pain, and similar studies in pain patients, offer strong support for the hypothesis that noxious stimulation activates limbic structures. Changes in rCBF index neuronal activity in specific brain regions.

The partial review that follows targets studies designed to capture the complex central processing associated with pain. Collectively, they have puzzled and challenged pain researchers with classical, sensory neurophysiology perspectives. While not perfectly consistent, they demonstrate beyond any doubt that massive parallel distributed processing occurs in the brain following tissue damage. Processing includes, but is not limited to, sensory pathways. The most striking feature of this massive, parallel distributed processing is that a great deal of it occurs in limbic brain. This provides supporting evidence for the contention that pain has an affective dimension.

Just what does increased rCBF in a brain area mean? Glib interpretation is tempting but dangerous. It seems naive to presume that the brain works on a rope and bell basis: pull the rope in the periphery and you ring a bell in a specific brain center. It is quite clear that the brain as a whole, and especially the limbic brain, operates as a system with complex feed-forward and feed-back mechanisms. Notions of "centers" for one or another function have largely disappeared from the landscape of contemporary brain research. Consequently, attempts to chase nociception-specific messages to nociception-specific centers are probably doomed to fail. Also, there is little justification for assuming that increased rCBF reflects sensory information processing in distributed neural networks in the brain. Affective processing may involve such processes as arousal and anticipation. Additionally, uncertainty exists in deciphering whether or not increased rCBF reflects corresponding activation/excitation or inhibition of neural structures.

Because of these uncertainties, clear interpretation of massively parallel rCBF indicators of brain activation still eludes us. Moreover, in the end, we have no idea how any of the areas feeds forward into the contents of consciousness. Despite these limitations, the following studies indicate a striking consistency, and they strongly implicate limbic structures in the construction of the pain experience.

Pioneering studies examined both normals and patients. Jones and colleagues (Jones *et al.* 1991) applied heat via a Peltier thermode to the hands of six normal volunteers. They contrasted the rCBF findings across three stimulus intensities ranging from noxious to innocuous. Pain-related changes in rCBF appeared in contralateral thalamus, lenticular nucleus and cingulate cortex. The same team studied rCBF in five cancer patients with pain before and after percutaneous, ventrolateral cervical cordotomy (DiPiero *et al.* 1991). They compared patients before pain with normals and then compared patients with themselves before and after neurosurgical intervention. The comparison of patients with normals revealed significantly less blood flow in three out of four of the individual quadrants of the hemithalamus contralateral to the side of pain in the cancer patients. Cordotomy abolished the differences. Cordotomized patients demonstrated decreased rCBF in the dorsal anterior quadrant of the thalamus contralateral to the side of pain, but no changes were evident in either primary somatosensory cortex or prefrontal cortex.

The lenticular nucleus, or lentiform nucleus, resides lateral to the thalamus and within the internal capsule. It comprises two parts, the larger putamen and (medial to it) the smaller globus pallidus, which is separated from the thalamus by the posterior limb of the internal capsule.

Talbot and associates (Talbot *et al.* 1991) stimulated the forearms of six normal volunteers with noxious heat from a contact thermode. Pain-related rCBF changes appeared in contralateral cingulate gyrus and in primary and secondary somatosensory cortex. Coghill and colleagues (Coghill *et al.* 1994) followed this with a PET study comparing rCBF changes in normal volunteers during painful heat stimulation and vibrotactile stimulation. With painful stimulation, subjects demonstrated rCBF changes in contralateral thalamus, primary and secondary somatosensory cortices, anterior cingulate cortex, insula, and frontal cortex. With vibrotactile stimulation, changes appeared in contralaterally in primary somatosensory cortex and bilaterally in secondary somatosensory cortex and insula. Both types of stimuli activated primary and secondary somatosensory cortical areas, but painful stimuli had a significantly greater effect on insula and in general produced a more widely dispersed effect.

Casey and colleagues (1994) delivered noxious and innocuous heat pulses to the forearms of volunteers during PET analysis of rCBF. Significant rCBF increases occurred contralaterally during painful stimulation in thalamus, cingulate cortex, primary and secondary somatosensory cortex, and insula. Ipsilaterally, secondary somatosensory cortex, thalamus, medial dorsal midbrain and cerebellar vermis also showed rCBF increases.

In a later study, Casey *et al*, (1996) sought to detect rCBF increases in 27 normal humans as they discriminated differences in the intensity of noxious and innocuous thermal stimulation applied to the nondominant (left) arm. They divided subjects into three groups of 9 each: repetitive contact heat stimuli (40 and 50 degree C of thermode stimuli), cold pressor, and warmth discrimination (36 and 43 degree C of thermode stimuli). Significant increases in rCBF to the 43 degrees C stimuli occurred in the contralateral ventral posterior thalamus, lenticular nucleus, medial prefrontal cortex (Brodmann's areas 10 and 32), as well as cerebellar vermis. The painful stimuli elicited more extensive brain activity. Significant rCBF increases to 50 degrees C stimuli appeared contralaterally in the thalamus, anterior cingulate cortex, premotor cortex, and secondary somatosensory (S2) and posterior insular cortices. Significant activity also appeared within the region of the contralateral anterior insula and lenticular nucleus. The ipsilateral premotor cortex and thalamus, and the medial dorsal midbrain and cerebellar vermis, also showed significant rCBF increases. In the heat pain and cold pain conditions, five areas responded consistently: cerebellar vermis, ipsilateral thalamus, contralateral premotor cortex, contralateral anterior cingulate cortex, and the contralateral insula/lenticular nucleus. Cold pain created a greater rCBF increase than did heat pain. This is consistent with evidence that subjects normally judge cold pain to be more intense and aversive than they do heat pain. These observations support the interpretation that rCBF increases during pain reflect activity of both the sensory and affective processing of nociceptive signaling.

Vogt and coworkers (1996) studied the rCBF responses of seven normal subjects to noxious and nonnoxious heat stimulation. They used statistical parametric mapping for the group to identify regions of altered relative rCBF. In addition, they fitted the PET data on a subject-by-subject basis to magnetic resonance images of the brain. The mapping analysis of the group showed one site with elevated rCBF in the midcingulate cortex and one in the perigenual cortex predominantly contralateral to the side of stimulation. There were bilateral sites of reduced rCBF in the cingulofrontal transitional cortex and in the posterior cingulate cortex as well. Co-registered PET and magnetic resonance images for individuals showed that only one case had a single, large region of elevated rCBF, while the others had a number of smaller regions. This study helps demonstrate the noteworthy range of individual differences in rCBF responses during pain.

Hsieh *et al.* (1995) investigated rCBF in eight patients with neuropathic pain (lateralized mononeuropathy). They compared two conditions: normal ongoing pain experience and a condition in which the experimenters had temporarily blocked the pain via lidocaine block. The ongoing neuropathic pain produced activation of bilateral anterior insula, posterior parietal, lateral inferior prefrontal, and posterior cingulate cortices as well as the posterior sector of the right anterior cingulate cortex. The contralateral posterior thalamus demonstrated reduced rCBF. However, they found no significant change in rCBF in the somatosensory areas SI and SII. The investigators concluded that these findings point to the affective-motivational dimension of chronic neuropathic pain.

In a later study, Hsieh and colleagues (1996) explored the effects of minor dermal injury elicited by intracutaneous injection of a minute amount of ethanol on rCBF in four subjects. A saline injection served as control. The painful condition (ethanol) prominently activated the hypothalamus, the periaqueductal gray (PAG), the prefrontal cortex (PFC), the insula, the anterior cingulate cortex, the posterior parietal cortex, the primary motor/somatosensory areas, the supplementary motor area, and the cerebellum.

Silverman and associates (1997) looked at pain threshold rCBF in six patients with irritable bowel syndrome, contrasting them to six normals tested under identical conditions of noxious rectal distension. For the healthy subjects, a significant relationship existed between activity of the anterior cingulate cortex and actual or simulated delivery of the painful stimuli, but no response occurred for nonpainful stimuli. In patients no response occurred in anterior cingulate, and instead they demonstrated a significant activation of left prefrontal cortex during both activation and anticipation.

Jones *et al.* (Jones *et al.* 1994) examined rheumatoid arthritis patients with chronic inflammatory pain in a test of the hypothesis that such pain alters endogenous opioid binding at receptors in the brain. A high concentration of such receptors exists in periaqueductal gray, medial thalamus, lentiform nucleus, anterior cingulate cortex and insular cortex. If chronic pain is associated with increased production of endogenous opioids and increased binding at receptors, then an exogenously introduced opioid substance should find fewer binding sites in these areas. The investigators used PET scanning to tracer quantities of ^{11}C diprenorphine following its intravenous injection in four patients, in pain and after pain relief. They observed significant changes in superior and inferior frontal cortex, straight gyrus, anterior and posterior cingulate, superior and mid-temporal cortices.

Derbyshire and coworkers (Derbyshire *et al.* 1994) studied rCBF in six patients with atypical facial pain, applying noxious and innocuous heat stimuli to the back of their hands contrasting their regional blood flow patterns to those of normal controls. Both patients and controls showed marked rCBF differences between painful and nonpainful conditions in thalamus, anterior cingulate cortex, lentiform nucleus, insula, and prefrontal cortex. The patient group showed increased blood flow in anterior cingulate cortex but decreased blood flow in prefrontal cortex.

Do chronic and acute pain produced different patterns in rCBF? Mountz and colleagues studied women diagnosed with fibromyalgia syndrome, a disorder characterized by widespread chronic pain and fatigue (Mountz *et al.* 1995). They examined resting state rCBF in ten patients and compared their data with those of 7 normal women. Resting regional bilateral blood flow was significantly lower in the fibromyalgia patients than in normals at thalamus, at the head of the caudate nucleus and in cortex. The observation of lower rather than higher rCBF levels in the fibromyalgia patients led the authors to speculate that chronic pain may eventually reduce blood flow in certain brain areas. They postulated that a release of C fiber neuropeptides in response to chronic noxious stimulation together with diminished rCBF altered central nervous system sensitivity to normally mildly noxious stimulation in fibromyalgia patients. These findings open new hypotheses about central differences in acute and chronic pain and the role of potential compensatory processes.

Collectively, these PET brain-imaging studies reveal massive distributed processing, largely in limbic structures, and thereby support the hypothesis that pain involves an affective component. Figure 2 shows a saggital view of the human brain, identifying structures implicated in PET studies of pain. In addition, it identifies the locus coeruleus. Thalamus, anterior cingulate cortex, insula, and hypothalamus emerge with high consistency across studies of both normal volunteers and patients with pain. This response pattern corresponds to MacLean's thalamocingulate division of the limbic brain (MacLean, 1990). Figure 3 provides a coronal perspective. Note that the insula appears as an extensive area of invaginated cortex. Thalamic neurons project to the cerebral cortex via the internal capsule. In addition, descending fibers extend from the cerebral cortex to subcortical structures including the basal ganglia, thalamus, brainstem, and spinal cord. The internal capsule contains the lenticular nucleus. The results of the various PET studies are broadly consistent with the hypothesis that noradrenergic activation plays a major role in the affective component of pain. Importantly, PET studies of humans experiencing pain corroborate findings from animal and human studies that use other methods.

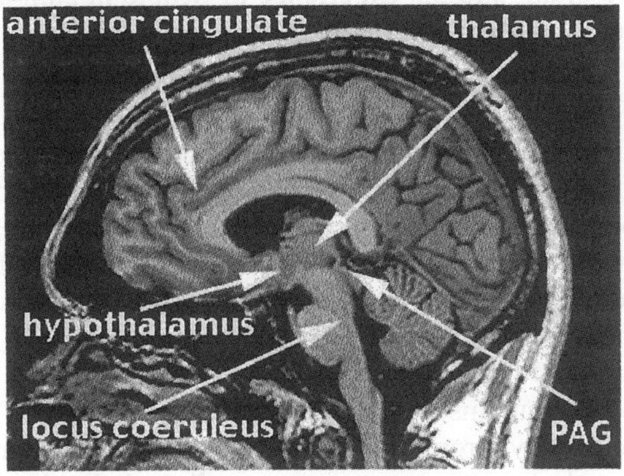

Figure 2. Key structures implicated in the affective dimension of pain.

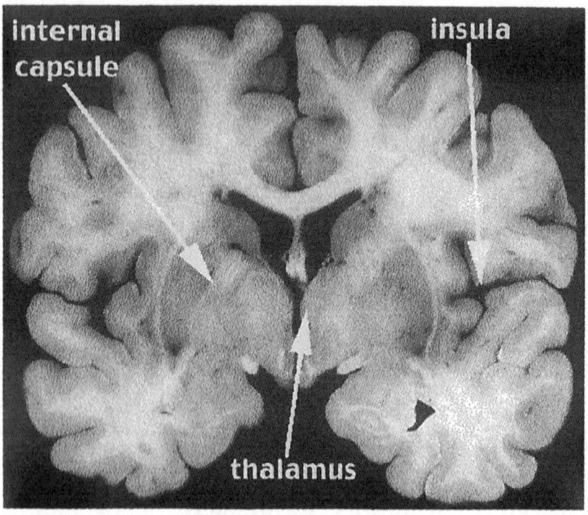

Figure 3. Coronal section displaying the insular cortex. The structures identified are active during pain.

Can one derive any strong interpretations about pain as a sensory modality (that is, can one think of pain as a separate channel function)? This line of research seems to demonstrate unequivocally that pain is not limited to a sensory modality. The patterns of central brain activation and arousal demonstrated thus far are not specific to pain as a sensory modality. Whether any of them are specific to threatening stimuli in general is a subject for future research. Central activation, not surprisingly, appears to correspond to higher order psychological processes. As more studies appear in various areas, it becomes increasingly clear that PET studies are telling us that such processes are a part of the complex experience of pain. This experience appears to stem from processing in emotion-linked areas of the brain, and the experience of people in pain confirms that pain involves strong emotional arousal.

5. Pain Asymbolia – When Pain Does Not Feel Bad

Pain asymbolia (PA) is a rare neurological condition caused by damage to a specific brain region - insular cortex. Patients with PA show normal behavior, but they react abnormally (i.e., indifferently) to potential threats and dangers presented to them (Berthier, Starkstein, & Leiguarda, 1988). In the absence of primary sensory deficits, PA patients showed a lack of withdrawal and absent or inadequate emotional responses to normally painful stimuli as well as both threatening gestures and verbal menaces.

Pain asymbolia is an instance of what Geschwind (1965) called sensory-limbic disconnection syndrome. In such syndromes, damage to specific anatomical sites (such as insular cortex) causes dissociation between normal pain perception and adequate emotional reaction, via the interruption of connections between the SII and the amygdala. Mesulam and Mufson (1985) demonstrated that posterior insula connects reciprocally to the sensory cortices (somatosensory, auditory, and visual cortex) and suggested that sensory-limbic (posterior insula-amygdala) interaction is critical for the affective-motivational content of perceptual experience.

Dissociation among brain function often follows physical and/or psychological trauma to the nervous system. Substantial neuropsychological evidence documents that dissociation of brain function can result from cerebral damage to different brain regions (e.g., hippocampal regions for explicit vs. implicit memory dissociation in amnesic patients and visual occipital cortex for conscious phenomenal seeing vs. intact visual information processing in blindsight). PA patients demonstrate dissociation of the normally integrated dimensions of pain, and they potentially suffer from the lack of adaptive responses to injury.

6. Pain Asymbolia and Qualia

Many consciousness researchers will want to know what sort of qualia PA patients experience with noxious stimulation. That is, how do they experience an injury that the rest of us perceive as painful? This is not the standard psychology question of how they would "behave" or "react" to noxious stimulation, but rather a question about their phenomenal experience of noxious stimulation. Evidence to date suggests that PA patients experience the sensory features of noxious stimulation (intensity, location, etc.), but they do not experience the normal affective or motivational arousal.

The proposal that pain is a multidimensional construct has received support and acceptance from pain clinicians and researchers since the inception of the multidisciplinary field. The IASP definition of pain discussed above reflects this. That pain is a multidimensional experience seems a safe assumption. These dimensions, translated into the philosopher's language, are qualia.

6.1 Dimensions of Pain and Superimposed Qualia

Taking this idea one step further, we submit that awareness of tissue trauma (i.e., pain) consists of two or more superimposed qualia. Roughly speaking, there are two dominant dimensions to qualia associated with pain: sensory and affective. Typically, the two dimensions are seamlessly integrated in our experience of pain, but they can separate under unusual circumstances, as they do in PA patients.

How do we account for the existence of qualia in the biological world? We assume that qualia would not have evolved if they did not serve any function at all (Cairns-Smith, 1998). Nonetheless, much of the debate over qualia in consciousness studies turns on the assumption that qualia serve no apparent function. Block's recently proposed distinction between phenomenal and access consciousness (Block, 1995) recapitulates the assumption that one can be aware of something that has no function in access consciousness. Phenomenal consciousness is simple experience, while access consciousness involves direct control. That is, a representation is access-conscious if it can contribute to direct control of reasoning, reporting and action.

Qualia associated with *conscious* sensation, perception, memory, and cognition do not readily lend themselves to a discernable functional characterization, at least in the ways that analytical philosophers conventionally conceptualize qualia. Conversely, one wonders if consciousness without qualia is possible. Abstract ideas and concepts can be intentional objects of which we are aware, and yet they do not seem to have strongly salient qualia associated with them. We contend that relations between consciousness and qualia remain problematic when we focus our discussion on primarily cognitive domains (perception of color, shape, objects, etc). When we move into domains more strongly associated with emotion and motivation, we see relations between consciousness and qualia differently. Qualia associated with affective dimension of pain, viewed from an evolutionary perspective, seem to serve the adaptive function of maximizing the survival of an injured organism.

Since several qualia are involved in the pain experience, for better or worse, we could imagine a situation where we lose a particular quale, in this case, pain's intrinsic aversiveness. How would this loss modify our experience accordingly? Should we say that we have become less aware of pain? The cases of PA patients suggest that there can be a breakdown of integrative processes underlying conscious experience of pain, but nonetheless, those PA patients who have lost pain's aversive qualia retain awareness of tissue trauma to some degree, at least in terms of pain's sensory dimensions. Most striking is the apparent lack of motivational and affective imperative to act on the part of these patients in response to an impinging source of tissue trauma. It seems that the elimination (or reduction) of the affective quale obviated one of the biological functions of pain – appraisal of the importance of the injurious event for the biological integrity of the organism. This leads us to briefly propose a potential role that qualia may play in the course of coordinating complex behavior.

6.2 What Can Qualia Do?

To many philosophers, it will seem strange to speak of a function for qualia. Nonetheless, we have recently seen some preliminary proposals concerning how to think about qualia within a mainstream scientific framework. In response to the question of "what can qualia do?" Banks (1996) commented on why functional theories neither account for qualia nor include qualia as a part of their theoretical constructs. He further discussed the inherent limits of any functional theory for understanding qualia. As he metaphorically put it, qualia are "hard to catch, lazy, and now excluded from the labor force altogether" (pg. 372-373). Banks concluded that qualia represent functionally relevant encodings in perception or imagery. Paul Churchland's recent attempt to explain color space using neural net models (Churchland, 1997) conforms to Banks' view on how scientific theories can address qualia. In the process of reaching this conclusion, Banks clarified several essential characteristics normally associated with qualia. These include: 1) perceptual qualia mirror the discriminations they reflect; 2) qualia are generated as a convenient summary of perceptual encodings that, while difficult to unpack, contains a great deal of information from multiple sources. Functionally speaking, qualia are the way perceptual systems serve up relevant information.

Despite the dominant notion that qualia serve no function, some scientists impute a significant functional role to qualia associated with conscious experience. Humphrey (1992) described five properties of sensations (his concept of sensations corresponds closely to the philosopher's notion of qualia). They characteristically : 1) belong to the subject, 2) are tied to a location in bodily space; 3) are modality specific; 4) are present-tense existing entities; and 5) are self-characterizing with respect to properties 1 to 4. Sensation provides several functions. Consider a circumstance in which you and I are in my office and I bang my knee inadvertently against the corner of my desk. One important function of qualia (sensations) is ownership. When the pain occurs, I know that it is mine and not yours. A second function of the qualia so generated is indexicality. If the sensation is pain, then I know that it belongs to the realm of the body and not to the realm of the external environment – the desk. In these ways, qualia are necessary for biological adaptation. Similarly, Gregory (1996a; 1996b) suggested in his editorial commentary for *Perception* that qualia serve to flag the present in order to separate it from memories and past knowledge. The present is uniquely important for survival. In his view, perceptions are analogous to the predictive hypotheses of science since both rely on knowledge (stored data, generalizations, assumptions). A critical difference between hypotheses of science and perceptual hypotheses may be that only perceptions have consciousness of qualia. Qualia represent aspects of the present selectively highlighted to facilitate real-time decisions essential for dealing with reality (Gregory, 1997).

Ramachandran and Hirstein (1997) recently argued that qualia differ from other brain states in that they have three *functional* characteristics. First, qualia are immutable; we cannot simply choose to start experiencing the sunset as green or to feel pain as if it were an itch. Second, qualia do not always uniquely constrain how you and I would subsequently behave; there is a potentially infinite set of possible actions and interactions resulting from a particular set of qualia. Third, qualia persist in immediate short-term memory so as to facilitate non-automatic decision-based action.

For the sake of argument, suppose we accept that qualia's principal function is to highlight information significantly associated with the present over any other potentially relevant information or knowledge (that remains non-conscious) that concurrently supports to guide our behavior in real-time. But if this were the case, one could argue that there must be a neural correlate for the qualia in question since evolutionary pressure has resulted in the selection of to-be-discovered physical mechanisms. Following this line of argument, one might well ask: why would we need phenomenal awareness of qualia? The hard problem again would seem to survive, even though some theoretical progress may be on the horizon.

Our approach to the question of why we have conscious experience at all is to treat this as a metaphysical question. We simply acknowledge the presence of consciousness and qualia associated with it and then try to come up with a functional characterization of what these constructs do within our theories/models. Of course, as Banks (1996) pointed out, conventional wisdom in science claims that qualia serve no function, so this question will lead back to a deadlock for most working scientists. In this regard, we think it is useful to clarify the exact nature of any hypothesized functions attributable to qualia. We note that qualia's functions are not necessarily *causal* but perhaps *enabling*; qualia can enable us to use information in the present for coordinating non-automatic decision-based action. (See Marcel, 1988, for a similar discussion of the function of consciousness.) As Ramachandran and Hirstein (1997) emphasized, perceiving a particular set of qualia does not always produce the same behavior.

Having discussed qualia and their functions for pain, we argue that qualia can vary in terms of the degree to which they constrain the production of subsequent behavior. Clearly, qualia associated with pain exert a stronger influence than those associated with, say, color in controlling how we interact with the world. Nonetheless, our views of qualia with respect to emotion and pain resonate those of Ramachadran and Hirstein (1997), Gregory (1996a; 1996b), and Humphrey (1992), all of whom suggested in one way or another that qualia flag the present and to make information in the present "salient" and "relevant" for non-automatic, decision-based action. The function of qualia is most evident in the case of a person with strong affective and motivational imperatives, such as relieving a severe pain. We contend that pain constitutes an ideal domain to pursue this and related questions in consciousness studies.

7. Conclusion

Classical thinking in neuroscience has characterized pain as a predominantly or entirely sensory experience. New evidence from animal studies and human PET studies makes it quite clear that tissue trauma initiates a complex pattern of central processing that involves both the thalamocortical sensory pathways and the limbic brain. Indeed, current knowledge would justify construing pain as an emotion with sensory features as opposed to the older notion of a sensory experience with emotional sequelae. We have argued that the phenomenal experience of pain involves at least two superimposed qualia: sensory and affective. Review of findings obtained from patients with damage to the insular cortex suggests that the affective quale of pain is critical in initiating adaptive/protective actions. We propose that the affective quale of pain represents the potential threat of an injurious event to the biological integrity of the individual. As such it contributes to defensive behavior and enables adaptive function.

References

Abercrombie, E. D. and B. L. Jacobs (1987a) "Single-unit response of noradrenergic neurons in the locus coeruleus of freely moving cats. I. Acutely presented stressful and nonstressful stimuli", *J. Neurosci.* 7:2837-2843.

Abercrombie, E. D. and B. L. Jacobs (1987b) "Single-unit response of noradrenergic neurons in the locus coeruleus of freely moving cats. II. Adaptation to chronically presented stressful stimuli", *J. Neurosci.* 7:2844-2848.

Alexander, J. and A. Black (1992) "Pain mechanisms and the management of neuropathic pain", *Curr. Opin. Neurol. Neurosurg.* 5:228-234.

Banks, W. P. (1996) "How much work can a quale do?", *Conscious. Cogn.* 5:368-380.

Barrett, J. A., M. B. Shaikh, H. Edinger and A. Siegel (1987) "The effects of intrahypothalamic injections of norepinephrine upon affective defense behavior in the cat", *Brain Res.* 426:381-384.

Bernard, J. F. and J. M. Besson (1990) "The spino(trigemino)pontoamygdaloid pathway: Electrophysiological evidence for an involvement in pain processes", *J. Neurophysiol.* 63:473-490.

Berthier, M., S. Starkstein and R. Leiguarda (1988) "Asymbolia for pain: a sensory-limbic disconnection syndrome", *Ann. Neurol.* 24:41-49.

Besson, J. M. and A. Chaouch (1987) "Peripheral and spinal mechanisms of nociception", *Psychol. Rev.* 67:67-185.

Bing, Z., L. Villanueva and D. Le Bars (1990) "Ascending pathways in the spinal cord involved in the activation of subnucleus reticularis dorsalis neurons in the medulla of the rat", *J. Neurophysiol.* 63:424-438.

Block, N. (1995) "On a confusion about a function of consciousness", *Behav. Brain. Sci.* 18:227-287.

Bonica, J. J. (1953) *The Management of Pain*, Philadelphia:Lea & Febiger.

Bonica, J. J. (1990) *The Management of Pain (2nd edition)*, Philadelphia:Lea & Febiger.

Boulton, A. (1992) "What causes neuropathic pain?", *J. Diabet. Complicat.* 6:58-63.

Boulu, P. and M. Benoist (1996) "Recent data on the pathophysiology of nerve root compression and pain", *Rev. Rhum. Engl. Ed.* 63:358-363.

Bowsher, D. (1976) "Role of the reticular formation in responses to noxious stimulation", *Pain* 2:361-378.

Bowsher, D. (1991) "Neurogenic pain syndromes and their management", *Br. Med. Bull.* 47:644-66.

Bremner, J. D., J. H. Krystal, S. M. Southwick and D. S. Charney (1996) "Noradrenergic mechanisms in stress and anxiety: I. Preclinical studies", *Synapse* 23:28-38.

Brody, H. (1973) "The systems view of man: implications of medicine, science, and ethics.", *Perspect. Biol. Med.* 17:71-91.

Burnstock, G. and C. H. V. Hoyle, eds. (1992) *Autonomic Neuroeffector Mechanisms*, Philadelphia: Harwood Academic Publishers.

Burstein, R., K. D. Cliffer and G. J. Giesler, eds. (1988) *The Spinohypothalamic and Spinotelecephalic Tracts: Direct Nociceptive Projections from the Spinal Cord to the Hypothalamus and Telencephalon*, New York: Elsevier.

Burstein, R., R. J. Dado, K. D. Cliffer and G. J. J. Giesler (1991) "Physiological characterization of spinohypothalamic tract neurons in the lumbar enlargement of rats", *J. Neurophysiol.* 66:261-284.

Butler, P. D., J. M. Weiss, J. C. Stout and C. B. Nemeroff (1990) "Corticotropin-releasing factor produces fear-enhancing and behavioral activating effects following infusion into the locus coeruleus", *J. Neurosci.* 10:176-183.

Cairns-Smith, A. G. (1998) "If qualia evolved", *Consciousness Research Abstract: Towards a Science of Consciousness 1998 Tucson III*, Thorverton, UK: Imprint Academic, p. 144.

Cannon, W. B. (1929) *Bodily Changes in Pain, Hunger, Fear and Rage*, New York: Appleton.

Casey, K. L., S. Minoshima, K. L. Berger, R. A. Koeppe, T. Morrow and K. A. J. Frey (1994) "Positron emission tomographic analysis of cerebral structures activated specifically by repetitive noxious heat stimuli", *J. Neurophysiol.* 71:802-807.

Casey, K. L., S. Minoshima, T. J. Morrow and R. A. Koeppe (1996) "Comparison of human cerebral activation pattern during cutaneous warmth, heat pain, and deep cold pain", *J. Neurophysiol.* 76:571-581.

Chapman, C. R. and M. Stillman. (1996) "Pathological Pain", in: *Pain and Touch* (2nd ed.), L. Kruger, ed, New York: Academic Press, pp. 315-342.

Chapman, C. R. (1996) "Limbic processes and the affective dimension of pain", in: *Towards the Neurobiology of Chronic Pain, Vol. 110*, G. Carli and M. Zimmerman, eds, New York: Elsevier, pp. 63-81.

Charney, D. S., S. W. Woods, L. M. Nagy, S. M. Southwick, J. H. Krystal and G. R. Heninger (1990) "Noradrenergic function in panic disorder", *J. Clin. Psychiat.* 51 Suppl A:5-11.

Charney, D. S. and A. Deutch (1996) "A functional neuroanatomy of anxiety and fear: implications for the pathophysiology and treatment of anxiety disorders", *Crit. Rev. Neurobiol.* 10:419-446.

Churchland, P. M. (1997) "How to measure similarity of implicit knowledge across structurally diverse brains?", Plenary talk presented at the *First Conference of the Association for the Scientific Study of Consciousnes*, Claremont, CA.

Coghill, R. C., J. D. Talbot, A. C. Evans, E. Meyer, A. Gjedde, M. C. Bushnell and G. H. Duncan (1994) "Distributed processing of pain and vibration by the human brain", *J. Neurosci.* 14:4095-5108.

Craig, K. D. and S. M. Weiss (1990) *Health Enhancement, Disease Prevention, and Early Intervention: Biobehavioral Strategies*, New York: Springer.

Craig, A. D. (1991) "Supraspinal pathways and mechanisms relevant to central pain", in: *Pain and Central Nervous System Disease: The Central Pain Syndromes*, K. L. Case, ed, New York: Raven Press, pp. 157-170.

Craig, A. D. (1992) "Spinal and trigeminal lamina I input to the locus coeruleus anterogradely labeled with Phaseolus vulgaris leucoagglutinin (PHA-L) in the cat and the monkey", *Brain Res.* 584:325-328.

Dahlstrom, A. and K. Fuxe (1964) "Localization of monoamines in the lower brain stem", *Experientia* 20:398-399.

Davar, G. and R. J. Maciewicz (1989) "Deafferentation pain syndromes", *Neurol. Clin.* 7:289-304.

Derbyshire, S. W., A. K. Jones, P. Devani, K. J. Friston, C. Feinmann, M. Harris, S. Pearce, J. D. Watson and R. S. Frackowiak (1994) "Cerebral responses to pain in patients with atypical facial pain measured by positron emission tomography", *J. Neurol. Neurosurg. Psychiat.* 57:1166-1172.

Devor, M. (1991) "Neuropathic pain and injured nerve: peripheral mechanisms", *Br. Med. Bull.* 47:619-30.

Di Piero, V., A. K. Jones, F. Iannotti, M. Powell, D. Perani, G. L. Lenzi and R. S. Frackowiak (1991) "Chronic pain: a PET study of the central effects of percutaneous high cervical cordotomy", *Pain* 46:9-12.

Elam, M., T. H. Svensson and P. Thoren (1985) "Differentiated cardiovascular afferent regulation of locus coeruleus neurons and sympathetic nerves", *Brain Res.* 358:77-84.

Elam, M., T. H. Svensson and P. Thoren (1986a) "Locus coeruleus neurons and sympathetic nerves: activation by cutaneous sensory afferents", *Brain Res.* 366: 254-261.

Elam, M., T. H. Svensson and P. Thoren (1986b) "Locus coeruleus neurons and sympathetic nerves: activation by visceral afferents", *Brain Res.* 375:117-125.

Elliott, K. J. (1994) "Taxonomy and mechanisms of neuropathic pain", *Semin. Neurol.* 14:195-205.

Fields, H. L. (1987) *Pain*, New York: McGraw-Hill.

Fillenz, M. (1990) *Noradrenergic Neurons*, Cambridge: Cambridge University Press.

Foote, S. L., F. E. Bloom and G. Aston-Jones (1983) "Nucleus locus ceruleus: new evidence of anatomical and physiological specificity", *Physiol. Rev.* 63:844-914.

Foote, S. L. and J. H. Morrison (1987) "Extrathalamic modulation of corticofunction", *Annu. Rev. Neurosci.* 10:67-95.

Fried, K., L. R. Govrin, F. Rosenthal, M. H. Ellisman and M. Devor (1991) "Ultrastructure of afferent axon endings in a neuroma", *J. Neurocytol.* 20:682-701.

Galer, B. S. and R. K. Portenoy (1991) "Acute herpetic and postherpetic neuralgia: clinical features and management", *Mt. Sinai J. Med.* 58:257-266.

Galer, B. S. (1995) "Neuropathic pain of peripheral origin: advances in pharmacologic treatment", *Neurology* 45:S17-25.

Gebhart, G. F. and T. J. Ness (1991) "Central mechanisms of visceral pain", *Can. J. Physiol. Pharmacol* 69:627-634.

Geschwind, N. (1965) "Disconnexion syndromes in animals and man. I", *Brain* 88:237-94.

Gray, J. A. (1982) *The Neuropsychology of Anxiety: An Enquiry into the Functions of the Septo-hippocampal System*, New York: Oxford University Press.

Gray, J. A. (1987) *The Psychology of Fear and Stress* (2nd ed.), Cambridge: Cambridge University Press.

Gregory, R. L. (1996a) "What do qualia do?", *Perception* 25:377-379.

Gregory, R. (1996b) "Peculiar qualia", *Perception* 25:755-6.

Gregory, R. L. (1997) "Knowledge in perception and illusion", *Philos. Trans. R. Soc. Lond. B. Biol. Sci.* 352:1121-1127.

Griep, E. N., J. W. Boersma and E. R. de Kloet (1993) "Altered reactivity of the hypothalamic-pituitary-adrenal axis in the primary fibromyalgia syndrome", *J. Rheumatol.* 20:469-474.

Heppelmann, B., K. Messlinger, H. G. Schaible and R. F. Schmidt (1991) "Nociception and pain", *Curr. Opin. Neurobiol.* 1:192-197.

Hsieh, J. C., M. Belfrage, E.-S. Stone, P. Hansson and M. Ingvar (1995) "Central representation of chronic ongoing neuropathic pain studied by positron emission tomography", *Pain* 63:225-236.

Hsieh, J. C., B.-M. Stahle, O. Hagermark, E.-S. Stone, G. Rosenquist and M. Ingvar (1996) "Traumatic nociceptive pain activates the hypothalamus and the periaqueductal gray: a positron emission tomography study", *Pain* 64:303-314.

Humphrey, N. (1992) *A History of the Mind*, New York: Simon and Schuster.

Isaacson, R. L. (1982) *The Limbic System* (2nd ed.), New York: Plenum Press.

Jänig, W. (1985) "Systemic and specific autonomic reactions in pain: efferent, afferent and endocrine components", *Eur. J. Anaesth.* 2:319-346.

Jänig, W. (1987) "Neuronal mechanisms of pain with special emphasis on visceral and deep somatic pain", *Acta Neurochir. Suppl. (Wien)* 38:16-32.

Jessell, T. M. and D. D. Kelly. (1991) "Pain and analgesia", in: *Principles of Neural Science* (3rd ed.), E. R. Kandel, J. H. Schwartz and T. M. Jessell, eds, New York: Elsevier, pp. 385-398.

Jones, A. K., W. D. Brown, K. J. Friston, L. Y. Qi and R. S. Frackowiak (1991) "Cortical and subcortical localization of response to pain in man using positron emission tomography", *Proc. Roy. Soc. Lon. Series B Bio. Sci.* 244:39-44.

Jones, A. K., V. J. Cunningham, S. Hakawa, T. Fujiwara, S. K. Luthra, S. Silva, S. Derbyshire and T. Jones (1994) "Changes in central opioid receptor binding in relation to inflammation and pain in patients with rheumatoid arthritis", *Brit. J. Rheum.* 33:909-916.

Kanosue, K., T. Nakayama, Y. Ishikawa and K. Imai-Matsumura (1984) "Responses of hypothalamic and thalamic neurons to noxious and scrotal thermal stimulation in rats", *J. Thermobiol.* 9:11-13.

Keefe, F. J. (1994) "Behavior therapy", in *Textbook of Pain*, P. D. Wall and R. Melzack, eds, Edinburgh: Churchill Livingstone, pp. 392-406.

Korf, J., B. S. Bunney and G. K. Aghajanian (1974) "Noradrenergic neurons: morphine inhibition of spontaneous activity", *Eur. J. Pharmacol.* 25:165-169.

Krukoff, T. L. (1990) "Neuropeptide regulation of autonomic outflow at the sympathetic preganglionic neuron: anatomical and neurochemical specificity", *Ann. N. Y. Acad. Sci.* 579:162-167.

Lacey, J. and B. Lacey. (1970) "Some autonomic-central nervous system interrrelationships", in: *Physiological Correlates of Emotion*, P. H. Black, ed., New York: Academic Press, pp. 205-227.

Lopez, J. F., E. A. Young, J. P. Herman, H. Akil and S. J. Watson. (1991) "Regulatory biology of the HPA axis: an integrative approach", in: *Central Nervous System Peptide Mechanisms in Stress and Depression*, S. C. Risch, ed, Washington, DC: American Psychiatric Press, pp. 1-52.

MacLean, P. D. (1990) *The Triune Brain in Evolution: Role in Paleocerebral Functions*, New York: Plenum Press.

Marcel, A. (1988) "Phenomenal experience and functionalism", in: *Consciousness in Contemporary Science*, A. Marcel and E. Bisiach, eds, Oxford: Oxford University Press, pp. p 121-158.

McMahon, S. and M. Koltzenburg (1990) "The changing role of primary afferent neurones in pain", *Pain* 43:269-272.

McNaughton, N. and S. T. Mason (1980) "The neuropsychology and neuropharmacology of the dorsal ascending noradrenergic bundle - a review", *Prog. Neurobiol.* 14:157-219.

Melzack, R. and P. D. Wall (1965) "Pain mechanisms: a new theory.", *Science* 150:971-979.

Merskey, H. (1979) "Pain terms: a list with definitions and a note on usage. Recommended by the International Association for the Study of Pain (IASP) Subcommittee on Taxonomy.", *Pain* 6:249-252.

Mesulam, M. and E. Mufson. (1985) "The insula of Reil in man and monkey: Architectonics, connectivity and function", in: *Cerebral cortex, Vol. 4*, A. Peters and E. G. Jones, eds, New York: Plenum, pp. 179-226.

Morilak, D. A., C. A. Fornal and B. L. Jacobs (1987) "Effects of physiological manipulations on locus coeruleus neuronal activity in freely moving cats. II. Cardiovascular challenge", *Brain Res.* 422:24-31.

Mountz, J. M., L. A. Bradley, J. G. Modell, R. W. Alexander, M. Triana-Alexander, L. A. Aaron, K. E. Stewart, G. S. Alarcon and J. D. Mountz (1995) "Fibromyalgia in Women: Abnormalities of Regional Cerebral Blood Flow in the Thalamus and the Caudate Nucleus are Associated with Low Pain Threshold Levels", *Arthr. Rheum.* 38:926-938.

Panksepp, J. (1986) "The anatomy of emotions", in: *Emotion: Theory, Research and Experience, Vol. 3*, R. Plutchik and H. Kellerman, eds, Orlando: Academic Press, pp. 91-124.

Papez, J. W. (1937) "A proposed mechanism of emotion", *Arch. Neurolog. Psych.* 38:725-743.

Perl, E. R. (1984) "Characterization of nociception and their activation of neurons in the superficial dorsal horn: first steps for the sensation of pain", in: *Advances in Pain Research and Therapy, Vol. 6*, L. Kruger and J. C. Liebeskind, eds, New York: Raven Press, pp. 23-52.

Peschanski, M. and J. Weil-Fugacza. (1987) "Aminergic and cholinergic afferents to the thalamus: experimental data with reference to pain pathways", in: *Thalamus and Pain*, J. M. Besson, G. Guilbaud and M. Paschanski, eds, Amsterdam: Excerpta Medica, pp. 127-154.

Ramachandran, V. and W. Hirstein (1997) "Three laws of qualia: What neurology tells us about the biological function of consciousness", *J. Consc. Stud.* 4:429-457.

Redmond, D. E. J. (1977) "Alteration in the functions of the nucleus locus coeruleus: a possible model for studies of anxiety", in: *Animal Models in Psychiatry and Neurology*, I. Hannin and E. Usdin, eds, New York: Pergamon Press, pp. 293-306.

Redmond, D. E. J. and Y. G. Huang (1979) "Current concepts. II. New evidence for a locus coeruleus-norepinephrine connection with anxiety", *Life Sci.* 25:2149-2162.

Roberts, W. J. (1986) "A hypothesis on the physiological basis for causalgia and related pains", *Pain* 24:297-311.

Saphier, D. (1987) "Cortisol alters firing rate and synaptic responses of limbic forebrain units", *Brain Res. Bull.*, 19: 519-524.

Sapolsky, R. M. (1992) *Stress, the Aging Brain, and the Mechanisms of Neuron Death*, Cambridge: MIT Press.

Sawchenko, P. E. and L. W. Swanson (1982) "The organization of noradrenergic pathways from the brain stem to the paraventricular and supraoptic neuclei in the rat", *Brain Res.* 273:275-325.

Selye, H. (1978) *The Stress of Life*, New York: McGraw-Hill.

Siddall, P. J., D. Taylor and M. J. Cousins (1995) "Pain associated with spinal cord injury", *Curr. Opin. Neurol.* 8:447-450.

Silverman, D. H., J. A. Munakata, H. Ennes, M. A. Mandelkern, C. K. Hoh and E. A. Mayer (1997) "Regional cerebral activity in normal and pathological perception of visceral pain", *Gastroenterology* 112:64-72.

Stanton-Hicks, M., W. Janig, S. Hassenbusch, J. D. Haddox, R. Boas and P. Wilson (1995) "Reflex sympathetic dystrophy: changing concepts and taxonomy", *Pain* 63:127-133.

Stone, E. A. (1975) "Stress and catecholamines", in: *Catecholamines and Behavior, Vol. 2*, A. J. Friedhoff, ed, New York: Plenum Press, pp. 31-72.

Sumal, K. K., W. W. Blessing, T. H. Joh, D. J. Reis and V. M. Pickel (1983) "Synaptic interaction of vagal afference and catecholaminergic neurons in the rat nucleus tractus solitarius", *J. Brain Res.* 277:31-40.

Svensson, T. H. (1987) "Peripheral, autonomic regulation of locus coeruleus noradrenergic neurons in brain: putative implications for psychiatry and psychopharmacology", *Psychopharmacology* 92:1-7.

Talbot, J. D., G. H. Duncan and M. C. Bushnell (1989) "Effects of diffuse noxious inhibitory controls (DNICs) on the sensory-discriminative dimension of pain perception", *Pain* 36:231-238.

Talbot, J. D., S. Marrett, A. C. Evans, E. Meyer, M. C. Bushnell and G. H. Duncan (1991) "Multiple representations of pain in human cerebral cortex", *Science* 251:1355-1358.

Vecht, C. J., d. B. H. J. Van and O. J. Wajer (1989) "Post-axillary dissection pain in breast cancer due to a lesion of the intercostobrachial nerve", *Pain* 38:171-176.

Villanueva, L., Z. Bing, D. Bouhassira and D. Le Bars (1989) "Encoding of electrical, thermal, and mechanical noxious stimuli by subnucleus reticularis dorsalis neurons in the rat medulla", *J. Neurophysiol.* 61:391-402.

Villanueva, L., K. D. Cliffer, L. S. Sorkin, D. Le Bars and W. D. J. Willis (1990) "Convergence of heterotopic nociceptive information onto neurons of caudal medullary reticular formation in monkey (Macaca fascicularis)", *J. Neurophysiol.* 63:1118-1127.

Vogt, B. A., S. Derbyshire and A. K. Jones (1996) "Pain processing in four regions of human cingulate cortex localized with co-registered PET and MR imaging", *Eur. J. Neurosc.* 8:1461-1473.

Wall, P. D. (1991) "Neuropathic pain and injured nerve: central mechanisms", *Br. Med. Bull.* 47:631-43.

Willer, J. C. *et al.* (1989) "Encoding of nociceptive thermal stimuli by diffuse noxious inhibitory controls in humans", *J. Neurophysiol.* 62:1028-38.

Willis, W, D. (1985) "Pain pathways in the primate", *Prog Clin Biol Res.* 176:117-133.

Willis, W. D. (1988) "Dorsal horn neurophysiology of pain", *Ann. New York Acad. Sci.* 531:76-"Mechanisms of somatic pain", in: *Current and Emerging Issues in Cancer Pain*, C. R. Chapman and K. M. Foley, eds, New York: Raven Press, pp. 67-81.

Willis, W. D. and K. N. Westlund (1997) "Neuroanatomy of the pain system and of the pathways that modulate pain.", *J. Clin. Neurophysiol.* 14:2-31.

Woolf, C. J. (1989) "Recent advances in the pathophysiology of acute pain", *Brit. J. Anaesth.* 63:139-146.

Woolf, C. J. and A. E. King (1990) "Dynamic alterations in the cutaneous mechanoreceptive fields of dorsal horn neurons in the rat spinal cord", *J. Neurosci.* 10:2717-2726.

PSYCHOPHYSIOLOGY OF EMOTIONAL PERCEPTION AND IMPLICATIONS FOR UNDERSTANDING EMOTION-MEMORY RELATIONSHIPS

MARGARET M. BRADLEY
NIMH Center for the Study of Emotion and Attention (Csea), Box 100165 HSC, University of Florida, Gainesville, Florida, 32610-0165, USA

ABSTRACT

Research collaborators at the Center for the Study of Emotion and Attention of the University of Florida have developed several standardized sets of visual, auditory, and word stimuli for the experimental study of emotional perception. Data from a large number of experiments in which these stimuli have been employed are consistent with the interpretation that behavioral, physiological, and self-report measures of emotion during perception reflect basic motivational dispositions that are either appetitive or defensive in orientation (emotional valence) and which vary in reactive intensity (arousal). These stimuli have also proven useful in studies of emotion-memory relationships, providing support for the "intensity" principle. Memory for these stimuli was found to be primarily sensitive to the arousal dimension, with differences in memory performance consistently obtained for emotionally arousing stimuli (either pleasant or unpleasant), compared to neutral stimuli.

1. Introduction

Emotions involve multiple response systems (behavioral, physiological, and self-report) and are highly variable in their psychophysiological composition. It is proposed that an important organizing factor in emotion is the individual's motivational state, determined by primitive defensive and appetitive circuits that have evolved to promote individual and species survival. From this perspective, emotional responses are significantly determined by stimulus valence (i.e., pleasant/ unpleasant; appetitive / defensive) and the intensity of the resulting activation.

In a number of recent studies (for review, see Bradley & Lang, 2000), we have explored emotion-relevant responses during perception. This context is useful not only because it naturalistically defines the onset of affect, but also because the individual is passive, motor interference is reduced, and a specific input event is the focus of current activity. Thus, the physiological and overt responses observed are primarily those that support perception, and those that are elicited by motivational parameters dictated by the stimulus. We have developed a number of different stimulus collections to study emotional perception in the laboratory, including sets of color photographs (the International Affective Picture System; IAPS, Lang, Bradley, & Cuthbert, 1999), digitized sounds (International Affective Digitized Sound system; IADS, Bradley & Lang, 1999a), and verbal stimuli (Affective Norms for English Words; ANEW, Bradley & Lang, 1999b). The prototypical paradigm involves presenting a series of emotional stimuli to participants, in the context of passive perception. Assessing the relationship between behavioral, physiological, and individual reports of emotional experience provides information regarding the organization of emotion in perception.

2. Psychophysiologic Studies of Emotion Stimulus Perception

Normal participants display impressive concordance between individual reports of emotional experience and psychophysiological reactivity in the perceptual situation (Bradley & Lang, 2000). Measures of facial electromyographic activity, including the corrugator (frown) and zygomatic (smile) muscles, consistently covary with reports of affective valence (i.e., pleasantness). Heart rate also reflects the hedonic valence of the perceived stimulus, leading to greater deceleration when processing unpleasant, compared to pleasant, materials. Electrodermal reactivity, on the other hand, which is presumed to reflect activity in the sympathetic nervous system, covaries linearly with reports of emotional arousal, increasing systematically with increases in rated arousal of the stimulus. Interestingly, cortical activity, measured either as discrete event-related potentials (e.g., P300) or as sustained slow wave activity, also co-varies systematically with rated arousal, with more cortical positivity measured when perceiving pleasant or unpleasant, compared to neutral, stimuli.

2.1 The Startle-Probe Methodology

Probe measures provide behavioral indices of emotional responding during perception. For example, presentation of a brief startling stimulus elicits an involuntary, reflexive eyeblink, whose function is primarily protective. The magnitude of the startle blink during emotional perception varies with the affective valence of the stimulus (Lang, 1995): Blinks are larger and faster when perceiving aversive or unpleasant stimuli, compared to pleasant materials. On the other hand, when a brief, non-startling tone probe is presented (to which a button press is required), these voluntary behavioral responses are slower in emotional perception, compared to when a neutral stimulus is processed. Taken together, the data are consistent with the idea that behavioral, physiological, and self-report measures of emotion during perception reflect basic motivational dispositions that are either appetitive or defensive in orientation (emotional valence) and which vary in reactive intensity (arousal).

3. Emotion and Memory

The role of emotion in memory continues to be elusive: When an emotional event occurs, what are the implications for memory performance? Clinical, anecdotal, and empirical evidence (for review see Bradley, 1994; Reisberg &

Heuer, 1995; Stein, Ornstein, Tversky & Brainerd, 1997) support practically the entire range of possible relationships relating emotion to memory. Better memory for unpleasant events is highlighted in "flashbulb" theories of memory (Brown & Kulik, 1977; Winograd & Neisser, 1992), which propose that exposure to traumatic events creates a very strong, almost veridical, memory representation. Conversely, "repression" theories maintain that memory is poorer for unpleasant events, sometimes even below the level of conscious awareness. A bias towards remembering pleasant events underlies the "pollyanna" hypothesis, which proposes that memory shows a preference for the positive. The "intensity" hypothesis ignores the parameter of hedonic valence and suggests that arousing events -- both pleasant and unpleasant -- are remembered better than those that are low in arousal. Taken together, past research provides a rich and varied set of both hypotheses and theoretical notions relating emotion to memory.

3.1 Empirical Studies of Emotion-Memory Relationships

Empirical evidence for each hypothesis arises from different domains of experimental study, which contributes to the diversity in these theories. For instance, flashbulb memory theories rely primarily on studies of memory for naturally occurring, culturally shared catastrophes (e.g., the assassination of a public figure; Pillemer, 1984). To systematically explore the relationship of emotion to memory in a controlled laboratory context, we have conducted a number of different studies assessing how people remember specific emotional stimuli that vary along dimensions of affective valence (pleasure) and arousal.

In these studies, we (Bradley, 1994; Bradley, Greenwald, Petry, & Lang, 1992) tested people's memory for affective pictures, sounds, or words that were previously presented in the context of a simple passive perception task. That is, during the encoding task, the participant was instructed to simply view or listen to a series of stimuli that varied in pleasure and arousal. An incidental memory paradigm was employed, in which participants were exposed to emotional stimuli, without knowing that their memory would be tested at a later point in time. In different experiments, memory was tested using free recall and recognition (either explicit and implicit) at short (i.e., 5 minutes), moderate (i.e., 1 day) and long (i.e. 1 year) delays. Gender and personality were also explored as they affect emotional memory performance.

Taken together, the data demonstrate general support for the "intensity" principle: Memory was primarily sensitive to the arousal dimension, with differences in performance consistently obtained for emotionally arousing stimuli (either pleasant or unpleasant), compared to neutral stimuli. Similar results from animal studies have led theorists such as Gold and McGaugh (1975; Gold, 1995) to propose that memory is specifically tuned to 'motivationally relevant' stimuli, preferentially storing events that are associated with highly negative or positive consequences. From a motivational perspective, a memory system that is differentially sensitive to events that are emotionally arousing is quite functional: These are the events that most threaten or support the organism's survival.

References

Bradley, M.M. (1994) "Emotional memory: A dimensional analysis", in: *Emotions: Essays on Emotion Theory*, S. van Goozen, N.E. Van de Poll, and J.A. Sergeant, eds, Hillsdale, New Jersey: Lawrence Erlbaum Associates, pp. 97-134.

Bradley, M.M., and P.J. Lang (1999a) *International Affective Digitized Sounds. Technical Manual and Affective Ratings*, Gainesville, FL: The Center for Research in Psychophysiology, University of Florida.

Bradley, M.M., and P.J. Lang (1999b) *Affective Norms for English Words (ANEW). Technical Manual and Affective Ratings*, Gainesville, FL: The Center for Research in Psychophysiology, University of Florida.

Bradley, M.M., and P.J. Lang (2000) "Measuring emotion: Behavior, feeling, and physiology", in: *Cognitive Neuroscience of Emotion*, R.D. Lane, L. Nadel, G.L. Ahern, J.J.B. Allen, A.W. Kaszniak, S.Z. Rapcsak, and G.E. Schwartz, eds, New York: Oxford University Press, pp. 24-61.

Bradley, M.M., M.K. Greenwald, M.C. Petry, and P.J. Lang (1992) "Remembering pictures: Pleasure and arousal in memory", *Journal of Experimental Psychology: Learning, Memory, and Cognition* 18:379-390.

Brown, R., and J. Kulik (1977) "Flashbulb memories", *Cognition* 5:73-99.

Gold, P.E. (1995) "Modulation of emotional and nonemotional memories: Same pharmacological systems, different neuroanatomical systems", in: *Brain and Memory: Modulation and Mediation of Neuroplasticity*, J.L. McGaugh, N.M. Weinberger, and G. Lynch, eds, New York: Oxford University Press, pp. 41-74.

Gold, P.E., and J.L. McGaugh (1975) "A single-trace, two-process view of memory storage processes", in: *Short-Term Memory*, D. Deutsch and J.A. Deutsch, eds, New York: Academic Press, pp. 355-390.

Pillemer, D.B. (1984) "Flashbulb memories of the assassination attempt on President Reagan", *Cognition* 16:63-80.

Lang, P.J. (1995) "The emotion probe: Studies of motivation and attention", *American Psychologist* 50:372-385.

Lang, P.J., M.M. Bradley, and B.N. Cuthbert (1999) *International Affective Picture System (IAPS): Technical Manual and Affective Ratings*, Gainesville, FL: The Center for Research in Psychophysiology, University of Florida.

Reisberg, D., and F. Heuer (1995) "Emotion's multiple effects on memory", in: *Brain and Memory: Modulation and Mediation of Neuroplasticity*, J.L. McGaugh, N.M. Weinberger, and G. Lynch, eds, New York: Oxford University Press, pp. 84-92.

Stein, N.L., P.A. Ornstein, B. Tversky, and C. Brainerd, eds (1997) *Memory for Emotional and Everyday Events*, Mahwah, New Jersey: Erlbaum and Associates.

Winograd, E., and U. Neisser, eds (1992) *Affect and Flashbulb Memories*, New York: Cambridge University Press.

IMAGERY AND EMOTION: INFORMATION NETWORKS IN THE BRAIN

PETER J. LANG
NIMH Center for the Study of Emotion and Attention (Csea), Box 100165 HSC, University of Florida, Gainesville, Florida, 32610-0165, USA

ABSTRACT

Emotions can best be conceptualized as action tendencies that serve immediate survival needs. Neural circuits subserving these action tendencies are largely subcortical, although connected to the cerebral cortex in humans to allow more elaborate processing of relevant information and more complex cognitive and behavioral output of emotional states. An associationist account of what more phenomenologically inclined theorists call "appraisal" is provided. This account avoids assumptions about subjective evaluation, and provides a mechanism that accounts for both the rapidity of emotional responses and their frequent "irrational" quality. In such a network view, emotions differ from other knowledge structures in being directly connected to subcortical appetitive and defense motivational systems.

1. Introduction

Emotions are action dispositions (Frijda, 1986; Lang, 1995). They evolved from reflexive somatomotor and vegetative reactions to appetitive or aversive stimulation, serving immediate survival needs (e.g., feeding, nurturance, sexual approach; fight, flight). The neural circuits supporting these simple behaviors (and related primary associative processes) are largely subcortical (see LeDoux, 1996). In human beings, however, these circuits are connected to large cerebral cortices that mediate the more elaborate information processing and complex cognitive and behavioral output of emotional states.

Emotions are instantiated when specific memory episodes (about context and behavior) are retrieved. Like other knowledge structures, emotional images are coded in memory as networks of mutually activating information units. In processing the network, activity in one unit is transmitted to adjacent units, and depending on the strength of activation, the entire structure may be engaged. The probability of network processing is increased with the number of units initially stimulated.

2. Emotion and Stimulus Representation

We presume that the fundamental emotion network is neural and that it is essentially opaque to consciousness. It can be thought of as a net of linked representations, which, in turn, might be individual neural sub-networks (Lang, 1984). The higher level representations (some of which certainly pass through awareness and are the stuff of affective reports) are of three basic types: stimulus, response, and meaning. Stimulus units reflect activation in sensory processors and are representations of perceived events. Response units code information mediating the three basic output procedures in emotion. Response units fall naturally into three general categories: (1) Language behavior, including both affective expression (e.g., distress calls, instrumental verbal aggression) and the description and evaluation of putative internal states; (2) Behavioral acts, including the many motor actions of emotion, such as facial expression and expressive posture, strength of approach or avoidance, as well as the modulation of secondary tasks (e.g., deficits in performance or control); and (3) Patterns of visceral and somatic activation (Bradley & Lang, 2000; Lang, 1993; Lang, Greenwald & Bradley, 1993). Meaning units refer to associated declarative (semantic) knowledge. This taxonomy is descriptively convenient; however, the actual neural sub-units may well cut across the proposed categories. For example, Hebb (1949) early described how visual stimulus representations might be based on the neural patterns instigated by eye movement responses.

2.1 Emotion and Imagery

Assumptions of the model are that emotion networks may be activated by any input that matches representations in its assembly (Lang, 1979); the greater the number and the verisimilitude of these matches, the greater the likelihood of network activation. It is further assumed that activation is facilitated when the associative strength of the net is high (greater coherence), in which case degraded cues readily instigate emotional processing. Thus, for example, a curled up garden hose readily prompts processing of a snake phobic's fear image network. Indeed, because of associative linkage, network activation does not depend on input from true events, e.g., actual danger or pain. Representations in the net may be broadly cued externally by language descriptions, moving and still pictures, diagrams, and other symbolic stimuli remote from the natural context; or internally, by semantic association, neuromuscular patterns and autonomic states (Vrana & Lang, 1990).

3. Conclusions

Readers will recognize the above model as an associationist account of what more phenomenologically inclined theorists call "appraisal." The network conception differs in avoiding assumptions about subjective evaluation, and in providing a mechanism that accounts for both the rapidity of emotional responses and their frequent "irrational" quality. In the network view, emotions differ from other knowledge structures in being directly connected to the subcortical appetitive and defense motivational systems (Lang, 1994). Thus, autonomic and somatic reflex activation is more probable and often more intense. Emotional images and memories, linked to the brain's basic survival system, are more persistent, more readily instigated by degraded cues and often refractory to instructional control.

References

Bradley, M.M., and P.J. Lang (2000) "Measuring emotion: Behavior, feeling, and physiology", in: *Cognitive Neuroscience of Emotion*, R.D. Lane, L. Nadel, G.L. Ahern, J.J.B. Allen, A.W. Kaszniak, S.Z. Rapcsak, and G.E. Schwartz, eds, New York: Oxford University Press, pp. 24-61.

Frijda, N.H. (1986) *The Emotions*, Cambridge: Cambridge University Press.

Hebb, D.O. (1949) *The Organization of Behavior: A Neuropsychological Theory*, New York: John Wiley and Sons.

Lang, P.J. (1979) "A bio-informational theory of emotional imagery", *Psychophysiology* 16:495-511.

Lang, P.J. (1984) "Cognition in emotion: Concept and action", in: *Emotions, Cognition, and Behavior*, C. Izard, J. Kagan, and R.B. Zajonc, eds, New York: Cambridge University Press.

Lang, P.J. (1993) "The three system approach to emotion", in: *The Organization of Emotion*, N. Birbaumer and A. Öhman, eds, Toronto: Hogrefe-Huber, pp. 18-30.

Lang, P.J. (1994) "The motivational organization of emotion: Affect-reflex connections", in: *The Emotions: Essays on Emotion Theory*, S. Van Goozen, N.E. Van de Poll, and J.A. Sergeant, eds, Hillsdale, New Jersey: Lawrence Erlbaum Associates, pp. 61-93.

Lang, P.J. (1995) "The emotion probe: Studies of motivation and attention", *American Psychologist* 50:372-385.

Lang, P.J., M.K. Greenwald, M.M. Bradley, and A.O. Hamm (1993) "Looking at pictures: Affective, facial, visceral, and behavioral reactions", *Psychophysiology* 30:261-273.

LeDoux, J. (1996) *The Emotional Brain: The Mysterious Underpinnings of Emotional Life*, New York: Simon and Schuster.

Vrana, S.R., and P.J. Lang (1990) "Fear imagery and the startle-probe reflex", *Journal of Abnormal Psychology* 99:189-197.

HEMISPHERIC ASYMMETRIES IN REPRESENTATION AND CONTROL OF EMOTIONS: EVIDENCE FROM UNILATERAL BRAIN DAMAGE

GUIDO GAINOTTI

Institute of Neurology, Catholic University of Rome, Largo A. Gemelli, 8,00168 Rome, Italy

ABSTRACT

Investigations of hemispheric asymmetries in representation and control of emotions have followed either the componential or the hierarchical model of emotions. After the first empirical studies, which showed that emotions are asymmetrically represented at the hemispheric level (giving rise to the first theoretical models of emotional lateralization), several studies have taken individually into account one or few specific components of emotions. Most of these investigations have focused attention on the communicative (sensory and expressive-motor) components of emotions, studying the perception or the production of facial or vocal emotional expression in patients with unilateral brain damage. Some proponents of this line of research have suggested that the right hemisphere might play a major role in functions of non-verbal communication, rather that in emotional behavior per se. However, later studies have shown that the right hemisphere superiority concerns not only the communicative, but also (and perhaps mainly) the vegetative components of emotions. These findings are inconsistent with the hypothesis that the right hemisphere superiority concerns non-verbal communication rather that emotional behavior per se. In more recent years, some authors have shifted attention from the componential to the hierarchical organization of emotions, assuming that both hemispheres may be involved in emotional functions, but that each of them may be mainly involved in a specific hierarchical level of emotions. Two models, belonging to this line of thought, have been proposed. The first model assumes that the hemispheric specialization may concern two different categories of emotions. The right hemisphere might mainly subserve the most primitive (survival related) categories of emotions, whereas the left hemisphere might play a major role in phylogenetically more recent social forms of emotions. The second model, that I prefer, assumes that the hemispheric specialization concerns two different levels of emotional processing, rather that two different categories of emotions. The lower emotional level, corresponding to the level of the automatically elicited spontaneous emotions, could be mainly represented in the right hemisphere, whereas the higher emotional level, subserving the conscious analysis and the intentional control of the emotional discharge, could be mainly represented in the left hemisphere.

1. Introduction

The problem of the hemispheric asymmetries in the representation of emotions is a rather recent one. More than a century had, in fact, elapsed between the first studies showing a left hemisphere dominance in the representation of language (Dax, 1836, quoted by Dax, 1865; Broca, 1865) and the first investigations showing clear laterality effects in the representation of emotions (Terzian & Cecotto, 1959; Gainotti, 1969, 1972).

During the last 30 years, however, the number of studies dealing (from the clinical, experimental and theoretical points of view) with the problem of the hemispheric asymmetries in the representation and control of emotions has steadily increased. After the first empirical studies, a clear tendency to approach the problem from theoretically motivated lines of research has emerged. These lines of research have taken into account two main features of the functional architecture of the emotional system, namely its componential nature and its hierarchical organization.

In the first part of this chapter, I will, therefore, briefly discuss some of the theoretical issues that have oriented the development of research in this area. This review will take into account: (a) the main analogies and differences between the emotional and the cognitive systems; (b) the principal components of emotions; (c) the hierarchical structure of the emotional system. After this introduction, I will more analytically review: (a) the results of the first empirical studies which have shown that emotions are not symmetrically represented at the hemispheric level, giving rise to the first theoretical models of emotional lateralization; (b) the outcome of studies conducted in brain-damaged patients to investigate hemispheric asymmetries in the representation of specific components of emotions; (c) the development of models aiming to link emotional lateralization to the hierarchical structure of the emotional system.

2. Main Characteristics of the Emotional System

2.1. Similarities and Differences between the Emotional and the Cognitive System

Most authors have considered the emotional and the cognitive systems as phylogenetically advanced adaptations, based on the integrated activity of a number of components. These components are roughly similar in both systems. They have the common functions of (a) scanning the external milieu, focusing attention on the most relevant stimuli, (b) analyzing these stimuli and computing their meaning, (c) providing an efficient and appropriate response, and (d) memorizing the most relevant data (stimulus characteristics, organismic response and outcome of this response). However, beyond these structural similarities, important differences also exist between the emotional and the cognitive systems. The general logic of these systems is different and their components must therefore have partly different characteristics.

According to Oatley and Johnson-Laird (1987), the organism employs two different operative systems to face a partially unpredictable environment and to select the most appropriate behavioral response. The first is the *emotional system,* considered as an emergency system able to interrupt the action occurring with an urgency procedure, and able to rapidly select a new operative scheme. The second is the *cognitive system,* considered as a more adaptive and evolved system, able to exhaustively analyze complex situations and to elaborate plastic and varied plans, but requiring much more time to carry out its work. This model assumes that the elementary and phylogenetically primitive emotional system may base its functioning on a limited number of modules (automata). The modules rapidly and automatically process a restricted number of signals and trigger an immediate response, selected from a small number of innate operative patterns, corresponding to the basic needs of the species in question.

2.2 Main Components of Emotional Behavior

The characteristics attributed by most authors to the main components of emotional behavior (summarized in Table 1) are essentially in agreement with Qatley and Johnson-Laird's (1987) interpretation of the logic of this system. Thus, with regard to the analysis of sensory information (listed in Table 1 as "Emotional computation of raw sensory data"), almost all authors recognize that what is required to evaluate the pleasant or dangerous significance of an external situation can be, at least in some cases, global, rapid and unconscious.

Table 1. Main Components of Emotional Behavior

		- Orienting of attention	
	Emotional Arousal		
		- Readiness to action	
LOW LEVEL AUTOMATIC COMPONENTS	Emotional Computation of Raw Sensory Data		
		Motor Reaction	Communicative Components
		- Facial expression	
	Emotional Response	- Vocal	
		- Bodily movements	
		Autonomic Reaction	
	Spontaneous Emotional Learning	(conditioned)	
	----------------------Emotional Experience----------------------		
	Cognitive Appraisal of Complex Emotional Situations		
HIGH LEVEL CONTROLLED COMPONENTS	Intentional Control of the Emotional Response		
	Controlled Learning of Emotionally Relevant Information (Declarative Memory)		

In agreement with the same model, most authors also acknowledge that the action schemata activated by the evaluation of an emotional stimulus are probably innate. These schemata correspond to a small number of basic emotions and reflect the most important interactive schemata of the human species at the communicative level or at that of proneness toward action. Darwin (1872) had already stressed the importance of the communicative aspects of emotions. Darwin had rightly pointed out that in man and in other social animals the facial and vocal expression of emotions are innate action patterns provided of high survival value and widely generalized across the human species.

Another emotional component is represented by the autonomic-vegetative response. This component plays a critical role in emotional behavior, both as a marker and an elicitor of the subjective experience of emotions (James, 1884; Levenson, 1992; Schacter, 1970). The autonomic-vegetative component is also a strong determinant of the efficacy of the behavioral response (Cannon, 1929). Both the communicative and the vegetative components of emotions have, therefore, been taken into account in studies dealing with the lateralization of human emotions.

The last point that I would just note in this section concerns the fact that, even when attention is focused on the componential nature of emotions, it is difficult to ignore the difference between elementary and more complex components of emotions. The elementary ones are linked to the spontaneous elicitation of automatic emotional responses, whereas the complex ones are related to the conceptual analysis of emotional stimuli and to the intentional control of the emotional response. This last point leads us to shift attention from the componential to the hierarchical structure of emotions.

2.3. The Hierarchical Structure of Human Emotions

Both anatomical and psychological models have stressed the hierarchical structure of human emotions starting from different phylogenetic and ontogenetic considerations. The anatomical models have drawn on the acknowledgement that the neural organization of emotions spans multiple phylogenetically different structures of the brain. For example in the lower brainstem are represented elementary adaptive reflexes (such as the patterns of laughing and crying), whereas in the cortico-limbic networks of the temporal and frontal lobe are located the interface structures between the emotional and the cognitive systems. Psychological models, such as that proposed by Oatley and Johnson-Laird (1987), account for the simplest and more elementary emotions as well as for the most precocious phases of child development, but not for complex emotions such as remorse, vanity or nostalgia. Since most authors assume that these complex emotions may derive from the primitive ones (thanks to mechanisms of emotion blending and of increasing interactions between the emotional and the cognitive systems), it became necessary to construct hierarchical, ontogenetic models capable of explaining these processes. Leventhal (1979, 1987), who has proposed that human emotions may derive from the activity of a hierarchical multicomponent system, has formulated this type of developmental model. This model is composed of three hierarchically organized functional levels: (1) the sensorimotor level, (2) the schematic level, and (3) the conceptual level, whose main characteristics are described in Table 2:

Table 2. Characteristics of the functional levels of emotional processing included in Leventhal's Model

THE SENSORY-MOTOR LEVEL

consists of a set of innate neuro-motor programs, which are triggered automatically by a certain number of environmental stimuli and which include components of motor and vegetative activation.

THE SCHEMATIC LEVEL

is based on the activity of emotional schemata, i.e. of prototypes of emotional behavior, formed (on the basis of conditioning processes) by the association between the innate neuro-motor programs and situations linked to these programs in the individual experience.

These emotional schemata are automatically reactivated during situations similar to those previously associated with the corresponding neuro-motor program. This automatic reactivation is accompanied by the evocation of the subjective and the expressive-motor components of the corresponding schema and is experienced as a true emotion.

THE CONCEPTUAL LEVEL

is based on a mechanism of conscious learning and is mediated by cognitive processes rather that by conditioning processes. This level stores abstract and propositional notions about emotions and the social rules concerning their expression. The activation of these propositional representations is not accompanied by the experience of the corresponding emotion (as is observed during activation of the emotional schemata).

Both the anatomical models and Leventhal's psychological model are based on the notion of different levels of emotional processing and on the assumption of a control of the highest over the lowest functional levels. Some authors have, however, proposed that a different principle (namely the phylogenetic difference between different categories of emotions) rather than the different complexity of the emotional computation to be performed, may underlie the hierarchical organization of emotions. This viewpoint has been put forward by McLean (1961) in his pioneering studies of the limbic system. According to this author, the most primitive forms of emotional behavior, such as the fight-flight reactions that are present in phylogenetically old species (e.g., reptilians), could be subserved by the hypothalamus and by the related parts of the paleostriatum.

On the other hand, the family-related patterns of emotional behavior (including social attachment and play), which are characteristic only of mammals, could be subserved by the cingulate gyrus, which is the phylogenetically more recent part of the limbic system. Both the hierarchical models based on the notion of different levels of emotional processing and those based on phylogenetic differences between different categories of emotions have been used in the interpretation of data relevant to the problem of hemispheric asymmetries in emotional behavior.

3. Periods that can be Distinguished in the Study of Emotional Lateralization

3.1 *The Period of the Purely Empirical Studies*

I have said in the introductory section of this chapter that only in the second part of this century did some unexpected clinical observations raise the problem of possible hemispheric asymmetries in the representation of emotions. The first data pointing in this direction were gathered by authors who observed different emotional behavior in patients submitted to a pharmacological inactivation of the right or left hemisphere (Alemà & Donini, 1960; Rossi & Rosadini, 1967; Terzian & Cecotto, 1959). These authors reported that injection of sodium amytal into the left carotid artery produces a "depressive-catastrophic reaction," characterized by bursts of tears and by a sad and pessimistic attitude. In contrast, pharmacological inactivation of the right hemisphere is followed by a "euphoric-manic reaction," characterized by a relaxed attitude with tendency to joke and laugh. Since these emotional manifestations were not related to the conditions of examination, they were considered as resulting from disruption of neural mechanisms specifically underpinning opposite aspects of mood (with a major involvement of the left hemisphere in "positive" emotions and of the right hemisphere in "negative" affects).

Some years later, I could partly confirm these clinical observations, by analyzing the patterns of emotional behavior shown by right and left brain-damaged patients during neuropsychological examination (Gainotti, 1969, 1972). I could, indeed, confirm that a "catastrophic reaction" follows left hemisphere injuries and that an "indifference reaction" is typical of patients with right hemisphere damage. However, I found misleading the equivalence proposed by Terzian and Cecotto (1959), Alemà and Donini (1960) and Rossi and Rosadini (1967) between "catastrophic reaction" and endogenous depression and between "indifference reaction" and euphoric-manic state.

As a matter of fact, catastrophic reactions of left brain-damaged patients usually consisted of increasing anxiety or sudden bursts of tears, triggered by repeated, frustrating attempts of verbal communication observed in a context of severe expressive disorder and motor impairment of the right hand. These emotional displays were, therefore, considered as a dramatic, but psychologically appropriate form of reaction to the catastrophic effects of the brain lesion, rather than as a form of biological depression.

On the other hand, right brain-damaged patients with an indifference reaction did not seem excited or euphoric, but rather indifferent, apathetic and unduly jocular. Furthermore, these patients also showed other paradoxical behaviors, such as the tendency to deny or to minimize their disabilities, sometimes coexisting with exaggerated expressions of hatred toward the paralyzed limbs. Overall, these patterns of behavior seemed to suggest an abnormal and inappropriate emotional reaction to the consequences of the brain damage, rather than pointing to a shift of mood toward an euphoric state.

To explain the contrast between the dramatic but psychologically appropriate reaction of left brain-damaged patients and the abnormal reaction of patients affected by right hemisphere lesions, I advanced the hypothesis of a right hemisphere dominance for emotional functions (Gainotti, 1972). According to this hypothesis, the emotional reaction should be appropriate when the right hemisphere is intact, whereas it should be absent or inappropriate when an extensive lesion of the right hemisphere inactivates the parts of this hemisphere involved in emotional functions.

Thus, the first clinical observations have allowed the formulation of two alternative models of the relationships between emotions and hemispheric asymmetries. According to the first model, proposed by the authors of the amytal studies, each hemisphere could be specialized for a different dimension of emotions, the left hemisphere being critically involved in positive emotions and the right hemisphere in negative emotions. According to the second model, which I have proposed, the clinical data point more to a general dominance of the right hemisphere for emotional functions than to a different specialization of the left hemisphere for positive emotions and of the right hemisphere for negative emotions.

3.2. The Period of Studies Dealing with Specific Components of Emotions

After the period of the first clinical investigations, several studies were conducted in patients with unilateral brain lesions and in normal subjects. These studies were conducted to test the alternative interpretations of (a) a general dominance of the right hemisphere for various aspects of emotional behavior, vs. (b) of a different hemispheric specialization for positive and negative emotions. Even though several lines of research and different components of emotions have been taken into account, the largest number of investigations studied the communicative aspects of emotions. These communicative aspects include the comprehension of the emotion expressed by a facial or vocal display and the expression of emotions through the facial or the vocal channels of emotional communication. The reasons that investigators have focused their attention on communicative aspects of emotions include: (a) the rediscovery of Darwin's seminal work by Tomkins (1962, 1963) and Ekman (1973, 1984) and their assumption that a typical pattern of facial expression may exist for a number of basic emotions; (b) the parallel development of sophisticated techniques for analysis of non-verbal communication, such as the Facial Action Coding System (FACS) procedure, developed by Ekman and Friesen (1978); and (c) the tendency to assume that hemispheric asymmetries emerge in the more complex, rather than in the more elementary components of a given function.

Since a critical survey of studies conducted in normal subjects exceeds the scope of this contribution, I will limit myself to a review of research conducted in patients with unilateral brain damage, briefly summarizing results of studies conducted in normal subjects. The methodology of investigations conducted with focal brain lesion patients has usually included testing the capacity of right and left brain-damaged patients to point to a face or a voice expressing a given emotion, or to communicate a given emotion through facial movements or with the affective contours of speech.

Results have consistently shown that right brain-damaged patients are often impaired in recognizing emotions expressed through tone of voice (Blonder *et al.*, 1991; Heilman *et al.*, 1975, 1984; Ross, 1981; Tucker *et al.*, 1977) and in the identification of facial emotional expressions (Blonder *et al.*, 1993; Borod *et al.*, 1986; Bowers *et al.*, 1985; De Kosky *et al.*, 1980). Other authors have shown that patients with right hemisphere injury are also impaired in the capacity to express emotions with the prosodic contours of speech (Ross, 1984; Tucker *et al.*, 1977) or through expressive facial movements (Blonder *et al.*, 1993; Borod *et al.*, 1986).

On the other hand, investigations conducted in normal subjects have allowed a better testing of the hypothesis assuming a different specialization of the left and right hemisphere for positive and negative emotions respectively. Some authors (e.g., Borod and Caron, 1980; Sackneim & Gur, 1978; Schwartz *et al.*, 1979) have shown that the right hemisphere dominance for functions of emotional communication is stronger for negative than for positive emotions. But overall, these studies have not supported the hypothesis of an interaction between hemisphere and emotional valence. They have rather confirmed, in agreement with investigations conducted in brain-damaged patients, the hypothesis of general superiority of the right hemisphere for functions of emotional communication (see Borod, 1993; and Gainotti, 1989, 1997 for reviews).

This fact has led some authors to hypothesize that the right hemisphere may have a superiority in the area of non-verbal communication analogous to that shown by the left hemisphere for language functions. Following this line of thought, Ross (1981, 1984) has suggested that disorders of non-verbal communication might be the basic defect of right brain-damaged patients and that emotional disturbances usually observed in these patients only reflect their inability to produce and to comprehend emotional signals. The "indifference reaction" of right brain-damaged patients should, therefore, be considered not as an abnormal form of emotional behavior, but simply as the consequence of a basic inability to correctly evaluate and express emotional signals.

Two main objections can be made to this hypothesis: The first refers to the fact that the right hemisphere superiority in tasks of emotional comprehension and expression has probably been overestimated. Thus, in research exploring the receptive level of emotional communication, Gainotti (1989) and Weddel (1989) found no difference between right and left brain-damaged patients on tasks requiring the identification of facial emotional expressions. Similar results have been obtained by Bradvick *et al.* (1990) and by Cancelliere and Kertesz (1990), studying the recognition of emotions expressed through the prosodic components of speech. Analogously, in investigations conducted at the expressive level, Mammucari *et al.* (1988) have found no difference between right and left brain-damaged patients studying the facial expressions elicited by positive and negative emotional movies. Similar results have been obtained by Caltagirone *et al.* (1989) and by Weddel *et al.* (1990), studying the production of posed (rather than spontaneous) facial emotional expressions, and by Bradvick *et al.* (1990) and Cancelliere and Kertesz (1990), studying the expression of emotions through the emotional contours of speech.

The second objection is even more relevant with respect to the subject of this section, since it refers to research conducted on another important component of emotions, namely the vegetative component. Irrespective of the exact scope of the autonomic response, the Ross' hypothesis predicts that this component of emotions should be intact in right brain-damaged patients, and this becomes particularly true if we accept with Schacter (1970) and Levenson (1992) that the autonomic component plays and important role in the generation of the emotional experience.

This prediction, however, is at variance with results of investigations that have studied the electrodermal response or other indices of autonomic activation in right and left brain-damaged patients, following presentation of emotional stimuli. This subject is discussed in greater detail within another chapter of this volume (Gainotti, this volume). At present, I will limit myself to noting that an important reduction of the vegetative response to emotional stimuli has been observed by several authors in right but not in left brain-damaged patients (see Gainotti, 1997, this volume; and Wittling, 1995 for reviews). Taken together, these data suggest that the emotional indifference of right brain-damaged patients is real and might be at least in part due to a reduced capacity to react with an appropriate vegetative response to emotionally laden stimuli.

Data consistent with the hypothesis of an important role of the right hemisphere in generating the vegetative components of emotions have also been obtained by Wittling and Roschmann (1993) and by Spence *et al.* (1996) during lateralized presentation of emotional stimuli to normal subjects. In both these studies, the autonomic response was higher during presentation of the emotional material to the right than to the left hemisphere.

3.3 The Period of Models Linking Hemispheric Asymmetries to the Hierarchical Structure of Emotions

In recent years, some authors have shifted attention from the componential to the hierarchical organization of emotions, assuming that both hemispheres play an important role in emotional functions, but that each of them may be specifically involved in a different level of emotional processing or of emotional representation.

Two main interpretations of the hemispheric asymmetries, which make reference to more general models of the hierarchical organization of emotions, have been proposed. According to the first interpretation, proposed by Ross *et al.* (1994), the right and left hemispheres might subserve two different categories of emotions. In the right hemisphere could be represented the most primitive (survival related) categories of emotions, i.e. those categories of emotions that, according to McLean (1961) are already present in phylogenetically old species, such as reptilians.

The left hemisphere, in contrast, could play a major role in phylogenetically more recent social forms of emotions. According to Ross *et al.* (1994), this interpretation could be consistent both with the right hemisphere hypothesis (assuming a general dominance of this hemisphere for emotional functions) and with the valence hypothesis (assuming a different specialization of the right hemisphere for negative emotions and of the left hemisphere for positive emotions). This statement is based on the claim that primitive emotions (lateralized to the right hemisphere) constitute the majority of the emotional schemata and have generally a negative valence, whereas social emotions (lateralized to the left hemisphere) have usually a positive valence.

It must be acknowledged, however, that empirical data supporting this interpretation are very scanty and that even the assumptions linking the primitive emotions with negative valence and the social emotions with positive valence seem only partly justified.

According to the second interpretation, which was originally proposed by Buck (1984) and Rinn (1984), and more recently developed by Gainotti *et al.* (1993), the hemispheric asymmetry might concern two different levels of emotional processing, rather than two different categories of emotions. In the most recent of these models, Gainotti *et al.* (1993), making reference to the Leventhal's (1979, 1987) conceptualization and terminology, proposed that the right hemisphere may preferentially subserve the "schematic level" and the left hemisphere the "conceptual level" of emotional processing. Since data showing a right hemisphere dominance for the autonomic components of emotions are clearly consistent with the hypothesis of a major involvement of this hemisphere in the 'schematic level' of emotional processing, I will focus my attention here on the role of the left hemisphere in the "conceptual level." In particular, I will take into account the problem of the leading role that this hemisphere could have in functions of emotional control, which constitute an important and dissociable aspect of the "conceptual level" of emotional processing.

Two main arguments, consistent with the hypothesis of a major role of the left hemisphere in functions of emotional control, have been advanced. The first refers to the possibility that both the expressive-motor and the autonomic components of the emotional response may be overexpressed by left brain-damaged patients. This possibility is supported by two sets of data: (a) clinical findings reported by Gainotti (1972), Buck and Duffy (1980) and House *et al.* (1989), who have noted that sudden outbursts of tears and other instances of increased facial emotional displays are often observed in left brain-damaged patients, and in particular in Broca's aphasic patients; and (b) data reported by Heilman *et al.* (1978) and by Meadows and Kaplan (1994), who, studying the autonomic response to emotional stimuli, have observed an increased reactivity in patients with left sided lesions in comparison with normal controls.

The second argument refers to a reinterpretation of data obtained by Sackeim and Gur (1978), Schwartz *et al.* (1979) and Borod and Caron (1980) studying the difference between the right and left half face in the expression of positive and negative emotions in normal subjects. These authors (as already mentioned) had observed a greater expressivity of the left (in comparison with the right) half face only for negative emotions, but not for smiling or other positive emotions. These findings had been interpreted within the context of the hypothesis assuming a different specialization of the left and right hemisphere for positive and negative emotions. However, Etcoff (1986) has rightly pointed out that smiling differs from the other emotional facial displays not only because of the positive valence of the emotion it usually conveys, but also because it constitutes the facial "emotional" expression more often intentionally used for social communication purposes.

If we assume a left hemisphere dominance for functions of emotional control, then the difference between left and right half face in the expression of negative and positive emotions can be viewed as the result of an interaction between the general superiority of the right hemisphere in spontaneous expression of emotions and that of the left hemisphere in control of the intentional facial expressive apparatus. As already proposed by Buck (1984) and Rinn (1984), the greater asymmetry between the right and left half face in the expression of negative emotions could be due to the greater inhibition exerted by the left hemisphere on the right half face in the overt expression of these socially non-communicable emotions. On the other hand, the lesser degree of asymmetry shown by smiling could be due to the contrast between the greater "natural" expressivity of the left half face and the greater intentional control of the left hemisphere over the expressive apparatus of the right half face.

4. Some Concluding Remarks About the Relations Between Hemispheric Asymmetries and Conscious Emotional Experience

The model that I have presented in the last part of this contribution has some implications as for the problem of the relationships between emotions, hemispheric asymmetries and conscious experience. If the level of emotional processing represented in the right hemisphere is mostly the schematic one, whereas the level represented in the left hemisphere corresponds to the conceptual one, then it is possible to predict that only an emotional stimulation of the right hemisphere (activating the level where emotional schemata are automatically and unconsciously aroused) should provoke an unconscious emotional experience.

Data consistent with this prediction have been recently reported by Ladavas *et al.* (1993) and by Spence *et al.* (1996). The former group studied the cognitive evaluation and autonomic response to subliminal and above-threshold presentation of emotional and non-emotional stimuli in a split-brain patient. The latter group investigated the cognitive evaluation and autonomic reaction to emotional and neutral scenes, briefly lateralized to the right and left hemisphere, in normal subjects. Both studies have shown: (a) that only the right hemisphere is able to selectively produce an appropriate autonomic response to the presentation of emotional material; and (b) that in the right hemisphere the production of the appropriate vegetative response can be dissociated from the cognitive evaluation of the eliciting stimulus.

Obviously, these data do not demonstrate that a full emotional experience can be unconsciously activated by the appropriate stimulus only in the right hemisphere, since the observed autonomic response is only a fragment of an emotional experience. Nevertheless, these data are clearly consistent with the hypothesis assuming that the level of emotional processing represented in the right (but not in the left) hemisphere can be described as the level where the emotional schemata are automatically and unconsciously activated.

References

Alemà, G. and G. Donini (1960) "Sulle modificazioni cliniche ed elettroencefalografiche da introduzione intracarotidea di iso-amil-etil-barbaturato di sodio nell'uomo", *Boll. Soc. Ital. Biol. Sperim.* 36:900-904.

Blonder, L.X., D. Bowers and K.M. Heilman (1991) "The role of the right hemisphere in emotional communication", *Brain* 114:1115-1127.

Blonder, L.X., A. Burns, D. Bowers, *et al.* (1993) "Right hemisphere facial expressivity during natural conversation", *Brain Cogn.* 21:44-56.

Borod, J.C. (1993) "Cerebral mechanisms underlying facial, prosodic and lexical emotional expression: a review of neuropsychological studies and methodological issues", *Neuropsychology* 7:445-463.

Borod, J.C. and H. Caron (1980) "Facedness and emotion related to lateral dominance, sex, and expression type", *Neuropsychologia* 18:237-241.

Borod, J.C., E. Koff, M. Perlman-Lorch and M. Nicholas (1986) " The expression and perception of facial emotion in brain-damaged patients", *Neuropsychologia* 24:169-180.

Bowers, D., R.M. Bauer, H.B. Coslett and K.M. Heilman (1985) "Processing of faces by patients with unilateral hemisphere lesions: Dissociation between judgments of facial affect and facial identify", *Brain Cogn.* 4:258-272.

Bradvik, B., C. Dravins, S. Holtas *et al.* (1990) "Do single right hemisphere infarcts or transient ischemic attacks result in aprosody?", *Acta. Neurol. Scand.* 81:61-70.

Broca, P. (1865) "Sur la faculte' du langage articule"', *Bull. Soc. Antropol.* 6:377-393.

Buck, R. (1984) *The Communication of Emotion*, New York: Guilford Press.

Buck, R. and R.J. Duffy (1980) "Nonverbal communication of affect in brain-damaged patients", *Cortex* 16: 351-362.

Caltagirone, C., P. Ekman, W. Friesen, G. Gainotti, A. Mammucari, L. Pizzamiglio and P. Zoccolotti (1989) "Posed emotional expression in unilateral brain damaged patients", *Cortex* 25:653-663.

Cancelliere, A.E.B. and A. Kertesz (1990) "Lesion localization in acquired deficits of emotional expression and comprehension", *Brain Cogn.* 13:133-147.

Cannon, W.B. (1929) *Bodily Changes in Pain, Hunger, Fear, and Rage (2nd ed)* New York: Appleton.

Darwin, C. (1872) *The Expression of the Emotions in Man and Animals*, London: Murray. (Reprinted Chicago: University of Chicago Press, 1965).

Dax, M. (1865) "Lesions de la moité gauche de l'encéphale coincidant avec l'oubli des signes de la pensée", *Gaz. Hebd. Med. Chir.*. 2:259-260.

DeKosky, S., K.M. Heilman, D. Bowers and B. Valenstein (1980) "Recognition and discrimination of emotional faces and pictures", *Brain Lang.* 9:206-214.

Ekman, P. (1973) "Darwin and cross-cultural studies of facial expression", in: *Darwin and Facial Expression,* P. Ekman, ed, New York: Academic Press, pp.1-83.

Ekman, P. (1984) "Expression and the nature of emotion", in: *Approches to Emotion,* K. Scherer and P. Ekman, eds, Hillsdale, NJ, Erlbaum.

Ekman, P. and W.V. Friesen (1978) *The Facial Action Coding System*, Palo Alto, CA: Consulting Psychologists Press.

Etcoff N.L. (1989) "Recognition of emotions in patients with unilateral brain damage", in: *Emotions and the Dual Brain,* G. Gainotti and C. Caltagirone, eds, Heidelberg: Springer-Verlag, pp.168-186.

Gainotti, G. (1969) "Réaction "catastrophiques" et manifestations d'indifferénce au cours des atteintes cérébrales", *Neuropsychologia* 7:195-204.

Gainotti, G. (1972) "Emotional behavior and hemispheric side of lesion", *Cortex* 8:41-55.

Gainotti, G. (1989) "The meaning of emotional disturbances resulting from unilateral brain injury", in: *Emotions and the Dual Brain,* G. Gainotti and C. Caltagirone eds., Heidelberg: Springer-Verlag, pp.147-167.

Gainotti, G. (1997) "Emotional disorders in relation to unilateral brain damage", in: *Behavioral Neurology and Neuropsychology*, T.E. Feinberg and M. Farah, eds, New York: Mc Graw-Hill.

Gainotti, G., C. Caltagirone and P. Zoccolotti (1993) "Left/right and cortical-subcortical dichotomies in the neuropsychological study of human emotions", *Cognition and Emotion,* 7:71-93.

Heilman, K.M., D. Bowers, L. Speedie and H.B. Coslett (1984) "Comprehension of affective and nonaffective prosody", *Neurology* 34:917-921.

Heilman, K.M., R. Scholes and R.T. Watson *(1975)* "Auditory affective agnosia", *J. Neurol.Neurosurg. Psychiatry* 38:69-72.

Heilman, K.M., H.D. Schwartz and R.T. Watson (1978) "Hypoarousal in patients with neglect and emotional indifference", *Neurology* 28:229-232.

House, A., M. Dennis, A. Molyneux, C. Warlow and K. Hawton (1989) "Emotionalism after stroke", *British Medical Journal* 298:991-994.

James, W. (1884) "What is an emotion ?", *Mind* 9:188-205.

Làdavas E., D. Cimatti, M. Del Pesce and G. Tuozzi (1993) "Emotional evaluation with and without conscious stimulus identification: evidence from a Split-brain patient", *Cognition and Emotion,* 7: 95-114.

Levenson, R.W. (1992) "Autonomic nervous system differences among emotions", *Psychol. Sci.* 3:23-27.

Leventhal, H. (1979) "A perceptual-motor processing model of emotion", in: *Perception of Emotion in Self and Others, vol 5*, P. Pliner, K. Blakenstein and I.M. Spiegel, eds, New York: Plenum.

Leventhal, H. (1987) "A perceptual motor theory of emotion", in: *Advances in Experimental Social Psychology vol.17*, L. Berkowitz, ed, New York: Academic Press.

McLean, P.D. (1961) "Le systeme limbique du point de vue de la self-protection et de la conservation de l'espece", in: *Physiologie et Pathologie du Rhinencephale,* T. Alajouanine, ed, Paris: Masson, pp.111-126.

Mammucari, A., C. Caltagirone, P. Ekman *et al.* (1988) "Spontaneous facial expression of emotions in brain-damaged patients", *Cortex* 24:521-533.

Meadows, M.E. and R.F. Kaplan (1994) "Dissociation of autonomic and subjective responses to emotional slides in right hemisphere damaged patients", *Neuropsychologia* 32:847-856.

Oatley, K. and P. Johnson-Laird (1987) "Toward a cognitive theory of emotions", *Cogn. Emot.* 1:29-50.

Rinn, W.E. (1984) "The neuropsychology of facial expression: A review of the neurological and psychological mechanisms for producing facial expression", *Psychol. Bull.* 95:52-77.

Ross, E.D. (1981) "The prosodias: functional-anatomical organization of the affective components of language in the right hemisphere", *Arch. Neurol.* 38:561-569.

Ross, E.D. (1984) "Right hemisphere's role in language, affective behavior and emotion", *Trends Neurosci.* 7:342-346

Ross, E.D., R.W. Homan and R. Buck (1994) "Differential hemispheric lateralization of primary and social emotions: implications for developing a comprehensive neurology for emotions, repression, and the subconscious", *Neuropsych. Neuropsychol. Behav. Neurol.* 7:1-19.

Rossi, G.F. and G. Rosadini (1967) "Experimental analysis of cerebral dominance in man", in: *Brain Mechanisms Underlying Speech and Language,* C.J. Millikan and F.L. Darly, eds, New York, Grune and Stratton.

Sackeim, H.A. and R.C. Gur (1978) "Lateral asymmetry in intensity of emotional expression", *Neuropsychologia* 16:473-481.

Schachter, 5. (1970) "The assumption of identity and peripheralist-centralist controversies in motivation and emotion", in: *Feelings and Emotions: The Loyola Symposium,* M.B. Arnold, ed., New York: Academic Press, pp.111-121.

Schwartz, G.E., G.L. Ahern and S.L. Brown (1979) "Lateralized facial muscle response to positive and negative emotional stimuli", *Psychophysiol.* 16:561-571.

Spence, S., D. Shapiro and E. Zaidel (1996) "The role of the right hemisphere in the physiological and cognitive components of emotional processing", *Psychophysiol.* 33:112-122.

Terzian, H. and S. Cecotto (1959) "Su un nuovo metodo per la determinazione e lo studio della dominanza emisferica", *Giornale di Psichiatria e Neuropatologia* 87:889-924.

Tomkins, S.S. (1962) *Affect, Imagery, Consciousness: vol.1, The Positive Affects*, New York: Springer.

Tomkins, S.S. (1963) *Affect, Imagery, Consciousness: vol.2, The Negative Affects*, New York: Springer.

Tucker, D.M., R.T. Watson and K.M. Heilman (1977) "Discrimination and evocation of affectively intoned speech in patients with right parietal disease", *Neurol.* 27:947-950.

Weddel, R.A. (1989) "Recognition memory for emotional facial expressions in patients with focal cerebral lesions", *Brain Cogn..* 11:1-17.

Weddel, R.A., D.J. Miller and C. Trevarthen (1990) "Voluntary emotional facial expression in patients with focal cerebral lesions", *Neuropsychologia* 28:49-60.

Wittling, W. *(1995)* "Brain asymmetry in the control of autonomic-physiologic activity", in: *Brain Asymmetry,* R.J. Davidson and K. Hugdahl, eds, Cambridge: MIT Press, pp.305-357.

Wittling, W. and R. Roschmann (1993) "Emotion-related hemisphere asymmetry: subjective emotional response to laterally presented films", *Cortex* 29: 31-448.

HEMISPHERE ASYMMETRIES FOR AUTONOMIC FUNCTIONS: EVIDENCE FROM NORMAL SUBJECTS AND BRAIN-DAMAGED PATIENTS

GUIDO GAINOTTI
Institute of Neurology, Catholic University of Rome, Largo A. Gemelli, 8, 00168 Rome, Italy

ABSTRACT

Independent lines of research have shown that the division of labor between the right and left hemisphere could concern not only the cognitive and emotional functions, but also autonomic activities, which are one of the main components of emotional behavior. Some of these investigations were part of research programs addressing questions about hemispheric asymmetries in representation and control of emotions. These studies have been conducted either in brain-damaged patients or in normal subjects, during tasks of selective emotional stimulation of the right or left hemisphere. Other clinically oriented studies have been motivated by epidemiological and neurophysiological data suggesting that hemispheric asymmetries might exist in the autonomic control of the heart. All these investigations have consistently shown that autonomic functions are lateralized in the human brain and that the right hemisphere plays a preeminent role from this point of view. However, both the question of the exact pattern of lateralization of the autonomic functions and the question of the relationships between autonomic and emotional asymmetries remain open and require further investigations. As for the first issue, two alternative models have been proposed. The first model assumes a right hemisphere superiority for every kind of autonomic function, whereas the second model posits a different specialization of the right hemisphere for sympathetic activities and of the left hemisphere for parasympathetic functions. As for the second issue, some authors assume an interdependence between autonomic and emotional cerebral asymmetries, whereas other authors maintain that no clear relationship exists between these two facets of brain lateralization. If we assume that asymmetries for complex behavioural activities probably emerge as a by-product of more basic interhemispheric differences, then an asymmetric representation of autonomic functions could be the prerequisite for the lateralization of the emotional system, considered as an emergency system, devised to respond rapidly and efficiently to situations relevant for the basic needs of the individual.

1. Introduction

The hypothesis of a right hemisphere dominance for autonomic functions is recent. It has been put forward in the last part of the 20th century, during investigations designed to clarify the precise nature of the right hemisphere dominance for emotions (see Gainotti, this volume, for review).

To be sure, this hypothesis had already been advanced, within the same context, by some classical authors, following Babinki's (1914) observation that anosognosia and anosodiaphoria (i.e. unawareness and lack of concern for hemiplegia) are usually found in patients with right hemisphere lesions. Discussing and extending these observations, Hirschl and Potzl (quoted by Schilder, 1935) had, in fact, advanced the hypothesis of a right hemisphere dominance for vegetative functions. This hypothesis, however, did not capture the attention of other scientists, since the prevailing line of thought was that only language and other cognitive functions were asymmetrically represented at the hemispheric level, whereas emotions and autonomic functions were not. Only after clinical studies showed that emotions are asymmetrically represented in the human brain, was the question of a possible lateralization of autonomic functions raised again and submitted to empirical testing.

The first study explicitly designed to address this issue in patients with unilateral brain lesions was reported by Heilman *et al.* (1978), who studied the galvanic skin response to painful stimuli applied to the hand ipsilateral to the damaged hemisphere. These authors reported a flattened vegetative response among patients with right hemisphere lesions, and in particular among those showing a unilateral neglect syndrome and signs of emotional indifference.

After this pioneer investigation, several other studies were conducted both in brain-damaged patients and in normal subjects. The first clinical investigations focused attention on the relationships between emotions and vegetative functions, studying the psychophysiological correlates of emotional activation among patients with unilateral brain lesions. More recent lines of research have investigated hemispheric asymmetries for autonomic functions, independently from emotional tasks. In recent years, a sizeable number of clinical studies have been devoted to questions concerning hemispheric asymmetries in cardiac autonomic control. Epidemiological and neurophysiological data seem, indeed, to suggest that different kinds of heart disorders could be linked to lesions of the right and left hemisphere. Finally, a very recent line of research has been conducted in normal subjects, experimentally studying the psychophysiological correlates of selective emotional stimulation of the right and left hemisphere. In the next sections of this chapter, I will, therefore, separately review results obtained in these different lines of research, before discussing, in the last part of this paper, the main questions that these studies have raised.

2. Clinical and Experimental Investigations

2.1. Psychophysiological Correlates of Emotional Activation in Patients with Unilateral Brain Lesions

Morrow *et al.* (1981), Zoccolotti *et al.* (1982) and, more recently, Meadows and Kaplan (1994) have studied galvanic skin response to emotional and neutral slides in patients with unilateral brain injuries and normal controls. Analogous investigations have been conducted by Zoccolotti *et al.* (1986) and by Caltagirone *et al.* (1989) studying heart rate changes and galvanic skin response to emotional (positive and negative) and neutral short films. All these studies have shown an important reduction of the galvanic skin response and of other indices of autonomic activation in right brain-damaged patients. This defect of vegetative response was usually specifically linked to the presentation of emotional material and was not significantly related to a poor cognitive evaluation of the emotional stimuli. Much less consistent has been the autonomic response of left brain-damaged patients. In some studies (e.g. Heilman *et al.,* 1978; Meadows & Kaplan, 1994) an autonomic response greater than that of normal controls has been found. In other investigations (e.g. Caltagirone *et al.*, 1989; Morrow *et al.*, 1981; Zoccolotti *et al.*, 1982, 1986) the autonomic response of left brain-damaged patients has been found to be lower than that shown by control subjects, but higher than that shown by right brain-damaged patients.

2.2. Studies on Hemispheric Asymmetries in Autonomic Heart Control

The brain control of cardiovascular activity is based on the outflow of sympathetic and parasympathetic pathways. The former originates from the vasomotor centers of bulb and medulla oblongata, which project to the preganglionic neurons, located in the intermediolateral cell columns of the spinal cord, and innervate the heart through cells lying in the stellate ganglion (Levy & Martin, 1979). The latter originates from neurons of the nucleus ambiguous and of the dorsal motor nucleus of the vagus (both located in the medulla oblongata) and course down through the vagal nerve to intracardiac ganglia in the wall of the heart (Levy & Martin, 1979). The outflow of the sympathetic and parasympathetic systems is modulated by various hierarchically organized brain structures. These range from the periaqueductal gray to the paraventricular hypothalamus and the central nucleus of the amygdala, and to various regions of the cerebral cortex, which include insula, anterior cingulate gyrus, orbital frontal cortex and parts of the somatic motor and sensory cortex (Cechetto & Saper, 1990; Oppenheimer & Hopkins, 1994).

Since the seminal papers of Mizeres (1958) and of Levy *et al.* (1966) it has been known that, at the level of the peripheral autonomic structures, both sympathetic and parasympathetic innervation of the heart are strongly lateralized (see Wittling, 1995, 1997 for reviews). In more recent years, however, it has become increasingly clear that not only the peripheral autonomic innervation, but also the hemispheric modulation of the heart vegetative control is lateralized. This claim is supported by investigations in human subjects that have examined: (a) the cardiac effects of unilateral brain lesions; (b) the consequences of unilateral hemispheric inactivation, and (c) the cardiac effects of unilateral hemispheric stimulation.

Table 1 provides methodological information and the main results of studies that have investigated the consequences of unilateral brain lesions on the autonomic control of the heart.

Table 1. Cardiac effects of unilateral hemispheric lesions in humans.

Authors	**Experimental Procedures and Outcome Measures**	**Main Results**
Zoccolotti *et al.* (1986) and Caltagirone *et al.* (1989)	Heart rate changes observed while viewing short emotional (negative) or neutral films	NC and LBDP showed a greater heart rate deceleration than RBDP in response to emotional films.
Yokoyama *et al.* (1987)	Heart rate changes during an attention-demanding task	RBDP showed a reduced heart rate response.
Lane *et al.* (1992)	Incidence of various kinds of arrhythmias in RBDP and LBDP	Supraventricular tachycardia was more frequent in RBDP.
Baron *et al.* (1994)	Various components of the power spectrum of heart rate variability	RBDP showed a reduction of the spectral power in the domain of sinus arrhythmias.
Sander and Klingelhöfer (1995)	Variability in circadian blood pressure and in cardiovascular measures	RBDP showed a reduced circadian blood pressure variability and a higher incidence of cardiac arrhythmias.
Naver *et al.* (1996)	Bedside tests reflecting sympathetic and parasympathetic influence on heart rate and blood pressure	RBDP showed a reduced heart rate response to respiratory changes.

Legend: NC = Normal Controls; RBDP = Right Brain Damaged Patients; LBDP= Left Brain Damaged Patients.

Taken together, data summarized in Table I show that autonomic control of the heart is disturbed more by right than by left brain lesions. In particular, these data show a pervasive reduction of heart rate variability in response to emotional stimuli (Caltagirone *et al.*, 1989; Zoccolotti *et al.*, 1986) or to attention demanding tasks (Yokoyama *et al.*, 1987), during increased respiratory activity (Naver *et al.*, 1996) or studying the circadian variability of blood pressure or of other cardiovascular parameters (Sander & Klingelhofer, 1995). This lack of heart rate variability is certainly abnormal since it is associated with an increased risk of sudden death in patients with and without a history of cardiac infarction (Johnson & Robinson, 1988; Kleiger *et al.*, 1987).

The consequences of a hemispheric unilateral inactivation on autonomic heart control have been studied by Rosen *et al.* (1982), Zamrini *et al.* (1990) and Yoon *et al.* (1997) in patients given an intracarotid amytal injection. Pharmacological inactivation of the left hemisphere was associated with an increased heart rate (considered as related to the stress provoked by the consequences of the hemispheric inactivation), whereas a similar response was not observed after injection of amytal into the right carotid artery (Rosen *et al.*, 1982; Zamrini *et al.*, 1990). Yoon *et al.* (1997), on the other hand, studied the power spectrum of heart rate variability before and after intracarotid amytal injection, to investigate the balance between sympathetic and parasympathetic activation. They observed a shift toward sympathetic predominance after left hemisphere inactivation, but no significant change of this balance after right hemisphere inactivation, and considered these data as indicative of a right hemisphere dominance for sympathetic activity.

Results obtained by Oppenheimer *et al.* (1992) studying heart rate and blood pressure changes during electrical stimulation of the right and left insular cortex are consistent with these conclusions. Sympathetic effects, with tachycardia and increased blood pressure were, in fact, mainly elicited by stimulation of the right insular cortex.

2.3. Psychophysiological Correlates of Selective Emotional Stimulation of the Right and Left Hemisphere in Normal Subjects

The psychophysiological correlates of the selective emotional stimulation of the right and left hemisphere have been recently studied in normal subjects, with different experimental procedures and contrasting results, by Wittling and coworkers (Wittling, 1990, 1995, 1997; Wittling *et al.*, 1998) and by Spence *et al.* (1996). Wittling and coworkers used lateralized film presentation, with a special experimental technique, to investigate blood pressure changes (Wittling, 1990), the power spectrum of heart rate variability (Wittling *et al.*, 1998) and various indices of myocardial activity (Wittling, 1997). The high frequency band of the spectral power was taken as an index of parasympathetic activity, whereas measures of myocardial performance were used to evaluate the sympathetic influence on the heart. The authors observed a greater increase of blood pressure and of the measures of myocardial performance during the right hemisphere film presentation and a significant increase of the high frequency component of the spectral power during left hemisphere viewing of the film. They interpreted these findings as indicative of a differential specialization of the right hemisphere for sympathetic activity and of the left hemisphere for the parasympathetic control of the heart. No interaction was found between the (emotional or neutral) quality of the film, the visual field stimulated, and the presence of psychophysiological asymmetries. The authors therefore concluded that the observed asymmetries were a direct consequence of the mono-hemispheric stimulation and were not related to the emotional nature of the task.

Spence *et al.* (1996), on the other hand, showed emotional and neutral slides briefly lateralized to the right and left visual half field, to normal subjects, using heart rate and pulse volume as measures of autonomic activation. They also asked their subjects to categorize each slide as emotional or neutral and examined reaction time as a measure of cognitive processing. The largest psychophysiological responses, including both the sympathetically mediated vasoconstriction and the parasympathetically mediated heart rate deceleration, were obtained after presentation of the emotional slides to the right hemisphere. Reaction times, on the contrary, failed to show a right hemisphere perceptual processing superiority. According to the authors, these results are at variance with the Wittling's conclusions, since they show: (a) that both the sympathetically mediated vasoconstriction and the parasympathetically mediated heart rate deceleration are predominantly related to a right hemisphere activation; and (b) that these autonomic responses are specific to emotional material, rather than reflecting a general state of activation of one cerebral hemisphere.

3. Open Questions in the Study of Hemispheric Asymmetries for Autonomic Functions

The contrast between the conclusions reached by Wittling *et al.* (Wittling, 1995, 1997; Wittling *et al.*, 1998) and by Spence *et al.* (1996), in the studies that I have just summarized, clearly indicate the main questions that remain open in the study of the relationships between hemispheric asymmetries and autonomic functions. The first question concerns the exact pattern of asymmetrical representation of the vegetative functions. All the authors agree on the prominent role of the right hemisphere in the modulation of sympathetic activities, but the hemispheric representation of parasympathetic activities remains controversial.

The second question concerns the relationship between hemispheric representation of emotions and of autonomic functions. Some authors (e.g. Spence *et al.*, 1996) assume a strong relationship between lateralization of emotions and of vegetative functions, while other authors (e.g. Wittling, 1995, 1997) assume an anatomical contiguity but a functional independence between lateralization of emotional and of autonomic functions.

3.1. How Are the Autonomic Functions Lateralized in the Human Brain?

A strong similarity exists between the models recently advanced to explain the lateralization of the autonomic functions and the first models advanced some decades ago to explain the lateralization of emotional functions in the human brain (see Gainotti, this volume). In both cases, one model assumes a general right hemisphere superiority for emotions (Gainotti, 1972) or for autonomic functions (Spence *et al.*, 1996), whereas a competing model assumes a different specialization of the right and left hemisphere for negative and positive emotions (Rossi & Rosadinl, 1967; Terzian & Cecotto, 1959) or for the sympathetic and parasympathetic components of vegetative functions (Wittling, 1995, 1997).

Furthermore, in the case of the lateralization of emotions, the right hemisphere dominance for negative emotions is supported by almost all the available data, whereas the left hemisphere prevalence for positive emotions is supported only by some studies. Similarly, the right hemisphere superiority for sympathetic functions is better supported than the left hemisphere dominance for parasympathetic activities. Almost all the data reported in the previous sections of this chapter show, in fact, a right hemisphere dominance for sympathetic functions, whereas the lateralization of parasympathetic activities is more controversial. On one hand, Wittling (1997) and, to a lesser extent, Zamrini *et al.* (1990) and Yoon *et al.* (1997) have reported data consistent with the hypothesis of a left hemisphere superiority for parasympathetic functions. On the other hand, results obtained by Zoccolotti *et al.* (1986), Yokoyama *et al.* (1987), Caltagirone *et al.* (1989), Naver *et al.* (1996), and Spence *et al.* (1996) seem to point to a right hemisphere dominance not only for sympathetic, but also for parasympathetic activities.

So, it seems safe to conclude that an asymmetry exists between the strong right hemisphere lateralization of the sympathetic functions and the weak hemispheric lateralization of the parasympathetic activities. This pattern of lateralization of the autonomic functions is perhaps not inconsistent with the model of lateralization of emotional functions that I have proposed in previous papers (Gainotti *et al.*, 1993; Gainotti, 1997, this volume). If the emotional system is (as proposed by Oatley & Johnson-Laird, 1987) basically an emergency system and if the spontaneous, automatic functioning of this system (schematic level) is mainly mediated by the right hemisphere, then it seems logical to expect that the energetic component of the autonomic system (namely the sympathetic section) may also be lateralized to the right hemisphere. Less compelling are the biological reasons that could suggest a right hemisphere lateralization of parasympathetic functions, since these functions are usually considered as the control mechanism of the autonomic system, devised to counterbalance the costs of the sympathetic response, restoring the organism's energetic resources. If we accept, as my model proposes, that the left hemisphere may mediate the conceptual level of emotional processing (and therefore control the functioning of the right hemisphere's schematic level) then there could also be some reasons to expect a weak left hemisphere lateralization of parasympathetic activity.

3.2. Relationships between the Lateralization of Emotions and of Autonomic Functions

According to Wittling (1995) three main lines of evidence suggest an independence between asymmetries concerning emotions and those concerning the autonomic functions. The first line of evidence consists in the observation that autonomic asymmetries have been found by Yokoyama *et al.* (1987) and by Wittling and Schwiger (1993) in conditions in which cerebral activation had been provoked by cognitive effort, rather than by emotional situations. The second consists in the fact that vegetative asymmetries have been repeatedly reported in humans in situations uninfluenced by emotional or cognitive factors, such as unilateral hemispheric inactivation by intracarotid amytal injection (Rosen *et al.*, 1982; Yoon *et al.*, 1997; Zamrini *et al.*, 1990) or electrical stimulation of the cerebral cortex (Oppenheimer *et al.*, 1992). The last line of evidence consists in the fact that when psychophysiological responses and subjective emotional response or cognitive evaluation of the emotional stimuli were assessed in the same experimental situation, no significant correlation was found between asymmetries in different response systems.

In my opinion, these data clearly show that there is a relative independence between emotional and autonomic systems, but do not necessarily prove that these two systems are independently lateralized. To be sure, the first two lines of evidence prove that the lateralized autonomic functions are not simply a component of the emotional system, but form an autonomous system. This system certainly plays a very important role in emotional behavior, but also intervenes in other kinds of energy demanding (e.g., motor or cognitive) tasks. On the other hand, the third line of evidence does not necessarily prove that emotional and autonomic functions are independently lateralized, because no instance of double dissociation (Shallice, 1988; Teuber, 1955) is reported by Wittling between asymmetries observed in different response systems. As a matter of fact, in all instances quoted by Wittling (1995) in which an asymmetry was observed in one, but not in another response system (Wittling, 1990, 1995; Wittling & Ptuger, 1990), this dissociation was a simple one. More precisely, the asymmetry concerned only the psychophysiological measure, but not the subjective experience or the cognitive evaluation of the emotional stimulus. The same statement also applies to other investigations, not quoted by Wittling (1995) but in which a similar asymmetry in only one response system was observed (e.g. Gainotti, 1989; Ladavas *et al.*, 1993; Mammucari *et al.*, 1988; Meadows & Kaplan, 1994; Spence *et al.*, 1996). If a double dissociation clearly indicates an independence between two phenomena, a simple dissociation can also suggest a quantitative difference (in terms of sensitivity or of difficulty) between two sets of measures, without necessarily proving their independence.

An alternative explanation of data reviewed by Wittling (1995) could, therefore consist in recognizing that the autonomic activities are a basic and sensitive component of the emotional functions, but are not exclusively involved in emotional activities, since they also contribute (although to a lesser extent) to non-emotional cognitive or motor functions. Furthermore, if we assume that hemispheric asymmetries for complex behavioral activities emerge as a function of more basic inter-hemispheric differences, then it is likely that an asymmetric representation of vegetative functions may be the antecedent of the asymmetric hemispheric representation of emotions. This simple model could allow us to understand: (a) why autonomic asymmetries can also be observed in tasks requiring a cognitive effort (and not only in emotional situations); (b) why autonomic asymmetries also result from the lateralized stimulation or inactivation of structures, such as the insular cortex, crucially involved in autonomic functions (Oppenheimer *et al.*, 1992); and (c) why studies considering both psychophysiological and subjective/emotional or cognitive/evaluative responses, have found greater asymmetries in the simple and sensitive autonomic functions, than in the complex cognitive or subjective (conscious) emotional response systems.

References

Babinski, J. (1914) “Contribution à l’étude des troubles mentaux dans l’hémiplégie organique cérébrale (Anosognosie)”, *Rev. Neurol.* 27:845-848.

Baron, S.A., Z. Rogovski and J. Hemli (1994) “Autonomic consequences of cerebral hemisphere infarction”, *Stroke* 25:113-116.

Caltagirone, C., P. Zoccolotti, G. Orginale, A. Daniele and A. Mammucari (1989) “Autonomic reactivity and facial expression of emotion in brain-damaged patients”, in: *Emotions and the Dual Brain,* G. Gainotti and C. Caltagirone, eds, Berlin: Springer, pp.204-221.

Cechetto, D.F. and C.B. Saper (1990) “Role of the cerebral cortex in autonomic function”, in: *Central Regulation of Autonomic Functions,* A.D. Loewy and K.M. Spyer, eds, New York: Oxford University Press, pp.208-223.

Gainotti, G. (1972) “Emotional behavior and hemispheric side of the lesion”, *Cortex* 8:41-55.

Gainotti, G. (1997) “Emotional disorders in relation to unilateral brain damage”, in: *Behavioral Neurology and Neuropsychology,* T.E. Feinberg and M. Farah, eds, New York: McGraw-Hill.

Gainotti, G. (1989) “The meaning of emotional disturbances resulting from unilateral brain injury”, in: *Emotions and the Dual Brain,* G. Gainotti and C. Caltagirone, eds, Berlin: Springer, pp.147-167.

Gainotti, G., C. Caltagirone and P. Zoccolotti (1993) “Left/right and cortical-subcortical dichotomies in the neuropsychological study of human emotions”, *Cognition and Emotion* 7:71-93.

Heilman, K.M., H.D. Schwartz and R.T. Watson (1978) "Hypoarousal in patients with neglect and emotional indifference", *Neurology* 28:229-232.

Johnson, R.H. and B.J. Robinson (1988) "Mortality in alcoholics with autonomic neuropathy", *J. Neurol. Neurosurg. Psychiatry* 51:476-480.

Kleiger, R.E., P. Miller, J:T. Bigger and A.J. Moss (1987) "Multicenter post infarction group: decreased heart rate variability and its association with increased mortality after acute myocardial infarction", *Am. J. Cardiol.* 59:256-262.

Làdavas, E., D. Cimatti, M. Del Pesce and G. Tuozzi (1993) "Emotional evaluation with and without conscious stimulus identification: evidence from a Split-brain patient", *Cognition and Emotion* 7:95-114.

Lane, R.D., J.D. Wallce, P.P. Petrosky, G.E. Schwartz and A.H. Gradman (1992) "Supraventricular tachycardia in patients with right hemisphere strokes", *Stroke* 23:362-366.

Levy, M.N. and P.J. Martin (1979) "Neural control of the heart", in: *Handbook of Physiology, Sect. 2: The Cardiovascular System, voL 1,* Bethesda, MD: American Physiological Society, pp.581-620.

Levy, M.N., M.L. Ng and H. Zieske (1966) "Functional distribution of the peripheral cardiac sympathetic pathways", *Circulation Res.* 19:650-661.

Mammucari, A., C. Caltagirone, P. Ekman, W. Friesen, G. Gainotti, L. Pizzamiglio and P. Zoccolotti (1988) "Spontaneous facial expression of emotions in brain-damaged patients", *Cortex* 24:521-533.

Meadows, M.E. and R.F. Kaplan (1994) "Dissociation of autonomic and subjective responses to emotional slides in right hemisphere damaged patient", *Neuropsychologia* 32:847-856.

Mizeres, N.J. (1958) "The origin and course of the cardioaccelerator fibers in the dog", *Anatomical Record* 132:261-279.

Morrow, L., B. Vrtunski, Y. Kim and F. Boller (1981) "Arousal response to emotional stimuli and laterality of lesion", *Neuropsychologia* 19:65-71.

Naver, H.K., C. Blomstrand and G. Wallin (1996) "Reduced heart rate variability after right-sided stroke", *Stroke* 27:247-251.

Oatley, K. and P. Johnson-Laird (1987) "Toward a cognitive theory of emotions", *Cognition and Emotion* 1:29-50.

Oppenheimer, S.M., A. Gelb, J.P. Girvin and V.C. Hachinski (1992) "Cardiovascular effects of human insular cortex stimulation", *Neurology* 42:1727-1732.

Oppenheimer, S.M. and D.A. Hopkins (1994) "Suprabulbar neuronal regulation of the heart", in: *Neurocardiology,* J.A. Armour and J.L. Ardell, eds, New York: Oxford University Press, pp.309-341.

Rosen, A.D., R.C. Gur, N. Sussman, R.E. Gur and H. Hurtig (1982) "Hemispheric asymmetry in the control of heart rate", *Abstracts Soc. Neurosci.* 8:917.

Rossi, G.F. and G. Rosandini (1967) "Experimental analysis of cerebral dominance in man", in: *Brain Mechanisms Underlying Speech and Language,* C.H. Millikan and F.I. Darley, eds, New York: Grune & Stratton Inc., pp.167-174.

Sander, D. and J. Klingelhöfer (1995) "Changes of circadian blood pressure patterns and cardiovascular parameters indicate lateralization of sympathetic activation following hemispheric brain infarction", *J. Neurol.* 242:313-318.

Schilder, P. (1935) *The Image and Appearance of the Human Body*, London: K. Paul.

Shallice, T. (1988) *From Neuropsychology to Mental Structure,* Cambridge: Cambridge University Press.

Spence, S., D. Shapiro and E. Zaidel (1996) "The role of the right hemisphere in the physiological and cognitive components of emotional processing", *Psychophysiology* 33:112-122.

Terzian, H. and C. Ceccotto (1959) "Su un nuovo metodo per la determinazione e lo studio della dominanza emisferica", *Giornale Psichiatr. Neuropatol.* 87:889-924.

Teuber, H.L. (1955) "Physiological psychology", *Ann. Rev. Psychol.* 9:267-276.

Wittling, W. (1990) "Psychophysiological correlates of human brain asymmetry: blood pressure changes during lateralized presentation of an emotionally laden film", *Neuropsychologia* 28:457-470.

Wittling, W. (1995) "Brain asymmetry in the control of autonomic-physiologic activity", in: *Brain Asymmetry,* R.J. Davidson and K. Hugdahl, eds, Cambridge, MA: MIT-Press, pp.305-357.

Wittling, W. (1997) "Brain asymmetry and autonomic control of the heart", *European Psychologist* 2:313-327.

Wittling, W., A. Block, S. Genzel and E. Schwiger (1998) "Hemisphere asymmetry in parasympathetic control of the heart", *Neuropsychologia* 36:461-468.

Wittling, W. and M. Pflüger (1990) "Neuroendocrine hemisphere asymmetries: salivary cortisol secretion during lateralized viewing of emotion-related and neutral films", *Brain and Cognition* 14:243-265.

Yokoyama, K., R. Jennings, P. Ackles, B.S. Hood and F. Boller (1987) "Lack of heart rate changes during attention-demanding tasks after right hemisphere lesions", *Neurology* 37:624-630.

Yoon, B.W., C.A. Morillo, D.F. Cechetto and V. Hachinski (1997) "Cerebral hemispheric lateralization in cardiac autonomic control", *Arch. Neurol.* 54:741-744.

Zamrini, E.Y., K.J. Meador, D.W. Loring, F.T. Nichols, G.P. Lee, R.E. Figueroa and W.O. Thompson (1990) "Unilateral cerebral inactivation produces differential left/right heart rate responses", *Neurology* 40:1408-1411.

Zoccolotti, P., C. Caltagirone, N. Benedetti and G. Gainotti (1986) "Perturbation des réponses végétatives aux stimuli émotionnels au cours des lésions hémisphériques unilatéralesæ, *Encéphale* 12:263-268.

Zoccolotti, P., D. Scabini, D. and V. Violani (1982) "Electrodermal responses in patients with unilateral brain damage", *J. Clin. Neuropsychol.* 4:143-150.

HIERARCHICAL ORGANIZATION OF EMOTIONAL EXPERIENCE AND ITS NEURAL SUBSTRATES[1]

RICHARD D. LANE
Department of Psychiatry, University of Arizona College of Medicine, Tucson, Arizona 85721, USA

ABSTRACT

The ability to recognize and describe emotion in oneself and others, called emotional awareness, may be conceptualized as a cognitive skill that undergoes a developmental process similar to that which Piaget described for cognition in general. The five levels of emotional awareness in ascending order are physical sensations, action tendencies, single emotions, blends of emotion, and blends of blends of emotional experience. The first two levels are implicit in that they constitute sensori-motor representations that may not be considered conscious emotional experiences per se. The latter three levels constitute explicit mental representations of experience. These five levels put unconscious and conscious processes on a continuum characterized by progressively increasing degrees of differentiation and complexity of the schemata used to process emotional information. They are hierarchically related in that functioning at each level adds to and modifies the function of lower levels. Psychometric and behavioral data supporting this conceptual framework are presented. Next, the work of other investigators is discussed, demonstrating the role of subcortical structures in implicit processing of emotional information. Functional neuroimaging studies of the neural correlates of emotional experience are presented which suggest that subregions within the anterior cingulate cortex may play a differential role in phenomenal and reflective conscious awareness of emotion, respectively. Parallels in the hierarchical organization of function at the psychological and neuroanatomical levels are discussed.

1. Introduction

In this chapter an approach to the study of emotional experience is presented, along with results of studies exploring this approach with psychometric, behavioral and functional neuroimaging methodologies. A framework for conceptualizing the relationship between emotional phenomena at different levels of functional organization is also presented. Based on the present author's findings and those of other investigators, a neuroanatomical model is described that addresses the distinction between implicit and explicit emotional processes. The general goal is to provide a unifying framework that will potentially

[1] Substantial portions of the present chapter are from: Lane, R.D. (2000) «Neural correlates of conscious emotional experience», in: *Cognitive Neuroscience of Emotion*, R.D. Lane, L. Nadel, G.L. Ahern, J.J.B. Allen, A.W. Kaszniak, S.Z. Rapcsak and G.E. Schwartz, eds., New York: Oxford University Press, pp. 345-370.

contribute to a theory of emotion that includes both unconscious emotional processes and conscious emotional experience.

2. Levels of Emotional Awareness

Emotions may be defined as information about the extent to which an individual is successful in achieving goals in interaction with the environment (Ortony, Clore, & Collins, 1988). This definition potentially encompasses both implicit and explicit emotional phenomena. Implicit phenomena include information in the form of facial expressions, gestures and posture, which convey messages to others in the environment. Explicit phenomena, such as conscious emotional experience, somatic experience, or recall of past emotional experiences, constitute internally-directed information.

Damasio (1994) notes that an advantage of conscious awareness of emotion is that it allows emotional information to be integrated with cognitive processes. If emotions were always nonconscious it would not be possible to voluntarily control emotional responses. If emotions are conscious, it is possible to think ahead, plan and generalize to similar but unfamiliar situations. Planning ahead requires drawing on past experiences as reference points. Thus, consciousness extends time from the present into both the past and future. Conscious awareness therefore affords flexibility of present response based on the history of an individual's unique interactions with the environment. This flexibility includes the capacity for emotional control.

To the extent that conscious and nonconscious aspects of emotion can be conceptually and empirically dissociated, specific neural correlates of conscious emotional experience would be anticipated. In order to search for such neural correlates, a method for measuring awareness of emotions is needed. Assessment of emotional experience typically involves asking subjects to rate the intensity (Izard, 1972; Watson, Clark, & Tellegen, 1988) or frequency (Larsen & Diener, 1987) of a given emotion on an ordinal scale. Although useful in certain contexts, this approach is prone to error for a variety of reasons, including other-deception, self-deception (Paulhus, 1985) or some other distortion or failure in retrospective memory. Thus, measures involving emotional state are not optimal. An alternative approach is to measure the trait ability to be aware of emotions in a way that does not rely on the accuracy of self-reports. An advantage of a between-subject or individual differences approach is that it is potentially applicable to a variety of clinically relevant phenomena in the domains of mental and physical health (Lane & Schwartz, 1987).

Lane and Schwartz (1987) proposed that an individual's ability to recognize and describe emotion in oneself and others, called emotional awareness, is a cognitive skill that undergoes a developmental process similar to that which Piaget described for cognition in general. A fundamental assumption of this model is that individual differences in emotional awareness reflect variations in the degree of differentiation and integration of the schemata used to process emotional information, whether that information comes from the external world or the internal world through introspection. To the extent that awareness of emotional information is adaptive, it follows that the more information one has about one's emotional state, the greater the potential to use this information in achieving adaptation success.

Lane and Schwartz (1987) posit five «levels of emotional awareness» which share the structural characteristics of Piaget's stages of cognitive development. The five levels of emotional awareness, in ascending order, are physical sensations, action tendencies, single emotions, blends of emotion, and blends of blends of emotional experience (the capacity to appreciate complexity in the experience of self and other). These levels describe the organization of experience. They describe traits, although they may also be used to describe states. The levels are hierarchically related in that functioning at each level adds to and modifies the function of previous levels but does not eliminate them. For example, level 4 experiences should be associated with more differentiated somatic sensations (level 1) than level 2 experiences. A given emotional experience can be thought of as a construction consisting of each of the levels of experience up to and including the highest level attained.

The term «structural characteristics» refers to the degree of differentiation and integration of the cognitive schemata used to process emotional information. This is a different use of the term «structure» from that of Ortony and colleagues (Ortony *et al.*, 1988), who use it to refer to the determinants of the specific kind of emotion activated. In the present context the term «structure» is used to refer to the degree of complexity of the emotion cues which can be perceived, or the nature of the cognitive processing of an emotional experience, once it has been activated. The development of schemata is driven by words or other representation modes that are used to describe emotion. This perspective draws on the symbol formation work by Werner and Kaplan (1963), who maintained that things in the world become known to an observer by virtue of the way in which they are symbolically represented. Thus, the nature of conscious emotional experience, and the ability to appreciate complexity in one's own experience and that of others, is influenced by what one knows about emotion, which itself is based on how emotion has been represented in the past.

Wine tasting can be used to illustrate this association between language and conscious awareness. Solomon (1990) compared novice and expert wine tasters in their ability to describe wines, discriminate between wines, and match written descriptions of the wines. Expert tasters used more descriptors and dimensions to describe wine than the novices. Furthermore, experts were successful in rank ordering wines based on sweetness, balance (proportion of sugar to acid) and tannin (astringency), while novices successfully rank-ordered wines based on sweetness only. In addition, descriptions of wines by experts were more successfully matched to the wines themselves when read by other experts than by novices. These results are consistent with the position that a given wine is experienced differently by an expert than a novice taster. The greater precision that experts demonstrate in describing wines may not just reflect their greater knowledge, but may also contribute to their more precise discriminative performance.

It is important to note that just as a person can still taste (experience) wine despite a complete lack of words to describe it, language is not necessary for conscious experience. It is known, for example, that intelligent thought is possible in prelinguistic children (Mehler & Dupoux, 1996), in adults without language, as -in so-called "linguistic isolates" who have grown up without language (Schaller, 1992), and in certain patients with aphasia and intact intellectual faculties (Ross, 1993). However, language can help to structure and establish concepts. These concepts can modify the allocation of attentional resources and thus the contents of conscious experience. Language, therefore, can enhance discriminative performance, whether it involves wine tasting or identifying emotions. To the extent that similar conceptual and attentional processes are involved, the parallels between increasing emotional and cognitive complexity can be readily understood.

3. Psychometric Characteristics of The Levels of Emotional Awareness Scale

The Levels of Emotional Awareness Scale (LEAS) is a written performance measure that asks the subject to describe his or her anticipated feelings and those of another person in each of twenty scenes described in two to four sentences (Lane, Quinlan, Schwartz *et al.*, 1990). Scoring is based on specific structural criteria aimed at determining the degree of differentiation in the use of emotion words (the degree of specificity in the terms used and the range of emotions described) and the differentiation of self from other. The details of scoring are described elsewhere. The following provides an example of a scene from the LEAS and responses that are scored at each level :

> You and your best friend are in the same line of work. There is a prize given annually to the best performance of the year. The two of you work hard to win the prize. One night the winner is announced: your friend. How would you feel? How would your friend feel?
>
> Examples of responses at each level:
>
> 0. I don't work hard to win "prizes." My friend would probably feel that the judges knew what they were doing.
> 1. I'd feel sick about it. It's hard for me to say what my friend would feel - it would all depend on what our relationship was like and what the prize meant to her.
> 2. I'd probably feel bad about it for a few days and try to figure out what went wrong. I'm sure my friend would be feeling

really good.

3. We would both feel happy. Hey, you can't win 'em all!
4. I would feel depressed - the friend in this light is just like any other competitor. I would also begrudgingly feel happy for my friend and rationalize that the judges had erred. My friend would feel very gratified but would take the prize in stride to save the friendship.
5. I'd feel disappointed that I didn't win but glad that if someone else did, that person was my friend. My friend probably deserved it! My friend would feel happy and proud but slightly worried that my feelings might be hurt.

The scoring requires essentially no inference by raters. Thus, the LEAS can be thought of as a performance measure of the ability to put feelings into words, which, based on the theoretical considerations reviewed above, should reflect the complexity of experience. Furthermore, since the scoring system evaluates the structure of experience and not its content, subjects cannot modify their responses to enhance their score, as is the case with some self-report instruments.

To date, eight separate psychometric studies have been conducted with the LEAS. The first study of Yale undergraduates (n=94) enabled us to examine the reliability of the LEAS and its correlation with other psychological tests (Lane *et al.*, 1990). The second study involved students at Chicago Medical School (CMS) (n=57) and focused on the correlation with the Levy Chimeric Faces Test (Lane, Kevley, DuBois *et al.*, 1995). The third study in Arizona and Minnesota (n=385) established norms for the scale (Lane, Sechrest, Riedel *et al.*, 1996). A fourth study with University of Arizona undergraduates (n=215) involved additional psychometric and psychophysiologic assessments. The fifth and sixth studies have been conducted in collaboration with Dr. Lisa Feldman-Barrett at Boston College. In addition, two international studies have been conducted: a study of 331 German students (Wrana, Thomas, Heindichs *et al.*, 1998) and a Canadian study of 30 subjects with borderline personality disorder and 40 control subjects (Levine, Marziali, & Hood, 1997). The findings from these studies are selectively reviewed below.

The LEAS has consistently been shown to have high inter-rater reliability and internal consistency (Lane, Reiman, Axelrod *et al.*, 1998). An adequate assessment of test-retest reliability of the LEAS in the general population has not been undertaken. Norms for age, sex and socioeconomic status have been established based on the study completed in Arizona and Minnesota.

In the Yale study we administered two instruments which, like the LEAS, are cognitive-developmental measures based on Piaget's model: the Sentence Completion Test of Ego Development by Loevinger (Loevinger & Wessler, 1970a; Loevinger, Wessler, & Redmore, 1970b) and the cognitive complexity of the description of parents by Blatt and colleagues (Blatt, Wein, Chevron, &

Quinlan, 1979). The LEAS correlated moderately (r=.37 and r=.36, respectively) and significantly (p<.01) in the predicted direction in both cases. These results support the claim that the LEAS is measuring a cognitive-developmental continuum and that the LEAS is not identical to these other measures.

One might question whether the LEAS is simply another measure of verbal ability. In the Yale sample the LEAS correlated r=.36 (p<.00l) with the vocabulary subtest of the WAIS. In the CMS study the LEAS correlated r=.17 (NS) with the Shipley Institute of Living Scale (Shipley, 1940), a multiple choice measure of verbal ability. These data suggest that verbal ability may contribute to LEAS performance. However, several studies have now been conducted demonstrating that when verbal ability is controlled significant effects are still observed.

For example, LEAS scores in men and women could be compared in all eight studies. In three of these studies measures of verbal ability, including the WAIS vocabulary subtest and the Shipley Institute of Living Scale, were also obtained. In each study women scored higher than men on the LEAS (p<.01), even when controlling for verbal ability (p<.05) (Barrett, Lane, Sechrest, & Schwartz, in press) . Thus, the finding that women score higher than men on the LEAS is a highly stable and generalizable finding. These data suggest that on average women are more sensitive to emotion cues in themselves and others than are men. This greater sensitivity has clear advantages in the realm of interpersonal relations and problem-solving, but may also contribute to the finding that women are approximately twice as likely to suffer from affective and anxiety disorders than are men (Gater, Tansella, Korten *et al.*, 1998; Breslau, Davis, Andreski *et al.*, 1997).

Lisa Feldman-Barrett administered the LEAS and the Weinberger Adjustment Inventory to 63 subjects at Penn State and 55 subjects at Boston College. In both samples the LEAS correlated significantly (p<.05, 2-tailed) with self-restraint, one of three superordinate dimensions of the scale. The LEAS also correlated significantly with impulse control, (r-- .35, p< .01, 2-tailed and r= .30, p< .05, 2-tailed), a component of self-restraint that involves the tendency to think before acting. Self-restraint refers directly to suppression of egoistic desires in the interest of long-term goals and relations with others. This replication in independent samples indicates that greater emotional awareness is associated with greater self-reported impulse control, and is consistent with the theory that functioning at higher levels of emotional awareness (levels 3-5) modulates function at lower levels (actions and action tendencies at level 2).

Evidence for the discriminant validity of the LEAS is provided by data from the Norms study and the Arizona undergraduate study. In both studies (n=385 and n--215, respectively) the Affect Intensity Measure (Larsen *et al.*, 1987), a trait measure of the tendency to experience emotions intensely, did not correlate significantly with the LEAS despite the large sample sizes. Thus, inadequate statistical power cannot explain the lack of correlation. The LEAS also does not correlate significantly with measures of negative affect, such as the Taylor Manifest Anxiety Scale and the Beck Depression Inventory. These results are consistent with the view that the LEAS measures the structure or complexity and not the intensity of affective experience.

4. The Levels of Emotional Awareness Scale: Behavioral Findings

A key assumption in this work on emotional awareness is that language promotes the development of schemata for the processing of emotional information, whether that information comes from the internal or external world. Furthermore, once the schemata are established they should affect the processing of emotional information whether the information is verbal or nonverbal. Thus, the LEAS should correlate with the ability to recognize and categorize external emotional stimuli. Furthermore, this correlation should hold whether the external stimulus and the response are purely verbal or purely non-verbal.

These hypotheses were tested in the Norms study by use of the Perception of Affect Task (PAT), a set of four emotion recognition tasks (35 items each) developed by Jim Rau and Alfred Kaszniak at the University of Arizona (Rau, 1993). The first subtask consists of stimuli describing an emotional situation without the use of emotion words. For example, «The man looked at the photograph of his recently departed wife.» The response involves choosing one from an array of seven terms (happy, sad, angry, afraid, disgust, neutral, surprise) to identify how the person in question was feeling. The fourth subtask is purely nonverbal. The stimuli consist of photographs of faces developed by Ekman (1982), each of which depicts an individual emotion. The response consists of selecting one from an array of seven photographs depicting emotional scenes without faces (e.g. two people standing arm-in-arm by a grave with their backs to the camera) . The other two subtasks involved a verbal stimulus (sentence) and a non-verbal response (from an array of seven faces) and a nonverbal stimulus (face) and a verbal response (from an array of seven words)

Across the entire scale, the correlation between the LEAS and the PAT was highly significant ($r=.43$, $n=385$, $p<.001$), accounting for about 18% of the variance. Furthermore, significant correlations were observed between the LEAS and each of the PAT subtasks. When dividing the sample into upper (high), middle and lower (low) thirds on the LEAS, the high LEAS subjects scored higher on each of the PAT subtasks than the low LEAS subjects. Thus, high LEAS scores were associated with better emotion recognition no matter whether the task was purely verbal or purely nonverbal (Lane *et al.*, 1996). Furthermore, when combining results for each of the seven emotion categories across the four subtasks (there were five stimuli of each emotion type in each subtask), the same findings for high, moderate and low LEAS subjects were observed (Lane, Shapiro, Sechrest, & Riedel, 1998). These findings support the claim that the LEAS is: 1) a measure of the schemata used to process emotional information, whether the information is verbal or nonverbal; 2) a measure of the complexity of experience; 3) not simply a measure of verbal ability.

A study performed at Chicago Medical School was our first attempt to relate the LEAS to brain function. Given that the LEAS is a psychological measure of an individual difference variable, we were interested in determining whether the LEAS correlated with individual differences in an aspect of brain function associated with the processing of emotional information. We selected the right hemispheric dominance (among right-handers) in the perception of facial emotion, in part because it has been consistently observed and in part because there are individual differences in the degree of lateralization of this function which are not well understood (Levy, Heller, Banich, & Burton, 1983a)

The measure of hemispheric dominance in the perception of facial emotion which we chose was the Levy Chimeric Faces Test (Levy, Heller, Banich, & Burton, 1983b) . This test consists of 36 chimeric or composite faces depicting a smiling half-face juxtaposed to a neutral half-face from the same subject. This composite is paired with its mirror image in a vertical array. The only difference between the two composites is whether the smile is in the left or the right visual field. The subject is asked to indicate whether the «strange picture» on the top or the bottom looks happier. Other studies have shown that the right hemispheric dominance (a preference for selecting the composite with the smile in the left visual field) on this task is consistently observed whether the stimuli are presented in free field in a group format, individually in a booklet format, or individually by tachistoscope. Furthermore, the right hemispheric advantage has been demonstrated using composite photographs consisting of sad as well as happy half-faces.

The results showed that the LEAS correlated significantly with the degree of right hemispheric advantage in performance of the LCFT ($r=.36$, $p<.05$) . Interestingly, the correlation between the degree of right hemispheric dominance and the LEAS improved ($r=.44$, $p<.003$) when restricting the sample to native English speakers (presumably because a measure completed in English is a more accurate measure of underlying schemata if completed in the subjects' native language) and when controlling for verbal ability using the Shipley Institute of Living Scale (Shipley, 1940). These data are consistent with the hypothesis that people who are more highly emotionally aware tend to preferentially use the hemisphere that is specialized for the detection of emotion cues.

5. Neural Correlates of Emotional Awareness

To further explore the functional neuroanatomy of emotional awareness, we administered the LEAS to subjects participating in a positron emission tomography (PET) study of emotion (Lane, Reiman *et al.*, 1998). Subjects included twelve right-handed female volunteers who were free of medical, neurological, or psychiatric abnormalities. The LEAS and other psychometric instruments were completed prior to PET imaging. Happiness, sadness, disgust, and three neutral control conditions were induced by film and recall of personal experiences (12 conditions). Twelve PET images of blood flow were obtained in each subject using the ECAT 951/31 scanner (Siemens, Knoxville, TN), 40 mCi intravenous bolus injections of ^{15}O-water, a 15 second uptake period, 60 second scans, and an interscan interval of 10 minutes.

To examine neural activity attributable to emotion generally, rather than to specific emotions, one can subtract the 3 neutral conditions from the 3 emotion conditions in a given stimulus modality (film or recall) . This difference, which can be calculated separately for the 6 film and 6 recall conditions, identifies regions of the brain where blood flow changes specifically attributable to emotion occur. These blood flow changes, which are indicative of neural activity in that region, can then be correlated with LEAS scores to identify regions of the brain that are associated with emotional awareness during emotional arousal.

Findings from this covariate analysis revealed one cluster for film-induced emotion with a maximum located in the right mid-cingulate cortex (BA 23; coordinates of maximum = [16, -18, 32]; $z=3.40$; $p<.001$ uncorrected) . For recall-induced emotion, the most statistically significant cluster was located in the right anterior cingulate cortex (BA 24; coordinates of maximum = [16, 6, 30]; $z=2.82$; $p<.005$ uncorrected) . A conjunction analysis was performed next to identify areas of significant overlap between the two covariance analyses. With a height threshold of $z=3.09$, $p<.001$, and an extent threshold of 5 voxels, a single cluster was observed in the right anterior cingulate cortex (BA 24) maximal at coordinates [14, 6, 30] ($z=3.74$, $p<.001$; $p= 9.2 \times 10^{-5}$; uncorrected). The point of maximum change was located in white matter adjacent to the anterior cingulate cortex. Given that blood flow changes in white matter are unlikely, the imprecision in anatomical localization associated with image normalization, the extension of the area of significant change into the anterior cingulate cortex, and the absence of other grey matter structures in the immediate vicinity, the likeliest location of this cluster is the anterior cingulate cortex (Lane, Reiman *et al.*, 1998).

The anterior cingulate cortex is a complex structure with numerous functions which «are difficult to quantify or even describe» (Vogt, Finch, & Olson, 1992). Traditionally the anterior cingulate cortex was thought to have a primarily affective function (Papez, 1937; Vogt *et al.*, 1992). However, in addition to emotion, it is now recognized to play important roles in attention, pain, response selection, maternal behavior, vocalization, skeletomotor function, and autonomic control (Vogt & Gabriel, 1993). The multiple functions of the anterior cingulate cortex likely contribute to the significant changes in activation that have been observed in a variety of studies. How can these different functions be reconciled with the present findings involving emotional awareness?

One answer might be that these various functions of the anterior cingulate cortex reflect its superordinate role in executive control of attention and motor responses (Lane, Reiman *et al.*, 1998). According to this view, emotion, pain or other salient exteroceptive or interoceptive stimuli provide moment-to-moment guidance regarding the most suitable allocation of attentional resources for the purpose of optimizing motor responses in interaction with the environment. The conscious experience of emotion could occur concomitantly and automatically as attention gets redirected by emotion. As such, a role of the anterior cingulate cortex in the conscious experience of emotion fits well with its other functions, but suggests that this role is not exclusive to emotion. To the extent that people who are more emotionally aware attend more to internal and external emotion cues, the cognitive processing of this information can contribute to ongoing emotional development.

6. Attention to Emotional Experience

In his seminal work on blindsight, Weiskrantz (1986) distinguishes between the fundamental constituents of a network mediating a function and the neural structures involved in commenting on or reflecting upon it. Blindsight is a condition caused by a lesion in the primary visual cortex in which patients are not consciously aware of visual stimuli but demonstrate behaviorally that the stimulus is perceived (Weiskrantz, 1986). The lessons learned from the study of blindsight patients in the aware and unaware states are potentially applicable to the understanding of the neural substrates of emotional awareness.

Weiskrantz (2000) raises the question of whether the dorsal anterior cingulate cortex is the final output or commentary stage at which awareness of emotion can be expressed and registered. A recent functional magnetic resonance imaging (fMRI) study of blindsight in the aware and unaware states (Sahraie, Weiskrantz, Barbur *et al.*, 1997), for example, reveals that dorsolateral prefrontal cortex is preferentially activated during visual processing in the aware state. We conducted a PET study that generated new data concerning the neural correlates of attention to emotional experience.

In this study (Lane, Fink, Chua, & Dolan, 1997b) we examined the pattern of neural activation associated with attending to one's own emotional experience. We used a selective attention paradigm based on the rationale that selective attention heightens activity in those regions that mediate a particular function (Corbetta, Miezin, Dobmeyer *et al.*, 1990; Fink, Halligan, Marshall *et al.*, 1996a). Thus, we reasoned that attention to one's own experience would activate those brain regions that are preferentially invloved in the conscious experience of emotion. To confirm that subjects were allocating their attention as we instructed, we had them indicate on a keypad how each emotion-evoking picture made them feel. In essence, we were examining an aspect of conscious experience involving commentary on that experience.

We studied ten healthy men as they viewed twelve picture sets, each consisting of pleasant, unpleasant and neutral pictures from the International Affective Picture System (Lang, Bradley, & Cuthbert, 1995). Pictures were presented for 500 msec every 3.0 seconds. Twelve PET-derived measures of cerebral blood flow were -obtained in each subject, one for each picture set. During half the scans subjects attended to their emotional experience (indicating on a keypad whether the picture evoked a pleasant, unpleasant or neutral feeling); during the other half they attended to spatial location (indicating whether the scene depicted was indoors, outdoors, or indeterminate). Across subjects, picture sets were counterbalanced across the two attention conditions.

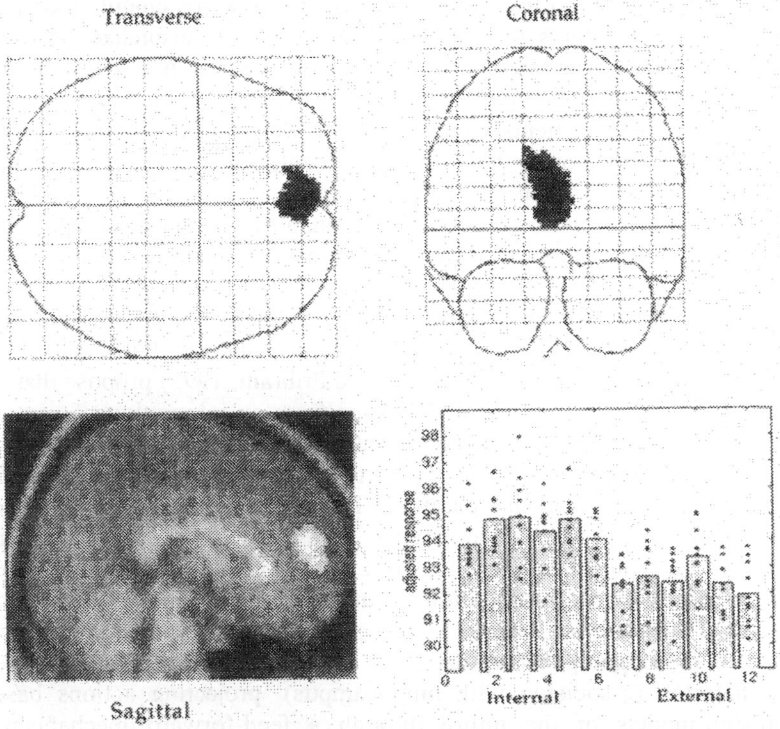

Figure 1. A statistical parametric map (SPM) showing significant cerebral blood flow increases in anterior cingulate cortex (BA32)/medial prefrontal cortex (BA9) during selective attention to subjective emotional responses (minus activations specific to the external focus condition). The figures in the left and right upper portion are projection images in the transverse and coronal planes, respectively. The sagittal view in the lower left depicts the spatial distribution of the activation in the internal focus condition ($z = 6.87$, $p < 0.001$, corrected) superimposed on the average structural MRI of the 10 male subjects. The figure in the lower right demonstrates blood flow values in each condition (internal: 1 – 6; external: 7 – 12).

During attention to subjective emotional experience, increased neural activity was elicited in rostral anterior cingulate cortex (BA32) and medial prefrontal cortex (coordinates: 0,50,16; $Z=6.74$, $p<.001$, corrected) (see Figure 1), right temporal pole, insula and ventral cingulate cortex (all $p<.001$, corrected) . Under the same stimulus conditions when subjects attended to spatial aspects of the picture sets, activation was observed in parieto-occipital cortex bilaterally ($Z=5.71$, $p<.001$, corrected), a region known to participate in the evaluation of spatial relationships.

Our interpretation of these findings is that the rostral anterior cingulate/medial prefrontal activation may reflect a representation of emotional state. Several lines of evidence support this view. This region is densely connected with the amygdala, orbitofrontal cortex, other sectors of the anterior cingulate cortex, and other paralimbic structures such as the insula (Price, Carmichael, & Drevets, 1996). It thus clearly receives information about the emotional significance of stimuli. Second, lesions in this area produce a blunting of emotional experience (e.g., in schizophrenic patients who underwent prefrontal leukotomy; Hoffman, 1949). Third, the neighboring dorsolateral prefrontal cortex clearly participates in working memory, keeping information temporarily «on-line» for use in cognitive operations (Goldman-Rakic, 1987). It is reasonable to hypothesize that a similar function may exist for interoceptive emotional information in a neighboring sector of prefrontal cortex.

Additional support for this hypothesis is provided by considering the strategic location of this area. Drawing on the work of Sanides (1970) and Goldberg (1987), Tucker and collegues (Tucker, Luu, & Pribram, 1995) propose that frontal lobe control of motivational impulses results from an integration of ventral and dorsal corticolimbic pathways. The ventrolateral pathway is derived from paleocortex (associated with olfactory cortex). The amygdala and orbitofrontal cortex are key structures in this pathway. This system links motor sequences to perceptual objects in a responsive manner. It restricts and monitors motivational impulses through a feedback mechanism. Lesions of the orbitofrontal cortex are typically associated with disinhibition, as in the case of Phineas Gage (Damasio, Grabowski, Frank *et al.*, 1994). Motor planning is articulated with specific reference to the ongoing perceptual input. The mediodorsal system is derived from archicortex (associated with hippocampus), projecting actions based on probabilistic models of the future through a feed-forward mechanism. The hippocampus and anterior cingulate cortex are key structures in this system. Action is based on a preexisting model rather than ongoing feedback about the course of action in the situation. Lesions in this area are associated with apathy and indifference. Thus, there may be reactive-based and planning-based motivational systems that participate in the regulation of behavior by the frontal lobe (see also Mega, Cummings, Salloway, & Malloy, 1997).

The area of rostral anterior cingulate/medial prefrontal cortex that we identified appears to be precisely situated between these two systems. Clearly, a representation of current emotional state facilitates guidance of current behavior and planning future behavior. The observation that greater emotional awareness is associated with greater impulse control may be particularly relevant in this context, as impulsiveness involves a failure to consider the future in the guidance of current behavior. A lack of impulse control is certainly evident in patients with frontal lobe lesions (cf. Phineas Gage) and is consistent with the findings of Morgan and colleagues (Morgan, Romanski, & LeDoux, 1993) that extinction of conditioned fear is greatly prolonged with lesions of the ventromedial prefrontal cortex (but see Gewirtz, Falls, & Davis, 1997). Cytoarchitectural studies reveal a gradual change, from caudal to rostral, in laminar characteristics from limbic periallocortex toward isocortical areas in medial prefrontal cortex (Barbas & Pandya, 1989), suggesting that a rigid distinction between rostral anterior cingulate cortex and medial prefrontal cortex is misleading. The midline location is consistent with other evidence that responses generated from internal cues are associated with activation of midline structures and responses generated from external cues are associated with activation of lateral structures (Chen, Thaler, Nixon *et al.*, 1995).

The findings from this study can therefore be interpreted as follows. When attending to one's own emotional state, several brain areas are activated including those involved in: 1) establishing a representation of the emotional state (rostral anterior cingulate/medial prefrontal cortex); 2) processing visceral information (anterior insula) (Augustine, 1996); 3) performing complex visual discrimination, possibly including retrieval of emotion-laden episodic memories (right temporal pole) (Fink, Markowitsch, Reinkemeier *et al.*, 1996b); and 4) regulating autonomic responses (ventral cingulate) (Vogt *et al.*, 1992).

7. Phenomenal Versus Reflective Conscious Awareness

It is interesting to consider the possibility that two different areas of the anterior cingulate cortex may be participating in different aspects of conscious emotional experience. What functions may be served by these two areas?

A fundamental distinction in the study of consciousness is that between primary and secondary consciousness (Farthing, 1992). Primary consciousness refers to phenomenal experience, the direct experience of an emotion, the taste of wine or the touch of a hand. Secondary consciousness refers to cognitive operations performed on the contents of primary consciousness (e.g., attending to or reflecting upon the contents of phenomenal awareness). This type of consciousness has also been referred to as metacognition (awareness of awareness) (Jarman, Vavrik, & Walton, 1995) or reflective conscious awareness (Farthing, 1992). It is clear that in order to reflect upon something one must have something (e.g. a representation) to reflect upon.

It is hypothesized that the correlation between dorsal anterior cingulate cortex and level of emotional awareness reflects phenomenal awareness of emotion, and that rostral anterior cingulate cortex participates in reflective awareness of emotion. It is noteworthy that Rainville and colleagues (Rainville, Duncan, Price *et al.*, 1997) demonstrated that dorsal anterior cingulate cortex participates in the affective component of pain, entirely consistent with the first part of this formulation. It is also noteworthy that an area of medial prefrontal cortex very close to that identified in our attention to emotional experience study has been implicated in the representation of the mental state of others, using so-called «theory of mind» tasks (Happe, Ehlers, Fletcher *et al.*, 1996). It is likely that the capacity to establish representations of one's own emotional state in infancy is closely linked with the perception and representation of the emotional state of others, including mother (Gergely & Watson, 1996). Given the similarity of the cognitive process involved, it is also likely that the representations of one's own state and that of others are established in neighboring and interconnected regions.

Much work remains to be done to confirm these hypotheses. Such future research would serve an integrative function. Stuss (1991a, 1991b) has discussed how the prefrontal cortex serves a self-monitoring and regulatory function. Damasio (1994) has discussed how the sense of self may derive in part from the somatovisceral sensations associated with emotion that are integrated with the higher cognitive functions of prefrontal cortex. It will be important to explore the extent to which the rostral anterior cingulate cortex/medial prefrontal cortex serves an exclusively emotional function, or like the dorsal anterior cingulate, appears to serve a superordinate function that may be greatly influenced by but is not necessarily exclusively dedicated to emotion. Such a conclusion is certainly possible in light of the plasticity of higher cortical functions based on the interaction between genetics, environmental experience and habitual modes of behavior (Elman, Bates, Johnson *et al.*, 1996). If this were observed, it would help to explain how emotion-related individual differences arise, and, indeed, contribute to an understanding of the neural substrates of unique individual personalities.

8. Summary, Conclusions, and Clinical Implications

The levels of emotional awareness model proposes that emotional phenomena in humans, at each level of functional organization, are potentially associated with awareness of some type. Thus, the lowest levels of organization, e.g. the autonomic activation (level 1) and action tendencies (level 2) associated with emotional arousal, as well as the higher levels of organization (levels 3-5), are each associated with a type of conscious experience. Level 1 and 2 phenomena, viewed in isolation, would not necessarily be considered indicators of emotion, and self-report inventories of emotional experience, such as the Positive and Negative Affect Scales (PANAS; Watson *et al.*, 1988), do not include many terms indicative of level 1 or 2 phenomena. Thus, in the context of actual level 1 or 2 emotional responses such inventories might falsely indicate that conscious emotional experience is not present. The levels of emotional awareness framework therefore puts conscious and unconscious processes on the same continuum, and at the same time distinguishes between types of unconscious (level 1 vs. level 2) and conscious (level 3 vs. 4 vs. 5) processes.

The major theories of emotion can be classified according to the level of processing upon which they focus. The James-Lange theory of emotion (James, 1884), which hypothesizes that somatic state is a determinant of emotional experience, is a Level 1 theory. Level 2 theories, which are action-oriented, include Darwin's theory that emotional displays serve adaptive functions (Darwin, 1965), or theories such as that of Lang (1993a) that focuses on appetitive/aversive motivational systems. Tomkins theory of individual emotions (Tomkins, 1962) is a Level 3 theory. Level 4 theories focus on blends of emotion, such as Izard's differential emotions theory (Izard, 1972) and Ekman's theory involving patterns of emotions (Ekman, 1982). Cacioppo's theory of bivariate evaluative space (Cacioppo & Berntson, 1994), in which both positive and negative responses can occur to varying degrees in response to a stimulus, is another example of a level 4 theory. Each theory addresses a coherent level of organization. To the extent that this is true, it is quite possible that each level of organization corresponds to an identifiable neurobiological state.

The evidence reviewed above provides the basis for a rudimentary neuroanatomical model of emotional awareness that distinguishes between implicit and explicit levels of function. Following Ladavas (Ladavas *et al.*, 1993), Levels 1 and 2 involve implicit processes that are automatic, modular and cognitively impenetrable. It would appear that subcortical structures participate in the automatic generation of emotional responses associated with absent or diffusely undifferentiated awareness. It may be speculated that the neural substrates of level 1 include the thalamus, hypothalamus, midbrain and brainstem. At Level 2, the sensorimotor enactive level, crude distinctions between globally positive and globally negative states can be made. Given that decorticate cats can demonstrate fear and pleasure reactions (Bard & Rioch, 1937), it is likely that the thalamus participates at this level. The amygdala appears to be preferentially activated in association with aversive stimuli (Tranel, 1997), and the ventral striatum, including the nucleus accumbens, is preferentially activated by appetitive or reward stimuli (Koob & Goeders, 1989). The outputs from this stage of processing are widespread. Emotions at this level are represented in actions such as gestures and other movements that have an either/or quality. Much evidence suggests that the basal ganglia participate in the automatic behavioral displays of emotional gestures and expression (Gray, 1995; Rolls, 1990). Orbitofrontal cortex activity appears to be associated with the perception of somatic sensations that bias behavior either toward or away from a stimulus (Damasio, 1994), and affects behavior by overriding automatic processes in the amygdala and participating in extinction, among other functions (Emery & Amaral, 2000). A key tenet of this model is that structures at this level, such as the amygdala, (Ledoux 1996; Cahill & McGaugh, 1998) are essential for implicit processing, and contribute to, but are not sufficient for, the explicit experience of discrete emotions or combinations thereof (i.e. levels 3-5)

Levels 3-5 involve explicit processes that are influenced by higher cognitive processes, including prior explicit knowledge. They are hypothesized to be mediated by the above structures and, in addition, based on the evidence presented above, by paralimbic structures including the anterior cingulate cortex and insula, and the medial prefrontal cortex. The rostral anterior cingulate cortex/medial prefrontal cortex appears to be necessary for the representation of emotion used in conscious cognition.

The hierarchical nature of this anatomical model, and the parallel hierarchical structure of the psychological model, should be evident. The hierarchical nature of brain structure and function has been recognized for many years. For example, Hughlings Jackson (1932) described the release of lower level functions by lesions higher in the neuraxis. Yakovlev (1948) proposed three levels of nervous system function, including a primitive inner core devoted to arousal and autonomic function, surrounded by a middle layer including the limbic system and basal ganglia and an outer layer, the most recent to emerge phylogenetically, including the neocortexand pyramydal system. This model was further elaborated by MacLean (1990) in his model of the triune brain, involving sequential evolution of the reptilian, palleomammalian and neomammalian brains. The challenge in the years ahead will be to generate a more differentiated and specific model based on the full range of neuroscientific methods, especially functional neuroimaging. Just how cortical and subcortical structures interact to produce these different levels of function remains to be elucidated.

It is interesting to speculate whether the process of representational redescription, as described by Karmiloff-Smith (1992), is mediated at least in part by structures hypothesized to be involved in explicit processing at levels 3-5. It may be precisely because these structures are not *uniquely* devoted to the processing of emotion that makes their emotion-related functions cognitively penetrable. Indeed, the level of emotional awareness of a given individual may be a function of the degree to which these very structures are or are not devoted to processing of emotional information from the internal and external worlds.

Elucidating the present model more fully is likely to have important clinical implications. The observation that patients with psychosomatic disorders had difficulty verbalizing feelings was the guiding clinical problem that led MacLean to expand on the Papez model of emotion (MacLean, 1949). MacLean's thesis was that interference with communication between limbic (visceral brain) and neocortical areas contributed to physical disease. The phenomenon to which MacLean referred is probably best captured currently by the clinical entity called alexithymia (Taylor, Bagby, & Parker, 1991), or «lack of words for emotion.» We have argued elsewhere that alexithymia is associated with emotional arousal in the absence of conscious awareness (Lane, Ahern, Schwartz, & Kaszniak, 1997a). Since greater conscious awareness of emotion is theoretically associated with progressively greater regulatory control of lower level processes, the relative absence of such awareness may be associated with autonomic and neuroendocrine dysregulation (Lane *et al.*, 1997a; Thayer & Lane, in press). The best recent evidence supporting this view comes from the observations that group psychotherapy designed to promote the awareness and expression of emotions (e.g. confronting fears directly) is associated with enhanced survival in patients with recurrent breast cancer (Spiegel, Bloom, Kraemer, & Gottheil, 1989) and malignant melanoma (Fawzy, Fawzy, Hyun *et al.*, 1993).

Alexithymia may be conceptualized as a failure of cognitive elaboration of modular emotion output. The challenge for the psychotherapist is to enable alexithymic individuals to make the transition from implicit to explicit processing of emotional arousal. Alexithymic patients are typically very difficult to treat. If they come for treatment they often do so at the urging of others. Successful therapy requires first educating them about the nature of their problem (Krystal, 1979). The next step is to help them overcome whatever motivational barriers exist in attending to and recognizing their own emotional experiences.

Just as wishes, expectancies and motivational states can influence the processing of exteroceptive stimuli, a history of psychological trauma can lead to motivated avoidance of dealing with one's own emotions that results in a deficit in the capacity for conscious emotional experiences. A safe and supportive interpersonal environment is essential for success in psychotherapy. It is therefore interesting to note that PTSD has been associated with decreased activity in the dorsal anterior cingulate relative to controls (Shin, Kosslyn, McNally *et al.*, 1997). Preliminary findings indicate that successful treatment of PTSD is associated with a return to normal levels of activity in this area (van der Kolk, *et al.*, 1997).

Once emotions are consciously acknowledged and experienced, the process of cognitive elaboration of emotion can then occur so that the origin and meaning of painful and distressing emotions can be understood, elaborated and used to promote adaptive behavior. If not brought to conscious awareness, emotional distress such as anger will be translated into action and/or a peripheral physiological response that may be maladaptive. The capacity for such explicit processing of emotion may indeed modulate the activity of those structures mediating implicit emotional processes. It will be important in the years ahead to explore the functional neuroanatomy of this process, the changes that occur in effective connectivity between brain regions (Buechel & Friston, 1997), and their influence on mental and physical health. In this regard, functional neuroimaging techniques will be useful in examination of the neural mechanisms by which labeling emotions verbally modifies activity of those structures involved in the conscious experience of and conscious reflection upon one's own emotional states.

References

Augustine, J.R. (1996) «Circuitry and functional aspects of the insular lobe in primates including humans», *Brain Research Reviews* 22:229-244.

Barbas, H. and D.N. Pandy (1989) «Architecture and intrinsic connections of the prefrontal cortex in the rhesus monkey», *Journal of Comparative Neurology* 286:353-375.

Bard, P. and D.M. Rioch (1937) «A study of four cats deprived of neocortex and additional parts of the forebrain», *Bulletin of the Johns Hopkins Hospital* 60:73-147.

Barrett, L.F., R.D. Lane, L. Sechrest and G.E. Schwartz (in press) «Sex differences in emotional awareness», *Personality and Social Psychology Bulletin*.

Blatt, S.J., S.J. Wein, E. Chevron and D.M. Quinlan (1979) «Parental representations and depression in normal young adults», *Journal of Abnormal Psychology* 88:388-397.

Breslau, N., G.C. Davis, P. Andreski, E.L. Peterson and L.R. Schultz (1997) «Sex differences in post traumatic stress disorder», *Archives of General Psychiatry* 54:1044-1048.

Buechel, C. and K.J. Friston (1997) «Characterising functional integration», in: *Human Brain Function*, R. S. J. Frackowiak, K. J. Friston, C. D. Frith, R. J. Dolan and J. C. P. Mazziotta, eds., San Diego: Academic Press, pp. 127-140.

Cacioppo, J.T. and G.G. Berntson (1994) «Relationship between attitudes and evaluative space: a critical review, with emphasis on the separability of positive and negative substrates», *Psychological Bulletin* 115:401-423.

Cahill, L. and J.L. McGaugh (1998) «Mechanisms of emotional arousal and lasting declarative memory», *Trends in Neurosciences* 21(7):294-299.

Chen, Y.-C., D. Thaler, P.D. Nixon, C.E. Stern and R.E. Passingham (1995) «The functions of the medial premotor cortex. II. The timing and selection of learned movements», *Experimental Brain Research* 102:461-473.

Corbetta, M., F.M. Miezin, S. Dobmeyer, G.L. Shulman and S.E. Petersen (1990) «Attentional modulation of neural processing of shape, color, and velocity in humans», *Science* 248:1556-1559.

Damasio, A.R. (1994) *Descartes' Error: Emotion, Reason, and the Human Brain.* New York: G.P. Putnam's Sons.

Damasio, H., T. Grabowski, R. Frank, A.M. Galaburda and A.R Damasio, (1994) «The return of Phineas Gage: clues about the brain from the skull of a famous patient», *Science* 264:1102-1105.

Darwin, C. (1965) *The Expression of Emotions in Man and Animals* (2nd edition), Chicago: University of Chicago Press.

Ekman, P. (1982) *Emotion in the Human Face* (2nd ed.), New York: Cambridge University Press.

Elman, J.L., E.A. Bates, M.H. Johnson, A. Karmiloff-Smith, D. Parisi and K. Plunkett (1996) *Rethinking Innateness: a Connectionist Perspective on Development*, Cambridge, MA: MIT Press.

Emery, N.J. and D.G. Amaral,(1999) «The role of the amygdala in social cognition», in: *Cognitive Neuroscience of Emotion*, R.D. Lane, L. Nadel, G.L. Ahern, J.J.B. Allen, A.W. Kaszniak, S.Z. Rapscak and G. E. Schwartz, eds., New York: Oxford University Press, pp. 156-191.

Farthing, G.W. (1992) *The Psychology of Consciousness*, Englewood Cliffs, NJ: Prentice Hall.

Fawzy, F.I., N.W. Fawzy, C.S. Hyun, R. Elashoff, D. Guthrie, J.L. Fahey and D.L Morton (1993) «Malignant melanoma. Effects of an early structured psychiatric intervention, coping, and affective state on recurrence and survival 6 years later», *Archives of General Psychiatry* 50:681-689.

Fink, G.R., P.W. Halligan, J.C. Marshall, C.D. Frith, R.S.J. Frackowiak and R.J. Dolan (1996a) «Where in the brain does visual attention select the forest and the trees?», *Nature* 382:626-628.

Fink, G.R., H.J. Markowitsch, M. Reinkemeier, T. Bruckbauer, J. Kessler and W.D. Heiss (1996b) «Cerebral representation of one's own past: neural networks involved in autobiographical memory», *Journal of Neuroscience* 16:4275-4282.

Fodor, J.A. (1983) *The Modularity of Mind*, Cambridge, MA:MIT Press.

Gater, R., M. Tansella, A. Korten, B.G. Tiemens, V.G. Mavreas and M.O. Olatawura (1998) «Sex differences in the prevalence and detection of depressive and anxiety disorders in general health care settings: report from the World Health Organization Collaborative Study on Psychological Problems in General Health Care», *Archives of General Psychiatry* 55:405-413.

Gergely, G. and J.S. Watson (1996) «The social biofeedback theory of parental affect-mirroring: the development of emotional self-awareness and self-control in infancy», *International Journal of Psychoanalysis* 77:1181-1212.

Gewirtz, J.C., WA. Falls and M. Davis (1997) «Normal conditioned inhibition and extinction of freezing and fear potentiated startle following electrolytic lesions of medial prefrontal cortex in rats», *Behavioral Neuroscience* 111:712-726.

Goldberg, G. (1987) «From intent to action: evolution and function of the premotor systems of the frontal lobe», in: *The Frontal Lobes Revisited,* E. Perecman, ed., New York: IRBN Press, pp. 273-306.

Goldman-Rakic, P.S. (1987) «Circuitry of the primate prefrontal cortex and regulation of behavior by representational memory», in: *Handbook of Physiology. Sect. 1. The nervous system. Vol. 5. Higher functions of the brain, Part 1,* F. Plum, ed., Bethesda, MD: American Physiological Society, pp. 373-417.

Gray, J.A. (1995) «A model of the limbic system and basal ganglia: applications to anxiety and schizophrenia», in: *The Cognitive Neurosciences,* M. S. Gazzaniga, ed., Cambridge, MA: MIT Press, pp. 1165-1176.

Happe, F., S. Ehlers, P. Fletcher, U. Frith, M. Johansson, C. Gillberg, R. Dolan, R. Frackowiak and C. Frith (1996 «Theory of mind in the brain. Evidence from a PET scan study of Asperger syndrome», *Neuroreport* 8(1):197-201.

Hoffman, J.L. (1949) «Clinical observations concerning schizophrenic patients treated by prefrontal leukotomy», *New England Journal of Medicine* 241:233-236.

Izard, C.E. (1972) *Patterns of Emotions*, New York: Academic Press.

Jackson, J.H. (1932) «Relations of different divisions of the central nervous system to one another and to parts of the body», in: *Selected Writings of John Hughlings Jackson*, J. H. Jackson, ed., London: Hodder and Stoughton, Ltd, pp. 422-443.

James, W. (1884) «What is emotion?» *Mind*, 4:118-204.

Jarman, R.F., J. Vavrik and PD. Walton (1995) «Metacognitive and frontal lobe processes: at the interface of cognitive psychology and neuropsychology», *Genetic, Social, and General Psychology Monographs* 121(2):153-210.

Karmiloff-Smith, A. (1992) *Beyond Modularity: A Developmental Perspective on Cognitive Science*, Cambridge, MA: MIT Press.

Koob GF and N.E. Goeders (1989) «Neuroanatomical substrates of drug self-administration», in: *Neuropharmacological Basis of Reward*, J.M. Liebman and S.J. Cooper, eds., New York: Oxford University Press, pp. 214-263.

Krystal, H. (1979) «Alexithymia and psychotherapy», *American Journal of Psychotherapy* 33:17-31.

Kunst-Wilson, MR. and RB. Zajonc (1980) «Affective discrimination of stimuli that cannot be recognized», *Science* 207:557-558.

Ladavas, E., D. Cimatti, M. del Pesce and G. Tuozzi (1993) «Emotional evaluation with and without conscious stimulus identification: evidence from a split-brain patient», *Cognition and Emotion* 7:95-114.

Lane, R.D., G.L. Ahern, G.E. Schwartz and A.W. Kaszniak (1997a) «Is alexithymia the emotional equivalent of blindsight?», *Biological Psychiatry* 42:834-844.

Lane, RD., G.R. Fink, P.M.L. Chau and R.J Dolan, (1997b) «Neural activation during selective attention to subjective emotional responses», *Neuroreport* 8(18):3969-3972.

Lane, R.D., L.S. Kevley, M.A. DuBois, P. Shamasundara and G.E Schwartz (1995) «Levels of emotional awareness and the degree of right hemispheric dominance in the perception of facial emotion», *Neuropsychologia* 33:525-528.

Lane, R.D., D.M. Quinlan, G.E. Schwartz, P.A. Walker and S.B. Zeitlin (1990) «The Levels of Emotional Awareness Scale: A cognitive-developmental measure of emotion», *Journal of Personality Assessment* 55(1&2):124-134.

Lane, R.D., E.M. Reiman, B. Axelrod, L.-S. Yun, A. Holmes and G.E. Schwartz (1998) «Neural correlates of levels of emotional awareness: evidence of an interaction between emotion and attention in the anterior cingulate cortex», *Journal of Cognitive Neuroscience* 10:525-535.

Lane, R.D., L. Sechrest, R.G. Reidel, V. Weldon, A.W. Kaszniak and G.E.

Schwartz (1996) «Impaired verbal and nonverbal emotion recognition in alexithymia», *Psychosomatic Medicine* 58:203-210.

Lane, RD. and G.E. Schwartz (1987) «Levels of emotional awareness: a cognitive-developmental theory and its application to psychopathology», *American Journal of Psychiatry* 144:133-143.

Lane, R.D., D.E. Shapiro, L. Sechrest and R. Riedel (1998) «Pervasive emotion recognition deficit common to alexithymia and repression», *Psychosomatic Medicine* 60:92.

Lang, P.J. (1993a) «The motivational organization of emotion: affect-reflex connections», in: *The Emotions: Essays on Emotion Theory,* S. Van Goozen, M. E. Van de Poll and J. A. Sergeant, eds., Hillsdale, NJ: Erlbaum, pp. 61-93.

Lang, P.J. (1993b) « The Three-System Approach to Emotion», in: *The Structure of Emotion: Psychophysiological, Cognitive and Clinical Aspects,* N. Birbaumer and A. Öhman, eds., Seattle, WA: Hogrefe and Huber Publishers, pp. 18-30.

Lang, P.J., M.M. Bradley and B.N. Cuthbert (1995) *The International Affective Picture System (IAPS): Photographic Slides*, University of Florida: The Center for Research in Psychophysiology.

Larsen, R.J. and E. Diener, (1987) «Affect intensity as an individual difference characteristic: a review», *Journal of Research in Personality* 21:1-39.

Lazarus, R. (1984) «On the primacy of cognition», *American Psychologist* 39:124-126.

LeDoux, J.E. (1996) *The Emotional Brain,* New York: Simon and Schuster.

Levine, D., E. Marziali and J. Hood (1997) «Emotion processing in borderline personality disorders», *Journal of Nervous and Mental Disease* 185(4):240-246.

Levy, J., W. Heller, MT. Banich and LA. Burton (1983a) «Are variations among right-handed individuals in perceptual asymmetries caused by characteristic arousal differences between hemispheres?», *Journal of Experimental Psychology: Human Perception and Performance* 9:329-359.

Levy, J., W. Heller, M.T. Banich and LA. Burton (1983b) Asymmetry of perception in free viewing of chimeric faces», *Brain and Cognition* 2:404-419.

Loevinger, J. and R. Wessler (1970a) *Measuring Ego Development, Vol. I. Construction and Use of a Sentence Completion Test,* San Francisco: Jossey-Bass.

Loevinger, J., R. Wessler and C. Redmore (1970b) *Measuring Ego Development, Vol. II. Scoring Manual for Women and Girls,* San Francisco: Jossey-Bass.

MacLean, P.D. (1949) «Psychosomatic disease and the 'visceral brain': Recent developments bearing on the Papez theory of emotion», *Psychosomatic Medicine* 11:338-353.

MacLean, P.D. (1990) *The Triune Brain in Evolution: Role in Paleocerebral Functions,* New York: Plenum Press.

Mega, MS., J.L. Cummings, S. Salloway and P. Malloy (1997) «The limbic system: an anatomic, phylogenetic, and clinical perspective», *Journal of Neuropsychiatry and Clinical Neurosciences* 9:315-330.

Mehler, J. and E. Dupoux (1996) *What Infants Know: The New Cognitive Science of Early Development*, Cambridge, MA: Blackwell Publishers.

Morgan, MA., L.M. Romanski and J.E. LeDoux (1993) «Extinction of emotional learning: contribution of medial prefrontal cortex», *Neuroscience Letters* 163:109-113.

Morris, J.S., A. Öhman and R.J. Dolan (1998) «Unconscious processing of aversively conditioned stimuli by the human amygdala», *Nature* 393:467-470.

Murphy, S.T. and R.B. Zajonc (1993) «Affect, cognition and awareness: affective priming with optimal and suboptimal stimulus exposures», *Journal of Personality and Social Psychology* 64:723-739.

Ortony, A., G.L. Clore and A. Collins (1988) *The Cognitive Structure of Emotions*, New York: Cambridge University Press.

Öhman, A., A. Flykt and D. Lundqvist (2000) «Unconscious emotion: evolutionary perspectives, psychophysiological data and neuropsychological mechanisms», in: *Cognitive Neuroscience of Emotion*, R.D. Lane, L. Nadel, G.L. Ahern, J.J.B. Allen, A.W Kaszniak, S.Z Rapscak and G. E. Schwartz, eds., New York: Oxford University Press, pp. 296-327.

Papez, J.W. (1937) «A proposed mechanism of emotion», *Archives of Neurology and Psychiatry* 38:725-734.

Paulhus, D.L. (1985) «Self deception and impression management in test responses», in: *Personality Assessment via Questionnaire*, A. Angleitner and J. S. Wiggins, eds., New York: Springer-Verlag, pp. 143-165.

Price, J.L., ST. Carmichael and W.C. Drevets (1996) «Networks related to the orbital and medial prefrontal cortex; a substrate for emotional behavior?», *Progress in Brain Research* 107:523-536.

Rainville, P., G.H. Duncan, D.D. Price, B. Carrier and M.C. Bushnell (1997) «Pain affect encoded in human anterior cingulate but not somatosensory cortex», *Science* 277:968-971.

Rau, J.C. (1993) «Perception of verbal and nonverbal affective stimuli in complex partial seizure disorder», *Dissertation Abstracts International [B]* 54:506-B.

Rolls, E.T. (1990) «A theory of emotion, and its application to understanding the neural basis of emotion», *Cognition and Emotion* 4(3):161-190.

Ross, E.D. (1993) «Acute agitation and other behaviors associated with Wernicke aphasia and their possible neurological bases», *Neuropsychiatry, Neuropsychology, and Behavioral Neurology* 6:9-18.

Sahraie, A., L. Weiskrantz, J.L. Barbur, A. Simmons, S.C.R. Williams and M.J. Brammer (1997) «Pattern of neuronal activity associated with conscious and unconscious processing of visual signals», *Proceedings of the National*

Academy of Sciences U.S.A. 94:9406-9411.

Sanides, F. (1970) «Functional architecture of motor and sensory cortices in primates in the light of a new concept of neocortex evolution», in: *The Primate Brain: Advances in Primatology,* C. R. Noback and W. Montagna, eds., New York: Appleton-Century-Crofts, pp. 137-201.

Schaller, S. (1992) *A Man Without Words*, New York: Summit Books.

Shin, L.M., S.M. Kosslyn, R.J. McNally, N.M. Alpert, W.L. Thompson, S.L. Rauch, M.L. Macklin and R.K. Pitman (1997) «Visual imagery and perception in posttraumatic stress disorder: A positron emission tomographic investigation», *Archives of General Psychiatry* 54:233-241.

Shipley, W. (1940) «A self-administering scale for measuring intellectual impairment and deterioration», *Journal of Psychology* 9:371-377.

Solomon, G.E.A. (1990) «Psychology of novice and expert wine talk», *American Journal of Psychology* 103:495-517.

Sommers, S. (1981) «Emotionality reconsidered: the role of cognition in emotional responsiveness», *Journal of Personality and Social Psychology* 41:553-561.

Spiegel, D., J.R. Bloom, H.C. Kraemer and E. Gottheil (1989) «Effect of psychosocial treatment on survival of patients with metastatic breast cancer», *Lancet* 2:888-891.

Stuss, D.T. (1991a) «Disturbance of self-awareness after frontal system damage», in: *Awareness of Deficit after Brain Injury: Clinical and Theoretical Issues*, G.P. Prigatano and D.L. Schacter, eds., New York: Oxford University Press, pp. 63-83.

Stuss, D.T. (1991b) «Self, awareness, and the frontal lobes: a neuropsychological perspective», in: *The Self: Interdisciplinary Approaches*, J. Strauss and G.R. Goethals, eds., New York: Springer-Verlag, pp. 255-278.

Taylor, G.J., R.M. Bagby and J.D.A. Parker (1991) «The alexithymia construct: a potential paradigm for psychosomatic medicine», *Psychosomatics* 32:153-164.

Thayer, J.F. and R.D. Lane (in press) «A model of neurovisceral integration in emotion regulation and dysregulation», *Journal of Affective Disorders*.

Tomkins, S.S. (1962) *Affect, Imagery, Consciousness, Vol. 1: The Positive Affects*, New York: Springer.

Tranel, D. (1997) «Emotional processing and the human amygdala», *Trends in Cognitive Sciences* 1:46-47.

Tucker, D.M., P. Luu and K.H. Pribram (1995) «Social and emotional self-regulation», in: *Structure and Functions of the Human Prefrontal Cortex,* J. Grafman, K.J. Holyoak and F. Boller, eds., New York: The New York Academy of Sciences.

Van der Kolk, B.A., J.A. Burbridge and J. Suzuki (1997) «The psychobiology of traumatic memory – Clinical implications of neuroimaging studies», *Annals of the New York Academy of Sciences* 821:99-113.

Vogt, B.A. and M. Gabriel (1993) *Neurobiology of Cingulate Cortex and Limbic Thalamus,* Boston: Birkhauser.

Vogt, B.A., D.M. Finch and C.R. Olson (1992) «Functional heterogeneity in cingulate cortex: the anterior executive and posterior evaluative regions», *Cerebral Cortex* 2:435-443.

Watson, D., L.A. Clark and A. Tellegen (1988) «Development and validation of brief measures of positive and negative affect: the PANAS scales», *Journal of Personality and Social Psychology* 54:1063-1070.

Weiskrantz, L. (2000) «Blindsight: Implications for the conscious experience of emotion», in: *Cognitive Neuroscience of Emotion,* R.D Lane, L. Nadel, G.L. Ahern, J.J.B. Allen, A.W Kaszniak, S.Z. Rapscak, and G.E. Schwartz, eds., New York: Oxford University Press, pp. 277-295.

Weiskrantz, L. (1986) *Blindsight,* Oxford: Oxford University Press.

Werner, H. and B. Kaplan (1963) *Symbol Formation: An organismic-Developmental Approach to Language and the Expression of Thought,* New York: John Wiley and Sons.

Whalen, P.J., S.L. Rauch, N.L. Etcoff, S.C. Mclnerney, M.B. Lee and M.A. Jenike (1998) «Masked presentations of emotional facial expressions modulate amygdala activity without explicit knowledge», *Journal of Neuroscience* 18:411-418.

Wrana, C., W. Thomas, G. Heindichs, M. Huber, R. Obliers, A. Koerfer and K. Köhle (1998) «Levels of Emotional Awareness Scale (LEAS): ein Beitrag zur empirischen Überprüfung von Validität und Reliabilität einer deutschen Fassung», Postervortrag bei der *47 Arbeitstagung des Deutschen Kollegiums für Psychosomatische Medizin,* Leipzig, March, 1998.

Yakovlev, P.I. (1948) «Motility, behavior, and the brain», *Journal of Nervous and Mental Disease* 107:313-335.

MENTAL REPRESENTATIONS, THE RETICULAR ACTIVATING SYSTEM AND EMOTIONS

BARBARA CABOTT
1238 N. W. Glisan, Suite C, Portland, OR 97209 U.S.A.

ABSTRACT

This paper introduces an overview of mental representations and their reconstruction through experiential, creative and multimodal methods in Psychotherapy. I will emphasize the all-important role of the arousal. Arousal is assumed to be dependent upon the Reticular Activating System (RAS) lying deep in the brainstem, with its rich interconnections to all of the cortex, thalamus, and limbic system, as providing the necessary conditions to create and re-create mental patterns. It is also important to note that emotions weave their ubiquitous thread throughout this arousal system and throughout the tapestry of mental representations. I will explain how multi-modal therapy methods mirror the work of the RAS. My approach is holistic and my methods are designed to access the central nervous system as a complete unit, working not only with behavior and intellect, but with emotions, movements, senses, perceptions and interpersonal interactions. I have synthesized the theories of Jean Piaget, Alexander Luria, Donald Hebb, Karl Pribram and Daniel Stern. This paper is a compilation of their theories, joined in support of the therapy I have developed for patients in my clinical practice.

1. Introduction

This paper introduces an overview of mental representations, and their reconstruction through experiential and multi-modal methods in psychotherapy. I will emphasize the role of the arousal, assumed dependent upon the Reticular Activating System (RAS) deep in the brainstem, as providing the necessary conditions to create and re-create mental patterns. Emotions weave their ubiquitous thread through this system and throughout the tapestry of mental representations.

1.1 Mental Representations

To date, the concept of mental representations defies definition, yet its origins date back to Renee Descartes' "1'idee representative." Since then, neuroscientists, psychoanalytic theorists, and cognitive psychologists have wrestled with the concept.

Paivio, as recently as 1986, claimed that defining mental representations is perhaps the most complex problem in all of science because it raises questions about the nature of thought, behavior, brain activity, developmental origins, environmental issues, and other difficult topics. He states, "...because of this complexity, we lack agreement on how to approach the problem, theoretically and even empirically" (Paivio, 1986, p 3).

Additionally, the study of mental representations does not provide conclusive evidence about the mechanisms that underlie them. David Hubel (1979) observes,

> In brief, there is an input: man's only way of knowing about the outside world. There is an output: man's only way of responding to the outside world and influencing it. And between input and output there is everything else, which must include perception, emotions, memory, thought and whatever else makes man human (Hubel, 1979, p. 8).

Perhaps we can fathom the complex world between "input" and "output" as the components that comprise mental representations, our internal pictures of the world whose elements reside in perceptions, emotions, memory, thought, and as Hubel said, "whatever else makes us human."

When elements within our nervous system link with stimuli from our environment, these combinations shape mental representations. These representations become the filters through which we perceive the world and also drive emotions and behavior. Acting like behind-the-scenes players, the separate elements join in much the same way that individual scenes make up a movie. Only of course, the theater where the movie plays is within our mind and environmental cues are the gears that propel the pictures onto the screen. Conscious mental representations are internal pictures, a panorama of sensations, movements, emotions, relationships, and places, which join together into a whole unit, or system, whenever a similar memory or cue sparks one of their segments.

While theorists disagree on an exact definition of mental representations, I maintain that they are linked to particular aspects of the nervous system and are clusters formed from a person's history. These representations, entangled with emotions, help explain the variety of human behavior, action and thought. They also weave a tangled web from their sensory-motor entryway to higher cortical functions, where they form into complex thoughts and representations of actual experiences. By the time representations are housed in the cortex, they are quite removed from the actual stimulus that formed them originally. Over time, they become as habitual and as "wired in" as the original reflexes that were at the root of their conception (Pribram, 1960).

Stern (1985) explains that while each unit is a reconstruction of real experience, it is not the actual experience. He points out that often what we believe to be memories are actually a distorted inner world. Unfortunately, these distortions can provide the fertile soil for neurotic mechanisms and negative feelings. These distortions usually increase over time, unless there is a means to change or "update" them.

Mountcastle (1975) points out another aspect of this when describing the perceptual system:

> Our brain is linked to what is "out there" by a few million fragile sensory nerve fibers, our only information channels, our lifelines to reality. Afferent nerve fibers are not high-fidelity recorders... the central neuron is a story-teller with regard to the nerve fibers, and it allows distortions... sensation is an abstraction, not a replication, of the real world (Mountcastle, 1975, p. 109).

Psychotherapy's first role is to bring these unconscious, abstract, habitual distortions into conscious experience, where their elements can be uncovered, observed, and changed. The emotional components that surround them are often the main access route to making them conscious. The role of the RAS in consciousness and thought is essential to this process.

1.2 The Reticular Activating System

The RAS has a crucial role in updating and transforming mental representations. This system, located low in the brain stem, has an important function in thought, consciousness and emotion. Constructed like a nerve net, it allows excitation to spread gradually. This gradual change can modulate the whole nervous system (Cabott, 1989).

It was in 1949 that Magoun and Maruzzi discovered that the Reticular Formation, located low in the brain stem, has a role in the mediation of attention and consciousness. We learned that the very existence of consciousness depended upon the integrity of the subcortex and brainstem rather than in the cortex alone. Initially, ascending fibers were found that ran upwards from the Reticular Formation, projecting to the limbic-hypothalamic system, the thalamus, and widely throughout the cortex, to stimulate the brain into wakefulness, or arousal. Later, descending connections were discovered, along with aminergic neuronal transmitters most closely correlated with arousal, attention, motivation, and emotion. These interconnected functions became known as the Reticular Activating System, a system Hebb (1980) described as subserving wakefulness, or the state of conscious alertness.

It was Alexander Luria (1973), the eminent Russian neuropsychologist, who elucidated the functional organization of the brain, spelling out three principal synergistic units and clearly emphasizing that reciprocity between the three units was required for organized conscious activity. The first unit is the RAS, a unit for regulating tone or waking. The second unit, comprising the sensory and posterior association cortices is dedicated obtaining, processing and storing information arriving from the outside world. The third unit, comprising the frontal lobes subserves programming, regulating and verifying mental activity.

In this interactive system, upper levels of the cortex descend, inhibit, help organize, regulate and modulate activities of the lower levels. The upper levels recruit the systems of the reticular formation. In this reciprocal relationship the most complex forms of conscious activity are possible. Luria noted that, "Man's mental processes in general, and his conscious activity in particular, always take place with the participation of all three units" (Luria, 1973, p. 45).

It was Hebb (1980) who helped us understand the relationship of the RAS to emotions. With the arousal functions closely entwined with the limbic system, he maintained there is an "intimate" connection with both emotion and memory. The RAS is involved in emotional behavior because it is the arousal system that puts the energy into an emotional response (Hebb, 1987), and emotions that put energy into the arousal system.

It seems that there is an optimal level of arousal. This level requires the close cooperation of RAS and the frontal lobes. This ideal "window" of arousal provides the best conditions for both modulation and activation. Too much "noise" in the system raises the arousal level and interferes with adaptive information processing. During weak or monotonous stimulation, arousal is lowered, making it difficult to get a meaningful signal through. Positive and negative emotions both activate the system. Consequently, I maintain the key to a successful therapeutic relationship is discovering the client's optimal arousal level and adequately modulating his/her emotions.

The primary source of excitation of the system is varied and multi-modal sensory input. Also, what is novel and fascinating actually heightens brain activity and is an important precondition for updating mental representations (Donchin, 1981). The RAS, when properly modulated and activated is a major ally in bringing elements of mental representations into consciousness that are "wired in" and thus otherwise intractable. As they are felt, experienced, and consciously observed, they have a chance to reform and align with adaptive thinking, feelings and behavior.

2. Case Study

Mental representations, once formed, are often large clusters of deeply layered components that reside within a person's memory and subjective inner world. Their parts are interweavings of the human nervous system, their subjective nature is a mental or cognitive picture that responds as a whole unit when interfaced with the environment. Their elements, mostly abstract and unconscious, often center around a major theme. Their construction in response to the outer world is usually a distortion, which can drive maladaptive thinking, feelings, and behavior. The feelings woven into the representation are usually the markers into the system.

Such was the case with Katrina, aged 41, whose major representation centered around being "dumb and stupid." She entered therapy presenting symptoms of anxiety and depression, recurrent nightmares, fear of success, and "blanking out" cognitively when asked almost any question. The elements of her mental representation were like wide inner antennae that pulled in and skewed any cues from her environment that aroused feelings of being "dumb and stupid."

Together, Katrina and I discovered her feelings that clustered around this belief that was imprinted when she was kept back in second grade. During this time, her father's drunken and violent outbursts kept her in a high state of arousal, likely interfering with the optimal level of her RAS, and impairing her thinking at school. When she had to repeat second grade her father yelled at her and called her "dumb and stupid." This became an inner representation that was joined with violence, fear, and humiliation. Over the years, it became more entrenched. Every experience that funneled into that theme raised her anxiety and closed down her arousal system. She had little awareness of the elements driving her negative thoughts, uncomfortable feelings, and inability to organize her life. She was not conscious of this all-pervading theme and the processes underlying it.

If we return to Luria's three units, we can conjecture that Unit Two was deeply imprinted with long-held memories of being "dumb and stupid." The fear and anxiety that were continually generated by this thought tipped her arousal system (Unit One) over the upper level of modulation, so that she could not summon the necessary energy to make plans, or follow through on goals as described in Luria's Unit Three. Thus, she was unable to quiet her feelings, change her view of herself, and organize her thoughts.

As Luria (1973) said, "It is known that in a state of lowered cortical tone the normal relationship between excitation and inhibition is disturbed, and the mobility of the nervous system, so necessary for mental activity to pursue its normal course is lost" (Luria, 1973, p. 45). This Katrina exemplified.

Her therapy involved accessing and modulating her arousal system using hypnosis and relaxation, along with using many combinations of varied sensory images. New concepts were introduced through imagery and art, and interpersonal trust and communication were established through a method of "conscious communication," where verbal communication was changed from being a "task demand" to a mutual sharing between patient and therapist from a calm state.

She was helped to refocus intentions and complete goals, by aligning with the process of her success as an athlete. Art, as a multilevel communication, conveyed meaning on several levels at the same time, allowing many components at once to become conscious and reworked. During therapy, she "aced" a licensing exam in physical therapy, quieted her nightmares completely, and no longer viewed herself as "dumb and stupid." Her emotions regulated, social interactions became easier and more complete, and she reached a long term goal of starting her own business.

3. Conclusions

I strive to understand the inner subjective life that lies between the "input" and "output," referred to by David Hubel (1979), and the major themes clustered around mental representations that often drive clients' maladaptive thoughts and emotions. It is essential for me in clinical practice to comprehend how mental representations are formed and re-formed, and the rules by which they operate within the central nervous system.

My therapy methods are designed to make representations and their components concrete, conscious, palpable, and closer to the cues that activated them. I believe that unwinding and re-forming these representations is at the essence of psychotherapy and the essence of cognitive, behavioral and emotional change.

References

Adrian, E.D., F. Bremer and H.H. Jasper, eds (1954) *Brain Mechanisms and Consciousness*, Springfield, Ill.: Charles C. Thomas.

Cabott, B. (1989) *The Functional Neuroanatomy of Mental Representations*, Unpublished dissertation.

Donchin, D. (1981) "Surprise, Surprise!", *Psychophysiology* 18:493-510.

Freud, S. (1953) *On Aphasia*, New York: International Universities Press.

Greenfield, N.S. and W.C. Lewis (1965) *Psychoanalysis and Current Biological Thought*, Madison, WI: University of Wisconsin Press.

Hebb. D.O. (1949) *The Organization of Behavior: A Neuropsychological Theory*, New York: Wiley.

Hebb, D.O. (1958) *A Textbook of Psychology*, Philadelphia: W.B. Saunders.

Hebb, D.O. (1980) *Essay On Mind*, Hillsdale, Hillsdale, NJ: Lawrence Erlbaum Associates.

Hubel, D.H. (1979) "The Brain in The Brain", *Scientific American* 2-15.

Kandel, E.R. and J.H. Schwartz (1985) *Principles of Neuroscience*, New York: Elsevier.

Kolb, B. and I.Q. Wishaw (1980) *Fundamentals of Human Neuropsychology*, San Francisco: H.W.H. Freeman and Company.

Luria, A.R. (1966) *Higher Cortical Functions in Man*, New York: Basic Books.

Luria, A.R. (1966) *Human Brain and Psychological Processes*, New York: Harper and Row.

Luria, A.R. (1969) "The Origin and Cerebral Organization of Man's Conscious Action", Lecture given to the *19th International Congress of Psychology*, London, England.

Luria, A.R. (1973) *The Working Brain*, New York: Basic Books.

Luria, A.R. (1976) *The Nature of Human Conflicts on Emotion, Conflict, and Will*, New York: Liveright.

Miller, G., E. Galanter and K. Pribram (1960) *Plans and Structure of Behavior*, New York: Henry Holt and Company.

Mountcastle, V.B. (1975) "The View from Within: Pathways to the Study of Perception", *The Johns Hopkins Medical Journal* 136:109-131.

Piaget, J. (1954) *The Construction of Reality in the Child*, New York: Basic Books.

Piaget, J. (1970) *Structuralism*, New York: Basic Books.

Piaget, J. (1971) *Biology and Knowledge: An Essay on the Relations between Organic Regulations and Cognitive Processes*, Chicago, Ill.: The University of Chicago Press.

Pribram, K.H. (1960) "A Review of Theory in Physiological Psychology", *Annual Review of Psychology* 11:1-40.

Pribram, K.H. (1962) "The Neuropsychology of Sigmund Freud" in: *Experimental Foundations of Clinical Psychology*, A. J. Bachrach, ed, New York: Basic Books, pp. 442-468.

Pribram, K.H. and M.M. Gill (1976) *Freud's Project Revisited*, New York: Basic Books.

Rossi, L.R. (1986) *The Psychobiology of Mind-Body Healing*, New York: W.W. Norton and Company.

Stern, D.N. (1985) *The Interpersonal World of the Infant*, New York: Basic Books.

Stern, D.N. (1994) Unpublished lectures from *Cape Cod Symposium*.

THE ROLE OF AUTONOMIC BALANCE IN EXPERIENCING EMOTIONS

BRANKA ZEI and MARC ARCHINARD
University Hospital of Geneva, Liaison Psychiatry,
51 Blvd. De la Cluse, CH 1205 Geneva, Switzerland

ABSTRACT

This research explores the role of the physiological component of emotional arousal. The concept of autonomic balance is presented theoretically and operationalized through measurement of heart rate variability (HRV). The role of the latter is examined in its relation to emotional arousal, as reflected in both subjective feeling and non-verbal vocal expression. Extraversion, as personality trait, and state anxiety, are included in the experimental design. The results lend support to the hypothesis that subjects with low HRV experience flattening of emotional reactions mainly in vocal expression, but also in subjective feeling. Implications of the findings are discussed in terms of the influence of HRV on interoception and emotional awareness.

1. Introduction and Theoretical Framework

Emotions have been characterised as psycho-physiological phenomena that include cognitions, visceral, humoral and immunological reactions, vocal and other non-verbal expressive displays, as well as activation of behavioral dispositions. The latter are supported by the autonomic nervous system (ANS).

Most studies on the ANS component of emotional reactions have focused on the sympathetic activation (for an extensive survey, see Cacioppo, Klein, Berntson, & Hatfield, 1993). However, the role of the parasympathetic branch of the ANS has not received equal attention in research involving adults. Within the framework of developmental psychology (Porges *et al.*, 1994), research has demonstrated that the base-level of vagal tone (defined as the amount of inhibitory influences on the heart by the parasympathetic nervous system) influences the expression and regulation of emotion as well as behavioral patterns in children (Porges 1992, 1995; Porges *et al.*, 1994). The base-level vagal tone has thus been related to autonomic responsivity in general. The latter has more recently been conceptualised in terms of autonomic balance (Friedman & Thayer, 1998) which is reliably quantified through measurement of heart rate variability (HRV). Although higher HRV is associated with normal emotional reactions, low HRV appears to be related to a series of affective and cognitive disturbances (Eysenck, 1985; Friedman & Thayer 1996; Klein, Cnaani, Harel, Braun, & Ben-Haim, 1995; Yeragani, Balon, & Pohl, 1990; Yeragani, *et al.*, 1995).

In order to explore the neurological underpinnings of emotions, we considered it meaningful to study the relationship between vagal tone, as indicator of autonomic balance, and emotional reactions reflected in vocal arousal and subjective feeling. We also assumed that the degree of awareness of subjective feeling would be linked to autonomic arousal.

The influence of autonomic activation on vocal behavior was first modeled by Williams and Stevens (1972) in terms of direct causal relationships between dominantly sympathetic or dominantly parasympathetic activation on the one hand, and voice intensity, vocal cord vibration and timing of speech on the other. We thus defined *vocal arousal* as a set of speech characteristics related to an emotional state. Porges *et al* (1994) provided a precise description of the link between vagal tone and vocal expression of emotion. Following the vocal feedback hypothesis (Hatfield, Hsee, Costello & Denney, 1995) whereby the degree of subjective experience is influenced by the proprioceptive and auditory feedback, we assumed that the degree of emotional awareness could also be correlated with the degree of vocal arousal.

2. Hypotheses

The basic vagal tone (expressed in HRV index units) influences emotional reactions in that the subjects with low HRV were predicted to display:

- A general flattening of vocal arousal and weaker vocal differentiation of emotions. More specifically, vocal arousal being predicted as higher in anger than in sadness (Scherer & Zei, 1988; Scherer, Banse, Wallbott & Goldbeck, 1991; Banse & Scherer, 1995), vocal differentiation of high versus low vocal arousal states would be diminished in subjects with low vagal tone.
- Lower levels of subjective emotional feeling.
- The awareness of an emotional state would be positively correlated with the degree of vocal arousal.

3. Methods

3.1 Subjects

Forty diabetic patients (18 female and 22 male), varying in age (range = 31-73 years, mean = 56; sd = 9), and duration of illness (range = 1-36 years ; mean =15 ; sd = 11), served as subjects for the study. Some of the patients had low levels of vagal tone due to lesions of the ANS, known as autonomic neuropathy.

3.2 Physiological and Psychological Measures

Two standard tests of the autonomic function (Vita *et al.*, 1986) were applied. They both measure the heart rate variability in two conditions: (1) Heart rate difference in deep breathing and (2) Lying-to-standing heart rate ratio. The results were age adjusted and combined into a composite HRV index. Psychological measures included the Speilberger State Anxiety Scale and the Eysenck Personality Inventory.

3.3 Induction of Emotions and Vocal Data

The subjects were asked to verbally recall their personal emotional experiences of joy, anger, and sadness. At the end of each recall they were asked to pronounce, in a mood congruent tone, the sentence "ALORS TU ACCEPTES CETTE AFFAIRE" ("So you accept the deal"). The sentence was presented in writing, without punctuation, so as not to suggest any tone of voice. The subjects were then asked whether they had subjectively felt (and to what degree) the emotion described during their recall. The results were coded on a scale of 0-3 (ranging from "not at all" to "very much"). The subjects' voices were recorded on a DAT recorder. The distance of the microphone to the mouth was kept constant

3.4 Acoustical Analyses

One hundred and twenty samples of the standard sentence were acoustically analysed. Three categories of vocal arousal indicators were extracted:

(1) *Fundamental frequency (F0) of vocal cord vibrations computed from the signal digitised at 44 kHz*. The following F0 parameters were extracted from the pitch curves and expressed in Hz.

- Mean, median, mode
- Range between 5th - 95th percentile, 5th percentile
- Maximum/minimum ratio, sd, coefficient of variation.

(2) *Acoustic energy computed from the raw signal values*. The following energy parameters were extracted from the amplitude envelopes and expressed in pseudo-decibel units:

- Maximum voiced energy
- Mean voiced energy
- Voiced energy range. The measurement was done at mid-point values of vowel nuclei .

(3) *Speed of delivery*, expressed in the number of syllables uttered per second. Prior to the measurement of the total signal length, all inter-syntagmatic pauses had been edited out. The speed of delivery thus corresponded more closely to articulation speed. The latter was expected to be slower in sadness than in anger. All the acoustical analyses were done by means of a Macintosh platform software "Signalyze" (Keller 1995).

3.5 Data Transformations and Creation of New Variables

In order to make the data directly comparable on a common scale, z-scores were calculated for all vocal parameters. Autonomic tests results were age adjusted and normalised against external reference values from healthy subjects (Vita *et al.*, 1986). A cumulated score on both tests was taken as the HRV index for each patient.

On the basis of curve-fitting, and upon inspection of partial correlations with the HRV index (controlling for age, anxiety state and extroversion), as well as linear multiple regression analyses, three vocal parameters appeared as significantly related to the HRV index. These were: F0 max/min ratio, voiced energy range, and the rate of delivery. We then calculated a summary score reflecting the overall degree of vocal arousal (Vocal Arousal Index) for each condition (anger, joy, sadness). We justify cumulating the three parameters into a composite score by the fact that while each of them can vary independently, they often maintain trading relationships and appear in configurations representing the speaker's personal way of signalling affect. Some speakers use mainly pitch parameters, while others use mainly energy parameters, speed of delivery, or any combination of the three basic dimensions of prosody.

Since we expected, the subjects with high HRV index to exhibit higher Vocal Arousal Index in anger than in sadness, we then calculated the delta between the vocal arousal index obtained in expressing anger and that obtained for sadness. Each subject was thus characterised by his/her Vocal Arousal Differential Index (ΔdB+Δmax/mIN+Δrate), reflecting the degree of his/her vocal differentiation between anger and sadness.

4. Results

4.1 Vocal Arousal

We performed linear multiple regressions (stepwise method) with Vocal Arousal Differential Index as the dependent variable and HRV index, demographic and psychological variables as independent variables. The results of the regressions show a highly significant effect for HRV index ($T = 7.189$; $p < .0001$) and a much lesser effect for state anxiety ($T = -2.052$; $p = .0470$). The HRV index alone explained 58% of data variance with the multiple $R = .79$. None of the other variables contributed significantly. From these results we can conclude that vocal differentiation of emotions is related, above all, to the HRV and marginally to anxiety state.

4.2 Self-Reported Subjective Feeling

Seventy-five percent of subjects reported felt anger (mean = 1.7; sd = 1.24), 97.5 % reported felt joy (mean = 2.5; sd = .78) and 95 % of the subjects reported felt sadness. Mann-Whitney U tests, with groups obtained by median split on HRV index, showed significant differences in the degree of felt sadness ($\underline{Z}$ = - 3.3; P=.0009), and anger ($\underline{Z}$ = - 2.4; P = .02). The groups with higher HRV reported a higher degree of subjective feeling for both sadness and anger than did those with lower HRV. By contrast, the correlations between Vocal Arousal Index and the degree of subjective feeling (controlled for demographic and psychological variables) did not show any significant correlation.

An unexpected finding concerned weeping episodes. Seventy-seven percent of subjects wept during the recall of sadness. The degree of weeping was coded from 0-3 with: 0 = absence of visible weeping; 1 = noticeable tears in the eyes; 2 = tears running down the face; 3 = tears running down the face accompanied by speaking difficulties. The correlation between the degree of crying and HRV index (controlled for anxiety, extroversion, gender and age) was calculated, revealing a highly significant relationship (r = .56, P = .000).

5. Discussion

Our hypothesis 1 was confirmed, in that HRV index, as indicator of autonomic balance, was found to be related to emotional arousal. The subjects with lower HRV exhibited a flattening of emotional reactions in two domains: vocal arousal and subjective emotional feeling. More specifically: (1) vocal differentiation between anger and sadness was smaller in subjects with low HRV compared with those with higher HRV, and (2) the degree of self reported subjective feeling was proportional to degree of HRV.

As for the unexpected finding concerning the degree of weeping being proportional to the HRV index, we had two complementary interpretations: (1) in neuropathic subjects, the destruction of the parasympathetic nerves causes diminished tearing, and (2) the emotional experience of sadness is altogether lesser in subjects with low HRV.

Our hypothesis 2 was not confirmed, in that the degree of subjective feeling was not found to be related to the degree of vocal arousal.

Our results concerning the flattening of emotional reactions agree with those of Andreasen and colleagues (Andreasen *et al.*, 1981), whose experiment demonstrated that affective flattening is reflected in a diminished variance in both amplitude and fundamental frequency of speech. The authors consider the acoustic analysis of voice patterns as an objective means of evaluating flatness of affect.

As for the results concerning subjective feeling, it appears meaningful to consider an explanation whereby higher levels of HRV may enhance the interoception of one's own cardiac response to emotional stress and consequently result in a higher degree of emotional awareness. Such a hypothesis would be in agreement with the findings of Davis *et al.* (1986), where subjects with high heart rate variability displayed more accurate perception of their own heart rates. In view of these findings it appears meaningful to assume that awareness of the strength of a subjective emotional feeling covaries with the degree of autonomic arousal and its interoception. The latter thus appears to be related to HRV index as indicator of basic non-emotional autonomic responsivity and/or autonomic balance.

References

Andreasen, N.C., M. Alpert and M.J. Martz (1981) "Acoustic analysis", *Archives of General Psychiatry* 38:281-285.

Banse, R. and K.R. Scherer (1996) "Acoustic profiles in vocal emotion expression", *Journal of Personality and Social Psychology* 70:614-636.

Cacioppo, J.T., D.J. Klein, G.G. Berntson and E. Hatfield (1993) "The psychophysiology of emotion", in: *Handbook of Emotions*, M. Lewis and J. Haviland, eds, New York: Guilford. Press, pp.119-141.

Davis, M.R., A.W. Langer, J.R. Sutter, P.D. Gelling and M. Marlin (1986) "Relative discriminability of heartbeat-contingent stimuli under three procedures for assessing cardiac perception", *Psychophysiology* 23:76-81.

Eysenck, H.J. (1970) *The Structure of Human Personality*, London: Methuen.

Eysenck, H.J. and M.J Eysenck (1985). *Personality and Individual Differences*, New York: Plenum Press.

Friedman, B.H. and J.F. Thayer (1996) "Spectral characteristics of heart period variability in shock avoidance and cold face stress in normal subjects", *Clinical Autonomic Research* 6:147-152.

Friedman, B.H. and J.F. Thayer (1998) "Autonomic balance revisited: Panic anxiety and heart rate variability", *Journal of Psychosomatic Research* 44:133-151.

Hatfield, E., C. K. Hsee, J. Costello, B. Schalekamp, M. Weisman and C. Denney (1995) "The impact of vocal feedback on emotional experience and expression", *Journal of Social Behavior and Personality* 10:293-312.

Klein, E., E. Cnaani, T. Harel, S. Braun and S.A. Ben-Haim (1995) "Altered heart rate variability in panic disorder patients", *Biological Psychiatry* 37:18-24.

Porges, S.W. (1992) "Autonomic regulation and attention", in: *Attention and Information Processing in Infants and Adults*, B.A. Campbell, H. Hayne and R. Richardson, eds, Hillsdale, NJ: Lawrence Erlbaum Associates.

Porges, S.W., J.A. Doussard-Roosevelt and A.K. Maiti (1994) "Vagal tone and the physiological regulation and emotion", *Monographs of the Society for Research in Child Development* 59:167-186.

Porges, S.W. (1995) "Cardiac vagal tone: A physiological index of stress", *Neuroendocrine and Biobehavioral Reviews* 19:225-233.

Scherer, K.R., R. Banse, H.G. Wallbott and T. Goldbeck (1991) "Vocal cues in emotion encoding and decoding", in: *Motivation and Emotion, vol. 15*, A.M. Isen, ed., New York: Plenum.

Scherer, K.R. and B. Zei (1988) "Vocal Indicators of affective disorders", *Psychotherapy and Psychosomatics* 49:179-186.

Vita, G., P. Princi, R. Calabro, A. Toscano, L. Manna and C. Messina (1986) "Cardiovascular reflex tests", *Journal of the Neurological Sciences* 75:263-274.

Williams, C.E. and K.N. Stevens (1972) "Emotions and speech: Some acoustical correlates", *Journal of the Acoustical Society of America* 52:1238-1250.

Yeragani, V.K., R. Pohl, K. Srinivasan, R. Balon, C. Ramesh and R. Berchou (1995) "Effects of isoproterenol on heart rate variability in patients with panic disorder", *Psychiatry Research* 56:289-293.

Yeragani, V.K., R. Balon and R. Pohl (1990) "Decreased R-R variance in panic disorder patients", *Acta Psychiatrica Scandinavica* 81:554-559.

PSYCHOPHYSIOLOGICAL ANALYSIS OF THE NONLINEAR DYNAMICS AND COMPLEXITY RELATED TO ATTENTIONAL CONFLICTS AND AFFECTIVE STATES

PATRICE RENAUD* and JEAN-PIERRE BLONDIN[0]

*Département de psychoéducation et de psychologie, *Université du Québec à Hull, 283 Boulevard Alexandre-Taché, Hull, Québec J8X 3X7, Canada*

[0]*Département de psychologie, Université de Montréal, 90 avenue Vincent-d'Indy, Quebec H3C 3J7, Canada*

ABSTRACT

The purpose of this study was to examine performance and cardiovascular data in the light of parameters expressing the complexity and dynamic instability of their process, and in the context of the Stroop task attentional effort expenditure. Dysphoric emotional state was also linked with performance and cardiovascular dynamics, and with the level of attentional conflict. Results indicate that competition between conflictual dimensions of a stimulus diminishes dynamic instability of the response process. Also, the affective state preceding performance seems to act as a modulator of dynamic instability and complexity of the organismic response process.

1. Introduction

Nonlinear dynamics and complexity, as they are related to attentional processes, refer to that safety margin needed by adapting organisms. Behaviorally speaking, the action patterns reflecting cognitive-perceptual adaptation might be translated into more or less foreseeable fluctuations taking place in time (Kelso, 1997, pp.187-224). Complexity, as the intricacy of ordering factors, speaks about the shock absorption capacity of a particular system without it being endangered in its structural integrity (Kaplan, 1991; May, 1991).

Increased competition among perceptual features contributes to stabilize neural dynamics (Hogg & Huberman, 1991). Stability corresponds, in that context, to the convergence toward a dominant representation that will take over the response output process (Prueitt *et al.*, 1995).

Emotional states act as a global priming factor of the attentional processes (Milner, 1996). They tune and modulate the psychophysiological dynamics engaged in actively mastering incoming information.

2. Material

2.1 Physiological Recording

Heart rate (HR) was monitored continuously on a Grass 7P44 cardiotachograph by means of two Ag/AgCl miniature electrodes placed on either side of the chest with the ground on the non-dominant wrist (i.e., the hand not used in pressing the response keys). By extrapolation, data were expressed in beat per minutes, at each second.

2.2 The Stroop Task

The attentional testing employed a computerized version of the Stroop task (Renaud & Blondin, 1997; Stroop, 1935), devised so that two types of stimuli could be presented: (1) Stroop stimuli (STR), which consisted of three color names printed in incongruent and conflictual colors, and (2) control stimuli (CI), which were strings of three Xs printed in either one of the three colors used for the Stroop stimuli.

The stimuli were presented one by one on the screen. Trials had a maximum duration of 2000 ms. Subjects were instructed to identify the color in which the stimuli were printed, and to do so as quickly and as accurately as possible using a response keyboard. Additionally, they had to wait for a sound signal to be heard, either at 400 ms or at 700 ms after the arrival of the stimulus, before responding to it. Each block of trials lasted five minutes. Reaction time (RT) was measured in milliseconds from the onset of the stimulus.

3. Method

3.1 Subjects and Procedure

Twenty-four subjects, varying in age between 19 and 30 years, constituted the experimental sample. They received the instructions and then had the opportunity to accustom themselves with the task. Before that part, they were asked to complete the Multiple Affect Adjective Checklist (MAACL; Zuckerman, 1960), which includes anxiety, depression and hostility scales, as a base level measure of dysphoric emotional state. The MAACL was also administered at two other occasions - after the CI and the STR conditions.

3.2 Statistical, nonlinear and complexity analysis

The parameters expressing the nonlinear properties were obtained through numerical simulations done with a program built to estimate, from time-series data, the dominant Lyapunov exponents (LE) of noisy nonlinear systems (Ellner, Nychka, & Gallant, 1992). This program, relying on neural nets, performed a global nonlinear regression for each individual RT and HR vector, in each condition. LE[1], which represent an operationalization of the concept of dynamic instability, and the number of hidden units (HU)[2] recruited by the calculus

---

[1] The more positive LE are, the more they refer to an unstable dynamical system.
[2] Higher complexity is indexed by a higher number of hidden units.

process, which is an index of complexity, were extracted from each simulation.

From a 2 x 2 counterbalanced factorial scheme, the impact of time pressure (400 ms or 700 ms delay) and attentional conflict resolution (CI or STR) on LE and HU was assessed. The data were analyzed using multivariate analyses of variance for repeated measures designs; univariate tests were used in the presence of a significant multivariate effect. Pearson's correlation coefficients were also computed to better understand the variables' interplay. The criterion for statistical significance was $p<.05$.

4. Results

4.1 Performance Dynamics

Table 1 presents the average LE found in each condition. Statistical analysis shows that STR condition produced a more stable performance dynamics than the CI condition, that is LE of a lesser amplitude for the RT dynamics when subjects had to deal with conflictual stimuli ($F(1,20)=4,49$, $p=.047$).

Table 1. Average Lyapunov exponents of the RT found in each condition

	CI	STR
400 ms	-0.542	-0.67
700 ms	-0.54	-1.075

4.2. Cardiovascular Dynamics

Analysis of the heart rate results indicated that STR condition generated less HU in the cardiovascular dynamics than the CI condition ($F(1,14)=5,69$, $p=.032$). Cardiovascular complexity seems to have been lesser when the metabolic requirement of the attentional task was higher because of perceptual competition. See Table 2 for the average HU found in each condition.

Table 2. Average hidden units of the HR found in each condition

	CI	STR
400 ms	2.00	2.067
700 ms	2.933	2.00

4.3. Dysphoric Emotional States

The STR condition produced a more pronounced DES ($F(1,23)=4.98$, $p=.048$ for anxiety; $F(1,23)=4.98$, $p=.036$ for depression) than the CI condition. Furthermore, base level DES scores obtained on the three emotional scales of the MAACL are significantly lower than those reported consecutively in the conflictual and the non-conflictual conditions ($F(1,23)=21.63$, $p<.001$ for anxiety;

$F(1,23)=25.72$, $p<.001$ for depression; $F(1,23)=20.09$, $p<.001$ for hostility).

4.4. Correlations

Pertaining to the cardiovascular dynamics, DES base level negatively correlates with HU in the STR condition at 400 ms on one side ($r=.61$, $p<.01$ for anxiety; $r=.62$, $p<.01$ for depression), and with LE in the CI condition at 700 ms on the other side ($r=.65$, $p<.01$ for anxiety).

5. Discussion

LE in both conflictual and non-conflictual tasks are indicative of dynamic processes at the edge of chaotic instability. Perceptual competition between the conflictual dimensions of the Stroop stimuli appears to diminish the dynamic instability of the response process. Emotional states preceding performance seem to act as modulators of dynamic instability and complexity of the organismic response process; more precisely, dysphoric emotions deplete the psychophysiological preparedness of the organism.

Future experiments will give the opportunity to reconsider the temporal aspects of the task, in order to better characterize the unfolding of attentional dynamics. Also, emotional states will be systematically controlled to conclusively test their impact on attention and its psychophysiology.

References

Ellner, S., D.W. Nychka and A.R. Gallant (1992) "LENNS, a program to estimate the dominant Lyapunov exponent of noisy nonlinear systems from time series data", *Institute of Statistics Mimeo Series #2235 (BMA Series #39)*, Statistics Department, North Carolina State University.

Hogg, T. and B.A. Huberman (1991) "Controlling chaos in distributed systems", *IEEE Transactions on Systems, Man and Cybernetics* 21:1325-1332.

Kaplan, D.T., M.I. Furman, S.M. Pincus, S.M. Ryan, L.A. Lipsitz and A.L. Goldberger, (1991) "Aging and the complexity of cardiovascular dynamics", *Biophysics Journal* 59:945-949.

Kelso, J.A.S. (1997) *Dynamic Patterns. The Self-Organization of Brain and Behavior*, Cambridge, MA: The MIT Press.

May, R.M. (1991) "Le chaos en biologie", *La Recherche* 232:588-598.

Milner, P.M. (1996) "Neural representations: Some old problems revisited", *Journal of Cognitive Neuroscience* 8: 69-77.

Prueitt, P.S., D.S. Levine, S.J. Leven, W.W. Tryon and F.D. Abraham (1995) "Introduction to artificial neural networks", in: *Chaos Theory in Psychology*, F.D. Abraham and A.R. Gilgen, eds, Westport, CT: Greenwood Press, pp. 195-263.

Renaud, P. and J.-P. Blondin (1997) "The stress of Stroop performance:

Physiological and emotional responses to color-word interference, task pacing, and pacing speed", *International Journal of Psychophysiology* 27:87-97.

Stroop, J.R. (1935) "Studies of interference in serial verbal reactions", *Journal of Experimental Psychology* 18:643-662.

Zuckerman, M. (1960) "The development of an affect adjective checklist for the measurement of anxiety", *Journal of Consulting and Clinical Psychology* 24:457-462.

AFFECTIVE NEUROSCIENCE AND EXTENDED RETICULAR THALAMIC ACTIVATING SYSTEM (ERTAS) THEORIES OF CONSCIOUSNESS

DOUGLAS F. WATT
Clinic for Cognitive Disorders, Quincy Hospital, Quincy, Massachusetts, 02169, USA

ABSTRACT

In the burgeoning literature about the neural basis of consciousness, affect is generally relegated to the back of the bus as an interesting "coloration" to the "hard problem" of consciousness. Most current theories of consciousness neglect evidence that emotion is a central organizing process for consciousness, probably one of its necessary and sufficient conditions. There are deep and intrinsic interpenetrations of global state functions that we have largely segregated, such as pain, affect, attentional functions and executive functions (as "slices" of the consciousness pie). Without central representation of value available "on-line," executive and attentional functions are collapsed at their base. Paralleling their extensive functional interpenetration, global state functions have vast overlap in putative neural substrates. Regarding neural correlates for emotion, broadly defined, the "limbic system" is so widely distributed that it has very unclear limits. This is derivative of the failure to clearly distinguish between emotion as a prototype or "primitive" vs. the much broader problems of emotional meaning, conditioning, and learning, as these relate to the global representation of value, which is interpenetrant with much of CNS activity. Even defining emotion in terms of its "primitives" or prototype affects yields differential but highly distributed-hierarchical neural substrates. Affect is elusively multi-dimensional, with patterned autonomic, endocrine, motor-executive, subjective pain/pleasure (valence), social/signaling, and cognitive (other/self appraisal) integrations. Emotional "primitives" are organized largely in diencephalic and midbrain structures ignored in most work on emotion, where most focus on telencephalic structures that support "valence tagging" (emotional learning and association) but that cannot underwrite valence itself. Basic connectivities between affective systems and the core systems of ERTAS underline the likely importance of these same primitive midbrain systems for consciousness: 1) connectivities between the midbrain reticular formation (MRF) and periaquaductal gray (PAG); 2) connections of thalamic intralaminar nuclei (ILN) to midbrain periaqueductal gray (PAG), various limbic, and basal ganglia (BG) systems; 3) predominant limbic modulation of thalamic nucleus reticularis thalami (nRt) "gatelets" by nucleus accumbens, paralimbic cortices, BG, and dorsomedial (DM) thalamus-prefrontal regions. Severe damage to PAG (a clearinghouse in the diencephalon-midbrain for primitive value operators with crucial projections to monoamine nuclei, ILN and MRF), profoundly impairs consciousness. PAG interactions with other ventral systems in SC and deep tegmental regions may form substrates for a primitive and basic neural representation of the self. But there can be only modest specificity at this point about the fundamental relations of emotion and consciousness, and many basic questions remain. At the end, some of these are reviewed, along with suggestions for future research to outline PAG's role and the role of "valence" or primary emotion in consciousness.

1. Introduction – Framing the Problem(s)

Consciousness and emotion are ancient topics as old as culture, still in their scientific infancy, and both slowly emerging into full respectability after decades of systematic neglect by science. Despite a modest resurgence in interest in the subject, emotion probably remains the most neglected and least understood subject relative to its importance in human life and in the whole of neuroscience. This is mostly likely over-determined. One aspect may be left over from Lange-James perspectives in which in which the richness of experienced emotion was reduced to an epiphenomenon, a sensory feedback from autonomic and motor efferents, a kind of phenomenologically compelling but ultimately irrelevant "neural mirage" or "after image" of the "real action" of emotion in autonomic and motor efferentation. Additionally, the explosion of cognitive neuroscience, in concert with the extensive discrediting of much of psychoanalytic thinking, has left emotion in a largely secondary role, despite dramatic lessening of the stranglehold that behaviorism had over thinking in psychology. Cognition is very much in ascendance these days, with some even assuming its foundations are fundamentally independent from affect, a position for which there is little evolutionary or neurological evidence. Finally, the relative disregard for emotion (until recently) in neuroscience may have major contributions from the intrinsic scientific and methodological difficulty of the subject itself:

1) Affect is elusively multi-dimensional, a complex composite of disparate elements;
2) There are formidable terminological and nosological issues, as emotion can be defined quite broadly (as emotional meaning, or emotional learning, which is vast and virtually interpenetrant with almost every higher activity in the CNS) or narrowly (the prototype emotional states of fear, rage, sadness, lust, etc.);
3) Emotion, broadly defined, in humans is spread out through many neocortical, paleocortical, subcortical, diencephalic, midbrain and brainstem systems, eluding neat localization in any "limbic system" unless the boarders of that are very broad;
4) There are differences between emotion as a conscious event (the activation of a strong feeling), and various unconscious types of emotional processing (i.e., unconscious valence assignments or affective behaviors such as avoidance) that have further divided the focus within the emotions research community: should we focus on feelings, or are they just a scientific distraction, while the real action of emotion is largely unconscious.

Recent work by LeDoux (1996), Panksepp, (1998) and Damasio (1994; 1998) have jointly moved emotion back onto center stage as a topic in neuroscience. However, within the burgeoning literature about the neural basis of consciousness, there are trends strongly paralleling this historical neglect of emotion, probably for the same basic reasons. Cognitive models dominate consciousness theory and research, while affect has been largely relegated to the back of the bus as an interesting coloration to the "hard problem" of consciousness (Chalmers, 1996), with a few exceptions. Cognition and affect are generally not conceptualized as intimately related, and emotion, within consciousness circles, is often seen as just an interesting type of "qualia" among many other types of qualia. Affect is a disadvantaged poor sister qualia at that, competing with better mapped visual awareness, which several adherents (notably Koch and Crick, 1995) offer as a best available neural network model for consciousness itself. Thus, the point of intersection of consciousness studies and emotion studies is "reduced" to the problem of the neural substrates of *conscious vs. unconscious emotional processes*, what I will call (after Chalmers, 1996) the "easy problem" about emotion and consciousness. Curiously, this very same conceptualization has been advanced in primary emotion research by the best known researcher in affective neuroscience (LeDoux, 1996). LeDoux is of the opinion that the main point of intersection of "affective neuroscience" and a science of consciousness is only the limited domain in which emotion enters experience through representation in consciousness mechanisms. This is seen as a small part of the big picture of emotional processing, much of which LeDoux sees (accurately) as going on unconsciously. Thus, overall, little consideration is given in "affective neuroscience" (excepting Panksepp, 1998), or in the rapidly expanding body of consciousness theory, to any potential role that emotion might have in underpinning consciousness, or to their potential intrinsic relations. The dominant assumption is that they are two fundamentally orthogonal processes.

The present review will attempt to present evidence for a somewhat broader view of the relationship between emotion and consciousness. This chapter will not address the "easy" problem of defining *neural substrates for conscious vs. unconscious emotion*, a question already reviewed by Öhman (1999), Damasio (1994), LeDoux (1996), and recently empirically investigated by Lane (2000) and Kaszniak *et al.* (1999), among others. Although the neural correlates of conscious (and unconscious) emotion are very important (and not really easy) problems, a less considered "hard" problem is whether emotion is simply one among many types of qualia, or is a necessary condition for consciousness itself. In other words, might consciousness require emotion, or are they truly "orthogonal"? This treatment will necessarily suffer from overly broad brushstrokes. The goal here will be to review evidence that emotion is a central organizing process for consciousness, and that any theory of consciousness must have a theory of emotion as one of its linchpins, which would suggest that many current theories of consciousness may have key foundations missing.

2. Terminological Problems - Just What Goes into Emotion Anyway?

Emotion is no "easy" problem. Just from the standpoint of simple definition, it is illusively, stubbornly, multi-dimensional. There are formidable terminological and nosological issues here, as emotion can be defined quite broadly (as emotional meaning, or emotional learning, which is vast and virtually interpenetrant with almost every higher activity in the central nervous system) or narrowly (the prototype emotional states of fear, rage, sadness, lust, etc.). Emotion in humans seems to bind together autonomic, endocrine, facial motor and global motor readiness activations, a poorly understood pain/pleasure valence, social signaling aspects, and higher cortical encodings (the high level other/self and social context encodings emphasized by many appraisal theorists) into a composite structure. From this perspective, emotion binds together virtually every type of information that the brain can encode. This composite ("supramodal") nature of affect has been a central factor in the morass of controversy and confusion in the various literatures on emotion. Because it has so many disparate features bound together, the study of emotion has often resembled the three blind men inspecting different portions of the elephant. Adding to the confusion and complexity, many of these various features can be, in varying degrees, either conscious or unconscious, and Öhman (1999) points out (in agreement with LeDoux, 1996) that "valence tagging" (assigning emotional meaning) can go on quite unconsciously. This has further divided the focus within the emotions research community: should we focus on feelings, or are they just a scientific distraction, while the real "action" of emotion is largely unconscious.

Although this complexity of features is frustrating (particularly for those looking for simple answers), this integration argues for a neglected point: that emotion might be related to, or a part of, the glue that holds the whole system together. In view of this, the generally low degree of tangency between emotion studies and consciousness studies is all the more surprising. This is particularly so, given the increasing emphasis by several prominent theorists (notably Baars, 1996; Newman, 1997; Taylor, 1999) on concepts of global access and global control (regarding competition between modular processors), and on the connectivities and networks that putatively support global integration of neural "information." Chalmers (1996) has argued that the most viable bridging principle for a theory of consciousness is the global integration or global availability of information in the central nervous system (CNS). This line of analysis suggests that possibly intrinsic connections between consciousness and emotion may have been generally poorly appreciated (a point developed further in the next section on Global State Functions). There are parallel blind spots within the emotions research literature, in that in both the "harder" neuroscience investigations (and in softer cognitive science approaches as well) most emotion research has been largely focused on the "top" of the processing hierarchy (and analogously the cognitive literature has focused largely on appraisal). There has been much less appreciation of the fundamental integration of higher cognition with basic social-biological value, which I would argue is the core scientific challenge for mapping emotion in humans, although Lane (2000), Damasio (1998), and several others are recently arguing for such hierarchical-distributed approaches. Such an emphasis on more integrative-distributed models for emotion might make some formal similarities with distributed models for consciousness more apparent.

While no definition of affect is going to satisfy all perspectives or theorists, there is some consensus that affect at least in humans (where its structure is frustratingly complex) involves at least (most of the time) a composite of the following elements:

1) *A precipitating event*, stressor or trigger that can be *external* (social event, threat, or affective expression or behavior of another in social context) or *internal* (a thought, memory, fantasy or other affect) or have related internal and external triggers); precipitants for affect run the gamut of possible stimuli.
2) *An assessment ("appraisal") of the precipitating event's meaning*, or some degree of cognitive processing that can come either *just before the experience of the affect* (a vital component of the precipitating process), *just after the activation of the affect* (a post hoc appraisal), or exist at both positions. This appraisal can be fleeting or detailed, deeply realistic and empathic, or profoundly distorted, or have complex admixtures of both realistic appraisal and distortion); processing of some events may have minimal "cognitive components" (e.g. predatory threats, primary loss experiences), and much of this meaning attribution can be quite unconscious.
3) *Subjective experiences along an intrinsic pain/pleasure axis (the crucial and poorly understood "valance" dimension of affect)* associated with various perceptions, ideas, sensations, actions, or representations of the precipitating event/trigger/stressor. Öhman (1999) points out that valence can also be assigned unconsciously or "subliminally").
4) *Motor, especially facial-motor changes, and differential motor "readiness" activations. These reflect the crucial adaptive "priming" of the executive systems by affect typically showing some version of approach vs. avoidance*, such as defensive, withdrawn, submissive, aggressive, seductive, affectionate/playful behavior; may show marked behavioral inhibition (freezing, blunting of expression) in the context of fear states, with considerable variability in terms of motor activity. In common English, these reflect our "personal intention," towards the situations/persons/events associated with affect.
5) *Complex autonomic - physiological changes (the crucial "visceral" aspect of emotion)*, with the most commonly studied being various cardiopulmonary

parameters, skin conductance, and various muscle tonic issues, but this aspect also could include endocrine and immune system changes.

3. The Functional Evidence – The Intrinsic Interpenetration of Global State Functions

In general, theories of consciousness have almost completely neglected distinctions between global state functions and channel functions (i.e., perception or any other modular thalamocortical processing channel) of the type proposed by Mesulam (1985), to their consistent detriment. It is as though all psychological functions somehow stand on democratically equal footing in the congress of consciousness, an intuitively appealing but questionable assumption. The distinction between global state functions and channel functions appears to be particularly relevant and informative regarding the conceptual difficulties that have been encountered by attempting to construct a general theory of consciousness solely from a theory of vision/visual awareness, as vision is a channel function. No adequate theory of consciousness can be constructed solely from an understanding of a channel function such as vision without mapping global state aspects such as visual working memory, the executive aspects of vision, etc. Recent work by Crick and Koch (1995) stops short of using any explicit version of global workspace theory. However, they do state that "the function of visual awareness is to produce the best current interpretation of the visual scene, in the light of past experience, and to make it available, for a sufficient time, to the parts of the brain that contemplate, plan and execute voluntary motor outputs (of one sort or another)." This is an acknowledgement that neuroscientifically valid notions of "visual awareness" cannot be constructed without reference to global state variables, in this case an explicit reference to executive functions supported in prefrontal systems.

Self-representation, pain, affect, attention, and executive functions (volition)) probably constitute the “big five” global state functions (GSF) that must all be linchpins in any viable theory of consciousness. They are deeply interpenetrating, both functionally, and in terms of their neural substrates. The maps we have currently for the neural architecture of these global state functions are heavily overlapping, and very highly distributed. Intriguingly, non-specific thalamic systems (various ILN systems and nRt) may be neglected key “players” in the neural architecture of all of these GSF. All notions referencing GSF share: (1) a central aspect of phenomenology as a starting point (i.e., feelings, attending, volition, pain, selfhood); (2) foundations of that aspect of consciousness mapped to deeply unconscious processes; (3) highly distributed networks running from ventral to thalamocortical regions, all involving reticular and non-specific thalamic systems; (4) basic interactions with other GSF, interactions currently poorly mapped in terms of neural substrates. Emotion modulates the higher aspects of global state in complex ways: by priming preparatory action sets, by directly influencing attentional and executive functions concurrent with its activation, as well as "downstream" influences (forming new emotional associations that are typically globally available (e.g., fear conditioning) and modifying basic habit systems) (see LeDoux and Panksepp chapters, this volume). Current hypotheses concerning the neuroanatomical distribution of other GSF include:

1) Affective functions (emotion broadly defined) are "localized" to a very diffusely distributed "limbic system" that seems to include just about every area of the brain excepting idiotypic cortex - "extended" notions about the limbic system include a host of prefrontal, paralimbic, telencephalic basal forebrain and subcortical gray matter systems (including the ventral basal ganglia, septal regions, and amygdala) many diencephalic regions, particularly anterior thalamus and hypothalamus, crucial midbrain areas, and monoaminergic brainstem core;
2) Attentional functions have been "localized" to RAS-MRF-thalamic loops, several other thalamic regions, prefrontal regions/associated basal ganglia, and several paralimbic, parietal, and heteromodal right hemisphere systems;
3) Executive functions have been "localized" to three parallel prefrontal-striatal-thalamic loops centered in dorsolateral, orbital and medial prefrontal regions.

These three mappings reference nothing that one could consider discretely separated regions or networks, and interestingly, all three are thought to be more crucially dependent upon right hemisphere systems (Mesulam, 1985). Lesions studies redundantly support this putative overlap in neural substrates: CNS lesions affecting attentional functions and executive functions (typically in prefrontal, basal ganglia, paralimbic, limbic, thalamic or brain stem regions) usually produce affective and personality changes. This suggests that affective, attentional and executive functions should be conceptualized as different kinds of poorly understood "global integration architectures," different slices of the consciousness pie. Put differently, consciousness might be the largest semantic "umbrella" subsuming many global state variables, each of them being part of a multicomponent neural envelope for consciousness. There is deep functional interpenetration (paralleling their neural architectural overlap) of the global state functions of affect, attentional function, and executive function that we have been taught (mostly) to conceptually separate:

a) The most critical aspect of attention relates to its executive aspects (what a person decides to focus upon, or what "grabs" attention), as these frames help define the content of working memory and are much more behaviorally relevant than its simpler "buffer" aspects (Baddeley, 1986);
b) Goals (invariably wish/fear based), implicit or explicit, conscious or unconscious, inform the frames for working memory (WM); these goals (implicit or explicit, conscious or unconscious) show *embedded value* and the

affective significance of virtually all WM frames – what is emotionally important and relevant has a major if not virtually determinant impact on defining foci of attention (the frames for working memory);

c) Affective activations critically influence and modulate executive functions, having the strongest impact on learning new paradigms for behavior – affects are the great internal reinforcers, a point neglected by most classically behaviorist points of view to their great (and eventually fatal) impoverishment, as they had to ideologically ignore internal processes about affect to prevent "inner life" from being smuggled in through the back door (these affective processes were relegated to what Skinner called the "black box" of the brain);

d) Executive function is geared globally towards the maximizing of pleasurable affect and the minimizing of painful affect (however adaptively or poorly this is conceived and executed, and despite enormous variations across individuals for what is rewarding vs. aversive);

e) Diseases that alter affective experience invariably affect motivation, underlining the specious nature of any distinctions between motivation (conceptualized as a core aspect of executive function by most) and emotion.

Perhaps the major "global derivative" of affective activation is the creation, elaboration and modification of the global representation of value (Watt, 1997), instantiated through long-term potentiation (LTP) mechanisms only partially mapped (see LeDoux and Panksepp's chapters in this volume. Representation of value has both conscious and unconscious elements, infiltrating virtually every aspect of personality and behavioral organization. Assigning or representing value is not primarily a cortical process or cognitive exercise dependent upon high level symbolic operations (although these elaborate value in vital ways for humans). Some appraisal literature has not considered that appraisal per se is not enough to explain emotion, only its frequent top-down activation by higher cortical encodings. Global changes in the representation of value probably rest in the kinds of complex co-activations fostered by LTP mechanisms linking thalamocortical regions with more ventral brain systems (networks involving amygdala and other basal forebrain structures, hypothalamic, midbrain, and brainstem systems). Idiotypic, unimodal and heteromodal cortices (isolated from midbrain, diencephalic, and subcortical systems) appear to be fundamentally devoid of mechanisms for defining biological or social value.

4. The Architectural Evidence - Neural Correlates of Emotion

It is axiomatic these days to think of emotion and memory of as "limbic system functions" but this is a problematic notion. Current schemes emphasize a division of the limbic system into a *paleocortical* evolutionary trend (amygdaloid-centered) and *archicortical* evolutionary trend (hippocampal-centered). Underlining the limitations of any unitary concept of the "limbic system," short term memory in the hippocampal-archicortical trend is seen currently as more allied with cognitive functions and cognitive mapping. These two evolutionary trends support "episodic memory" (a transmodal serial linkage of cortical sensory-motor encodings with spatiotemporal coordinates to enable short term memory) and "emotional memory" (the linkage of cortical and thalamic sensory-motor encodings with "valenced" activations of autonomic and other systems (best known for fear).

The borders of the "limbic system" are vague and have been extended decade by decade like the erosion of a vast neural shoreline (see Fig. 1). Various "extended" notions about the limbic system include a host of paleocortical paralimbic, basal ganglia, thalamic and hypothalamic, basal forebrain and other subcortical systems, including even monoaminergic portions of the RAS brainstem core (Derryberry & Tucker, 1992). Some models of the limbic system suggest that the entire prefrontal systems and heteromodal association cortex in the right hemisphere could be considered part of the limbic system – as its extended association cortex. One wonders what is left – what is not "limbic system" beyond idiotypic or primary cortex, and regions such as Broca's or Wernicke's areas. In mammals, neural correlates for even the prototype states (from Panksepp, 1998) - namely fear, rage, lust, separation distress, play/joy - are spread out through many systems in the basal forebrain, diencephalon and midbrain. This is consistent with an assumption that all global state functions have dense (and interpenetrant) roots in ventral brain systems, and from those roots, recruit many regions (increasingly differentially up top). There few borders to the "limbic system" if one fails to make clear distinctions between primitive or primary emotion and "emotional meaning," "secondary emotions" and "emotional learning." These are very broad corticolimbic functions, affected by sum total of long term changes in many systems mediated by synaptic plasticity mechanisms following the activation of primary affect. This "spread" of the "limbic system" suggests that the brain is not non-specifically dedicated to the processing of "information," but to the processing of events in terms of an interpenetrating hierarchy of biological, social and personal-subjective values, dedicated to the deepening interpenetration of "value operators" at many levels of the brain. This is also consistent with the assumption that hierarchical distributed models are critical to the modeling of all global state functions.

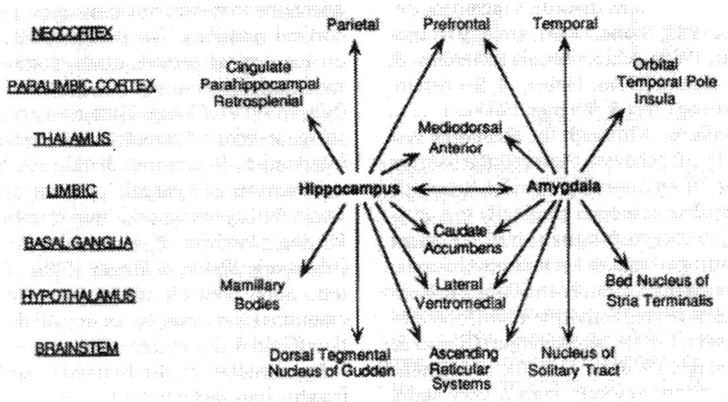

Figure 1. The “Limbic system” is highly distributed ("fully distributed"?). Highly parallel and globally distributed nature of limbic connectivities running from the brainstem (including portions of ERTAS) to the highest neocortical centers. Omitted from this schema are critical limbic inputs into non-specific thalamic systems (NRT/ILN – the reticular and intralaminar nuclei) from basal ganglia, prefrontal systems, accumbens, and paralimbic cortices, and crucial regions in the midbrain (from Derryberry &Tucker, 1992).

A difficult problem has been understanding the biphasic nature of affect (the "plus or minus" nature of all affective valence). This primary feature of affect has been appreciated for as long as there has been human culture: that we have loves and hates, likes and dislikes, attractions and aversions. We are thoroughly ambivalent creatures in our relationships with significant others and we struggle with our ambivalence from birth to death. Perhaps the most famous instantiation of this ancient principle was Freud’s dual instinct theory, and appreciation for the depth of human ambivalence is perhaps the beginning of emotional wisdom. Yet as fundamental as this is, its neural architecture is not at all clear. Work on emotion and startle probe investigations (Lang, 1993; Kagan, 1992) suggests that the bipolarity of affect must be grounded in two systems that would have to be push-pull and mutually inhibitory. There is also work relating these differential affective valences to the two hemispheres (the right biased towards the experience of negative affects, and left biased towards benign or positive affect), and also to prefrontal and parietal cortices (Davidson, 1992). These are empirically fairly robust correlations that tempted Heilman (1997) to suggest that modules for positive affect vs. negative affect are organized in left vs. right hemispheres (or in parietal vs. prefrontal regions). I criticized this line of thinking (Watt, 1998), as a "cortico-centric" conceptualization of affect, which neglected evidence that higher cortical zones by themselves cannot supply valence or value at all. Such a conceptualization ignores evidence that stimulation at various levels of distributed subcortical systems (brainstem, different portions of midbrain and hypothalamus, septum vs. amygdala) can elicit painful or pleasurable affect, while neocortical stimulation does relatively little to generate affect (Panksepp, 1996). This cortical "lobar" or "laterality" view of valence also ignores how metabolic imaging tasks intrinsically over-represent cortical regions, as they are inherently more active metabolically (Panksepp, 1998). Instead, correlations of valence with activations of large lobar and even hemispheric regions may just be endpoints or global activation tendencies derivative of a complex neurodevelopmental course not yet charted. This neurodevelopmental course builds highly distributed networks involving counterbalanced complementary structures at multiple levels of the neuroaxis: multiple brainstem regions in the reticular core, midbrain and diencephalic regions that appear to provide “affective prototypes” or “primes” (discussed below), several basal forebrain regions, several paralimbic regions (esp. cingulate vs. orbital frontal) and the highest neocortical regions (parietal vs. prefrontal, left vs. right hemisphere). Just as Broca’s areas gets control over bilateral motor neurons, specific right frontal regions probably get control over "dysphoria neurons" that may be highly distributed. Given that the cortex may be the “playground of the emotions” (Panksepp, 1998) and that cortical systems are essential for the common top-down activation of emotion via complex appraisals, a cortico-centric theory of valence is easy to defend. However, this is not consistent with what is known about the prototype states, and it does not ground emotion in evolutionarily important “rewards” and “punishments” that encode rewards for actions/events that promote survival and procreation and aversive consequences for those that promote the opposite. Such a theory must have a much more primitive base than one placing positive and negative valence in complex heteromodal and paralimbic regions (like prefrontal and parietal lobes). Such a neural system for value or valence must integrate species and individual survival/biological need issues (hunger, thirst, sexual-reproductive issues, etc.) with defense issues, and with the complex problems of attachment and social relations to conspecifics (Watt, 1998). Attachment and all affective issues are hugely interpenetrating, increasingly so as one climbs the phylogenetic tree into the primate and then hominid lines (Bowlby, 1969).

There is certainly major evidence for the role of the amygdala in “valence tagging,” or establishing valence (fear conditioning to be more precise) for certain classes of stimuli, due to much elegant work by LeDoux (summarized in LeDoux, 1996). This structure seems to act as a high level correlator associating complex stimuli from thalamus and various cortical regions with autonomic, endocrine and behavioral activations via its efferent outflows from the central nucleus to various targets in hypothalamus, brain stem, and PAG midbrain. One theoretical conundrum in all this, generally not acknowledged, is that a high level correlator structure such as the amygdala cannot by itself supply primitive biological or social value or valence, only associate more primitive value or valence activations with various encodings from the more dorsal portions of neuroaxis such as cortex and thalamus. Supporting this line of analysis is empirical evidence (Panksepp, 1998 for summary) suggesting that coherent fear or rage states ***can*** be activated by structures underneath the amygdala, ***not*** simply "pseudo-affect," or autonomic and other "fragments" of true affect. In other words, the long standing assumption that such affects are "sham," (e.g. "sham rage" from hypothalamic stimulation) may be a "sham" itself.

5. The Architectural Evidence (Part II): "Unpacking" Valence and Value into Affective Prototypes in the Midbrain

The work of Jaak Panksepp (1996; 1998) is contributory in "unpacking" valence into its fundamental modularity (different kinds of primitive + and -) in terms of core prototypes of social-biological relationship to other species and conspecifics, with these likely fundamental to all mammalian brains. This work has been neglected in most reviews of the neural correlates of emotion. Panksepp (1998) summarizes a group of core networks in the midbrain-diencephalon that support prototype affects or emotional "primitives." Each distributed circuit (Table 1) appears to modulate a prototype "grade A" primary emotion: attachment/bonding, nurturance, sadness/separation distress, fear, rage, play/joy, and a seeking/expectancy system that supplies non-specific motivational arousal and probably much of the primitive preconscious substrate for the fundamental experience of *hope*. Each circuit is heavily (but not exclusively) neuropeptide mediated, and projects to periaquaductal gray (PAG).

Table 1. Distributed Midbrain-Diencephalic-Basal Forebrain Chemoarchitectures for Emotional Primitives (Prototype Emotions) (Extracted from Panksepp, 1998)

Affective Behavior	**Distributed Neural Networks and Major Structures**	**Neuro-modulators**
Non-Specific Motivational Arousal – Seeking and Exploratory Behavior	Ventral Tegmental Area (VTA) to more dorsolateral hypothalamic to periaqueductal gray (PAG), with diffuse mesolimbic and mesocortical "extensions." Nucleus accumbens as crucial basal ganglia processor for emotional "habit" systems.	DA (+), glutamate (+), many neuropeptides including opiods, neurotensin, CCK
Rage/Anger – ("Affective Attack")	medial amygdala to bed nucleus of stria terminalis (BNST) to anterior and ventromedial and perifornical hypothalamic to more dorsal PAG	Substance P (+) (? Ach, glutamate (+) as nonspecific modulators?)
Fear	central & lateral amygdala to medial and anterior hypothalamic to more dorsal PAG to nucleus reticularis	Glutamate (+) and neuropeptides (DBI, CRF, CCK,

	pontine caudalis	alpha MSH, NPY)
Sexuality	BNST & corticomedial amygdala to preoptic & ventromedial hypothalamus to lateral and ventral PAG	Steroids (+), vasopressin and oxytocin
Nurturance/ maternal care	Anterior cingulate to bed nucleus of stria terminalis (BNST) to preoptic hypothalamic to VTA to more ventral PAG	Oxytocin (+), prolactin (+), dopamine, opiods
Separation Distress/ Social Bonding	Anterior cingulate/anterior thalamus to BNST/ventral septum to midline & dorsomedial thalamus to dorsal preoptic hypothalamic to more dorsal PAG (close to circuits for pain)	Opiods (-/+) oxytocin (-/+), prolactin (-/+) CRF (+) for separation distress, ACh (-)
Play/Joy/ Social Affection	Parafascicular/centromedian thalamus, dorso-medial thalamus, posterior thalamus, projecting to ventral PAG (septum inhibitory re: play)	Opiods (+ in small - mod. amounts, - in larger amounts), ACh (+)
? Social Dominance	Not clear if separate from activation of play systems and inhibition of fear systems??	

Legend: (–) inhibits prototype; (+) activates prototype; PAG = periaqueductal gray; CCK = choleocystokinin; CRF = corticotrophin releasing factor; ACTH = adrenocorticotropic hormone; DBI = diazepam binding inhibitor; ACh = acetylcholine; DA = dopamine; MSH = melanocyte stimulating hormone; NPY = neuropeptide Y.

6. The Architectural Evidence (Part III): Connectivities Between the "Limbic System" and ERTAS Systems

There are many sites for intersection between the basic architecture of value/emotion with the ERTAS proposed by several reviewers as the mostly likely neural network architecture for consciousness (Baars and Newman, 1993; Newman, 1997; Baars, Newman & Taylor, 1998; Taylor, 1999). These critical sites for important interactions are seen at each level of the ERTAS, starting with the reticular core and progressing all the way up to the prefrontal systems: much of the reticular core, particularly its monoaminergic portions, which we have been taught to associate with cortical arousal, was originally aimed at primitive limbic forebrain arousal (with cortical arousal evolutionarily "tacked on"):

1) There are important afferents from PAG to both MRF and several ILN/ midline thalamic systems. PAG shows both ventral (*vl*) and dorsal (*dl*) pathways projecting (differentially) to non-specific thalamic ILN systems (including the centrolateral (*dl*), centromedian (*vl*), parafascicular (*vl*), and paraventricular (*dl*) midline systems) as well as feeding back differentially

onto different portions of hypothalamus and the various monoamine "spritzers" (Cameron, 1995 a/b);

2) the nucleus reticularis thalami (nRt) shows dominant limbic, paralimbic and heteromodal control of nRt gating, with modulation of nRt "gatelets" by nucleus accumbens, paralimbic cortices, BG, and DM thalamus-prefrontal regions (Cornwall, et al 1990); more specifically, there is NA gating of HC and prefrontal afferents to nRt (Newman and Grace, 1998). Taylor (1999) emphasizes that the limbic system input dominates the anterior portions of nRt without reciprocal inhibitory controls (see figure 2);

3) reciprocal connections of ILN to reticular core, midbrain PAG, limbic, BG, and many cortical systems (Newman, 1997); There are extensive prefrontal projections to ILN, as part of the top down control of ERTAS, including from paralimbic prefrontal systems (cingulate and orbital frontal) (Newman and Baars, 1993).

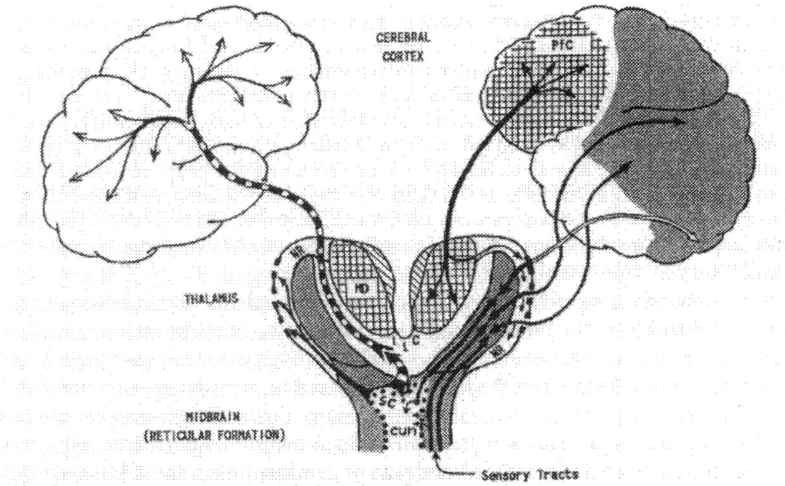

Figure 2. Representation of Current ERTAS Global Workspace Theory, showing a coronal section through the midbrain and thalamus, illustrating projections (arrows) between them, and with the cerebral cortex. The shaded areas represent: classical sensory pathways (in the midbrain); the ventral nuclei of the thalamus; and the areas of cortex (right side) with which these nuclei share projections. The crosshatched areas designate the medial dorsal (MD) nucleus and pre-frontal cortex (PfC). The unshaded areas in the thalamus and midbrain constitute the reticular core responsible for the global activation of a Tangential Intracortical Network via projections (dashed arrows) from the midbrain Reticular Formation (sc/cun) and intralaminar complex (ILC). The heart of this extended activation system is the nucleus reticularis (NR). (from Newman and Baars, 1993).

Taken together, these argue that "non-specific" nRt/ILN regions crucial to the extended reticular thalamic activating system (ERTAS) have rich, even dominant limbic, paralimbic and heteromodal connectivities, which provide a neural basis for: 1) emotionally relevant stimuli to influence attentional content; 2) for different aspects of emotion to enter consciousness; and 3) for limbic inputs to facilitate binding in widely distributed networks via their influence on non-specific thalamic systems (binding and synchronous activation of widely distributed networks being one of the leading mechanisms thought to possibly underpin the integration of features in qualia (Llinas, et al, 1994; Engel et al, 1997). These dense limbic and paralimbic connectivities of the non-specific thalamic systems thought to subserve gating and binding functions suggest that affect must be more than simple "coloration." Gating and binding may depend in part on affect's global "valance tagging" of all other encodings. Gating appears more dependent on the higher paralimbic, prefrontal and limbic inputs into nRt, while binding and arousal may depend on the lower level PAG projections to ILN and midline thalamic systems. It is currently unclear what specific role the PAG – ERTAS connectivities might play in the generation of consciousness, or in conscious vs. unconscious emotion, but it is impossible that the projections of all the prototype affective systems into PAG are basically functionally trivial. Some possible correlates are:

1) *facilitation of "intention integration" functions supported in ILN* (Symthies, 1997), which appear to be required for coherent "intentional" content in ERTAS ("agency"), and *facilitation of arousal functions supported in MRF*;
2) possible unconscious contributions via *a tuning or priming role roughly analogous to what Newman (1997) outlines for SC (superior colliculus) role vis a vis cortical structures* in ERTAS that are involved in higher perceptual–motor functions, with SC supplying a primitive ambient mapping of the spatial envelope around the organism essential for higher cortical structures to function coherently. This analogy would make PAG a global low resolution "tuner" of higher limbic systems, priming them in certain affective directions via a global value mapping;
3) Implicit in its function as a clearinghouse for the prototype affects outlined in Panksepp (1998), PAG might run ongoing inhibitory competitions between them (possibly in its radial topography between its four longitudinal columns), roughly analogous to what nRt is doing higher up in the neuraxis with a much more informationally dense and complex thalamocortical system. Since there would be less to do, it should be a faster structure for competitions than nRt, consistent with the lower presentation thresholds for affective valence assignments than for perceptual qualia. This nRt analogy is speculative, but hardly without evolutionary sense, and not addressed anywhere in the literature on PAG that I am aware of. One might wonder how (except via a clearinghouse structure that is running on-going internal competitions) could one instantiate a neural system for rapid, centrally

administrated inhibition of any kind of sexual or play activations, when the brain gets any kind of signal from the rage/fear systems that the organism must go into a defensive mode. Lateral inhibition at the telencephalic level would probably be much too slow. This type of competitive process between potential affective "attractors" could not be supported in amygdala, given its minimal/ambiguous role in play, its central role in rage, limited role in attachment, etc., (i.e., just too much variability in its participation). It is also too high in the system to function as a convergence zone for primary affective activations. Instead, work on amygdala suggests a high level correlator that allows linkage of both higher and lower resolution perceptual encodings with basic defensive responses, responses that must be instantiated by more ventral systems in hypothalamus and PAG (the output targets for the amygdala's central nucleus (LeDoux, 1996)).

A PAG-instantiated competition might provide substrates for the differential agonism and antagonism of the prototypes in adaptive functioning, underpinning much of "adaptive common sense" in basic emotional functioning. For example, low to moderate degrees of activation of the play system would agonize the sexuality systems, and low to moderate degrees of activation of the fear systems would agonize the rage systems, while large activations of the rage systems might inhibit the fear systems, etc. This running of competitions between the prototypes would insure very rapid central inhibition of play, sexual and other positive states that animals could ill afford to have "hanging around" in the context of the need to rapidly mobile fear and rage systems in survival situations. Adaptive functioning would not be consistent with anything other than very fast shut-down of the positive affective systems (attachment, play, sexuality) when survival is at stake, as it is far worse to "botch" defense against a lethal predatory threat than to miss out on a sexual or "fun" encounter. Additionally, large activations of the negative affective systems would have the ability to shut down the positive systems for quite some time, but not visa versa. Thus, the dampening balance in PAG is probably against the positives, possibly consistent with the literature on PAG emphasizing its role in defensive functions.

Clearly, there are many unanswered basic theoretical issues here and a dearth of empirical materials to address them. Are the thresholds for conscious affects mediated directly in PAG, or is it in nRt (as it may be for more cognitively mediated contents) or elsewhere? What are most basic neural activation elements for emotional qualia (feelings)? This is where things get murky (this controversial issue of what constitutes "feelings"), although the traditional answers of visceral and other sensory feedback, and LeDoux's recent emphasis (1996) on working memories about affective events (triggers?) are prominent in the literature. Panksepp's work suggests the possibility that the core of "feelings" are provided by intrinsically pleasurable or unpleasurable action primes ventrally organized in

PAG. But even this does not address the issue of what makes these activations rewarding or aversive. Why do emotional activations feel good or bad? There is nothing learned or particularly modifiable in this crucial and still mysterious issue of valence, and this core of feeling "good" or "bad" does not seem to require much leadership from cortex, although cortex clearly can participate via complex working memory systems that can hold complex content on line. Being enraged or fearful is not rewarding, unless it becomes paired with experiences that lead from those states to positive states, and then of course, it is the anticipation of the positive states that becomes rewarding, as those states are intrinsically desirable and positive.

In this sense, the older visceral feedback notions from James and the newer working memory dimensions emphasized by LeDoux both still leave out any explanation for the intrinsic dimension of painful or pleasurable state or organismic value, which frames emotion as a global state function with close ties to pain/pleasure. Clearly, there are sensory-motor correlates to emotional states, but these don't adequately explain valence, at least not by themselves. (For another perspective on this troublesome problem of intrinsic valence, a problem that will not "go away", see Chapman's treatment of pain in this volume). Perhaps a basic model for this is provided by hypothalamic set-point detection. When internal physiological states are outside a desirable range, both visceral sensations and action dispositions (thirst, and pursuit of fluids) are activated. But phenomenal states of rage, separation distress, fear must have similar mechanisms, that these are "*not OK*" departures from ideal organismic baselines, activating defensive responses, while play and affection, sexual stimuli, etc., must encode or activate the opposite, setting in motion basic appetitive mechanisms. These are central and not peripheral aspects of affect. Events that modify this crucial valence aspect alter the whole experiential picture, and not just "pieces" or fragments of affect (e.g., administration of small dose of opiates to animals in separation distress). It isn't that the animals continue crying, but "just don't really mean it anymore," ("sham distress") or that they are still sad but are "just not expressing it anymore" (behavioral inhibition). The entire coherent emotional response drops out and is suppressed by the opiates. Any assumption here of the primacy of working memory over valence (or that WM completely explains valence) is putting the dorsal (cortical) cart before the ventral (limbic) horse. What is most disconcerting about the working memory and visceral feedback explanations is a curious "sensory-centered" and "cogno-centric" bias, as these offer essentially a "no comment" on this crucial valence dimension of emotion, its intrinsic reward value, and probably fundamental ties to pain and pleasure. In summary, I would argue that the notion of valence cannot be modeled without some concept of action priming. We know that something is negatively or positively valenced because genetically coded action paradigms tell us that our organism wants more or less of it. Without that, there is simply no way to model intrinsic value.

Other evidence for this experiential "primacy" of valence and its functional independence from working memory comes from the phenomenology of patients with deliriums, which I have argued (Watt, 1995) is a relatively pure disorder of working memory (particularly its framework aspects, its thematic coherence and "streaming" (Newman and Grace, 1998). Delirium is typically associated with Alzheimer's disease (AD) and anticholinergic drugs, probably because of the important cholinergic mediation of the ERTAS. The only focus that these patients can typically maintain is on some emotionally loaded internal directive (typically to get the "____" out of the hospital, or to get out of the posey that often is restraining them, etc.) These are patients for whom the normal complex smooth sequencing of WM is largely collapsed (depending of course on the severity of the confusional state). WM loses its coherence in these delirious patients, except in terms of their dramatic perseverations driven by emotionally primitive states. I suspect that this is due to the relative collapse of the normal field of thalamocortical interactions supported in ERTAS, and the relatively primitive and limited states that the conscious system can now enter, with these being largely determined by emotionally primitive reactions: fear, rage, primitive needs states re: other people, separation distress, etc. There is nothing like the normal operation of a smoothly sequenced working memory in these patients, no coherently sustained complex cognitive content, and this is not simply a feature attributable to their baseline dementias which sometimes are quite mild. Delirious patients without baseline dementia also show the same collapse of working memory, and the same perseverative focus on basic affective themes, although delirium is much rarer outside of some stage of AD, and in these contexts it typically requires a more severe metabolic, structural or neuromodulatory disturbance for its generation. This suggests that emotion provides coherent activations to the ERTAS, and that in states of delirium those are mostly all that is left for the extended reticular thalamic activating system in terms of global chaotic attractors. The "fine grain" of more cognitively and cortically tuned consciousness is gone, and one is left with the much coarser emotional priming(s) that primitive subcortical systems, those presumably less dependent on the cholinergic modulation of ERTAS, provide.

We are still in the dark about this fundamental valence aspect of emotions, what makes them feel good or bad. However, this line of analysis suggests that valence may have to do with the global resonances initiated in PAG (and the differences between the defensive and appetitive systems activated in PAG and other ventral brain regions such as hypothalamus) which are then spread throughout ERTAS via the PAG to ILN and MRF connectivities at the base of ERTAS. Conscious primary or prototypic emotion (fear, rage, separation grief, etc.), may be a "trajectory" category for the whole of ERTAS. Clearly, sensory feedback about visceral and motor activity, episodic memories, working memories about trigger phenomena, and cognitive appraisals that act as top down drivers activating emotion, are all part of the primary conscious experience of emotion, but all of these still leave out the question of what makes feelings rewarding or aversive, their intrinsic ties to pain and pleasure, which I would argue is the core of "valence." To understand that, we need a deeper understanding of pain and pleasure as primitive qualia in consciousness, which in turn requires an understanding of primitive but integrated body and value mappings, subjects around which we currently have minimal established neuroscience. However, I would argue that PAG is clearly a candidate structure to provide such primitive value mappings, while amygdala is not, due to its dorsal position in the brain, its primary involvement in defensive functions, and its relative dependence upon activation of PAG to instantiate an emotional response.

Regarding this larger problem of conceptualizing the relations of emotion and consciousness, there seems little acknowledgement in the literature that since we know much more about cognition and language than we do about pain and pleasure that the possible foundations for consciousness in pain and pleasure are getting short shrift. The consciousness literature at times seems gravitationally drawn into basic assumptions that consciousness can be constructed almost value-free, a neat cognitive process that is largely independent of any "messy" foundations in organismic pain, pleasure, other biologically grounded values. Of course, such a vision of consciousness would make it fundamentally independent of primitive defensive and appetitive systems in the basal forebrain, diencephalon and midbrain that are themselves probably evolutionary extensions of more basic hypothalamic set-point homeostatic mechanisms. If so, these are highly questionable assumptions from the standpoint of evolutionary theory, which would virtually mandate explanations in which higher cortical processes are adaptive extensions or modifications of biologically successful "lower," more primitive mechanisms.

7. Summary: Consciousness Theory and Emotion

This PAG-MRF-ILN intersection between traditional ERTAS architectures and the midbrain-diencephalic architecture for emotional primitives underlines that traditional distinctions between a dorsal cognitive thalamocortical architecture and a ventral limbic architecture are misleading on two counts: so-called "non-specific" thalamic regions are centrally involved in both cognition and emotion, and midbrain PAG global value mappings appear essential for normal conscious functioning via their influence on MRF and ILN at the base of ERTAS. Although PAG lesions in rats (Panksepp, 1998, unavoidably involving SC) have been shown to virtually ablate consciousness, homologous cases in primates (and without the superior colliculus being involved) allowed a dim kind of twilight state but without any kind of behavioral spontaneity/intentionality. PAG – ERTAS connectivities may explain how global representation of value is essential for self and world to become "real," as PAG outputs may facilitate binding of features (perhaps particularly the binding of motor representations supported in various ILN systems) and other forms of binding supported in other non-specific thalamic systems. The ontogenetic prototype for this may be the binding of features into a coherent and emotionally positive image of the mother's face that marks the infant's first (intrinsically emotional) response to another, but we know virtually nothing about this at a neural systems level. Despite our ignorance of neurodevelopment, from a ontogenetic standpoint there is some evidence for this supposition that affective systems participate in feature binding: early conscious states appear to be prototypically affective. "Feelings" may be *proto-qualia*, or *proto-experiences* (what Panksepp (1998) calls "e-qualia").

Emotion (in the form of these midbrain-diencephalic primitives) defines biologically compelling prototypes for self-world relations based in "primes" for relations to other species and conspecifics, such as the confrontation with a predator (the fear system), the attachment to and/or loss of a mate, child or parent (the bonding and separation distress systems), the confrontation with an aggressor/rival (the rage system), or playful affection with a conspecific (the joyous engagement supported in the play system). Given the assumption (Metzinger, 1998) that consciousness depends on coherent self-world models, this set of primitives thus generates a group of primary "wetware instantiated" models for basic self-world relations that could be (further) cognitively developed. One is led to wonder here if these primes or prototypes operate as resonance points within the global ERTAS architecture for consciousness, and whether these primes are essential foundations for primitive qualia. These prototype affective states, by initiating various global resonances, may prime the "virtual reality" generation that many (see Revonsuo, 1995; 1998) see underpinning consciousness, as consciousness reflects a real time, ongoing "virtual self-world model." At this point we do not know if self-world models can exist if completely stripped of neural connection to these affective primes or basal neural prototypes for self-world relations, but the richest contents of consciousness suggest that the influence of these primes is deceptively pervasive. It is not some truly isolated "redness of red" that catches our eye, but rather the aesthetic (affective) redness of the rose that reminds us of things and people we love. It is not some isolated motor proprioception that we experience richly, it is the last agonizing and impossible stretch to the ball that scores the winning goal, or fails to.

Consciousness has to reflect global integration, as both agency and value are deeply embedded in what might initially appear to be passive sensory qualia. Some have argued persuasively (Panksepp, 1998; Cotterill, 1995), as agency is central to selfhood, that the most basal foundations for consciousness must rest on motor maps (which may have more stability than sensory maps). In any case, sensory-motor coherencies, and their connections to value mappings seem central to C: the connectivities between a primitive value map in PAG, a primitive sensory map largely in superior colliculus, and primitive motor maps in deep tectal and tegmental areas may form the most basal neural representation for self (Panksepp, 1998). It would be very hard to know that one existed if one could not correlate on-going sensory changes with activated action schematas, and both of these with value schematas that generate and predict inherent internal rewards and "punishments." These correlations may enable the most basic and primitive feeling that we exist, and that what we do matters, and has effects, good and bad. Without primitive value correlates (possibly largely contributed from PAG) interacting with the primitive sensory and motor mappings, sensory-motor correlations wouldn't mean anything by themselves. Although there are probably many "NCC" (neural correlates of consciousness), such integrations have to start subcortically and from there generate reiterations in highly distributed networks, particularly in the prefrontal systems, which a number of authors have implicated in the task of self-representation and self-awareness. In any case, this poorly appreciated midbrain integration of sense, value and action may form foundations for a primitive yet superordinate "self-model" that Metzinger sees as an essential foundation for consciousness (Metzinger, 1998), underpinning the normal "ownership" of qualia, a basic property of consciousness as yet unexplained.

8. Suggestions for Future Research

There is much still work to be done in understanding the prototypes for affect generated in the midbrain and diencephalon, and how these are reiterated throughout higher layers of neural architecture (such as septum, amygdala and paralimbic regions) dedicated to "valence tagging" of various higher encodings, forming progressively more complex emotional associations/actions. Every prototype will need a Joe LeDoux to understand how its particular version of long term emotional learning works in terms of structures, pathways and molecules, although there may be very similar molecular mechanisms for most LTP. But projection of all of the prototypes affective systems into PAG suggests cautions about adopting LeDoux's (1996) recommendation that we abandon notions of a "limbic system" under the assumption that there are many discrete emotional systems in the brain that have little substantive architectural integration. There may be greater "spread" of various discrete emotional systems "up top," but increasingly crowded lanes of traffic as one gets into the diencephalic systems, with final intersections in PAG. But there is not much established theoretical fabric here, and much ignorance about the fundamental relations between emotion and consciousness. Basic questions include:

1) How does one even define conscious? Are mammals in a real sense sentient beings? (There is clearly some intrinsic "Nagelian" uncertainty here, and we don't have any real outline for a minimally conscious neural architecture.) Is there just one type of consciousness, one that requires full awareness that one exists and high level linguistic-cognitive abilities, or is it better to think in terms of some version of a developmental hierarchy with a fuzzy transition to a periconscious substrate in earlier phylogenesis? Or is this too fuzzy?
2) Is the generation of affective valence - that stimuli are organismically positive or negative - and the arousal of the whole forebrain (towards consciousness) more or less fundamentally separate or "orthogonal?" Is arousal inherently motivational and hot, or "neutral - cool" and related primarily to the need to have cortical systems optimally tuned for sensory information processing and symbolic representation? Trickier yet, are these truly separate from a global neurodynamic point of view? In other words, is "cognitive neutrality" partially illusory? Or does optimal cognitive "openness" or maximally adaptive symbolic representation (even just of the natural world let alone of one's self) require a certain euthymic dynamic balance in affective systems?
3) If valence and arousal are basically "orthogonal," then how does valence enter consciousness? Since arousal is ventrally organized and by definition foundational (running the length of the reticular columns in the medulla to midbrain to thalamic ILN), if arousal and valence are fundamentally separate, how and where does valence get introduced into the forebrain systems? Since valence must integrate homeostasis ("biological needs"), and the "state-space" of conspecific relations in mammals ("social needs"), valence itself seems to mandate distributed hierarchical systems.)
4) In outlining the substrates of a complex function, is it more important to conceptualize the brain in terms of mostly discrete modular systems, or in terms of global integrative processes, or are these just the latest (neurodynamically misinformed) examples of a localizationist - mass action dispute? Is the current balance of thinking tilted away from adequately weighting the importance of global state functions (what central processes integrate the modular aspects), or not?
5) Is current theorizing about affect cogno- or cortico-centric? Or is emotion fundamentally dependent upon and driven by cortical and cognitive mechanisms? Or are cognitive-appraisal issues in fact even under-weighted? Is consciousness also fundamentally dependent on high level cognitive and cortical mechanisms, or do "foundational processes" more primitive than cognition simply recruit cognition as the latest layer on the onion?
6) Does "structural chauvinism" significantly color research and theory in neuroscience, with various theorists and researchers championing their favorite structure (such as paleocortex or amygdala or PAG)? Or is just this

intrinsic to the correct identification of the basic modularity of the system? A hierarchical distributed model, with PAG at the base and cingulate and perhaps much of the right hemisphere at the processing apex and amygdala right in the middle has much to recommend it empirically (see Lane's chapter). In relationship to such hierarchical models (which few quarrel with in principle), is there adequate appreciation of contributions from the base of the processing hierarchy (the deep ventral reticular - midbrain systems) to emotion and consciousness? Are the PAG - hypothalamic regions mostly passive output systems that just insure the right visceral/autonomic/motor buttons are pushed? Are these ventral systems more integrative? Related to all these questions, are the ventral members of any processing hierarchy more essential than the dorsal ones to the basic integrity of the functions enabled by the distributed network? Or are the "higher systems" more crucial?

7) Are conscious emotions or "feelings" basically dependent on working memory? Or is working memory in conscious emotional states more a cognitive "re-presentation" of complex aspects of valence and value that are more deeply felt? Do those deeply felt valences from the prototype emotional states require much cognitive function, or are they "pre-categorical" conscious states that reflect primary organismic valuing? If the latter, why and how do they feel "good" and "bad" (what gets valence into consciousness)? Does this just stem from learned (and semantically labeled) identifications of basic visceral-autonomic changes in the different affective states? Or is primary organismic value instantiated through more fundamental mechanisms, such as the model of hypothalamic set point detection might suggest?

8) Is "working memory" itself a fairly delimited higher cortical function, or a more global thalamocorticolimbic integration that requires a fair amount of background (tonic) affective information (given the heavy limbic and paralimbic connectivities of nRt, and the deep influence that emotion has on attentional functions).

These questions clearly have a basic overlap and map a vast theoretic landscape. We obviously want tests for these questions. But it is worth emphasizing that much current research in consciousness studies *and* emotion studies hasn't formulated that these are important questions *AT ALL*. In the absence of that, and the generation of research informed by these questions, how can one expect much progress? But if we grant that these are important questions, how might we investigate them? Three specific approaches are suggested here: one looking at global neurodynamic issues, a second approach based on examining thresholds for conscious emotion that fall underneath thresholds for working memory, and a third advocating for metabolic imaging technologies that compensate for relative cortical hypermetabolism.

First, there is the question of whether conscious emotions show global neurodynamic correlates. We don't know if conscious emotions truly reflect "global operators" and (if so) where and how are those global resonances initiated and distributed? John Smythies' (1998) retrieval of a fascinating experimental paradigm by Roy John and Keith Killam (1967, cited in Smythies, 1998) supports a methodology to better identify structures/connectivities that initiate onset of an emotional state. One could look more closely at the neural correlates for the appearance of global and coherent behavioral changes that are part of the emotion "gestalt," and to what extent activity in certain structures (carrying the neurodynamic "rhythm marker") initiates a specific global resonance signature coincident with onset of an emotion. Ideas advanced here would suggest that the onset of emotion (even for fearful stimuli) would not coincide with the initial processing of precipitating stimuli in the lateral amygdala, and not even with central amygdala outflows (although that would be closer in time) but with the consolidation/activation of operators in PAG. Although all these events would be very close in time, time-synched video recordings of animal behavior would probably offer enough temporal resolution when linked to neurophysiological monitoring of multiple sites. Relative independence (or sizable time lags) between thalamocortical operators appearing (with clear evidence of conscious emotional correlates), and the activation of ventral brain operators in hypothalamic - PAG networks (with minimal evidence of conscious correlates) would clearly support LeDoux's and Taylor's positions on these undecided questions, and cast doubt on Panksepp's and Damasio's. But in any case, the method John Smythies (1998) suggests would offer a path to explore all this much more in detail, if one could get the monitoring electrodes in enough of the right places.

A second empirical pathway would be to look at events in which brief presentations of emotional stimuli allow awareness of valence without recognition of content. Work by Bradley and Lang in this volume shows us that we can get "valence information" in consciousness with stimuli presented under the thalamocortical thresholds required by competitive nRt gatelets, so that no working memory can be established in cortex (no sustainable cortical attractors can be generated.) It is relatively easy for stimuli to fall underneath the thalamocortical thresholds (brief presentations that elicit priming but no conscious experience), but these strongly valenced stimuli apparently can still fall above what are briefer "affective valence thresholds." This suggests the obvious point that these systems are faster, more primitive, phylogenetically earlier, etc, but does not establish by itself that they are foundational for consciousness. How might one examine this further? Studies that look at stimulus events under the thalamocortical thresholds but above the affective thresholds, such as work done by Arne Öhman, Margaret Bradley, Peter Lang (summarized in this volume) provide a testing ground for investigating this crucial phenomenon of possible multidimensional valence without clear cortical encodings. The traditional point of view is largely that this "valence information" consists largely of autonomic feedback (? via the cingulate), but without any essential contribution from the ventral affective brain systems themselves, particularly PAG (although a neat distinction here is problematic due to important cingulate - PAG connectivities). However, it is not clear that autonomic feedback by itself allows differentiation of valence. It is possible that autonomic feedback is simply too non-specific, although this appears not yet solidly established either way. An alternative hypothesis is that the fundamental conscious experience of valence, that something is desirable or not, may have foundations in the PAG to ILN pathways, and that PAG afferents to the non-specific thalamic systems allow a fast pathway for valence information to influence the whole state of the thalamocortical umbrella.

Emotions theorists/researchers have almost totally neglected these two crucial aspects of valence: first, that it is more specific than just + and - ; second, that it might have its own competitive-inhibitory and "global distribution" mechanisms. There has been little thought given to the possibility that PAG's clearinghouse function suggests that it might run competitions like nRt, and that these might underwrite the much quicker time constants that the empirical literature has supported for activating valence (vs. activating a working memory). Following further down this path (using brief presentations under the nRt - thalamocortical thresholds), a potential psychological research strategy would be unpacking discrimination of simple positive/negative valence discriminations into discriminations of the full panoply of prototypes: can people distinguish brief presentations (again too brief for WM to get established via a sensory thalamocortical pathway) in terms of anger vs. fear stimuli, play vs. sexual stimuli, or separation distress vs. fear stimuli? If they cannot discriminate much above chance, this would suggest that whatever the pathways for information into GW (global workspace) that do not require a primary thalamocortical mechanism are very low resolution pathways indeed, supporting only the crudest of valence distinctions, and not the prototype affective discriminations enabled in PAG, where there is possible competitive-inhibitory gating of the primary affective states. This possible finding (no modular or prototype discrimination possible for brief affective stimuli) would weaken the hypothesis that PAG's outflows to MRF and ILN have much to do with conscious emotion, while the converse finding would strengthen the suspicion that PAG outflows to the ILN and reticular systems provide some fundamental substrates for conscious emotion. Full rostral-caudal lesions of PAG in humans would offer much towards the clarifying of PAG's role in emotion and consciousness, but this scenario is virtually impossible naturalistically and grotesquely unethical non-naturalistically. Currently, I am aware of no technologies for reversible chemical lesions at this level of the brain.

Finally, a third empirical approach would be based on metabolic imaging. In the interests of outlining processing hierarchies, I have asked leading neuro-imagers how they might devise compensatory weightings in the metabolic imaging of emotion for the ventral systems' inherently less active physiology. Since many theorists of emotion support hierarchical models (at least in principle if not in practice), this would "level the playing field" and allow the relative activation states of more ventral diencephalic and midbrain/brainstem systems to be more accurately imaged in various affective tasks.

Hopefully following the lead of Damasio (1998) and Panksepp (1998), there will be greater interest in exploring the ventral and primitive portions of the "limbic system" and their complex interactions with crucial subcortical telencephalic systems and paralimbic regions such as the cingulate. Such work would move us towards developing hierarchical distributed models for emotion and consciousness that do not suffer from "cortical chauvinism." This is not to suggest a "counter" midbrain-diencephalic chauvinism, but rather the more complex idea that these ventral limbic and dorsal cortical systems are profoundly interactive. If one does not understand the brain as a whole, one cannot truly understand the functions of its parts, even the supposedly discrete modular processors in cortex. Such hierarchical-distributed models for emotion will need to inevitably interdigitate with similarly hierarchical models for attention, pain, self-representation, and executive functions as part of the complex multi-component "neural envelope" for consciousness. Such a widely integrative theory of global state functions is in its infancy and is still being stitched together (erratically and ambivalently) from barely compatible fabrics. Clearly, these different functions do not represent "the same thing." But they all speak to Nature's fundamental integration of brain function which (I believe) is a principle that organizes the essential neural foundations for consciousness, whatever their final architectural, neurodynamic, and neuromodulatory foundations turn out to be. If this is true, then the concerted study of emotion (or any of the global state variables) will in the end underline the necessity of such an approach to consciousness, as more segmented approaches (assuming fundamental separations between attention, emotion, executive functions, self-representation, pain, etc. that ignore their essential borders and interactions) will eventually lead honest researchers to the critical borders emotion shares with other global state variables. All of these global state functions control the structure of and access to global workspace; to use Baar's (1996) theater metaphor, these functions are different members of the stage, writing and directing crews that jointly determine which actors get into the spotlight of consciousness.

References

Baars, B. (1988) *A Cognitive Theory of Consciousness*, Cambridge, MA: Cambridge University Press.

Baars, B.J. (1993) "How does a serial, integrated and very limited stream of consciousness emerge from a nervous system that is mostly unconscious, distributed, and of enormous capacity?", in: *CIBA Symposium on Experimental and Theoretical Studies of Consciousness*, G. Bock and J. Marsh, eds, London: John Wiley and Sons, pp. 282-290.

Baars, B.J. and J. Newman (1994) "A neurobiological interpretation of the Global Workspace theory of consciousness", in: A. Revonsuo and M. Kamppinen, eds, *Consciousness in Philosophy and Cognitive Neuroscience*, Hillsdale, NJ: Lawrence Erlbaum.

Baars, B.J. (1996) *In the Theater of Consciousness: The Workspace of the Mind*, New York: Oxford University Press.

Baars, B.J., J. Newman and J.G. Taylor (1998) "Neuronal mechanisms of consciousness: A relational global workspace framework", in: *Toward a Science of Consciousness II*, S.R. Hameroff , A.W. Kaszniak and A.C. Scott, eds, Cambridge, MA: MIT Press, pp. 269-277.

Baddeley, A. (1986) *Working Memory*, Oxford: Oxford University Press.

Barlow, H.B. (1980) "Nature's joke: A conjecture on the biological role of consciousness", in: *Consciousness and the Physical World*, B.D. Josephson and V.S. Ramachandran, eds, Oxford: Pergamon Press, pp. 81-94.

Bandler, R. and A. Depaulis, eds (1991) *The Midbrain Periaqueductal Gray*, New York: Plenum Press.

Blanck, G. and R. Blanck (1974) *Ego Psychology: Theory and Practice*, New York: Colombia University Press.

Bowlby, J. (1969) *Attachment and Loss, Vol 1: Attachment*, New York: Basic Books.

Bowlby, J. (1973) *Attachment and Loss. Vol. 2: Separation, Anxiety and Anger*, New York: Basic Books.

Cameron, A.A., I.A. Khan, K.N. Westlund and W.D. Willis (1995) "The efferent projections of the periaqueductal gray in the rat: a Phaseolus vulgaris-leucoagglutinin study. II. Descending projections", *Journal of Comparative Neurology* 351:585-601.

Cameron, A.A., I.A. Khan, K.N. Westlund, K.D. Cliffer and W.D. Willis (1995) "The efferent projections of the periaqueductal gray in the rat: a Phaseolus vulgaris-leucoagglutinin study. I. Ascending projections", *Journal of Comparative Neurology* 351:568-84.

Chalmers, D. (1996) "Facing up to the problem of consciousness", *Journal of Consciousness Studies* 2:200-219.

Cornwall, J., J.D. Cooper and O.T. Phillipson (1990) "Projections to the reticular thalamic nucleus in the rat", *Experimental Brain Research* 80:157-171.

Cotterill, R. (1995) "On the unity of conscious experience", *Journal of Consciousness Studies* 2:290-312.

Crick, F. and C. Koch (1995) "Are we aware of neural activity in primary visual cortex?", *Nature* 375:121-23.

Damasio, A. (1994) *Descartes' Error: Emotion, Reason, and the Human Brain*, New York: Putnam.

Damasio, A. (1998) "Emotion in the perspective of an integrated nervous system", *Brain Research Review* 26 (2-3):83-86.

Davidson, R.J. (1992) "Anterior cortical asymmetry and the nature of emotion", *Brain and Cognition* 20:125-151.

Derryberry, D. and D.M. Tucker (1992) "Neural mechanisms of emotion", *Journal of Consulting and Clinical Psychology* 60:329-338.

Engel, A.K., P. Fries, P.R. Roelfsema, P. König and W. Singer (1997) "Temporal binding, binocular rivalry, and consciousness", *Association for the Scientific Study of Consciousness*, On-Line Seminar, *http://server.phil.vt.edu/assc/esem.html*.

Fein, D., S. Joy, L.A. Green and L. Waterhouse (1996) "Autism and pervasive developmental disorders", in: *Neuropsychiatry: A Comprehensive Textbook*, B. Fogel, ed., New York: Harcourt Brace, pp 571-615.

Freeman, W. (1995) *Societies of Brains*, New York: Lawrence Erlbaum Associates.

Heilman, K.M. (1997) "The neurobiology of emotional experience", *Journal of Neuropsychiatry* 9:439-448.

John, E.R. (1962) "Some speculations on the psychophysiology of the mind", in: *Theories Of The Mind*, J. M. Scher, ed., New York: The Free Press, pp. 80-121.

Joseph, R. (1990) *Neuropsychology, Neuropsychiatry and Behavioral Neurology*, New York: Plenum Press.

Kagan, J. (1994) "On the nature of emotion", *Society for Research in Child Development* 59 (2-3): 7-24.

Kaszniak, A.W., S.L. Reminger, S.Z. Rapcsak and E.L. Glisky (1999) "Conscious experience and autonomic response to emotional stimuli following frontal lobe damage", in: *Toward a Science of Consciousness III*, S.R. Hameroff, A.W. Kaszniak and D.J. Chalmers, eds, Cambridge, MA: MIT Press, pp. 201-214.

Lane, R.D. (2000) "Neural correlates of conscious emotional experience", in: *Cognitive Neuroscience of Emotion*, R.D. Lane, L. Nadel, G.L. Ahern, J.J.B. Allen, A.W. Kaszniak, S.Z. Rapcsak and G.E. Schwartz, eds, New York: Oxford University Press, pp. 345-370.

Lang, P.J. (1993) "Emotion and psychopathology: a startle probe analysis", *Personality and Psychopathology Research* 16:163-199.

LeDoux, J. (1996) *The Emotional Brain. The Mysterious Underpinnings of Emotional Life*, New York: Simon and Schuster.

Llinas, R.R, U. Ribary, M. Joliot and X.-J.Wang (1994) "Content and context in temporal thalamocortical binding", in: *Temporal Coding in the Brain*, G. Buzsaki, R.R. Llinas and W. Singer, eds, Berlin: Springer Verlag.

Lozsadi, D.A. (1994) "Organization of cortical afferents to the rostral limbic sector of the rat reticular thalamic nucleus", *Journal of Comparative Neurology* 341:520-533.

MacLean, P.D. (1985) "Evolutionary psychiatry and the triune brain", *Psychological Medicine* 15:219-221.

Mesulam, M. (1985) "Patterns in behavioral neuroanatomy: association areas, limbic system, and hemispheric specialization", in: *Principles of Behavioral Neurology*, M. Mesulam, ed., Philadelphia: FA Davis Company, pp. 1-70.

Metzinger, T. (1998) "Being No One", Plenary Address, *ASSC Conference on Neural Correlates of Consciousness*, June 21, 1998, Bremen, Germany.

Newman, J. (1997) "Putting the puzzle together: Towards a general theory of the neural correlates of consciousness", *Journal of Consciousness Studies* 4(1&2): 47-66, 101-121.

Newman, J. and B.J. Baars (1993) "A neural attentional model for access to consciousness: A Global Workspace perspective", *Concepts in Neuroscience* 4:255-290.

Newman, J. and A. Grace (1998) "Newly elucidated circuitry subserving the selective gating of fronto-hippocampal systems contributing to the stream of consciousness: A model for the modulation of attention by affective states and episodic representations",*http://www.zynet.co.uk/imprint/Tucson neuroscience*.

Öhman, A. (1999) "Distinguishing unconscious from conscious emotional processes: Methodological considerations and theoretic implications", in: *Handbook of Cognition and Emotion*, T. Dalgleish and M. Power, eds, Chichester, UK: Wiley, pp. 321-352.

Panksepp, J. (1989) "The neurobiology of emotions: Of animal brains and human feelings", in: *Handbook of Psychophysiology*, T. Manstead and H. Wagner, eds., Chichester, UK: John Wiley and Sons, pp 5-26.

Panksepp, J. (1989) "The psychobiology of emotions: The animal side of human feelings", in: *Emotions and the dual brain: Experimental Brain Research Series 18*, G. Gainotti and C. Caltagirone, eds, Berlin: Springer-Verlag, pp. 31-55.

Panksepp, J. (1991) "Affective Neuroscience: A conceptual framework for the neurobiological study of emotions", in: *International Reviews of Emotion Research*, K. Strongman, ed., Chichester, UK: Wiley, pp 59-99.

Panksepp, J. (1993) "Neurochemical control of moods and emotions: amino acids to neuropeptides", in: *Handbook of Emotions*, M. Lewis and J.M. Haviland, eds, New York: Guilford Press, pp. 87-107.

Panksepp, J. (1996) "Affective Neuroscience: A paradigm to study the animate circuits for human emotions", in: *Emotions: An Interdisciplinary Approach,* R. Kavanaugh, B. Zimmerberg and S. Fine, eds, Hillsdale, NJ: Erlbaum, pp 29-60.

Panksepp, J. (1998) *Affective Neuroscience*, New York: Oxford University Press.

Papez, J.W. (1937) "A proposed mechanism of emotion", *Archives of Neurology and Psychiatry* 79:217-224.

Pope, K.S. and J.L. Singer (1976) "Regulation of the stream of consciousness: Toward a theory of ongoing thought", in: *Consciousness and Self Regulation: Advances In Research*, Vol. II, G. Schwartz and D. Shapiro, eds, New York: Plenum, pp. 101-138.

Putnam, F.W. (1992) "Discussion: Are alter personalities fragments or figments", *Psychoanalytic Inquiry* 12:95-111.

Revonsuo A. (1995) "Consciousness, dreams, and virtual realities", *Philosophical Psychology* 8:35-58.

Revonsuo A. (1998) "How to take consciousness seriously in cognitive neuroscience", *http://www.zynet.co.uk/imprint/Tucson/neuroscience.*

Schore, A. (1994) *Affect Regulation and the Origins of the Self: The Neurobiology of Affective Development*, Hillsdale, NJ: Lawrence Erlbaum Associates.

Smythies, J. (1997) "The functional neuroanatomy of awareness", *Consciousness and Cognition* 6:455-481.

Smythies, J. (1998) "A fascinating experimental paradigm", *http://server.phil. vt.edu/assc/watt/smythies4.html.*

Taylor, J.G. (1999) *The Race for Consciousness*, Cambridge, MA: MIT Press.

Tucker, D.M., P. Luu and K.H. Pribram (1995) "Social and emotional self-regulation", *Annuals of the New York Academy of Science* Dec 15:213-39.

Watt, D.F. (1986) "Transference: a right hemisphere event? An inquiry into the boundary between psychoanalytic metapsychology and neuropsychology", *Psychoanalysis and Contemporary Thought* 9(1):43-77.

Watt, D.F. (1990) "Higher cortical functions and the ego: the boundary between psychoanalysis, behavioral neurology and neuropsychology", *Psychoanalytic Psychology* 7(4):487-527.

Watt, D.F. (1995) "Central challenges to neural network theory: syndromes of diseased consciousness, the ontogenesis of limbic networks and the phenomenology of value", *Syllabus of the World Congress of Neural Networks*, July, 1995.

Watt, D.F. (1997) "Affect, corticolimbic connectivities, and the representation of value: neurodevelopmental issues for current extended reticular thalamic activating theories of consciousness", Presentation at the *Brain and Self Conference*, Elsinore, Denmark, August, 1997.

Watt, D.F. (1998) "Affect and the limbic system: some hard problems: Commentary", *Journal of Neuropsychiatry* 10:113-116.

Watt, D.F. (1998) "Emotion, Cognitive Neuroscience and Consciousness Studies: Is Emotion Really One of Easy Problems?", On-Line Seminar at the University of Arizona, *http://www.consciousness.arizona.edu/emotion.html*.

PSYCHOLOGICAL PERSPECTIVES

INTRODUCTION: PSYCHOLOGICAL PERSPECTIVES

ALFRED W. KASZNIAK
Center for Consciousness Studies, Departments of Psychology, Neurology & Psychiatry, University of Arizona, 1503 E. University, Tucson, Arizona 85721, U.S.A.

The study of emotion was a core interest in the earliest days psychology's independence from its parent discipline of philosophy (e.g., Hall, 1897; James, 1884, 1890). This interest was stimulated, in no small part, by the seminal observations of Darwin (1872/1965) on emotional expression in humans and other animals. Emotion was also a clear preoccupation in the early development of psychoanalysis (Breuer & Freud, 1925/1956). However, emotion received sparse attention during the 19th or early 20th century experimental psychology, in either Germany, America, or Great Britain (see Boring, 1950). Nonetheless, by 1927 there was sufficient psychological scholarship and empirical research on emotion to warrant organization of the Wittenberg Symposium on Feelings and Emotions (Reymert, 1928), which brought together 34 distinguished American and European psychologists who presented papers to an audience of several hundred. Examination of the theoretic issues and methodologic problems addressed at this gathering shows substantial overlap with those issues and problems that psychologists continue to grapple with today. The current view of emotion as manifesting in the five components of physiological arousal, cognitive appraisal, conscious experience or feeling, action tendency, and expressive behavior, can be seen as an extension of early conceptualizations, such as those of James (1890), who wrote:

> Objects of rage, love, fear, etc., not only prompt a man to outward deeds, but provoke characteristic alterations in his attitude and visage, and affect his breathing, circulation, and other organic functions in specific ways. When the outward deeds are inhibited, these latter emotional expressions still remain, and we read the anger in the face, though the blow may not be struck, and the fear betrays itself in voice and color, though one may suppress all other sign. (p. 442)

Such historical continuity should not, however, be interpreted as reflecting any lack of conceptual or empirical progress. Indeed, progress in the psychology of emotion has been particularly rapid in recent years, reflected in both the accelerating rate with which relevant new papers and books are published and in the degree to which consensus has developed regarding several previously debated issues. Even in the few short years between publication of the first (Lewis & Haviland, 1993) and the second (Lewis & Haviland-Jones, 2000) editions of the *Handbook of Emotions*, much significant growth in conceptual clarity, theoretic and methodologic sophistication, and the accrual of new data can be seen.

The authors of papers within this third section of the present volume provide current reviews of a representative cross-section of the psychology of emotion. Among the several key questions in this area addressed by these authors are: What is the nature (componential and temporal structure) of emotion? What is the basic structure that characterizes self-reports of emotional experience? What are the distinguishable processes of emotion, and what functional role do these processes serve? What comprises the phenomenal experience of emotion? What are the antecedents of emotion episodes, and what do these antecedents tell us about the functions of emotions? How do social contexts influence the elicitation and experience of emotion? How does the face contribute to emotion and development of a sense of self? How are facial expressions of emotion influenced by social norms and the particulars of a given social interaction? What contributes to the appreciation and experience of humor? What are the respiratory, vocal, facial, and body movement components of spontaneous versus contrived laughter? How can emotion be incorporated into the development of purposeful artificial (robotic) systems, and how might this inform our understanding of the functions of emotion? How can the complex emotion of romantic jealousy be understood? What do clinical disorders of emotional experience and expression, such as alexithymia, tell us about the mental representation of emotion? What functional role does emotion play in decision-making, problem solving, and creative activity?

Despite the apparent diversity of questions being addressed within this section, converging points of agreement emerge: From a psychological perspective, emotions cannot be conceived as monolithic, but rather as multicomponential processes, with different components (e.g., appraisal, experience, expression, physiological arousal) serving different functional roles in adaptive behavior and an ecology of mind. In addition, any adequate understanding of emotion elicitation and the experience of emotion must include the role played by social context. Finally, it is clear that emotion and cognitive processes are constantly interactive, with both simultaneously involved in how we make decisions, solve problems, or create the myriad of our personal and cultural artifacts.

References

Boring, E.G. (1950) *A History of Experimental Psychology* (Second Edition), New York: Appleton-Century-Crofts.

Breuer, J. and S. Freud (1925/1956) *Studies on Hysteria*, London: Hogarth Press and the Institute of Psychoanalysis.

Darwin, C. (1872/1965) *The Expression of the Emotions in Man and Animals*, Chicago: University of Chicago Press.

Hall, G.S. (1897) "A study of fears", *American Journal of Psychology* 8:147-249.

James, W. (1884) "What is an emotion?", *Mind* 9:188-205.

James, W. (1890) *The Principles of Psychology, Volume II*, New York: Henry Holt and Company.

Lewis, M. and J.M. Haviland, eds (1993) *Handbook of Emotions*, New York: Guilford Press.

Lewis, M. and J.M. Haviland-Jones, eds (2000) *Handbook of Emotions*, New York: Guilford Press.

Reymert, M.L. (1928) *Feelings and Emotions: The Wittenberg Symposium*, Worcester, Massachusetts: Clark University Press.

THE NATURE AND EXPERIENCE OF EMOTIONS

NICO H. FRIJDA
Department of Psychology, Amsterdam University, Roetersstraat 15, 1018 WB Amsterdam, Netherlands

ABSTRACT

Emotions are multicomponential responses to events that vary over time, and their duration varies from a few seconds to a day or more. The duration of the component processes also varies, necessitating a hierarchical description of the emotional responses. At the basic-process level, emotion's functionally distinct processes are to be distinguished: affect processes; appraisal processes; activated action dispositions; and regulation processes. Most specific for emotion phenomena are affect and shifts of behavioral and attentional control. Phenomenally (rather than functionally), one of the major emotion components is emotional experience. Experience, too, can be analyzed in terms of constituents. The only elementary, uanalyzable qualia are pleasure and pain, and the awareness of incomplete control over thought and behavior -- the shifts in control precedence that gave rise to the very name "affect" (meaning "being affected"). Beyond that, experience includes awareness of the event-a-appraised, awareness of current state of action readiness, and awareness of bodily state. These components may be assumed to be common to animals and humans. Humans differ from animals in thc richness of events-as-appraised and of the cognitions attached to all components. In addition, emotional experience includes cognitions about and evaluations of one's emotion, giving rise to an emotion's "significance" and moral evaluation, with subsequent regulation and secondary emotions.

1. What is "An Emotion"?

This chapter will discuss the nature and experience of emotions, from a psychological perspective. From that perspective, "emotion" is the name for the process or processes underlying multicomponential responses to events. Emotions are constructs derived from bundles of phenomenal components, and emotion names are names for particular bundles of such components. To what extent these bundles form coherent wholes --that is, stem from a unitary process-- is a matter for empirical examination and theoretical inference. Such theory will have to deal with the major fact that the various components appear to be only moderately correlated. I will come back to that.

At a phenomenal level, the major components are emotional experience, behavior and motivational signs, and physiological changes. Describing the components, however, only partly accounts for the nature of the responses. Each of the components, in each single response instance, shows a given development over time. In fact, although emotions are often referred to as "states", they are better described as processes over time. The course of the responses over time merits separate attention. Although emotions are often thought of as rapid and brief responses, their time course only rarely consist of a rapid rise in responding, culminating in a peak and then slowly decaying to base level. When asking subjects to report on past emotions, and asked to draw the course of the emotions over time on a computer screen, the one-peak graph was an exception. More common were courses with more or less gradual onset, and multiple peaks and valleys. Durations of a few seconds again were an exception. Durations ranged up to a day or even more, and the average was more than one hour (Frijda *et al*., 1991).

The descriptions of the incidents suggested that the various response components might each have quite different durations, some lasting over the entire period reported, others only for a brief time, or occurring intermittently. Evidently, the bundles of responses can and should be described at different levels of analysis, and at different levels of abstraction. The phenomena suggest a hierarchical model of what emotional responses consist of.

The highest level of description is a transaction or emotion episode level. An emotion can be viewed as a transaction between subject and environment concerning an event that is relevant to the individual and that involves a given "relational theme" (Lazarus, 1991) such as an achievement, a loss, a threat, a harm, or an offense. The transaction constitutes an emotion episode during which there are certain constancies: a continuous core appraisal (appraisal of the theme at issue), continuous personal involvement in the issue, and usually continuous attentional, cognitive, and physiological activation.

Viewed at a lower level, the "emotion level," different reactions succeed one another. There are successive periods, each characterized by specific appraisals (e.g., a threat appraised first as unexpected, then primarily as uncontrollable, then perhaps as impossible to overcome at all), and by particular changes in action readiness (stopping of behavior or "surprise," readiness to withdraw, or "fear," abandoning of taking a stance, or "despair"; Frijda, 1986). Each of these "emotions" consists of a set of processes at the basic emotion-process level, that each, in turn, may become manifest in sequences of executive processes such as variations in autonomic arousal, sequences of facial expressions, and the like.

The basic processes are best described not from a phenomenological but from a functional perspective. The major ones are affect processes, appraisal processes, activation processes, notably involving action dispositions, and regulation processes. I will briefly discuss each of them. Together, they form prototypical instances of emotions, but they may occur more or less independently.

2. Affect

The term "affect" in a restricted sense: to denote the experiences of pleasure and pain, and the processes underlying those experiences. The processes are interpreted as processes of stimulus or state acceptance and stimulus or state non-acceptance, which is what the experiences of pleasure and pain mean to the individual. The processes organize openness or closure with respect to the stimuli at hand, as well as the individual's acceptance or non-acceptance of its overall momentary state, as in moods. Positive affect or pleasure is the state that individuals seek to maintain and that on occasion allows disengagement from vital interests, in states of playfulness. Negative affect or pain is the state individuals seek to get away from and, if impossible, that leads to overall behavioral disorganization and, eventually, apathy and stereotyped behaviors, as shown by battery animals.

When exceeding a certain intensity, affect is a process that affects widely divergent responses, such as attention, resource distribution over various functions, the thresholds for behaviors that aim to strengthen or terminate interaction with a given stimulus event, and probably certain sensory thresholds such as that for pain. It is one of the core processes that make a response an emotional one and, indeed, forms one of the reasons to introduce a term or assume a "faculty" like emotion. The existence of affect introduces value in a world of fact (to borrow from Wolfgang Köhler). It accounts for the existence of preferences, for manifest or experienced palatability and aversiveness of stimuli, and for behavioral priorities other than those based upon habit strength. It thereby also is one of processes accounting for the characteristic of "passivity" of emotional experience and behavior, that stamped the early concepts of "passion" (behavior aims not resulting from active intent, so not "actions") and "affect" (those reasons for behavior that affected the person). Affect is the distinguishing mark (or one of the distinguishing marks) that single out experiences and behaviors to be called "emotional", rather than the much less specific response component of autonomic arousal.

Affect processes may become manifest in feelings of pleasure or pain. They may instead become manifest in likes or dislikes, defined either as experiences or as behavioral tendencies. They may also manifest themselves solely in enhanced sensitivity to particular stimuli, or in the enhanced affective response to other stimuli than those that elicited them, as evident from studies by Öhman (see Öhman & Wiens, this volume) and by Zajonc (e.g., Murphy & Zajonc, 1993).

Affect is the basis for dividing emotions into positive and negative, pleasant and unpleasant ones. There do exist mixed or unclear ones, when positive and negative affect coexist, or are both aroused by the same stimulus event, when that event has hedonically opposite implications (such as loss of a spouse meaning both loss of an intimate relationship and gain in novel opportunities). One may also distinguish neutral emotions, when emotions are not defined by the presence of affect but by a change in behavioral control, as in the cases of surprise and desire.

3. Appraisal

"Appraisal" refers to the information processes that transform an external stimulus event (or a memory or a thought) into an affectively valent event. The process may be as simple as that which turns the smell of food into a stimulus for getting hold of it, or as complex as that which turns reading a stock exchange figure into a trigger for despair.

In both cases, appraisal of an event so that it leads to an emotion (that is, to a multicomponential response) usually involves two different appraisal processes: one that pertains to the event's affective value, and another that pertains to what one can or cannot do to deal or cope with the event; the first is often referred to as "primary appraisal", and the second as "secondary appraisal" (Lazarus, 1991). Secondary appraisal consists of picking up context information that is relevant for the selection of one mode of emotional response or another (e.g., whether the event can be prevented to occur, or its effects be attenuated), or by drawing upon previous experience, and by using information about one's coping potential in making the selection. All this means that appraisal usually involves cognitive processes, to which I will come back in my chapter on emotion antecedents.

Appraisals may concern a particular object or event, or the world, the environment, as a whole; or they may concern the general state of one's resources (Morris, 1992). The latter states of appraisal are usually referred to as moods. They are unfocused appraisals (Frijda, 1993).

Appraisal manifests a specific and remarkable function of the human mind. It occurs automatically (e.g., Bargh & Pietromonaco, 1982). It allows relevant stimuli to activate affect or other aspects of emotion more or less regardless of the individual's direction of attention or action goals of the moment. Inversely, emotional sensitivities and concerns can be aroused, again more or less regardless of circumstances. A passing injurious remark may set off an emotion of distress or anger, whatever goal one is engaged in when hearing or overhearing it. Generally speaking, appraisal reveals the human (and, presumably, animal) capacity for automatic affective evaluation of events.

4. Activation Processes

Emotional responses include behaviors, among which facial expressions and vocal intonation changes, as well as locomotions, the handling of objects, and verbal activities. Several different of those behaviors, notably those that occur together in response to particular events, show functional equivalence. They have similar meaning for effecting changes in subject-object relationships, such as enhancing subject-object interaction, decreasing such interaction, obstructing an antagonist's interference. The inference from such equivalence is that emotions involve changes in motivational states regarding subject-object relationships. I call these changes "changes in action readiness". Emotion mechanisms include mechanisms for producing such changes in action readiness.

Motivational changes with respect to the establishment or modification of subject-object relationships is probably the best general formula for understanding emotional behavior and emotional urges. Joy, one can say, aims at openness and intercourse with the world. Anger aims at neutralizing an interference or offense. Fear aims at preventing or avoiding harm; etc. The motivational changes themselves can be understood as resulting from the activation of action dispositions present in memory, notably in neural dispositions. From the perspective of analyzing emotional phenomena, one may distinguish three general types of such dispositions.

First, there are general activation mechanisms. "Activation" is used here to mean "tonic readiness to act" (Pribram, 1981). Many emotions involve activation increase; some others are characterized by deactivation, presumably caused by blocking of these activation mechanisms or loss of resources, as in depressive apathy and exhaustion. "Activation", as a psychological construct, is inferred from variations in generalized behavioral readiness, as seen in muscle tone, persistence in time and over obstacles, frequency of behavioral manifestations and frequency of behavioral change. It usually, or perhaps always, is accompanied by changes in autonomic activation. Some emotional states are characterized by diffuse activation, for instance those called "excitement" or "being upset"; others also involve more specific action dispositions.

Second, there exist activations for relational activity at the general or strategic level: generalized readiness to interact or not to interact, the behavioral correspondents of pleasure and pain. As mentioned before, affect would seem to involve generalized activation for any response process involving enhanced and decreased interaction with the environment, respectively (Lang, this volume); more or less intense affect certainly does. The point of relevance here is that many emotions involve nothing else, at the response end, than affect or this strategic readiness, and perhaps enhanced general activation. Said plainly, many emotions are nothing but states of positive or negative affect, coupled to global readiness for acting in a relationship-enhancing or relationship-weakening fashion, and perhaps coupled to enhanced or decreased activation, and a representation of a given cognitive appraisal. So for many emotions labeled as instances of "sadness", of "hope", of "being moved" or of "distress".

Third, there exist activation of one or several specific action dispositions, each involving the motivation to achieve or maintain a particular subject-environment relationship. The dispositions are similar or identical to those identified by Panksepp (1998), Gray (1982), and others. Activation of such dispositions leads to activation or organization of motor dispositions, gross bodily patterns as well as facial expressions; the dispositions correspond to what are often thought of as "basic emotions" such as joy, anger, fear, active sadness or panic, surprise, and disgust or revulsion (e.g., Ekman, 1992; Oatley, 1992). However, psychological analysis suggests that there exist additional elementary mechanisms and additional basic forms of relational action readiness,-- additional, that is, to those basic emotions. They include the dispositions underlying social emotions (proximity-seeking or attachment, affective bonding, sympathy, submission), pronounced in motivational patterns called love or affection and in emotions like shame, humility, or dependency and *amae*. (Markus & Kitayama, 1991). They also include the disposition for having or achieving ego-object fusion and loss of the self, as this occurs in emotions like *awe*, in religious emotions, in infatuation, and again in emotions like *amae*.

Activation of any kind of these action dispositions may involve activation of what, from this perspective, may be called support mechanisms: attention deployment and attention allocation, and the energy mobilization processes manifest in autonomic changes. They can be seen as preparatory for, or supporting, motor behavior and the motivational readinesses indicated.

Each of the action dispositions, presumably, has its own particular sensitivity to particular patterns of incoming information. In view of what is known about the antecedents of emotions (see chapter on that topic), one can best view the activation of the various action dispositions to occur along two ways: external stimuli impacting directly upon their sensitivities, and information from memory or thought doing so. The two ways are, of course, equivalent to LeDoux's (1996) two routes. Thresholds of activation may vary, for instance by residual activation from previous events. The result is that, on occasion, information that corresponds only to a fragment of what the sensitivity is set for can elicit the corresponding emotion. Activation of an action disposition and of support mechanisms has the major implication of causing or facilitating a shift in goal priorities and behavioral control. That is, events that elicit activation changes as meant, as well as those eliciting affect of any intensity, tend to cause a shift in "control precedence," a high likelihood that control will switch towards attending to the emotional issue, cognitive preoccupation with it, the action readiness aroused or the resource utilization by autonomic arousal. Shift in control precedence is the other major emotion characteristic, and the other aspect of the "passivity" that tends to set emotions apart from instrumental or habitual behavior, and that is drawn in to understand the various consequent "irrationalities," performance decrements, single-mindedness, and cognitive range changes that are often described in connection with emotions. Note that I am talking about shifts in precedence. The shifts are not necessarily towards actual control of behavior or behavioral planning. Emotions can be attenuated or kept in check, or other behavior-controlling functions may retain top priority; but during emotions, thoughts and feelings about the emotional issue distract or jump to top priority when those other functions slacken.

As I said, the action dispositions underlying states of action readiness correspond to what are often considered as "basic emotions". This latter concept suggests that a fixed relationship between the various different response components exists, all being commanded by a single particular disposition. The internal structure of the action dispositions discussed is not very clear, however. The moderate correlations between components make such fixed links between components (between a particular type of action readiness, a particular autonomic pattern, a particular facial expression, a particular experience) unlikely. The situational dependence of each of those individual components also upsets the notion of fixed links or of a unitary disposition from which all components spring. Of course, alternative modes of organization are possible. One involves viewing the dispositions as motivational states of readiness to reach a particular type of relational change, each compiling the set of active components in accordance with the specific situation, the specific response repertoire, and the support requirements of the specific constellation. Much research will be needed to clarify the mode of organization involved and, thereby, how tenable is the notion of basic "affect programs" (Ekman, 1992).

Whatever their organization, activation of the different types of action disposition occurs separately. Some emotional responses contain full-fledged activation of tactical, specific action dispositions; they are the prototypical instances of joy, anger, fear etc. Others consist only of general activation increase, or only of strategic readiness or affect arousal, or only of those plus awareness of some event as appraised, as indicated above with the example of hope. Indeed, a category like "jealousy" refers to some appraisal plus negative affect plus, in many cases, one of the specific forms of action readiness that fit that negative affect. Sometimes that is anger, sometimes it is sadness, sometimes it is fear, and sometimes it is diffuse "distress" or nameless paralysis. Many actual instances of emotions like anger, joy, fear etc. also consist only of appraisal and affect and, perhaps, general activation. This analysis implies that it is erroneous to regard the totality of emotions as variants and mixtures of the "basic emotions". They are not basic in that sense. Emotion space is not neatly hierarchical, and certainly not neatly divided into the 6 or so basic emotion segments.

5. Regulation Processes

Emotion control is not an influence affecting the emotion system from without, because of rational considerations or cultural norms (Frijda, 1986, 1988). Emotion control stems, at least in part, from the anticipation of aversive consequences of the emotional reaction itself. It is effected, at least in part, by inhibition dispositions that also operate as such in shaping certain emotions, notably freezing and anxiety (Gray, 1982). Inhibition, moreover, is directly functional in regulating social interactions (Delgado, 1975; DeWaal, 1989, 1996). Inhibition represents a general and ubiquitous component of emotional responding, adjusting response so that it fits the situation, because all immoderate responses, whether animal or human, upset adaptation. Emotional responses, in all their components, are the outcome of the balance between activations and inhibitions reciprocally evoked by the activations. An emotional response generally may be due to some activation, some loss of regulation, or both.

6. Other Components

Reactions to emotional events may include further components. One is cognitive activity. Emotions give rise to, or include, mental rumination -- an important factor in extending the duration of emotion episodes (Rimé, 1996). Other cognitive activities include emotion-generated attributions that, for instance, turn an emotion of anger (attribution of an effect to actions of an actor) into one of hatred (attribution of an effect to the personality of an actor), or depressive sense of failure into shame, with concomitant change in action readiness and judgements of the emotion itself. They also include judgements about one's emotion which generate a secondary emotion, such as shame about one's fear. Emotions are often doubled by emotions aroused by having a particular emotion, and such secondary emotions are in part accountable for emotion regulation.

Changes in action readiness are usually triggered by environmental events as appraised, and they usually are attributed to such an event. Correspondingly, action is usually directed towards an object, whether external or in thought. The changes in action readiness, however, also exist without an object, or without a specific object but directed towards the environment as a whole, becoming manifest in threshold changes for actual responses towards objects or events. This again corresponds to what one calls "moods": changes in action readiness not directed towards a specific object. Describing a mood in terms of a generalized appraisal, a diffuse state of action readiness, and an objectless feeling are alternative options and represent, in all likelihood, correlated phenomena.

7. Emotional Experience

Experience is one of the components of emotion that can be distinguished at the phenomenal level. It is only one of the components. Stimulus events may evoke affect changes, appraisals, changes in action readiness and physiological responses without conscious awareness of the eliciting events nor of the reactions. But many or most emotions include emotional experience, and on occasion experience may be the only observable response to an emotional event.

In this section I will discuss the nature and function of emotional experience. Such discussion is not easy, in part because experience is private by definition, and accessible mainly by language. But the difficulties are deeper. Conscious experience is not a direct and unaltered reflection of underlying information processes, including sensory and bodily ones. As extensively argued by Marcel (1983; and by Lambie & Marcel, in preparation, with respect to emotional experience in particular): experience represents a transformation, the outcome of an operation, upon underlying information processes. It changes with the perspective taken of those processes, and the attention directed towards these. Emotional experience changes form and nature according to circumstances, just as visual perception does, where you can see objects or colors or forms. This applies even to the basic affective experiences of pleasure and pain.

Discussion is also made more difficult because of several preconceptions. One of these is the notion, taken from Hume, that emotional experience is a sort of sensation. It suggests that emotional feelings are irreducible qualia, like the sensations of red or green; or that emotional experiences are inexpressible, ineffable. However, none of the qualia of feelings, apart from pleasure and pain appears irreducible, let alone unanalyzable. More important is that emotional feelings have properties that disqualify considering them as sensations, whether as irreducible qualia or as body feelings, as perhaps James and others have held. Emotional experience has a number of characteristics that carries it way beyond sensations, and stamp it as "emotional," in distinction of sensations. These characteristics are closely linked to the processes described in the preceding sections. First, emotional experiences are intentional; they are *about* something. Second, emotional experiences are passions; one feels affected by them. Third, emotional experiences are experienced as having *meaning*, in part by affecting the self. Fourth, emotional experiences are *relational*. A few remarks on each of those follows.

As to the intentionality of emotional feelings: one is angry, happy, or sad about something. Recall that this does not mean that the experience is caused by some object or event, but that the experience itself bears upon an object or event and concerns the latter. Intentionality implies a nonsymmetric subject-object relationship. This is in some sense the case even for moods that were defined as states of appraisal or action readiness not bearing upon an object. This notwithstanding, mood experiences are not closed in themselves. They are experienced as states of the self or as states of the world as a whole. Anxious mood, depressed mood, happy mood are not disembodied free-floating qualia but concern one's state, one's "being in the world" as welcome or unwelcome.

The felt "aboutness" implies that the experience is felt to emanate from the object that it is about. It comes from apparent properties of the object, and these properties affect me, and theirs is the initiative.

The third feature of emotional experience: it involves meaning. In part this is implied in affect carrying value, or acceptance or non-acceptance of object or state. In part it is an implication of its being felt to affect the subject. Emotional experience has links to the self, to the self-as-actor, the felt source of spontaneity and of taking initiatives. It intrudes upon those initiatives. Note that this does not involve high-level cognition but low-level cognition, information-processing-in-action, like that which controls the stability of the visual field in a mammal that moves its eyes, and which stability depends upon recording self-initiated head or eye movements. Such meaning is also closely linked to body awareness, to the sensations arising from autonomic arousal. The role of body awareness in distinguishing emotional from non-emotional experience, and one emotion from another, is small, contrary to what James and Schachter and Singer supposed. However, it is the clearest sign of being affected. In Sartre's (1948) terms, it represents that emotional events or feelings are to be taken seriously. The body is gripped, which enters awareness both by body sensations and by the interruption of action and attention that these provide.

Emotional experiences are not merely intentional; they are also *relational*: the fourth feature. Emotions are felt as playing between me and the object, the world as a whole, or myself as a subject. They affect the subject not only by coming from the object, but also by imposing a particular relationship with it.

Phenomenally, emotional experiences are not necessarily subjective states, not states of oneself, but states of how things are between me, the subject, and the object. In a feeling of sadness or anger, things are not so good; in a feeling of joy they are smooth and accepted; and so forth. Emotional experiences are experiences of the state of commerce from the standpoint of one of the participants in the commerce. In principle, and at its most unreflected, emotions are experienced as perceptions: things are beautiful, events are threatening or awful or seductive. At its most elementary, emotional experience is how the world appears, as imbued with meaning (Sartre, 1946; Lambie and Marcel, in preparation).

Emotional experience, however, does not always correspond to this description. I indicated that every state of consciousness is more or less a construction based upon input and other information; and the nature of the description depends upon personal attitudes and direction of attention. When emotions are experienced as subjective states, felt within oneself, the intentional, relational aspect tends to disappear because the reflection abstracts from the object and focuses upon the evaluative aspect as such. This is the way in which emotional experience figures in classical introspection. As Titchener remarked, affect tends to disappear rapidly when attended to. Feelings, he wrote, are "evanescent", a statement the recent sufferer of bereavement would not easily subscribe to because the attention is towards the loss, not towards the feeling.

Emotional experience is built up from three partly independent aspects that correspond to the three major components discussed in the preceding sections. They fill out and specify the overall experiential aspect of being affected. First: emotional experience includes experience of how the object or event affects me: experience of affect. Second, it includes what the object means to me: what it might do to me, offer me, withhold from me or give me: experience of the object-as-appraised. Third: it includes experience of what I might do relative to it, want to do, not want to do or be unable to do, in order to deal or cope with what it might do or offer: experience of state of action readiness. I will briefly discuss these aspects of emotion under the perspective of experience.

7.1. Experience of Affect

Experiences of pleasure and unpleasantness occur in the feelings of like and dislike that may accompany sensory experience. However, they are better exemplified by their extreme instantiations in rapture and bliss, and in suffering and anguish. Strange to say, pleasure and pain (in the extended sense) have for a long time been almost absent from the psychology of emotion. It sometimes seems as if students of emotion have long suffered from professional anhedonia. The term pleasure does not figure in the indexes of major emotion texts (e.g., Mandler, 1984) or in Griffith's book (1997); it has only recently regained prominence, in descriptions of sensory pleasures, of sex, of flow experience produced by activity one is fully engaged in, and of emotions proper. Pleasure and pain is the core of what emotions really are.

The relational aspect of affect is evident. Phenomenally, pleasure and pain can nevertheless present themselves in different guises: as inner or subjective "feelings", as properties of an object that looks attractive or repulsive, or as an object that one likes or dislikes. And they carry their prolongation or claim for change within themselves and impose it upon the subject. At higher intensities, they dominate and impose control precedence. Pain and anguish grip the person, as in the unmitigated fires of just having lost a beloved or a child,--fires that burn and burn, that preempt attention, twist the body, wrap it in a poisoned gown, force it into seeking action. Pain *has* to end, and it has to end *now*. Bliss seduces, demands immersion, its anticipation suffers delay with difficulty, it demands immersion in the Lethe.

The fact that affect experience is "intentional," usually is about something does not imply that what it is felt to be about necessarily corresponds to what actually evoked the experience. One may err in what one feels pleasure or displeasure to be about. Experimentation with subliminal affect arousal provides illustrations, and it may well be frequent in daily life. There probably is some sort of elementary attribution processes involved in attaching affect to an object (Russell & Barrett, 1999). It is conceivable that the neural activity underlying affect becomes fully meaningful or intentional only after interaction with other processes, such as accessing readiness for approach and avoidance responses. Absence of those latter kinds of access may turn affective experience into inchoate, diffuse "feelings", of the kinds that seem to occur among the many emotions in epileptic auras (MacLean, 1993).

7.2. Appraisal Experience

Emotional experience under the angle of appraisal most clearly reveals that such experience is perceptual experience, or can be. Attraction is out there; beauty is out there, seductiveness is out there, and so are threats, insults, and painful losses. Events-as-appraised include experience of what the event might offer me, the subject, do to me or withhold from me or give me, and whether it is something I can or cannot do something with or about. Coping ability was mentioned as one of the factors involved in emotion arousal; it enters experience through the event-as-appraised: the event appears as powerful or contemptible, as seductive, as overpowering, and the like.

Experience of an event-as-appraised need not be articulate. Situations appear as "uncanny," "weird," or "catastrophical," without the subject being able to specify why or what she means. Weird is weird is weird, as far as experience goes. Note that, generally speaking, appraisal (experienced or not) always results from the interplay of external and internal information, even if experience does not show this. For non-emotional experience it is of course no different: look at perceived verticality.

Analytically, appraisal experiences can be accounted for by way of particular patterns of causal or component features. "Personal loss" implies that something of value has gone that was present before, etcetera. Research has shown clear links between emotion terms and experiences and such patterns of components. Cognitive appraisal theories see the experiences of different emotions (or the semantics of emotion labels) as projections of appraisal patterns composed of the values on a small set of discrete appraisal dimensions (Frijda, 1986; Scherer, 1984; Roseman *et al.*, 1996). The dimensions include valence or pleasantness, felt goal obstructiveness or conduciveness, controllability of the event or sensed coping potential, uncertainty about event outcome, causal agency or responsibility, and novelty . Appraisals then are among the elements distinguishing one emotional experience from another, thus solving the problem facing James and others: how is the feeling of one emotion distinct from another.

Appraisal extends beyond the meaning of the immediately given situation. The meanings of events usually extend in various directions: whether they may continue in time or will be over in a flash, for instance. Meanings include what may happen next; "threats" are events that are likely to develop in harmful fashion. They include other aspects of the temporal field: a nasty remark feels different when it is the tenth remark in a row, when unpleasantness leaves no respite, when the situation cannot be escaped from. Each emotional awareness is fitted in a field of possibilities that are experienced in more diffuse and inarticulate, or more definite and articulate fashion. Articulateness is a variable feature of experience.

As with what affect is about: experience of event-as-appraised is not necessarily isomorphic to the event-as-appraised that has triggered a change in action readiness, or any other response component. The qualities of experience do not neatly match the underlying information sources. "Weird" does not neatly tell what was it that made the event look weird; and it may not come from event properties at all but, for instance, from one's inability to handle or place the information (see my contribution on emotion antecedents). Experience provides hypotheses concerning the information sources, but no more.

7.3. Experienced State of Action Readiness

Emotional experience, finally, includes awareness of one's state of action readiness: one's impulse, desire, and the estimates of the likely success of one's actions. Experience of one's state of action readiness is what gives emotional experience its most specific flavor, together with affect. Activation, of course, is one of the two major dimensions of emotion self-report, in the studies by Russell (1979, 1980; Russell & Barrett, 1999), and Lang (this volume).

"Basic" emotion labels to a very large extent refer to modes of action readiness: the hostility, the desire to retaliate, in anger, the desire to self-protect in fear, the sense of being paralyzed in anxiety, the loss of interest and impetus in apathetic depression. That is, when asking subjects to indicate the urges or impulses or inclinations they feel during particular emotions, correspondences with major emotion categories are strong and unambiguous (Frijda, Capers &

Terschure, 1989), more or less regardless of language and culture (Frijda *et al.*, 1995).

Awareness of state of action readiness is the consciousness taken of one's actual state of action readiness, as present in action programming, motor preparation or execution, general activation, and in the various support mechanisms such as attention. States of action readiness may well become conscious independent of peripheral motor innervation; after all, one is aware of one's intentions and plans, so why not of one's urges. Feedback from peripheral activity may contribute, though, as it does according to "facial feedback theory".

Awareness of action readiness includes that of unreadiness, listlessness or apathy. It also includes the results of monitoring one's action planning and preparation. There are feelings of helplessness or hopelessness, implying that no coping action could be devised or will be possible in the future. There likewise are feelings of brashness, self-assuredness, being in command, that come from having relevant actions, if needed, at the tip of one's fingers. Again, these involve low-level cognitions that are also available to the victorious rat in a rat contest, or in the well-trained bull at the beginning of a bullfight. Of course, these are suppositions of what a rat or bull might feel; but they are inferences from features of behavior, and whosoever wants to explain these features differently go ahead.

7.4. Full Emotional Experience

In summary: emotional experience is a composite of affect, appraisal, felt intent or impulse, as well as feedback from bodily reaction. This implies that it exists regardless of reflection. It can be purely perceptual, immersed in meaning-as-perceived. It can be entirely noncategorial, without involving assigning an emotion name. It is experience-in-interaction with events and one's dealing with events.

Full human emotional experience includes more than the aspects discussed. I should mention the presence of results of reflection and of labeling. I only alluded to the appraisal of the event with its full temporal and implicational context. It encompasses awareness of the possible or likely responses of others to one's actions, whether actual or forthcoming ones. It thus includes the anticipated aversive or beneficial response consequences. That is: emotional experience includes some awareness of the instigators, of emotion regulation and control, and of one's responses as controlled or non-controlled.

Emotional experience also includes a further important aspect: that what I earlier (Frijda, 1986) described as the "significance" of one's emotion, the set of implications of one's emotions with regard to social effects and one's moral standards and ideals, and the emotional consequences of these implications. One may despise one's fear or pride oneself in one's shame. One may expect others to despise one's fear. One may be aware, while being angry, that God condemns it, or society at large, or both; and which makes a difference for experience. Emotional experience typically is embedded in a field of implications and consequences for oneself, one's interactants, one's social environment, that constitute the difference between one fear and another, one man's or woman's fear with another's, fear of the supernatural from fear of spiders. Those differences in experience are consequential, and may be so consequential as to lead to the invention of particular emotion names. Litosh is a grief that one accepts, wallows in, and secretly delights in (Czechs have it, according to Kundera); wrath is anger with a tinge of moral justification and sense of righteousness; and so forth. They are not merely names: they have behavioral implications. Litosh induces little effort to appease it, and wrath does not suffer from moral objections. Anger that God condemns may obtain added tenseness, added violence, and less pleasure, than anger for God's sake. Anger that breaks through a barrier, presumably, has a different time course from that which runs smoothly.

The event-as-appraised includes further potentialities, notably one's own, and in particular those of emotions that might be aroused by one's actions in response to it. I call these anticipated or imagined emotions "virtual emotions": emotions that one feels one might have if certain acts were performed. They are not emotions, but experientially and in origin they are like emotions.

What I have in mind are two major classes of emotional reactions: emotions that one would have if such and such conditions would obtain; and empathic emotions, emotions that someone else might feel or presumably does feel. They are important factors in the control of social behavior. Imprudent and rash behavior is restrained by glimpses of possible eventual shame or guilt. The function of shame and guilt emotion lies precisely in instigating efforts to avoid or forestall them. Harshness towards others is restrained by consideration for the other's eventual sadness. Virtual emotions probably are not unique to humans. Animals may act helpfully and with consideration, and they restrain their anger to prevent unwanted damage, as when restraining anger towards infants, or in view of maintaining the ongoing social relationship. DeWaal (1996) gives a wide array of illustrations of such acts and restraints in primates and elephants; the most likely candidate for that which instigates them is empathic feeling. Also, it would seem that virtual emotions are precisely what is lost after frontal brain damage, as described by Damasio (1994).

8. The Functional Role of Emotional Experience

Discussion of experience, emotional or otherwise, is made difficult because of the double face of human conscious awareness. In the first place, consciousness is consciousness of the world. It allows response, as when seeing a beautiful face activates approach and other behavior tendencies. Such consciousness has been termed "irreflexive consciousness" by phenomenological philosophers such as Sartre (1946). In the second place, human consciousness includes awareness of that consciousness of the world. We not only see, but we know that we see and what we see, which allows naming and conscious awareness in a more explicit sense. The two aspects are linked, except perhaps in ecstasy and during states of dissociation; during torture, for instance, the individual may feel that he or she looks down upon him/herself while writhing and shrieking in pain. The fact that the link can be broken indicates that taking stock of awareness is a distinct process from the awareness that is being taken stock of. There is little to distinguish irreflexive awareness from nonconscious information processing as it occurs in blind sight or, presumably, in right-hemisphere stimulation of split-brain patients, except its being reportable *post hoc*.

These considerations are relevant for evaluating the role and function of emotional experience in emotion. There is, for instance, an important distinction between emotional awareness and awareness of what the emotion is about, or what has caused it. When terrified, one may not know why one is terrified, while knowing that one is terrified, and what one is terrified of. We know when terrified by a scream in the dark, or by the siren upon city hall's roof. But we do not know why it terrifies us, while that still may be something in the information, and the associations to it. This, of course, is relevant for the questions about nonconscious emotion activation. "Nonconscious emotion activation" may mean several different things, that are also functionally different: lack of awareness of what caused the emotion, lack of awareness of which aspect of the object caused the emotion, or why, or lack of awareness of the emotional response itself. In Öhman's experiments, there may be lack of awareness in all these regards; in those on the persistence of conditioned fear in rats by LeDoux, only that regarding the why of the causation.

All this touches upon the main question: what difference does it make whether any of those aspects of the emotion process is conscious or nonconscious? What difference does it make whether or not a reportable emotional experience is aroused? What, if any, is the function of emotional awareness, in regard to stimulus, to stimulus meaning, or to response, considering the fact that affective processes exist that do not seem to presuppose awareness, or at least reportable awareness, such as the effects of nonconscious exposure to affective stimuli.

Those latter effects demonstrate that emotional reactions can occur without awareness. Presumably, all elementary "expressive" motor reactions can occur without emotional awareness, or at least independent of it. An aggressive reaction to an unexpected stimulus is triggered by the unexpected stimulus, but not necessarily by the awareness of being startled by that stimulus and, perhaps, even without awareness of that stimulus (as in Zajonc-type experiments).

But any more involved response -- elaboration of a reflex-like aggressive response into watchfulness, aggressive stance, attack -- all need the awareness of both the triggering event, its unacceptability or negative affective value, and one's "desire" to neutralize it, the action tendency.

Notably, the motivational aspect of emotions would seem to be dependent upon awareness. Pleasure and pain, as feelings, have urgency and persistence. They drive behavior till exhaustion or termination of the feelings. And indeed, their function would seem to be precisely that, and it is not clear how the function could be fulfilled in a different way. Affect, I think, is a central signal generated by concern relevance, of such a nature that it gains attention and keeps it tied, and that it simultaneously influences any response system of the organism. Peptides could do that for glandular and motor systems, but not for the cognitive system needed for planning and maintaining the plans over obstacles and interruptions, and reflection upon the meaning of the event and one's response to it..

This implies the supposition that emotional awareness is a prerequisite for sustaining action in relation to an emotional object over and beyond a first encounter, or in the physical absence of the object. "Readiness" that goes beyond a specific motor readiness (but which is a matter of continuing goal-orientation, being set to achieve a particular relational change) would seem not to be possible without awareness.

References

Damasio, A. (1994) *Descartes error. Emotion, Reason, and the Human Brain*, New York: Avon Books.

De Waal, F. (1989) *Peacemaking Among Primates*, Cambridge, MA: Harvard University Press.

De Waal, F. (1996) *Good Natured: The Origins of Right and Wrong in Humans and Other Animals*, Cambridge, MA: Harvard University Press.

Delgado, J.M.R. (1975) “Inhibitory systems and emotions”, in: *Emotions: Their Parameters and Measurement*, L. Levi, ed., New York, Raven Press, pp. 183-204.

Ekman, P. (1992) “An argument for basic emotions”, *Cognition and Emotion* 6:169-200.

Frijda, N. H. (1986) *The Emotions*, Cambridge,UK: Cambridge University Press.

Frijda, N.H. (1988) “The laws of emotion”, *American Psychologist* 43:349-358.

Frijda, N.H. (1993) “Moods, emotion episodes, and emotions”, in: *Handbook of Emotions*, J. Haviland and M.Lewis, eds, New York: Guilford Press, pp. 381-403.

Frijda, N.H., P. Kuipers and E. Terschure (1989) “Relations between emotion, appraisal, and emotional action readiness”, *Journal of Personality and Social Psychology* 57:212-228.

Frijda, N.H., S. Markam, K. Sato and R. Wiers (1995) “Emotions and emotion words”, in: *Everyday Conceptions of Emotion*, J.A. Russell, J.-M..Fernández-Dols, A.S.R. Manstead and J. Wellenkamp, eds,Dordrecht: Kluwer, pp. 121-144.

Frijda, N.H., B. Mesquita, J. Sonnemans and Van S. Goozen (1991) “The duration of affective phenomena, or emotions, sentiments and passions”, *International Review of Emotion and Motivation*, K. Strongman, ed., New York: Wiley, pp. 187-225.

Gray, J.A. (1982) *The Neuropsychology of Anxiety: An Enquiry Into the Functions of the Septo-Hippocampal System*, Oxford, UK: Oxford University Press.

Griffith, P.E. (1997) *What Emotions Really Are*, Chicago: University of Chicago Press.

Lambie, J.A. and A.J. Marcel (in preparation) "The varieties of emotion experience".

Lazarus, R.S. (1991) *Emotion and Adaptation*, New York: Oxford University Press.

LeDoux, J. (1996) *The Emotional Brain*, New York: Simon and Schuster.

MacLean, P.D. (1993) "Cerebral evolution of emotion", in: *Handbook of emotions*, M. Lewis and J.M. Haviland, eds, New York: Guilford Press, pp. 67-85.

Mandler, G. (1984) *Mind and Body: The Psychology of Emotion and Stress*, New York: Norton.

Marcel, A. (1983) "Conscious and unconscious perception: An approach to the relations between phenomenal experience and perceptual processes", *Cognitive Psychology* 15:238-300.

Markus, H.R. and S. Kitayama (1991) "Culture and the self: Implications for cognition, emotion, and motivation", *Psychological Review* 98:224-253.

Morris, W.N. (1992) "The role of mood in affective systems", in: *Emotion. Review of Personality and Social Psychology, Vol. 13*, M. Clark, ed., Beverly-Hills, CA: Sage, pp. 256-293.

Murphy, S.T. and R.B. Zajonc (1993) "Affect, cognition, and awareness: Affective priming with optimal and suboptimal stimulus exposures", *Journal of Personality and Social Psychology* 64:723-739.

Oatley, K. (1992) *Best Laid Schemes: The Psychology of Emotions*, Cambridge, UK: Cambridge University Press.

Pribram, K.H. (1981) "Emotions", in: *Handbook of Clinical Neuropsychology*, S.B. Filskov and T.J. Boll, eds., New York, Wiley, pp. 102-134.

Rimé, B. (1996) "Social sharing of emotion", in: *ISRE'96: Proceedings of the 9h Conference of the International Society for Research on Emotions,* N.H. Frijda, ed., Toronto: ISRE publications pp. 3-16.

Roseman, I.J., A.A. Antoniou and P.E. Jose (1996) "Appraisal determinants of emotions: Constructing a more accurate and comprehensive theory", *Cognition and Emotion* 10:241-278.

Russell, J.A. (1979) "Affective space is bipolar", *Journal of Personality and Social Psychology* 37:345-356.

Russell, J.A. (1980) "A circumplex model of affect", *Journal of Personality and Social Psychology* 39:1161-1178.

Russell, J.A. and L. F. Barrett (1999) "Core affect, prototypical emotional episodes and other things called *emotion*: Dissecting the elephant", *Journal of Personality and Social Psychology* 76:805-819.

Sartre, J.P. (1946) *The Emotions*, New York: Philosophical Library.

Scherer, K.R. (1984) "On the nature and function of emotion: A component process approach", in: *Approaches to emotion*, K.R. Scherer and P. Ekman, eds, Hillsdale, NJ: Erlbaum, pp. 293-317.

ANTECEDENTS AND FUNCTIONS OF EMOTION EPISODES

NICO H. FRIJDA
Department of Psychology, Amsterdam University, Roetersstraat 15, 1018 WB Amsterdam, Netherlands

ABSTRACT

Emotions are aroused by appearance or disappearance of pleasant and unpleasant events, or by anticipation of these contingencies. Pleasant and unpleasant events are events that directly and innately evoke an affective response, or are associated or conditioned to such primary affective stimuli; and stimulus events appraised as relevant to some concern. Arousal of emotions involves appraisal, that is, some process that transforms perceived events into events with affective value. These processes are sometimes simple (as with innately valent stimuli), but most often involve some processing of information. This implies that antecedents of emotions usually include cognitive processes. The cognitions concern features of the emotion-arousing event, its relation to concerns, the subject's response propensities, and his or her coping potential. The nature of these antecedents, and the nature of emotional reactions indicate the functions of emotions: signaling concern-relevance of events; and motivating and organizing actions to do something with or about those events. These functions are primarily realized through action dispositions that benefit adaptation in various different ways.

1. Emotion Antecedents

Emotions consist of affect, appraisals, and changes in action readiness and support mechanisms, part of which are reflected in emotional experience. What are their antecedents? What are the antecedents of emotions in general, and thus of any or all of those components? And what are the antecedents of a particular kind of emotion, that is, for different affects, different appraisals, different states of action readiness?

Events that arouse emotions show infinite variety. Also, many different kinds of event are capable of arousing each kind of emotion. How are we to reduce those multitudes to a manageable number, and to understand them?

Following a behaviorist lead, one can say that, by and large, emotions are aroused by pleasant and unpleasant events (or, in behaviorist parlance, by positive or negative reinforcements), or by anticipation of such events (see Fig. 1, derived from Mowrer, 1960). And what are pleasant and unpleasant events? There would seem to exist two kinds. The first kind consists of stimulus events that are pleasant or unpleasant as such, by nature or by associative learning. Bodily pain and sweet substances belong to the stimuli that are intrinsically pleasant or unpleasant; most aversive tastes and traumatic stimuli to those that obtained it by associative learning.

The second kind consists of stimulus events that are relevant to the satisfaction of concerns. Relevance may mean achievement of satisfaction or recuperation from loss or threat; and it may mean harm to satisfaction or loss of present satisfaction, as indicated in the scheme of Table 1.

Table 1. Emotion antecedents (after Mowrer, 1960).

	positive reinforcer	negative reinforcer
(signal for) increase	pleasant	unpleasant
(signal for) decrease	unpleasant	pleasant

2. Concerns

"Concerns" refers to motives, major goals, desires, values, and affective sensitivities. Together, they can be defined as states of affairs that an individual prefers to obtain (Frijda, 1986). Those states may ultimately involve affectively valent stimuli of the first kind (as they do when the concern is to obtain or retain pleasant sensory stimulation or to avoid evil smells). Many concerns, however, have to be understood differently (as with the concern to retain body temperature, which is defined by a set-point and not by some preset "pleasant temperature").

"Concern relevance" is assigned a central place in emotion arousal, for several reasons. It accounts for the dependence of emotions upon the personal meaning of events; personal meaning is most readily understood as grasping the implications of an event for the fulfillment of one's major goals, motives. and values. It also accounts for the fact that, to some extent, different people respond emotionally to different events, and that a given event may arouse an emotion now, and not tomorrow: concerns differ between people (I care for certain people that are indifferent to you), and within the same people from moment to moment (my loves may vary in strength and priority). Most importantly, it accounts for the multitude of events that may arouse emotions. Any individual has numerous concerns, for which different events are relevant. Also, many different events are relevant to any given concern, because satisfaction of each concern may be promoted or threatened in many different ways. In addition, there is hardly a limit to the number of events that can be interpreted as being relevant in some way or other to any or many concerns.

The separation between emotions and concerns is important, and considerably clarifies the analysis of emotions. Different emotions, on the whole, correspond to different contingencies with regard to the satisfaction or non-satisfaction of any concern (Frijda, 1986; Oatley, 1992; Scherer, 1984; I will go into some detail below). A given emotion --say sadness-- may occur when your beloved leaves you, when your paper is not accepted, when you are unable to buy a ticket for that opera performance, which all represent different concerns. Also, any concern may give rise to any emotion, when the corresponding contingency occurs. The fate of one's beloved arouses emotion because she or he is your beloved, and it may make you happy, sad, angry, or proud, according to what happens.

By and large, the variety of emotions is limited (when different emotions are defined by contingency or by mode of action readiness or activated action disposition; see my chapter on the nature of emotions), but that of concerns is not. The number of kinds of event that elicits some emotion is large, but it is transparent: it equals the number of different emotions multiplied by the number of concerns multiplied by the number of contingencies, and then by the number of events that can be interpreted as representing a given contingency with respect to a given concern!

Separating emotions from concerns frees understanding emotions from an undue fixation upon "stimuli". Most emotions by far are elicited not by stimuli but by constellations of a concern and some contingency regarding its satisfaction. The well-being of one's beloved is something one cares about, whatever the precise stimulus pattern that represents it, or harm to it. And for most sources of satisfaction or dissatisfaction, the precise stimuli are not easy to specify, if indeed they can be defined in terms of "stimuli".

It is useful to realize how large and varied the range of concerns in an adult human actually is! Among the concerns are elementary ones like sex, absence of pain or other aversive stimulation, but also equally elementary but stimulus-wise less simple ones like proximity of an attachment figure. They include many less tangible concerns such as status in the group, self-esteem, successfully exercising the abilities that form part of one's self-concept. Other more or less subtle ones are the elementary social concerns: having social interaction and exchange, belonging to a group, belonging to a group with high status in its own eyes and those of others, belonging to a larger unity. Some of those concerns are evolutionary perhaps more elementary than others, nevertheless, they all are concerns, the concerns are there, and emotions are aroused when they are at stake.

Each of those general or "source" concerns, or combinations of them, may give rise to acquired "surface" concerns, specific for a given individual: the proximity and well-being of a selected individual, for instance -- that one for you, this one for me; a particular life goal, a particular interest or a particular ideology. From the point of view of understanding emotions they function like the source concerns that they may derive from.

3. Emotion Arousal

Thus, emotions are aroused by events relevant to a concern. One has to modify: emotions are aroused by events appraised in a way that renders them relevant to a concern. I will later discuss the notion of appraisal as an aspect of emotion antecedents. Positive emotions result when the event is appraised as favorable to a concern, and negative emotions result when events are appraised as harmful or obstructive of concern satisfaction. Neutral emotions (surprise, alertness may arise when relevance is uncertain.

There is a complication. Most events are relevant for more than one concern. Loss of a spouse means loss of partnership as well as sex as well as income; but it also may mean increase in freedom and self-determination. An event, therefore, may evoke several emotions, and often affectively opposite ones. It also may evoke an emotional response that derives its strength from more than one source. Indeed, emotional intensity has been shown to be proportionate to both the number and the importance of the concerns that give the event its emotional meaning (Frijda *et al.*, 1992).

That intensity does not depend merely upon the concerns, but also upon how serious the harm or how extensive the satisfaction promises to be. It also depends upon the individual's ability or inability to cope with or profit from the event. An offender with a knife produces more fear than a bare handed one; but he evokes little fear in a jiu-jitsu expert. A neutral or harmless event may evoke anxiety in an individual who has lost all sense of coping competence, because under that condition every event may turn out to be relevant. Event magnitude and coping potential thus not only affect emotional intensity, but also whether or not an event arouses an emotion at all.

Generally speaking, by and large an emotion is aroused only when relevant events involve meeting an uncertainty about being able to deal with the event, or when they imply having overcome a previous uncertainty. When events --even highly relevant ones-- can be countered in routine fashion, no emotion occurs. When favorable events can be profited from in routine fashion, no emotion occurs either. Previous uncertainty is a condition for positive emotions like enjoyment, joy, hope, relief, or pride, and actual uncertainty is a condition for a sense of meeting a challenge.

Emotion antecedents thus include the individual's concerns and affective sensitivities, events relevant to those concerns and sensitivities, and his or her coping confidence.

4. Expectations

I said that emotions are aroused, by and large, by pleasant and unpleasant events. The reservation, "by and large", is due to the fact that certain reactions are called emotions by some investigators, and not by others, notably surprise, astonishment, and curiosity or interest. They are neither inherently positive nor negative, and they are elicited by events that deviate from expectations (Frijda, 1986; Mandler, 1984; Meyer *et al.*, 1991; Oatley, 1992). Relevant and unexpected events do not constitute separate classes, however. Deviation from expectancy may by itself be concern-relevant. It may be pleasant (as in interesting novelty and a surprise party) as well as unpleasant (when it is difficult to meet, as on occasion with a surprise party). Also, unexpectedness contributes to coping uncertainty. However, the condition should be added in view of the "cognitive" emotions.

5. Antecedents of Different Emotions

What causes the different pleasant and unpleasant emotions? The answer of course depends upon the criteria for what distinguishes one emotion from another. One criterion is what elicits it. Jealousy is the name for the emotion that is caused by seeing one's love object engaging in love-play with someone else, and envy for that caused by seeing someone else possessing what one would like to possess oneself. The antecedent is of course a useless criterion in the present context of asking what the antecedents of different emotions are. To answer that question, one needs a criterion that is independent of antecedents.

Such a criterion is found in evidence of states of action readiness. Major emotions, notably what are often considered "basic emotions", are in fact distinguished by major modes of action readiness and action dispositions, as discussed in the previous chapter. The question "what are the antecedents of different emotions?" is best understood as: what are the antecedents of the various states of action readiness or of the activation of the various action dispositions? There are further response components, such as patterns of autonomic response, like sympathetic arousal and the orienting reflex. I will refer to them in passing.

Basic emotions have sometimes been thought of as responses to particular specific stimuli. Sadness has been viewed as the response to pain, fear to loss of balance, to encounters with a predator, or to the anticipation of pain, and anger to frustration. There is no support for such a scheme; the various elicitors of each emotion are not easily subsumed under one such heading, not even when their range is extended by conditioning. The schema of Table 1 offers a more promising starting point. Different emotions are elicited by different contingencies involving pleasant and unpleasant stimuli, rather than by different stimuli. The schema of Table 1 can be extended by differentiating the contingencies beyond increase and decrease. The effect of increase (or its extreme, appearance) differs when the individual can avail herself of the opportunity, or meets difficulties in doing so. The effect or decrease or loss differs likewise with the presence or absence of coping potential (Gray, 1982). All this can be readily rephrased in terms of different contingencies with regard to achieving concern satisfaction (Frijda, 1986) or goal achievement (Oatley, 1992). Oatley distinguishes the conditions for various emotions as follows (Table 2):

Table 2 Emotion antecedents (modified from Oatley, 1992)

achievement of valued goal	happiness
failure of valued goal loss of an active goal	sadness
frustration of an active goal	anger
threat to self-preservation goal goal conflict	fear

Lazarus (1991) has described the contingencies in terms of the "core relational theme" that is involved in an event, in which, again, terms like achievement, threat, and loss occur. The contingencies can also be described analytically rather than as categories, in terms of patterns of "appraisal dimensions" (Smith & Ellsworth, 1985; Frijda, 1986) or "stimulus evaluation checks" (Scherer, 1984), which allows greater flexibility and variety. The appraisal components can be viewed as constituents of emotional experience, and they were mentioned in the previous chapter in that role. They can also be viewed as features of the emotion antecedents as appraised. Each emotion is elicited by an event representing, or appraised as involving, a particular pattern of values on appraisal dimension or evaluation check outcomes. Several proposals for plausible sets of components have been made; Table 3 reproduces that of Roseman *et al.* (1996).

Table 3. Appraisal dimensions (according to Roseman *et al.*, 1996)

unexpectedness	whether event violates one's expectations
situational state	motive-consistency or inconsistency
motivational state	minimize punishment vs. maximize reward
probability	uncertainty or certainty
agency	circumstances, other person, or self
control potential	nothing, or something one can do about the situation
problem type	instrumental or intrinsic

The various contingencies, themes or components are cognitive, in the sense that they involve utilization of information other than that of the emotion-eliciting stimulus *per se*. There may not even be an emotion-eliciting stimulus. "Loss" is not a stimulus, but a constellation including the absence of a stimulus that was valuable for some concern and was present before. The information processes concerned can be low-level and elementary. "Absence" may merely mean a thwarted expectation, and it receives its full emotional impact by low coping potential, that is, by the awareness that nothing can be done or has succeeded to restore the lost object. Viewed in this manner, even the panic of an operated rat exposed earlier to a traumatic stimulus in LeDoux's (1996) experiments results from a cognitive appraisal process. Not the affective value of the conditioned stimulus does so, but the panic response does, as the subject has no way to do anything whatever about the highly aversive stimulus, and in some sense "knows" this.

Differential utilization of information provides a plausible explanation of the elicitation of different emotions. That information utilization is frequently referred to as "cognitive appraisal". It need not consist of a distinct evaluation process, performed for the sake of emotion. It may consist, and no doubt usually consists, of picking up certain items of available information, and perhaps attending to them (while at other times not picking them up or not attending to them). No doubt it often also consists of viewing a given input in the light of available cognitive schemas. It does so when some non-information is viewed as "absence," or when a familiar and usually responsive object does not move and remains unresponsive in a way beyond one's coping potential. I allude to the fact that even animals can recognize a dead attachment object, and suffer accordingly (e.g. see Goodall, 1972; DeWaal, 1996 for similar perceptual feats).

The differential cognitive processes can be conceptualized as the construction of a representation of the antecedent-as-appraised. They can also be conceived, and perhaps more plausibly, as differential sensitivities or responsiveness of the various action dispositions to particular patterns of information (Frijda, 1993). Both interpretations imply processes of abstracting the information patterns from the informational inputs and aroused schemas. These processes often are elementary. Any mismatch between input and a "neuronal model" (Sokolov, 1963) arouses the orienting reflex mechanism. But often the processes draw in higher cognitive abilities.

6. Cognitive Antecedents

Concern relevance is only in part a property of the event-concern relationship, just as coping confidence is only in part a matter of actual coping ability. An event is relevant for a concern when it is so tagged by the individual. Also, secondary appraisal uses causal attributions and may involve uncertainties that the individual attaches to the input from the actual event. The events themselves may be cognitive: thoughts, fantasies, upsetting memories. Said in general fashion:

emotions are aroused by events appraised as relevant to some concern, under the light of coping potential and other context information as appraised.

Emotions result from appraisal processes that may be minimal in the case of affect arousal by intrinsically valent stimuli. Usually, however, the appraisal process is more complex and more a cognitive process. The cognitive processes may involve true inferences, as when Othello became disturbed by a stray handkerchief. They may involve cognitive schema's, as when a depressive individual gets deeply upset by a minor failure or just the expectation of failure. They may involve entirely symbolic meanings, as when the taste of pork evokes disgust in a Jew or Muslim, or when enraged Iranians burn the American flag.

Emotions thus often are aroused by the intervention of cognitive processes. To this statement a number of qualifications have to be made.

First, the quantifier "often" makes clear that cognitive processes are not a necessary condition for emotion arousal, witness the emotions aroused by severe pain or sexually seductive stimuli. Much confusion can arise from the fact that such emotions can be enhanced, prevented or inhibited by cognitive influences. The effect of such influences does not imply that they were operative in the emergence of emotion as such.

Second, even when cognitive processes do mediate emotion arousal, they still do not form a sufficient factor. Personal concerns are often essential. Death of a friend causes distress, but not death of a former friend, or not as much. In addition, many relevant cognitions fail to arouse emotions. Television images of the horrors caused by tornado Mich do not very much upset viewers far away. Cognitions have to involve access to previously evoked affect or to facilitated or obstructed action planning in order to be emotionally effective. They have to include something like cognitive schemas precipitated by personal encounters with events like those reported, or as generated by imagery (Lang, this volume); they have to consist of "schematic representations" (Power & Dalgleish, 1996; Teasdale & Barnard, 1992).

It should be stressed that "cognition" does not imply consciousness, either of the relevant information or of the process of using that information. Neither does it necessarily imply inference in any articulate sense. "Cognitive", as used in understanding emotion arousal, means utilization of information that goes beyond the information given, in whatever way. The cognitive processes can be very low-level, such as perceptual spontaneity and causality (involved in agency attributions upon which much anger rests), novelty assessment as underlies both startle, surprise, and the orienting reflex, self-reference as also underlies the stability of the visual field during eye movements.

The role of cognition in emotion arousal coexists with failures of cognitive processes. Such failures may be equally prominent and, in fact, are characteristic for emotion arousal. The cognitive processes involved in emotion arousal usually do not use all information available at the time of arousal. They do not even use all information that might be pertinent to evaluate emotional relevance or urgency.

Cognitive processes in emotion arousal show the functionally important features of being slanted towards using minimal information, evoking responses whenever a stimulus aspect fits response dispositions, and they appear to respond preferentially to information that is present, not to that which is not present, nor to information that is incompatible with presented affectively relevant information. They tend to respond to affectively valent sensory stimuli, to associations or other informational links to events with such valence, and to just those aspects of stimulus events for which this is the case. They tend to be less sensitive to information that is not affectively valent or that might signal affective innocence. These features allow rapid appraisal, and appraisal that is relatively independent of the informational complexity of the event or its associated information.

Finally, the information used by the cognitive processes in emotion arousal does not necessarily correspond to the cognitions that are part of conscious appraisal. Conscious appraisal may include attributions made after the fact of emotion arousal (Frijda, 1993).

7. Further Emotion Antecedents

The conditions actually eliciting various emotions suggest that emotions can be aroused for "insufficient reasons." Formerly (before the era of drugs), temporal lobe epileptics might fly into a lethal rage by a minor unexpected noise, just as anyone after a day of being harassed at work may blow up at home at being asked an innocent question, or as a hyper-startler might do when unexpectedly touched (Simons, 1996).

There appear to exist moment-to-moment fluctuations in the propensity to appraise events in a particular way, or in the sensitivity or arousability of the action dispositions (which of these ways to phrase the mechanism is the optimal one will be hard to decide). The threshold for activating certain action dispositions may even be so low that they become activated by only a fraction of the informational pattern that usually is needed to activate them. Action dispositions, in other words, may have their own state of activation or arousability that can make them more or less independent from the pick-up of information or from the meaning generally or normally carried by the corresponding events, or even from the meaning the subject would attach to them.

In any case, propensity for particular emotions can vary for several reasons other than the information that the eliciting event carries, whether directly or through associated information. Propensities vary with mood; happy moods make silly jokes seem funny, for instance. They vary with prior activation, as in the example of the person falling out with the dog after a day of harassment at the office. They may vary as a function of prior activation, as in the irritability after a quarrel, or through sensitivity or threshold changes with physiological causes. Sexual stimuli facilitate subsequent anger, and vice versa (Zillmann, 1983), which may well be due to increased testosterone level; and hormonal changes are probably responsible for enhanced premenstrual irritability (Van Goozen *et al.*, 1994). They also vary with personality disposition. Certain temperamental traits are best understood as propensities for particular emotions: extroversion is interpreted as a Positive Affect disposition (Watson & Clark, 1984), neuroticism as Negative Affect, Depressiveness as a propensity for sadness and self-aversion (Power & Dalgleish, 1996); there also exist anger propensity and trait anxiety. The traits imply that the frequency of the particular emotions is higher in some individuals than in others and that, by consequence, a given event is sooner appraised as exhilarating, frightening, depressing, angering, by some individuals than by others.

All this means that particular emotions are elicited either because a certain appraisal is rendered likely by the pattern of information contained in the eliciting event and its associated information, in interaction with the individual's coping potential; or because a momentary change has occurred in propensity to appraise or respond in a certain fashion; or because of a more stable propensity of the individual to do so.

8. The Functions of Emotions

The analysis of what arouses emotions, and the affect and action dispositions that emotions consist of, points clearly to the functions of emotions.

First of all, emotions detect and signal the occurrence of concern-relevant (as well as intrinsically pleasant or unpleasant) events. That, of course, is precisely what affect does. It signals those events both to the various information processing and action systems, and to awareness. In fact, affect is what makes motivation possible, in the sense of sustained seeking for particular objects or states. Motivational goals are recognized as goals because not having reached them as well as hitting upon them generates affect, notably in systems with only modestly developed representational capacities.

The provisions for emotional appraisal allow to do that detecting and signaling of relevant events without a need for extensive processing of the possible implications of and associations to the input information. I indicated the information-processing bias involved in emotional appraisal, and in the sensitivities of the various action dispositions. They seem to act upon a false alarm bias, that is only secondarily corrected by regulation functions, when time permits to do so.

By detecting concern-relevant events, emotion (or at least affect, in the sense used here) provides a basis for preferences and decisions, and it provides such a basis to prefer and decide within a relatively short time, in some "absolute" fashion. Events are liked or disliked, while relative liking-more or disliking-more is another matter. Affect allows preferences without extensive comparative assessments of merit, to the extent that "merit" makes sense without affect. Without affective appraisal, the individual may not be able to reach any decision in a given situation, or not within reasonable time, as Damasio (1994) has argued.

And then affect and other aspects of appraisal provide the motivation to do something with or about those relevant events. They activate ready-made response provisions, or they demotivate if doing something appears impossible or too costly. Emotions are an organism's provisions for concern-realization or, more, precisely -- recall the role of uncertainty in emotion arousal -- they are the provisions for concern realization when routine procedures do not suffice.

Doing something about relevant events of course means doing what activated action dispositions, the states of action readiness suggest, under the appropriate conditions. The appropriate conditions, by and large, are the contingencies to which the action dispositions are geared to respond. Emotions thus are functional in dealing with relevant events in two ways: by distinguishing the various contingencies -- the appraisal processes -- and by activating classes of behavior or, in the event, by suppressing behavior or abstaining from it. Appropriate conditions, I said, are "by and large" the various emotion-specific contingencies. The contingencies may include features that might make behavior inadvisable, such as the power of an angering opponent, or the absence of a distinct location of threat. These are the conditions for behavioral inhibition that at some level drastically modifies the contingency-emotion relationships. At the same time, they are features that tend to be neglected, in particular when activation is high or inhibition resources are low.

9. Functions of Emotional Behavior

How can emotions be effective in doing something about events, and thus safeguarding concern relevance? There are several ways.

Nonspecific mechanisms affecting behavior have been mentioned in the earlier chapter. General activation, attentional arousal, and several of the other support mechanisms such as autonomic arousal have applicability over a wide range of contingencies. Some aspects may have had useful functions in the phylogenetic past under restricted but still emotion-nonspecific circumstances. The usefulness of the skin conductance response, for instance, is still a matter of debate, since it is most pronounced where the sweating least serves thermoregulation (hand palms, foot soles); it may have served running on slippery soil or climbing trees. Motivation loss, as in depressive apathy has been suggested to serve energy conservation or facilitating the detachment from a now useless attachment, but it may just represent a side effect of how activation mechanisms operate.

The clearest function of the more specific action dispositions --those corresponding to fear, anger, disgust, surprise, desire-- is in dealing with a relevant event. They serve to modify the relationship, in direct confrontations with an event: to block the progress of an antagonist, decrease one's vulnerability to harmful agents, expel evil, or at least unpleasant, substances, better orient in the visual and auditory environment with regard to an object in an unknown location ("surprise"), get closer to desirable objects and avail oneself of obtained opportunities through the impulses of the emotions of lust, greed, and enjoyment. This obvious function of emotionally instigated behavior has been at the core of adaptive interpretations of emotions ever since Darwin, and elaborated by Plutchik (1980). The modes of action readiness indicated earlier can all be considered instrumental in producing a particular relational change, with each of them in principle capable of solving a particular predicament or availing of a particular kind of opportunity.

There are some subtle variants of such functionality in direct dealing. Freezing, inhibitory anxiety, would seem to serve to suppress behavior under conditions where it is unclear what risks whichever action might entail. That same cautionary function is served by inhibition in social situations as in embarrassment and stage fright: the person may act stupidly, but not wrongly. Joy, that is, the impulse towards expansive and playful behavior, produces diffuse interaction with the environment, notably the social one, thus enlarging the field of novel and unsought-for opportunities.

Interestingly, dealing with relevant events can be achieved in two different ways. Approach, avoidance, sensory exposure and withdrawal, as well as attack are affect the relationships directly; the body movements directly modify the relationships. However, many emotional actions modify the relationship with relevant events indirectly, by influencing the actions of a protagonist or antagonist. Emotions are, to a very large extent, geared to operate in a social environment. Many emotion manifestations have the function of influencing other individuals to do the work for you. Whereas aggression, deployment of force, may factually block unwanted approach, intimidation and threat displays make the antagonist stop approach of his or her own accord; it is obviously a cheaper alternative to fighting it out. The distress call is useful because others hear it and respond to it. Crying is effective because, or in so far as, others are moved to solve the distressing predicament for you. Smiling is effective in establishing friendly relationships because others understand it as signaling affinitive intent and absence of hostile intent, so that no aggression from their side follows, or friendly approach is evoked.

The duality between direct and social functionality can be recognized in facial expressions. Some expressions are best viewed as automatized instrumental actions, in the way that Darwin analyzed them. But others are best understood as social signals: threat displays, submission displays (the bent head in shame, remorse, and embarrassment), affiliation displays, nurturance-demanding displays (Fridlund, 1994; Frijda, 1986). Note that all these emotions "work" thanks to matching sensitivities in those who perceive the manifestations. When the calls for nurturance are rejected, crying only increases irritation.

The social consequences importantly expand the functionality of emotions. Jealousy is a case in point. As anger, it intimidates the partner as well as the rival and discourages their dallying. As manifest suffering it operates like blackmail, to the same effect. Jealousy is a functional emotion. It is not mere meaningless suffering or urge towards punishment of the partner or rival. It is not at all unlikely that more marriages are saved than hurt by jealousy.

In a similar fashion does envy spoil the other's pleasure, if only in the eyes of the subject, by denigrating the value of what the other has obtained (Klein, 1977). Spoiling that pleasure helps. It restores some of the inequality in possession as well as in happiness and social status linked to it. Such consequences of emotion go a long way in explaining their occurrence or even existence. Revenge seems puzzling because the original harm cannot be undone. However, it restores self-esteem and a sense of having the fate in one's own hands. These are gains that outdo the harms one may also suffer, and that notably revenge can lead to. From that perspective there is nothing irrational in it (Frijda, 1994).

10. Social-Regulatory Functions

In addition to such adaptive dealing, emotions fulfil an important role in maintaining or modifying interpersonal relationships. Sometimes it is in their very nature to do so; in other cases the effects on relationships seem secondary. Sadness is an instance of the latter. To the extent that sadness involves low coping potential with regard to some loss, one of its main elements is helplessness; helplessness is manifested by passivity, waiting for others to take initiatives, and by crying. Crying can be called the mode of social influence of the helpless or powerless (Frijda, 1986). Calling for help thus establishes a relationship of dependency of one with the other interactant, if only for the moment of sadness. Crying, indeed, is probably a mixture of a distress call and a submission signal (notably the tears might be), in which latter function it can serve as an aggression inhibitor, as indeed it does during marital and other quarrels.

Submission is still more prominent in several emotions other than sadness, and more clearly functional. It is a main element in shame, guilt feeling, remorse, humility, timidity, and embarrassment. The mode of operation of the submission appears different in each case: siding with the antagonist with respect to self-rejection in shame, subservience and taking responsibility in guilt-emotion, atonement in remorse, inconspicuousness in timidity and embarrassment, truly submitting in humility. All serve in some way to attenuate hostility, rejection, and social-emotional distance. Shame accepts the rejecter's point of view, timidity and embarrassment recognize one's place in an unequal status relationship or emotional position (as when being in love). Guilt feeling equalizes suffering. You cannot undo the deed but you can balance the relationship to the offended person (Baumeister, Stillwell, and Heatherton, 1994), as in shame you cannot undo the deed or shed the shameful property, but you can repair resulting social rejection.

Anger, too, has a prominent social-regulatory function. It may well be its main and major function, with self-defense in a second place. Anger would seem in the first place to serve to settle social power problems, among humans no less than among primates (Averill, 1982; DeWaal, 1982). Marital quarrels are more about relative power than about what they seem to be about (Frijda, 1987) .

The relational function is of course obvious in the truly social emotions like sympathy, affinity, tenderness and love. They involve forms of action readiness to establish and maintain particular personal relationships , which motivate behavior that elicits reciprocation. Tender feelings motivate caretaking behaviors that help to achieve the goal of well-being in the other while at the same time establishing a relationship that me either be asymmetrical or symmetrical.

Emotions that establish interpersonal relationships may do so for solving some emotional confrontation (as in distress crying), but they may also do so for those relationships' own sake, and for an indefinite durations. Emotions like friendly feeling, affection, and interpersonal interest all tend to establish a personal relationship, sometimes for a few seconds, sometimes for a lifetime; interpersonal interest is a potent instigator of affection and sympathy in the recipient. The smile is a carrier of the message of these emotions. A smile not merely modifies a relationship but may constitute one when it is answered; and it my modify or strengthen one, as in erotic contact. Mutual smiling *is* a form of relationship.

One of the functions of (certain) emotions thus is the establishment of interpersonal relationships. Affinitive emotions function in that way. It may even be one of their *raisons d'être.* Obviously, whether or not the affinitive emotions fulfil that function depends upon their being shared; it depends upon sensitivities in the target person that make him/her reciprocate. Affection is an emotion which makes emotions or makes one accept emotions.

Feelings of tenderness, respect, and admiration are further emotions that each dictate a particular type of relationship which may be sought for its own sake, that is, for satisfying concerns for social bondedness and self-loss, or for adopting a particular position in the social group. There exists a variety of positions that each have their satisfactions and that each are achieved or maintained by particular emotions. The hierarchical relationships of submission and dependency of course are generally useful to the person in the lower position (and in no way uniquely Japanese) for the advantages and benefits of receiving attention, care, protection and sharing in another's power and resources. The dependent position is prominent in the Japanese concept of *amae*. and in the same emotion, unnamed, in other people (Markus & Kitayama, 1991).

The emotions named sympathy, pity, being moved: they satisfy the concerns of interest in the welfare of others, of having or creating bonds, and of harmoniously interacting with others (Baumeister & Leary, 1995). And there are the emotions like awe, fusion with others, certain forms of love or desire to fuse with someone else, with the environment as a whole, or with space, prominent in religious experiences and in mass manifestations, as they were experienced by several million people at the occasion of the death and funeral of princess Diana of England. People all over the world felt they were together for a few moments, and delighted in it.

Obviously, emotions also establish relationships from the dominance end. Anger not merely fights for momentary dominance, --when not effectively countered it posits a power relationship, just as contempt and pride do. By being angry one tends to claim a right to it, by contempt one puts someone else down; by pride one lifts oneself up.

Joy provides a final example of how emotions establish relationships. Joy does establish the position of freedom with regard to vital needs and obligations. Joy is the action tendency underlying play behavior: laughing, hopping, singing, putting one's arm around a stranger's shoulders, doing other useless things. In joy, one temporarily adopts a gratuitous relationship with the environment as well as with other individuals. Its further functions in receiving information and learning social skills and acquiring social bonds have already been mentioned.

The constitutive role of interpersonal emotions is general. Emotions of trust and familiarity in interpersonal activities, permit the establishment of further relationships, such as sexual ones. There is no sex without such emotions, as has appeared from the research by Harlow on the detrimental effects of early isolation from peers upon later sexual desire and competence (e.g., Harlow, 1969)

11. Functions of Emotions in Continued Interaction

The social-regulatory function of emotions has interesting extensions. Many emotions derive their major functional role from their role in continued social interactions, or from the expectation of such continued interaction. They serve not primarily to solve an actual confrontation but to regulate future ones of a similar kind in the ecology. Many emotions, in fact, seem to be there to make their own future occurrence superfluous.

Revenge clearly illustrates the point. Even if the reason for wanting revenge cannot be undone, taking revenge shows that you cannot be trifled with at subsequent occasions (Frijda, 1994). It also represents an effort towards the restoration of power inequality, again useful in view of continued interaction. Vengefulness thus is functional in making revenge unnecessary in the future, as long as one is manifestly ready for it. These functions of course parallel the more immediate one of restoration of self-esteem that was mentioned in the preceding section.

Shame can be viewed in a similar way. Shame is painful, and thus motivates to prevent its future occurrence, either by changing behavior directly, or by a covert controlling mechanism, in the form of "virtual emotion". One glimpses the shame that might emerge if one behaved in the rejected fashion. Shame thereby is the motor for dynamic conformity (Scheff, 1988).

Guilt emotion, too, is painful, in particular because it implies loss of acceptance, love or respect, by others or by God. It motivates careful or considerate behavior towards others, or with respect to the values of society, again either as anticipation of the emotion or as a virtual emotion.

Anticipation of future occurrence may even be the major function of grief and sadness. Like vengefulness, sadness represents a well-known paradox: you cannot undo the loss, so why suffer sadness? But sadness may well serve mainly to prevent its own occurrence, under conditions that that is still possible. You may do what you can to prevent loss: to care for, be careful with, remain close to, not to alienate or not to lose the cared-for person. The paradigm for grief is not someone's death but the child losing sight of its mother.

A final example is that of regret. Regret is famously irrational, because the bad thing (or not having the good thing) is past and cannot be repaired; so why worry? But you can try to prevent it, by learning from your mistakes and, primarily, by being set to avoid it. Regret-imagery, anticipation of regret, functions to instill prudent behavior. It can indeed be considered the major motor behind advertising, the power of "sales", and of "direct marketing". One will miss all those advantageous offers if one does not grasp them while one can.

Frank (1988) advances the suggestion that many seemingly irrational emotions exist to assure others of one's commitment to particular lines of action. This allows those others to reckon with those future lines of action, to the benefit of the sender. Revenge illustrates the point. In Frank's view, evolution has taken hold of that function and made the emotions into innate ones for that benefit.

Emotions, finally, are functional in a wider context. They form the fabric of human (and perhaps nonhuman) interaction, and thus the maintenance of social groups and cooperation, both for the individual's purposes and the relationship with other groups. By their being interesting to other individuals, they support social bonding and group coherence. They also do so by their signal value: expression of emotions signal the presence of relevant events in the environment. How interesting emotions are to others appears from the ubiquity of social sharing (Rimé, 1996). About 90% of emotional incidents are shared with at least one other individual within one day after their occurrence. The larger part of those shared incidents are secondarily shared with third parties. The interest in shared incidents originates, I think, in an elementary interest in emotional interaction in both partners in the sharing. The emotional value of widely shared emotions, like those of the people attending the funerals of princess Diana, of Khomeiny, or of the million or so that attended the funeral of the famous Egyptian singer Oum Kalthoum would seem to have the same source. Emotions belong to the major cementing agents in human groups (and perhaps in animal groups), whether by empathy, by just being in it together, or by exchanging responses about the events.

12. Dysfunctional Emotions

There are limits to the functionality of emotions, and thus to their functional interpretation. Emotions are often irrational, in the sense that their absence would have been more favorable to the individual. They often also are dysfunctional, in that they decrease efficiency with regard to the emotional goal as well as with regard to other ones. Emotions often interfere with task performance and adaptation. Phobias, devastating and persistent grief, performance deficit in anxiety and eagerness, paralyzing anxiety and paralyzing shame provide examples. How can we explain this and reconcile function and dysfunction?

One explanation invokes evolutionary lags. Some emotion dispositions may have been formed in a different ecology from the one presently prevailing. It cannot be the major explanation because, as I tried to show, most emotions serve adaptation to the present-day environment. But the principle may be of use for understanding particular phenomena, such as the galvanic skin response, or generally the ubiquity of sympathetic disturbance in emotional situations not requiring physical effort.

More important for explaining dysfunctional emotional responses is the way cognition operates in connection with emotions. The emotion system is built for rapid reaction with a false alarm bias. It responds to partial information and, more important, to here-and-now information. It often lacks depth of cognitive exploration, particularly of exploration of reasons for not responding emotionally in a given fashion. Harmful consequences are not always anticipated because they are more remote, less immediately present. All this may be viewed as a deficit, because evolution often tinkers and not, as a rule, considers all implications of a given change. But it may not really concern a deficit but a reasonable option. More depth or range of processing of course would take more time and requires more resources.

Some nonfunctional or dysfunctional responses indeed result from resource limitations. Regulation is such a limited resource; it can be spent by exhaustion. Yet, the entire emotion system may be built for operation along with regulation processes, according to a shoot first, ask questions later strategy. There are further limited resources: for vigorous physical action (as in anger), for cognitive explorations, for attention deployment. Emotional events and coping activity, by their very nature, are often taxing; they arise in response to difficulties.

Then there exist concern conflicts. Many emotions result from conflicting interests that may be irresolvable. Also, frequently the concerns rather than the emotions are harmful or dysfunctional. Desire for heroin, for unlimited power and dominance, greed, desire for national glory, are examples. It is the desire for heroin that is dysfunctional, not the anger or despair when not having it. Emotions have merely "local rationality" (De Sousa, 1988). They are efficient in safeguarding satisfying concerns, regardless of the adaptiveness of the concerns, and often regardless of other concerns that might be jeopardized by reacting upon a given one.

Finally there exists social imbalance. Social influences may upset emotional function by weakening or enhancing regulatory control, or the valence and appraisal of particular antecedents.

References

Averill, J.R. (1982) *Anger and Aggression: An Essay on Emotion*, New York: Springer.

De Sousa, R. (1988) *The Rationality of Emotions*, Cambridge, MA: MIT Press.

De Waal, F.B.M. (1982) *Chimpanzee Politics. Power and Sex Among Apes*, London: Jonathan Cape.

DeWaal, F.B.M. (1996) *Good Natured.: The Origins of Right and Wrong in Humans and Other Animals*, Cambridge, MA: Harvard University Press.

Fridlund, A.J. (1994) *Human Facial Expression: An Evolutionary View*, New York: Academic Press.

Frijda, N.H. (1987) "Macht, matchtsstrijd en emoties: ruzies tussen partners", [Power, power struggle, and marital quarrels], in: *Macht en Beïnvloeding*, O. Wiegman and H.Wilke, eds, Deventer: Van Loghum Slaterus.

Frijda, N.H. (1986) *The Emotions*, Cambridge: Cambridge University Press.

Frijda, N.H. (1993) "The place of appraisal in emotion", *Cognition and Emotion* 7:357-388.

Frijda, N.H. (1994) "The Lex Talionis: On vengeance", in: *Emotions: Essays on Emotion Theory*, S.H.M.Van Goozen, N.E. Van de Poll and J.A. Sergeant, eds, Hillsdale, NJ: Lawrence Erlbaum, pp.263-290.

Frijda, N.H., A. Ortony, J. Sonnemans and G. Clore (1992) "The complexity of intensity", in: *Review of Personality and Social Psychology, Vol. 6*, M. Clark, ed., Beverley-Hills: Sage, pp. 60-89.

Goodall, J. (1972) *In the Shadow of Man*, New York: Dell Books.

Harlow, H.F. (1969) "Age-mate or peer affectional systems", in: *Advances in the Study of Behavior, Vol.2*, D.S. Lehrman, R.A.Hinde and E.Shaw, eds, New York: Academic Press, pp. 334-384.

Klein, M. (1977) "Envy and gratitude", in: *Envy and Gratitude and Other Works, 1946-1963*, M. Klein, ed., New York: Delacorte.

LeDoux, J. (1996) *The Emotional Brain*, New York: Simon and Schuster.

Meyer, W.U., M. Niepel, U. Rudolph and A. Schützewohl (1991) "An experimental analysis of surprise", *Cognition and Emotion* 5:295-311.

Mowrer, O.H. (1960) *Learning Theory and Behavior*, New York: Wiley.

Oatley, K. (1992) *Best Laid Schemes: The Psychology of Emotions*, Cambridge, UK: Cambridge University Press.

Plutchik, R. (1980) *Emotion: A Psychoevolutionary Synthesis*, New York, Harper and Row.

Power, M. and T. Dalgleish (1996) *Cognition and Emotion: From Order to Disorder*, Mawah: Erlbaum.

Rimé, B. (1996) "Social sharing of emotion", in: *ISRE'96: Proceedings of the 9h Conference of the International Society for Research on Emotions*, N.H. Frijda, ed., Toronto: ISRE publications, pp. 3-16.

Roseman, I.J., A.A. Antoniou and P.E. Jose (1996) "Appraisal determinants of emotions: Constructing a more accurate and comprehensive theory", *Cognition and Emotion* 10:241-278.

Scherer, K.R. (1984) "On the nature and function of emotion: A component process approach", in: *Approaches to Emotion*, K.R. Scherer and P. Ekman, eds, Hillsdale, NJ: Erlbaum, pp. 293-317.

Simons, R. C. (1996) *Booh! Culture, Experience, and the Startle Reflex*, Oxford, UK: Oxford University Press.

Smith, C.A. and P.C. Ellsworth (1985) "Patterns of cognitive appraisal in emotion", *Journal of Personality and Social Psychology* 48:813-838.

Sokolov, J.N. (1963) *Perception and the Conditioned Reflex*, Oxford, UK: Pergamon Press.

Teasdale, J. and P. Barnard (1993) *Affect, Cognition, and Change*, Mawah, NJ: Erlbaum.

Van Goozen, S.H.M., N.H. Frijda, V.M. Wiegant, E. Endert and N.E. Van de Poll (1996) "The premenstrual phase and reactions to aversive events: A study of hormonal influences on emotionality", *Psychoneuroendocrinology* 21:479-497.

Watson, D. and L.A. Clark (1984) "Negative affectivity: The disposition to experience aversive emotional states", *Psychological Bulletin* 96:465-490.

Zillmann, D. (1983) "Transfer of excitation in emotional behavior", in: *Social Psychophysiology: A sourcebook*, J.T. Caccioppo and R.E. Petty, eds, New York: Guilford Press, pp. 215-240.

TO THINK AND TO FEEL: NONCONSCIOUS EMOTIONAL ACTIVATION AND CONSCIOUSNESS

ARNE ÖHMAN and STEFAN WIENS
Psychology Section, Department of Clinical Neuroscience, Karolinska Insttutet, Karolinska Hospital, Z6, 171 76 Stockholm, Sweden

ABSTRACT

Folk psychology maintains that consciousness plays a decisive role in the control of human behavior. This conviction, which is consistent with the Cartesian Cogito ("I think, therefore I am"), has been challenged by evidence of nonconscious mediation of psychological processes. The present article reviews this evidence with particular emphasis on our own research using the technique of backward masking to study nonconscious activation of emotion. Even though subjects do not recognize emotional target stimuli when these are followed immediately by masking stimuli, subjects nevertheless show differential psychophysiological responses to the masked target stimuli. Brain imaging studies have shown that masked emotional stimuli specifically activate the amygdala via subcortical pathways. Research further suggests that information from masked emotional stimuli may be accessible to the cognitive system through feedback from autonomic responses. These findings suggest that nonconscious processes constantly monitor the surrounding world for stimuli of emotional significance. This view is consistent with the notion that bodily feedback affects feelings in core consciousness and that the role of consciousness in emotion is to interpret and make sense of feelings.

1. The Cartesian Cogito and the Primacy of Consciousness

"Cogito, ergo sum" (I think, therefore I am) Descartes' statement in "Meditationes de prima philosophia" from 1641, remains one of the most influential ideas in Western thought: The one thing you can't doubt is that you are conscious. Because of its undeniable truth, Descartes claimed that this statement provides the pillar upon which our understanding of the world must rest. Often the "Cartesian Cogito" has been interpreted not only as an ontological statement but as a normative one as well: Conscious insight and conscious control of action is what thou shalt strive for. From this perspective, it is not surprising that Western thinkers have been skeptical to intuitions, gut feelings, hunches, passions and other ill-defined phenomena that seem to challenge the supremacy of rational cognition in the control of human action. The view of humans as conscious, rational beings has pervaded not only philosophy but also everyday notions of human psychology, which typically has held consciousness to be the node upon which causes of conduct converge, and from which actions originate.

The emphasis on conscious thought promoted by the "cogito" puts emotion outside the center of human existence, as a potentially threatening alien force, lurking in the body. Emotion represents a lower existence of dark irrational forces that may corrupt or seduce the rational mind. The mark of wisdom is to let reason domesticate passion.

For the pioneers launching psychology as an empirical science in the late 19th century, consciousness was the self-evident object of their scientific pursuit, and introspection was its main methodological tool. William James (1884) incorporated emotion into the enterprise by proposing that its conscious component, the feeling, resulted from the perception of the bodily activation that he saw as a central ingredient of emotion. He viewed emotion as a bridge between body and mind, thus accepting emotion as part of the emerging science of psychology. However, his open-mindedness came to a halt when considering unconscious psychological processes. Allowing unconscious processes as an object of study, he once remarked, was "the sovereign means for believing what one likes in psychology and of turning what might become a science into a tumble ground for whimsies (James, 1890, p. 163).

Thus it was a zeitgeist of formidable strength that Freud (1894) challenged by postulating an unconscious level of mentation showing up, for example, in hysteric symptoms or in dreams. In fact, the zeitgeist was so strong that it could cope with Freud simply by ignoring him. Indeed, even a century later, folk psychology remains remarkably resistant against assigning an important role to nonconscious processes, in spite of a hundred years of psychoanalytic thought. In everyday life, as well as in the courtroom, people are thought to be transparent both to themselves and to observers, to have a free will, to know what they are doing, and to have conscious access to the determinants of their actions. However, in psychology as a science, the dogmas of introspection were effectively challenged by the Würzburg school a couple of decades following Freud's first publications. The work of this group showed that there were psychological phenomena that resisted introspection and that could not be reduced to the content of consciousness. The basic experimental result that refuted traditional introspective psychology was the demonstration of "image-less thoughts, – "obscure, intangible, unanalyzable, indescribable (mental) contents that are neither sensations nor ideas" (Boring, 1950, p. 403). Some years later the action in psychology moved across the Atlantic, first to turn functional and then behavioristic, and in the process, consciousness was banned as a respectable object for scientific scrutiny. The most influential behaviorist, B. F. Skinner argued that "Freud's argument that we need not be aware of important causes of conduct leads naturally to the broader conclusion that awareness of cause has nothing to do with causal effectiveness" (Skinner, 1954/1972, p. 247). In his view, the basic mistake that Freud shared with the introspectionists was that consciousness gave privileged access to the causal determinants of behavior.

The neglect of consciousness that academic psychology inherited from the behaviorists left a breach between scientific and everyday notions of the role of consciousness. For our introspective selves, psychological phenomena such as thoughts, feeling and actions always occur in temporal closeness with conscious experience, and thus there is a short step to attribute them (perhaps illusorily) to consciousness (Merikle & Daneman, 2000). Thus, folk psychology adheres to the intellectual heritage from Descartes in giving consciousness a key role in controlling human behavior. However, according to Gray (1995), nothing that we know about physiology, behavior, the evolution of brain and behavior, or in robotics and artificial intelligence, provides a compelling case for consciousness as a scientific hypothesis, and neither does the hypothesis of consciousness provide a useful scientific account of this knowledge. Therefore, Gray (1995) claimed that the only reason for advancing a hypothesis of consciousness is because it occurs as a datum in our own experience. In other words, even though the "qualia" of the "cogito" may not be the pillar of all knowledge that Descartes claimed it to be, it is still regarded by (at least some) contemporary scientists as the founding pillar upon which a science of consciousness must rest.

2. Nonconscious Processes in Psychology: The New Look in Perception

Curiously enough, although academic psychologists for most of the 20th century have neglected or denied a place for consciousness in their science, they have also been very skeptical to accepting its converse, that is, nonconscious psychological processes. One would have thought that if consciousness is not something worth worrying about, then there is little reason to take offense by the notion of nonconscious determinants of behavior, or nonconscious psychological processes. Nevertheless, claims to have demonstrated experimentally that psychological processes can be nonconsciously mediated were met with considerable resistance, and such claims had to withstand severe critical scrutiny to become accepted. Although some of the heat of these debates may be fueled by the implication that accepting nonconscious processes would confirm psychodynamic theorizing, some of it may also be attributed to the prevailing strength of the intellectual heritage placing consciousness at the center of the psychological arena (Merikle & Daneman, 2000).

The controversy of nonconscious processes, for example, became evident with the "New Look" of the early fifties, which claimed to have demonstrated nonconscious influences on perception. In particular, inspired by psychodynamic notions, it was suggested that perception was modulated by perceptual defenses, *i.e.*, the active blockage of threatening information from reaching consciousness (*e.g.*, McGinnies, 1949). To overcome some of the methodological problems plaguing earlier studies of nonconscious perception, Lazarus and McCleary (1951) developed a conditioning paradigm to demonstrate emotional responses to consciously non-recognized words, a phenomenon they called "subception". They conditioned emotional activation, as indexed by skin conductance responses (SCRs), to nonsense syllables by pairing these stimuli with an annoying electric shock. In a subsequent test session, shocked and non-shocked control syllables were tachistoscopically exposed at durations varying from below to above the threshold for recognition. When data were analyzed only from trials where the subjects failed to verbally recognize the stimulus, subjects showed larger SCRs to shock-associated than to non-shock-associated syllables. This was taken as methodologically sound evidence of nonconscious discrimination of emotionally relevant stimuli that the subjects were unable to discriminate consciously.

However, in an incisive analysis, Eriksen (1960) argued that the subception effect was a statistical artifact resulting from psychometric necessities. Eriksen argued that because both SCRs and verbal reports are only partly valid indicators of a common perceptual process, dissociations between the two measures are inevitable. Thus, partial correlations will occur to the effect that SCRs may distinguish between shock- and non-shock-associated words with verbal reports held constant (*i.e.*, trials on which the subjects failed to recognize the stimuli). Eriksen (1960) replicated the subception effect reported by Lazarus and McCleary (1951) in several studies from his own laboratory, but in no case did he find that presumably emotionally relevant measures such as the SCR were more sensitive than verbal reports. Eriksen´s (1960) experimental work and intriguing analysis of the literature laid the idea of nonconscious perceptual mechanisms to rest for decades, and critiques of new claims for nonconscious perception (*e.g.*, Marcel, 1983) were inspired by his analysis (*e.g.*, Holender, 1986). Nevertheless, some four decades later it is difficult to deny that there is a phenomenon of nonconscious perception (Merikle & Daneman, 2000).

3. Nonconscious Processes in Contemporary Cognitive Psychology

With the advent of a strong, scientifically based cognitive psychology in the sixties and the seventies, the debate of nonconscious processes took a new turn because emerging models of cognitive processing implied the existence of nonconscious processes. For example, Erdelyi (1974) argued that contemporary information processing models necessitated nonconscious processing stages. This conclusion was reaffirmed both by observers centrally located in the psychodynamic tradition (Dixon, 1981; Shevrin & Dickman, 1980) and by textbooks in main stream cognitive psychology (*e.g.*, Lachman, Lachman & Butterfield, 1979).

Several lines of research involving nonconscious processes were developed within the main stream of experimental psychology. First, as pointed out by James (1890), the concept of consciousness is obviously related to the concept of selective attention. In the mid-seventies, research on attention introduced a distinction between automatic versus conscious (Posner, 1978) or automatic versus controlled (Shiffrin & Schneider, 1977) information processing that was based on the concept of "limited cognitive resources". The cognitive resource concept was developed to account for the fact that the selectivity of attention is better described in terms of a flexible and strategic distribution of limited processing resources across stimuli and tasks than in terms of structural bottlenecks letting through only one of the potential inputs (Kahneman, 1973). Automatic processing was defined as resource independent, whereas controlled processing was defined as heavily dependent on resources (Schneider, Dumais, & Shiffrin, 1984). This implies that tasks handled by automatic processes can be carried out concurrently without mutual interference. The performance in consciously controlled tasks, on the other hand, is severely degraded by forced time-sharing with other similarly controlled tasks. This distinction between attentional control systems captures important differences between nonconscious and conscious mental processes.

Second, memory processes are related to conscious mental activity. For example, James´(1890) version of short-term memory, primary memory, was defined as the phenomenological present in the sense that its content had never left consciousness. Similarly, the modern concept of working memory (Baddeley, 1992) is obviously related to conscious processes (*e.g.*, Kihlstrom, 1987; LeDoux, 1996). In the early eighties, the concept of "implicit memory" was introduced to describe instances of nonconscious memory, that is, evidence of memory that subjects could not attribute to a consciously recalled learning episode (Graf & Schacter, 1985). For example, when subjects were shown a list of words, they were primed by this exposure on a subsequent word stem completion task even if they failed to remember having seen the words. Furthermore, brain damaged subjects with severe amnesia performed quite normally on implicit memory tasks, whereas their explicit recognition performance was very poor (Warrington & Weiskrantz, 1974; see Schacter, 1987, for an early review of research on implicit memory).

Third, interest in mechanisms related to nonconscious psychological processes also became evident in social psychology. Zajonc (1980) argued that we automatically assess stimuli in terms of whether we like them or not, whereas to determine what we think about them is a slower, more error-prone and deliberate process. To use his own phrase, "preference precedes inference". Part of the empirical support for this assertion came from studies of the mere exposure effect, *i.e.*, that subjects tend to like objects that they have been previously exposed to. After Kunst-Wilson and Zajonc (1980) exposed subjects to very brief (1 ms) visual stimuli (irregular geometric shapes), subjects were presented with these stimuli in pairs together with similar distractor stimuli, and asked to indicate which one they liked best and which one they recognized best. Even though recognition was at chance, subjects preferred previously exposed stimuli to the distractors. This finding showed that mere exposure effects do not require conscious mediation. Similarly, other investigators reported that nonconsciously presented trait information affects person impression (Bargh & Pietromonaco, 1982; Lewicki, 1986), and that such effects are mediated by complex but nonconscious rule learning (Lewicki, 1986).

Nonconsciously mediated psychological phenomena were also demonstrated in a neuropsychological context. Prosopagnosia, an inability to recognize people from their visual appearance, may result from damage to visual association cortices (Damasio, Tranel & Damasio, 1990). Two independent case studies showed that prosopagnosics recognize familiar faces in terms of enhanced SCRs even though these patients were completely unable to identify the familiar persons when asked to identify them verbally (Bauer, 1984; Tranel & Damasio, 1985).

In the mid-eighties, there was converging evidence of psychological systems that operate outside of consciousness, even though the terms "conscious" and "nonconscious" were seldom used explicitly. Marcel (1983), however, used these terms to explain findings of nonconscious priming in lexical decision and the Stroop color-word interference tasks, and in presenting an ambitious theory giving nonconscious processing a central role in perception.

However, even though the existence of nonconscious effects was widely accepted, the question was still open concerning their importance. To use Loftus and Klinger´s (1992) catchy phrase, how smart or dumb is the nonconscious? How complicated a system is it, and can it handle complex information? How flexibly can it deal with novel situations? And does it do what is best for us? Greenwald (1992) reviewed the available evidence to address these questions. He concluded that there is substantial evidence for nonconscious cognition but that its performance appears unimpressive: it can handle semantics of single words but not of sentences. But, the semantics of single words (or pictures) may be sufficient for controlling a key area of human functioning—that of emotion.

4. The Somatic Marker Hypothesis

In an influential work that explicitly challenged the Cartesian Cogito, Damasio (1994) revived the Jamesian idea, claiming that decision making processes are influenced by the feedback of bodily responses in emotion. Neuropsychological data from patients with lesions in the ventromedial frontal cortex suggested that damage to this area undermines adaptive functioning. Patients with such lesions showed little impairments in neuropsychological domains such as language or memory, but somehow their lives appeared to fall into pieces as a result of series of unfortunate and ill-informed decisions, even in persons with a history as successful decision-makers. In Damasio's (1994) interpretation, decisions of such patients were no longer informed by emotion because the lesion disrupted neural pathways between the prefrontal cortex and primary emotional areas such as the amygdala in the anterior temporal lobe. Damasio argued that feedback from autonomic and other bodily activity associated with emotion, called *somatic markers*, plays a crucial role in decision making. Thus, the autonomic arousal elicited by stimuli related to some of the options facing a decision-maker can be profitably used to limit the number of alternatives that have to be considered in forming a decision. For example, because they activate aversive emotions, a number of alternatives are eliminated, leaving fewer options for the decision-maker to ponder. In Damasio's own words, "somatic markers do not deliberate for us. They assist the deliberation by highlighting some options (either dangerous or favorable).... You may think of it as a system for automated qualification of predictions, which acts, whether you want it or not, to evaluate the extremely diverse scenarios of the anticipated future before you. Think of it as a biasing device" (Damasio, 1994, p. 174).

Because of the proposed importance of autonomic input in emotion, autonomic changes have to occur (or at least be initiated) very quickly after an eliciting stimulus to influence conscious cognition. In fact, to compensate for the general sluggishness of autonomic effectors, they would have to be activated virtually instantaneous with the stimulus, if they are to affect subsequent cognitive processing. Because conscious thought is too slow to accomplish this, theorists giving input from the autonomic nervous system a critical role in emotion and decision making must assume that autonomic responses can be activated nonconsciously, after only a minimum of processing of the eliciting stimulus. In other words, automatic and nonconscious stimulus processing must be sufficient for emotional activation to affect decision processes (Damasio, 1994). Thus, notwithstanding his distaste for nonconscious mechanisms, James' (1884) basic ideas of a role for bodily feedback in emotion at least implicitly assumes the existence of automatic, nonconscious stimulus processing routines.

Further, for somatic markers to be useful in decision making, Damasio's (1994) argument appears to require that autonomic responses can be more or less automatically conditioned to new stimuli without requiring awareness of the conditioning contingencies. For example, Bechara, Damasio, Tranel and Damasio (1997) showed that normals, but not patients with lesions in the ventromedial frontal cortex, showed elevated SCRs to risky alternatives in a gambling task, even before they could consciously identify and avoid the high risk alternatives. Thus, decision options that have (consciously or nonconsciously) been associated with bad outcomes in the past, elicit emotional responses that help the decision-maker to shy away from bad choices without losing time and efficiency by having to consult with advanced conscious levels of information processing.

5. Nonconscious Activation of Emotion to Masked Stimuli

5.1. The Amygdala Fear Circuitry

As reviewed above, there is good reason, both theoretically and empirically, to take the existence and impact of nonconscious psychological processes seriously. Indeed, Damasio's (1994) somatic marker hypothesis suggests that conscious cognition may be preceded and shaped by nonconscious emotional activity. Recent research on the neuroscience of emotion has demonstrated that emotional states such as fear can be activated by degraded stimulus input. LeDoux (1990b; 1996) has delineated the neuroarchitecture of this system, through a series of studies of the neural control of auditorily elicited conditioned emotional responses in the rat. He and his coworkers demonstrated that the amygdala was critical in the activation and efferent control of fear, in that neural pathways to hypothalamic, midbrain and brainstem nuclei control autonomic responses, behavioral responses (*e.g.*, freezing), and potentiation of defense reflexes (*e.g.*, startle) originate in the central nucleus of the amygdala. The amygdala can be activated from several cortical areas and also via a direct neural link from auditory nuclei in the thalamus (medial geniculate body) to the "significance evaluator" and "fear effector system" in the lateral and central amygdala, respectively. This monosynaptic link provides the amygdala with immediate information about gross features of emotionally relevant auditory stimuli. It bypasses the traditionally emphasized thalamo-cortical pathway, which gives full meaning to the stimulus, and the cortico-amygdala link, which is presumed to activate emotion. It is described as a "quick and dirty" transmission route. It "probably does not tell the amygdala much about the stimulus, certainly not much about Gestalt or object properties of the stimulus, but it at least informs the amygdala that the sensory receptors of a given modality have been activated and that a significant stimulus may be present" (LeDoux, 1990a, p. 172). In this way, the amygdala can initiate rapid activation of defense responses. This system is postulated to be adaptively biased toward false positives rather than false negatives. This is because it is less costly to abort falsely initialized defense responses than failing to elicit defense when the threat is real.

5.2. Backward Masking: A Method to Study Nonconsciously Activated Human Emotion

The LeDoux model of fear activation suggests that fear can be elicited without the stimulus necessarily having to be processed in cortical sensory areas. Applied to humans, this statement implies that fear can be activated by stimuli that are not consciously recognized. In an extensive series of studies, the first of which were reported by Öhman (1986), we have examined emotional responses to non-recognizable stimuli in human subjects (see reviews in Öhman, 1996; Öhman & Mineka, in press, and Öhman & Wiens, in press). These studies have relied on backward masking to present stimuli outside of conscious awareness.

When a brief visual stimulus is immediately followed by another visual stimulus and stimulus parameters are adjusted properly, the latter stimulus masks the former stimulus from being consciously perceived. For example, if the first stimulus is of short duration (*e.g.*, up to 30 ms), backward masking occurs in that the second stimulus, the *mask*, prevents perception of the first stimulus, the *target*. However, even though the target is blocked from conscious access, it is still processed to considerable depth (*e.g.*, Marcel, 1983). Thus, rapid, automatic processes may be directed toward the target stimulus before the mask disrupts more elaborate processing.

Esteves and Öhman (1993) studied backward masking of facial expression stimuli, and Öhman and Soares (1993) examined another class of fear-related stimuli, small animals such as snakes and spiders. Subjects were exposed to pairs of stimuli, in which a target stimulus either depicting a facial emotion (happy or angry; Esteves & Öhman, 1993) or one of four visual stimuli (snakes, spiders, flowers or mushrooms; Öhman & Soares, 1993) was followed by a second stimulus, which either portrayed a neutral facial expression or a randomly composed picture, respectively. The stimulus-onset-asynchrony (SOA) between these two stimuli was systematically varied from short (20 ms) to long (about 300 ms), whereas the mask was always presented for 30 ms. After each stimulus pair, subjects were asked to judge what the target was (*e.g.*, a happy or an angry face) and to rate their confidence in their answers. Across several experiments, the data showed that subjects required at least 100 ms exposure of the target for confident correct decisions. In contrast, when duration of target and SOA was 30 ms or less, subjects performed at chance and also reported low confidence in their ratings. This result is robust across experimental conditions (Esteves & Öhman, 1993; Öhman & Soares, 1993), including shock administrations during the perceptual task (Esteves, Parra, Dimberg, & Öhman, 1994).

5.3. Nonconscious Elicitation of Conditioned Responses to Facial Stimuli

Using backward masking, it is possible to examine if emotionally evocative stimuli activate psychophysiological responses even if the stimuli are not recognized consciously because they are backwardly masked. To make sure that stimuli such as faces actually elicit enhanced responses, Pavlovian conditioning procedures have been used to induce enhanced SCRs to stimuli that subsequently are presented under conditions of backward masking. For example, an angry face is briefly presented and followed by a mild electric shock to the fingers whereas another face is presented without the shock. The question is if subjects will respond differentially to these faces when the stimuli are presented masked. The 30 ms SOA between a target and a mask has been used in several studies examining SCRs to masked facial stimuli. Esteves, Dimberg and Öhman (1994) reported three experiments in which subjects were conditioned to angry faces with a shock UCS. After conditioning was established (documented by enhanced SCRs to the shock associated face, the CS+), the subjects were tested by masked presentations of the CS+ and the non shock-associated face, the CS-. In spite of the fact that the masking conditions prevented conscious perception of the stimuli, SCR data consistently showed larger responses to the CS+ than to the CS-.

In a study by Parra, Esteves, Flykt and Öhman (1997), subjects were presented with several exemplars of angry and happy faces. Some of the angry faces were followed by shock during an acquisition phase. In a subsequent extinction phase, the previously shocked faces were either presented as targets and masked by neutral faces, or served as masks of neutral faces (and were thus recognizable). Irrespective of whether the presentation was nonconscious or conscious (*i.e.*, the CS+ was presented as targets or masks), subjects showed reliable differential SCRs that did not differ between conditions. However, whereas familiarity ratings of previously presented faces were high when the CS+ served as masks, the ratings were low when the CS+ occurred as targets. Thus, there was a clear dissociation between SCRs and familiarity ratings. Whereas SCR conditioning was similar in both masking conditions, familiarity ratings differed greatly between masking conditions.

In another experiment, Parra *et al.* (1997) compared masked and nonmasked extinction after conditioning to nonmasked angry CS+ and happy CS-. In addition, subjects rated their shock expectancy in the 3.5 s interval between the CS and the UCS. Consistent with previous findings, a strong dissociation between shock expectancy and SCR was observed. That is, whereas skin conductance responses showed similar differential conditioning in both masking conditions, shock ratings showed an interaction between conditioning and masking, which indicate that the differentiation between CS+ and CS- was much larger in the nonmasked than in the masked condition.

Wong, Shevrin, and Williams (1994) used schematic negative and positive evaluated faces and included measurement of slow cortical potentials as an additional measure. After conditioning with the negative face serving as the CS+ and the positive face serving as the CS-, subjects were exposed to masked presentations of the stimuli below rigorously defined individual thresholds for recognition. In masked extinction, subjects showed a distinct slow negative potential after the CS+ that preceded the point of previous UCS presentations. This waveform was comparable to a previously described expectancy wave occurring in anticipation of an emotionally or motivationally relevant stimulus (*e.g.*, Lang, Öhman & Simons, 1978). This finding suggested to the authors that "an anticipatory process...can be elicited entirely outside awareness" (Wong *et al.*, 1994, p. 87). Taken together, these findings suggest that emotional processing, as indexed by psychophysiological measures, occurs outside of awareness and may be relatively dissociated from other processing.

5.4. Facial Responses to Masked Facial Stimuli

Dimberg (1982) reported that subjects tend to respond to pictures of emotional faces by mimicking the emotional expression of the stimulus faces in their facial muscles (as assessed by electromyography, EMG). Thus, when exposed to an angry face, subjects show increases in activity of the *corrugator supercilii* muscle mediating the frown, and little change in the *zygomatic major* muscle, which connects the corner of the mouth and the cheekbone and mediates a smile. With a happy face, however, the opposite pattern was found with increases in the zygomatic and no change or a decrease in the corrugator muscle.

To examine whether the patterned EMG response could be elicited to stimuli presented outside of awareness, Dimberg, Elmehed, and Thunberg (2000) measured facial responses to masked facial stimuli. Dimberg *et al.* (2000) presented angry, neutral and happy faces briefly (30 ms) and masked by a 5-sec presentation of a neutral face. Even though subjects could recognize consciously only the neutral masking face, they nevertheless showed differential facial EMG responses to the target stimuli. The masked happy face elicited a larger zygomatic response than the neutral or angry face, whereas the masked angry face elicited a larger corrugator response than the neutral or happy faces. These differences emerged quickly within 500 ms after stimulus onset. Taken together with our findings (Esteves *et al.*, 1994; Parra *et al.*, 1997), the results indicate that both unspecific autonomic responses (*i.e.*, SCRs) and specific muscular responses (*i.e.*, facial EMG) are elicited by masked emotional stimuli.

5.5. Fear Responses to Masked Phobic Stimuli

It could be questioned if the data on physiologic responses reviewed so far reflect genuine emotional responses. Even though angry faces elicited facial EMG responses in the study reported by Dimberg *et al.* (2000), the level of emotion induced by a pictorial representation of anger was probably low. Similarly, although the conditioning procedure used by Esteves *et al.* (1994) and Parra *et al.* (1997) induced enhanced responses to angry faces, the resulting level of emotional activation still was likely to be low. However, genuine fear can be studied by selecting individuals with phobic fears and then exposing these individuals with fear-relevant pictures while measuring psychophysiological responses. Studies using this design have shown that snake- and spider-phobic subjects show strongly elevated psychophysiological responses to visual presentations of their feared object (*e.g.*, Fredrikson, 1981, Hamm, Cuthbert, Globisch & Vaitl, 1997). For example, Globisch, Hamm, Esteves, and Öhman (1999) found that snake-fearful and spider-fearful subjects, in contrast to normal controls, show enhanced skin conductance responses, heart rate accelerations as opposed to decelerations, and blood pressure increases when presented with fear-relevant pictures. In addition, startle probe stimuli presented after onset of the fear-relevant pictures showed substantial potentiation of the startle blink reflex indicating escape and avoidance inclinations (cf., Lang, Bradley, & Cuthbert, 1990). These results demonstrate that fearful subjects show a pronounced emotional response to feared stimuli.

Using the backward masking technique, Öhman and Soares (1994) tested the hypothesis that phobic fear can be activated to masked phobic stimuli. They selected subjects who were highly fearful of either snakes or spiders (but not both) as well as nonfearful controls. Subjects were exposed to two stimulus series consisting of repeated presentations of pictures of snakes, spiders, flowers, and mushrooms. In the first series, target pictures were masked, whereas in the second series, targets were presented without masks. Skin conductance responses were recorded as an index of the physiological response component of fear. In addition, subjects were exposed to an extra series of pictures in which they were asked to rate their affect in terms of valence, activation, and control.

The results concerning skin conductance were that, irrespective of masking condition, subjects who were afraid of snakes showed elevated responding to snakes compared to spiders and neutral stimuli, subjects who were afraid of spiders showed specifically elevated responses to spiders, and nonfearful subjects did not differentiate among stimulus categories. Thus, results from the masked series were very similar to those from the non-masked series, which suggests that most of the response was nonconsciously recruited. These psychophysiological findings were paralleled in subjective ratings of affect. That is, even when stimuli were masked, snake fearful subjects reported more negative valence, more arousal and less control to the masked snakes pictures than to any other pictures. Similar results were obtained for spider fearful subjects to spiders, whereas nonfearful subjects did not differentiate among stimulus categories. Thus, emotional aspect of the stimulus content became available to the cognitive system even though conscious recognition was ruled out because of masking.

In concert, the experimental results reviewed in this section show quite conclusively that emotional responses can be elicited by masked stimuli. This is true regardless of whether subjects are exposed to emotional facial stimuli (Dimberg, 2000), to facial stimuli that elicit enhanced responses because of Pavlovian conditioning (Esteves *et al.*, 1994; Parra *et al.*, 1997), or to pictures portraying phobic content in phobic subjects (Öhman & Soares, 1994). Consistent with the LeDoux (1996) model of how the brain fear circuit is organized, these results suggest that emotional responses can be activated from degraded stimuli that do not reach consciousness because of backward masking.

6. Activating the Human Amygdala with Masked Stimuli

To determine in humans if results on nonconscious emotional activation to masked fear stimuli involve the amygdala, it is necessary to measure amygdala activation non-invasively. Morris, Öhman and Dolan (1998) used positron-emission tomography (PET) to measure changes in regional cerebral blood-flow to examine amygdala activation to masked fear-associated face stimuli. In a conditioning procedure that preceded PET scanning, participants were presented with pictures of four different male faces, two of which were angry and two neutral. One of the angry faces was paired with an aversive noise, and thus served as a CS+. The second angry face served as a CS-. Additional conditioning trials were presented in between PET scans. During PET scans target-mask pairs were presented with short inter-trial (5 sec) intervals with two different masking conditions. One of them involved the CS+ and the CS- as target stimuli, each presented for 30 ms, and each one immediately and consistently followed by one of the two neutral faces, presented for 45 ms as masks. In this condition, therefore, the CSs were effectively masked. For the second masking condition, the order of targets and masks was reversed. Thus, because the neutral faces were presented for 30 ms and were immediately followed by the CS+ or the CS- presented for 45 ms, the CSs could be clearly seen by the subjects. SCR data confirmed reliable differential responding to the CS+ and CS- in both the masked and the nonmasked extinction series during the PET scans. This design allowed a stringent test of the hypothesis that human fear conditioning involves amygdala activation. By contrasting images of regional cerebral blood flow produced by the masked CS+ and the masked CS-, a difference revealed the location in the brain of the fear response conditioned to the CS+ elicited by a nonreportable fear-relevant stimulus.

A similar contrast of the nonmasked CS+ and the CS- revealed the location in the brain of a fear response elicited by a consciously elicited stimulus.

Confirming the hypothesis of amygdala involvement, the overall contrasts between the CS+ and the CS- showed significant differential activations specifically in amygdala regions. This effect, however, showed an interaction with brain laterality. For the masked CSs, only the right amygdala was activated; for nonmasked CSs, only the left amygdala was activated. However, these effects were not perfect mirror images of each other. For recognized stimuli, activation was somewhat superior and posterior to that seen with nonreportable stimuli, even though both resided within the amygdala complex. In the present context, the important finding is that the human amygdala was specifically activated by conditioned fear stimuli, whereas the laterality effect needs to be addressed by future research.

Further data on fear responding to masked stimuli were reported by Whalen *et al.* (1998). These authors used functional magnetic resonance imaging (fMRI) to measure responses of the human amygdala to masked presentations of (nonconditioned) facial stimuli. When periods of repeated presentations of masked fearful faces were compared with periods of repeated presentations of happy faces, enhanced amygdala activation was found to the fearful faces. Thus, even though conscious recognition of the stimuli was prevented by backward masking, reliable activation of the central structure of the fear circuit, the amygdala, was observed to the fearful face.

In a follow-up to the study of Morris *et al.* (1998), Morris, Öhman and Dolan (1999) examined whether nonconscious activation of the amygdala by nonrecognized stimuli occurred via subcortical pathways, as may be expected from LeDoux's (1996) model. Thus, Morris *et al.* (1999) examined the neural connectivity between the amygdala and other brain regions when the amygdala was activated by masked stimuli. The specific hypothesis was based on the literature on blind sight (Weiskrantz, 1986; 1997). According to this literature, the remaining but nonconscious visual capacity sometimes seen in patients with damage to the primary visual cortices may be mediated by a parallel visual route through the superior colliculus and pulvinar nucleus of the thalamus served by large, rapidly conducting neurons. Morris *et al.* (1999) proposed that this route might be less susceptible to masking than the usual visual pathway through the lateral geniculate. In support of this conjecture, Morris *et al.* (1999) reported that activation of the (right) amygdala by masked stimuli could be reliably predicted from activation in the superior colliculus and the pulvinar, whereas such a relationship was not obvious when the (left) amygdala was activated by nonmasked stimuli. The research by Morris *et al.* (1998; 1999) demonstrates that a similar system to that previously delineated in the rodent brain (for review, see LeDoux, 1996) for fear conditioning appears also to be operating in the human brain.

7. The Concept of a Fear Module

The fact that human autonomic responses related to fear appear to be controlled by a neural circuit that appears similar to that in rodents suggests that the neural organization of fear is common to mammals and that it reflects a joint evolutionary origin. Thus, fear responses may reflect evolutionary coupling of autonomic function to behavioral demands to assure effective coping with emotional situations. Because of their ancient origin, these processes operate quite independently of the more recently evolved capacity for advanced computations in primates and, most notably, in the hominoid line of descent. As we have shown, there is strong experimental evidence that autonomic responses associated with fear can be activated from stimuli that are not consciously processed. Indeed, as proposed in LeDoux's (1996) animal model, and as indicated by results in humans reported by Morris *et al.* (1999), the central structure of the fear network, the amygdala, does not require cortical input for its activation by a fear stimulus. It appears, therefore, that the autonomic nervous system is continuously modulated by automatic appraisal processes that monitor the surroundings for threat, independent of conscious awareness. When these automatic perceptual processes encounter threatening stimuli, selective attention is switched to the threat to evaluate it more thoroughly, or to activate defense responses immediately if they are very close.

Several authors have taken this automaticity and the speed of emotional activation (as exemplified in fear) to suggest that there are "automatic appraisal mechanisms" or "inescapable" affective responses that are independent of higher cognition (*e.g.*, Robinson, 1998; Zajonc, 1980; see Griffith, 1997 for review). Öhman and Mineka (1999) postulated that fear activation can be understood in terms of a "fear module" (cf. Fodor, 1983; Griffith, 1997), that is, a relatively independent behavioral, psychophysiological, and neural system specifically tailored by evolution to solve adaptive problems related to potentially life-threatening situations. This fear module has four important characteristics. First, it is preferentially activated by evolutionarily relevant threatening stimuli in aversive contexts, often as a result of the learning of selective associations between fear-relevant stimuli and aversive events. Second, it can be automatically activated by fear stimuli with no need for conscious recognition of the stimulus. Third, it is encapsulated in the sense that it is relatively impenetrable to cognition. Finally, it reflects an independent and specific neural circuit that is centered on the amygdala and incorporates a series of subcortical structures that in concert control fear responses (*e.g.*, LeDoux, 1996; Fanselow & LeDoux, 1999).

7.1. Experiencing the Fear Module

The fear module is primitive in the sense that it evolved as a result of evolutionary contingencies millions of years ago. However, it now operates in a human brain capable of advanced thought, language, and the conscious experience of emotion. Humans can talk about emotions and they have emotional experiences. Awareness of an emotion requires not only that an emotional stimulus is recognized, but also originates in feedback from the emotional responses that are elicited by the stimulus. For example, experiencing a racing heart when a shadow appears from the dark alley contributes to the feeling of fear. This is consistent with James's (1884) idea that such feedback *is* the emotion. You feel the emotion when you experience its effect on your body. Thus, the feeling of fear is the experience of an activated fear module (LeDoux, 1996).

The Jamesian idea has stood the test of time remarkably well, in spite of merciless critics (*e.g.*, Cannon, 1927). Several decades ago, the role of bodily feedback in emotion was revived as a modulator of emotional intensity rather than as a determinant of emotional quality (Schachter & Singer, 1962). Later on, the effects of autonomic feedback provided the cornerstone to one of the most articulated cognitive theories of emotion (Mandler, 1975). More recently, as we have seen, Damasio (1994) argued that somatic markers have a crucial role in cognitive processes such as decision making.

7.2. Feedback from Autonomic Responses in Masked Fear Conditioning

Some unexpected results in a conditioning study reported by Öhman and Soares (1998) may be interpreted in terms of the somatic marker hypothesis. In this study, subjects were presented with a differential conditioning task in which one stimulus (the CS+, *e.g.*, a snake) was presented masked and followed after a 4 sec interval by a mild electric shock, whereas another stimulus (the CS-, *e.g.*, a spider) was also presented masked but never followed by shock. Although subjects were unable to discriminate between the masked CS on a forced-choice recognition task, they nevertheless showed differential SCRs to the CS+ and the CS-, thus confirming that emotional responses can actually be conditioned outside of awareness (see also Esteves, Dimberg, Parra & Öhman, 1994). Further, when a group of subjects was asked to rate their expectancy of shock immediately after each CS presentation, these subjects showed differential shock expectancies to masked images of CS+ and CS–. That is, subjects rated shocks as more likely to occur on CS+ trials than on CS- trials. Thus, even though subjects could not recognize the stimuli, they somehow appeared to have access to information resulting in differential shock expectancy ratings. Similarly, Parra *et al.* (1997) found that after subjects were conditioned to nonmasked CSs and shocks were no longer administered, subjects rated shocks as somewhat more likely to occur after the masked CS+ than after the masked CS–.

To account for these findings, Öhman and Soares (1998) invoked the somatic marker hypothesis, suggesting that participants may have been able to predict shocks based on autonomic cues associated with the conditioned fear response. This interpretation, however, is weakened by the failure to find reliable correlations between SCRs and shock expectancy ratings (Öhman & Soares, 1998; Parra *et al.*, 1997). However, Katkin, Wiens, and Öhman (2000) postulated that it might not be the mere occurrence of autonomic responses, but their sensation that allows shock prediction. That is, subjects in the Öhman and Soares (1998) and Parra *et al.* (1997) studies may have based their shock expectancy on their sensation of the conditioned autonomic response, and this sensation may have acted as a cue for impending shock. If so, then subjects who are sensitive to the visceral cues associated with the conditioned fear response would be more accurate in predicting the occurrence of shocks than subjects who are insensitive to visceral cues, despite similar levels of actual autonomic activity.

To test this hypothesis, Katkin *et al.* (2000) tested subjects on the same conditioning task as that in the study by Öhman and Soares (1998). In addition, they assessed separately subjects' general sensitivity to visceral activity, which was indexed by performance on a heartbeat detection task (Katkin, 1985). In this task, subjects were presented with tones that varied in the delay interval from ECG R-waves, and subjects judged the temporal relationship between the tones and their own heartbeat sensations. Because tones at different delay intervals had exactly the same rhythm and differed only in respect to the delay interval from the R-waves, only subjects who could detect their heartbeats had a reference point to discriminate between the delay intervals (Katkin, 1985). Therefore, subjects were classified as good or poor heartbeat detectors based on their ability to judge the temporal relationship between tones and heartbeats.

The results showed reliable skin conductance conditioning to masked CSs as well as greater expectancy of shock after the masked CS+, replicating the basic findings of Öhman and Soares (1998). However, shock expectancy ratings interacted with heartbeat detection. This interaction showed that only the good heartbeat detectors predicted shock beyond chance. In other words, good heartbeat detectors rated shock as somewhat likely on CS+ trials and as somewhat unlikely on CS- trials, whereas poor detectors did not differentiate the CS+ and the CS- in their shock ratings. Thus, the results of the Katkin *et al.* (1999) study suggest that the ability to predict shock may be limited to participants who are sensitive to their own autonomic activity. The findings are consistent with Damasio's (1994) somatic marker hypothesis and further support the notion that hunches, or "gut feelings," are based in part on the sensation of visceral cues.

8. Concluding Comments

We opened this article by appealing to the Cartesian Cogito: I think, therefore I am. Our main message has been that true enough, humans are conscious creatures, but that consciousness is limited in the sense that there are many important psychological processes to which it is denied access. Influential as the cogito has been, strict adherence to it leaves out large segments of human psychology. In particular, our research indicates that a nonconscious level of emotional activity monitors automatically the world around us quite independently from consciousness. If this level locates important events, a primary result is that consciousness is called upon, partly because the nonconscious level directs attention towards important stimuli, partly because the automatically activated emotional response is fed back to conscious awareness. The joint result of these two processes is the emergence of an emotion in consciousness, of perceiving the stimulus and having a feeling about it. In fact, a case could be made to the effect that the "I think" should be exchanged with the "I feel" as the basic datum in consciousness. Thus, the great Czech novelist Milan Kundera may have been right in his remark that the cogito is a statement by a man who underestimates toothache. And he is in good company. Humphrey's (1992) "raw sensations" and Damasio's (1999) "core consciousness" both involve feelings and provide similar fundamental theoretical concepts in otherwise quite different treatments of the problem of consciousness. Damasio (1999), in particular, is close to the present analysis in his insistence that feelings derive from a basic biological level of organism-environment interactions generating bodily responses that feed back to determine core consciousness.

But even though thinking may not be the basic fact as established in the cogito, it may, of course, still be very important as a short-hand for a level of conscious cognition that is distinctly human. As argued by Gazzaniga (2000), the human brain is a wonder of evolutionary design that is quite capable automatically to handle a very large part of the transactions between an individual and his or her environment. The role of consciousness is primarily to make sense, to interpret what is going on and organize it into a coherent story or narrative of our lives. This narrative becomes a theory or map of the world that is very handy to help us navigate through life and that can provide consistency and meaning to what we do. As many of us have realized, what the brain makes us do is not always what we want to do, or even that we think we are doing. Nevertheless, the story that emerges from the left hemisphere interpretive system may be an important vehicle for making us less vulnerable to capricious environmental stimuli and their effect on our body.

Acknowledgement

The research reviewed in the article was supported by grants to the first author from the Swedish Council for Research in the Humanities and Social Sciences.

References

Baddeley, A. (1992) "Working memory", *Science* 255:556-559.

Bargh, J. A. and P. Pietromonaco (1982) "Automatic information processing and social perception: The influence of trait information presented outside of conscious awareness on impression formation", *Journal of Personality and Social Psychology* 43:437-449.

Bauer, R. M. (1984) "Automatic recognition of names and faces in prosopagnosia: A neuropsychological application of the Guilty Knowledge Test", *Neuropsychologia* 22:457-469.

Bechara, A., H. Damasio, D. Tranel and A. R. Damasio (1995) "Deciding advantageously before knowing the advantageous strategy", *Science* 275:1293-1295.

Boring, E. G. (1950) *A History of Experimental Psychology*. 2nd ed., New York: Appleton-Century-Crofts.

Cannon, W. B. (1927) "The James-Lange theory of emotions: A critical examination and an alternative theory", *American Journal of Psychology* 39:106-124.

Damasio, A. R. (1994) *Descartes´ Error: Emotion, Reason, and the Human Brain*, New York: G. Putnam´s Sons.

Damasio, A. R. (1996) "The somatic marker hypothesis and the possible functions of the prefrontal cortex", *Philosophical Transactions of the Royal Society* B 351:1413-1420.

Damasio, A. R. (1999) *The Feeling of What Happens: Body and Emotion in the Making of Consciousness*, New York: Harcourt, Brace and Co.

Damasio, A. R., D. Tranel and H. Damasio (1990) "Face agnosia and the neural substrates of memory", *Annual Review of Neuroscience* 13:89-109.

Damasio, A. R., D. Tranel and H. Damasio (1990) "Individuals with sociopathic behavior caused by frontal damage fail to respond autonomically to social stimuli", *Behavioral Brain Research* 41:81-94.

Damasio, A. R., D. Tranel and H. Damasio (1991) "Somatic markers and the guidance of behavior: Theory and preliminary testings", in: *Frontal Lobe Function and Dysfunction*, H. S. Levin, H. M. Eisenberg and A. L. Benton, eds., New York: Oxford University Press, pp. 217-229.

Dimberg, U. (1982) "Facial reactions to facial expressions", *Psychophysiology* 19:643-647.

Dimberg, U., K. Elmehed and M. Thunberg (2000) "Unconscious facial reactions to emotional facial expressions", *Psychological Science* 11:86-89.

Dixon, N. F. (1981) *Preconscious Processing*, Chichester: Wiley.

Erdelyi, M. H. (1974) "A new look at The New Look: Perceptual defense and vigilance", *Psychological Review* 81:1-25.

Eriksen, C. W. (1960) "Discrimination and learning without awareness: A methodological survey and evaluation", *Psychological Review* 67:279-300.

Esteves, F. and A. Öhman (1993) "Masking the face: Recognition of emotional facial expressions as a functions of the parameters of backward masking", *Scandinavian Journal of Psychology* 34:1-18.

Esteves, F., U. Dimberg and A. Öhman (1994) "Automatically elicited fear: Conditioned skin conductance responses to masked facial expressions", *Cognition and Emotion* 8:393-413.

Esteves, F., C. Parra, U. Dimberg and A. Öhman (1994) "Nonconscious associative learning: Pavlovian conditioning of skin conductance responses to masked fear-relevant facial stimuli", *Psychophysiology* 31:375-385.

Fanselow, M. S. and J. E. LeDoux (1999) "Why we think plasticity underlying Pavlovian fear conditioning occurs in the basolateral amygdala", *Neuron* 23:229-232.

Fodor, J. (1983) *The Modularity of Mind*, Cambridge, MA: The MIT Press.

Fredrikson, M. (1981) "Orienting and defensive reactions to phobic and conditioned fear stimuli in phobics and normals", *Psychophysiology* 18:456-465.

Freud, S. (1894/1962) "The neuro-psychoses of defense", in: *The Standard Edition of the Complete Psychological Works of Sigmund Freud.* Vol. 3, J. Strachey, ed., London: Hogarth Press.

Gazzaniga, M. S. (2000) "Cerebral specialization and interhemispheric communication. Does the corpus callosum enable the human condition", *Brain*, 123:1293-1326.

Globisch, J., A. O. Hamm, F. Esteves and A. Öhman (1999) "Fear appears fast: Temporal course of startle reflex potentiation in animal fearful subjects", *Psychophysiology* 36:66-75.

Graf, P. and D. L. Schacter (1985) "Implicit and explicit memory for new associations in normal and amnesic subjects", *Journal of Experimental Psychology: Learning, Memory, and Cognition* 11:501-518.

Gray, J. A. (1995) "Contents of consciousness: A neuropsychological conjecture", *Behavioral and Brain Sciences* 18:659-676.

Greenwald, A. G. (1992) "New Look 3. Unconscious cognition reclaimed", *American Psychologist* 47:766-769.

Griffith, P. E. (1997) *What Emotions Really Are*, Chicago: The University of Chicago Press.

Hamm, A. O., B. N. Cuthbert, J. Globisch and D. Vaitl (1997) "Fear and the startle reflex: Blink modulation and autonomic response patterns in animal and mutilation fearful subjects", *Psychophysiology* 34:97-107.

Holender, D. (1986) "Semantic activation without conscious identification in dichotic listening, parafoveal vision, and visual masking: A survey and appraisal", *Behavioral and Brain Science* 9:1-66.

Humphrey, N. (1992) *A History of Mind. Evolution and the Birth of Consciousness*, New York: Simon and Schuster.

James, W. (1884) "What is an emotion?", *Mind* 9:188-205.

James, W. (1890/1950) *The Principles of Psychology*. Vol. 1, New York: Dover.

Kahneman, D. (1973) *Attention and Effort.*, Englewood Cliff, NJ: Prentice-Hall.

Katkin, E. S. (1985) "Blood, sweat, and tears: Individual differences in autonomic self-perception", *Psychophysiology* 22:125-137.

Katkin, E. S., S. Wiens and A. Öhman (1999) "Nonconscious fear conditioning and heartbeat detection", Paper submitted for publication.

Kihlstrom, J. F. (1987) "The cognitive unconscious", *Science* 237:1445-1452

Kunst-Wilson, W. R. and R. B. Zajonc (1980) "Affective discrimination of stimuli that cannot be recognized", *Science* 207:557-558.

Lachman, R., J. L. Lachman and E. C. Butterfield (1979) *Cognitive Psychology and Information Processing: An Introduction*, Hillsdale, NJ: Erlbaum.

Lang, P. J., M. M. Bradley and B. N. Cuthbert (1990) "Emotion, attention, and the startle reflex", *Psychological Review* 97:377-398.

Lang, P. J., A. Öhman and R. F. Simons (1978) "The psychophysiology of anticipation", in: *Attention and Performance VII*, J. Requin, ed., Hillsdale, NJ: Erlbaum, pp.469-485.

Lazarus, R. S. and R. A. McCleary (1951) "Autonomic discrimination without awareness: A study of subception", *Psychological Review* 58:113-122.

LeDoux, J. E. (1990a) "Fear pathways in the brain: Implications for a theory of the emotional brain", in: *Fear and Defense*, P. F. Brain, S. Parmigiani, R. J. Blanchard and D. Mainardi, eds., London: Harwood, pp. 163-178.

LeDoux, J. E. (1990b) "Information flow from sensation to emotion: Plasticity in the neural computation of stimulus value", in: *Learning and Computational Neuroscience: Foundations of Adaptive Networks*, M. Gabriel and J. Moore, eds., Cambridge, MA: Bradford Books/MIT Press, pp. 3-51.

LeDoux, J. (1996) *The Emotional Brain. The Mysterious Underpinnings of Emotional Life*, New York: Simon and Schuster.

Lewecki, P. (1986) *Nonconscious Social Information Processing*, Orlando, FL: Academic Press.

Loftus, E. F, and M. R. Klinger (1993) "Is the unconscious smart or dumb?", *American Psychologist* 47:761-765.

Mandler, G. (1975) *Mind and Emotion*, New York: Wiley.

Marcel, A. (1983) "Conscious and unconscious perception: An approach to the relations between phenomenal experience and perceptual processes", *Cognitive Psychology* 15:238-300.

McGinnies, E. (1949) "Emotionality and perceptual defense", *Psychological Review* 56:244-251.

Merikle, P. M. and M. Daneman (2000) "Conscious vs. unconscious perception", in: *The Cognitive Neurosciences*. 2nd ed., M. S. Gazzaniga, ed., Cambridge, MA: The MIT Press.

Morris, J., A. Öhman and R. J. Dolan (1998) "Modulation of human amygdala activity by emotional learning and conscious awareness", *Nature* 393:467-470.

Morris, J. S., A. Öhman and R. J. Dolan (1999) "A subcortical pathway to the right amygdala mediating "unseen" fear", *Proceedings of the National Academy of Sciences* 96:1680-1685.

Öhman, A. (1986) "Face the beast and fear the face: Animal and social fears as prototypes for evolutionary analyses of emotion", *Psychophysiology* 23:123-145.

Öhman, A. (1996) "Preferential preattentive processing of threat in anxiety: Preparedness and attentional biases", in: *Current Controversies in the Anxiety Disorders*, R. M. Rapee, ed., New York: Guilford, pp. 253-290.

Öhman, A. and S. Mineka (in press) "Fear, phobias, and preparedness: Toward an evolved module of fear and fear learning", *Psychological Review*.

Öhman, A. and J. Soares (1993) "On the automaticity of phobic fear: Conditioned skin conductance responses to masked phobic stimuli", *Journal of Abnormal Psychology* 102:121-132.

Öhman, A. and J. J. F. Soares (1994) "Unconscious anxiety: Phobic responses to masked stimuli", *Journal of Abnormal Psychology* 103:231-240.

Öhman, A. and J. J. F. Soares (1998) "Emotional conditioning to masked stimuli: Expectancies for aversive outcomes following nonrecognized fear-relevant stimuli", *Journal of Experimental Psychology: General* 127:69-82.

Öhman, A. and S. Wiens (in press) "On the automaticity of autonomic responses in emotion: An evolutionary perspective", in: *Handbook of Affective Sciences*, R. J. Davidson, K.R. Scherer and H. H. Goldsmith, eds., New York: Oxford University Press.

Parra, C., F. Esteves, A. Flykt and A. Öhman (1997) "Pavlovian conditioning to social stimuli: Backward masking and the dissociation of implicit and explicit cognitive processes", *European Psychologist* 2:106-117.

Posner, M. I. (1978) *Chronometric Explorations of Mind*, Hillsdale, NJ: Erlbaum.

Ring, C. and J. Brener (1992) “The temporal location of heartbeat sensations”, *Psychophysiology* 29:535-545.

Robinson, M. D. (1998) “Running from William James’ bear: A review of preattentive mechanisms and their contributions to emotional experience”, *Cognition and Emotion* 12:667-696.

Schachter, S. and J. E. Singer (1962) “Cognitive, social, and physiological determinants of emotional state”, *Psychological Review* 69:379-399.

Schacter, D. L. (1987) “Implicit memory: History and current status”, *Journal of Experimental Psychology: Learning, Memory, and Cognition* 13:501-518.

Schneider, W., S. T. Dumais and R. M. Shiffrin (1984) “Automatic and control processing and attention”, in: *Varieties of Attention*, R. Parasuraman and D. R. Davies, eds., New York: Academic Press, pp. 1-28.

Shevrin, H. and S. Dickman (1980) “The psychological unconscious: A necessary assumption for all psychological theory”, *American Psychologist* 35:421-434.

Shiffrin, R. M. and W. Schneider (1977) “Controlled and automatic human information processing: II. Perceptual learning, automatic attending, and a general theory”, *Psychological Review* 84:127-190.

Skinner, B. F. (1954/1972) “A critique of psychoanalytic concepts and theories”, in: *Cumulative Record. A Selection of Papers*. 3rd ed., B. F. Skinner, ed., New York: Appleton-Century-Crofts, pp 239-248.

Tranel, D. and A. R. Damasio (1985) “Knowledge without awareness: An autonomic index of facial recognition”, *Science* 228:1453-1454.

Warrington, E. K. and L. Weiskrantz (1974) “The effect of prior learning on subsequent retention in amnesic patients”, *Neuropsychologia* 12:419-428.

Weiskrantz, L. (1986) *Blindsight. A Case Study and Implications*, Oxford: Oxford University Press.

Weiskrantz, L. (1997) *Consciousness Lost and Found*, Oxford: Oxford University Press.

Whalen, P. J., S. L. Rauch, N. L. Etcoff, S. C. McInerney, M. B. Lee and M. A. Jenike (1998) “Masked presentations of emotional facial expression modulate amygdala activity without explicit knowledge”, *Journal of Neuroscience* 18:411-418.

Zajonc, R. B. (1980) “Feeling and thinking: Preferences need no inferences”, *American Psychologist* 35:151-175.

THE EXPERIENCE OF EMOTION: SITUATIONAL INFLUENCES ON THE ELICITATION AND EXPERIENCE OF EMOTIONS

URSULA HESS

Department of Psychology, University of Quebec at Montreal, P.O.Box 8888, Station A, Montreal, Québec., H3C 3P8, Canada

ABSTRACT

The experience of emotions and affects is a pervasive aspect of our life. We all feel well or moody on a particular day, are impatient or pleased with something we are doing and experience happiness at a compliment or anger at a slight. The present paper discusses the social context influences on the elicitation and experience of emotions from the point of view of appraisal theories. Appraisal theories of emotions posit cognitive evaluations of such aspects of the emotion eliciting event as its novelty and pleasantness, the degree to which it helps or hinders ongoing plans and goals, the degree to which individuals believe themselves able to cope with the event, the degree to which what happened appears just and unjust, etc. Yet, the outcome of these appraisals as well as the behavioral consequences of the outcomes are not the same for everyone and a number of situational influences have been found to be of importance in this context. In this paper I argue that these factors not only exert influence on the display and labeling of emotions but also modulate the appraisal process.

1. Introduction

The present paper discusses situational influences on the elicitation and experience of emotions. Specifically, such influences as cultural emotion norms and roles as well as the relationship between interaction partners in terms of power or gender are considered. These situational aspects may influence the emotion process at different points. First, they may influence the perception of the situation and thus act as a filter for those elements of the situation that enter the appraisal process. Second, they may influence the appraisal process as such. Third, they may influence the emotion display and the labeling of the subjective feeling state. The former two are influences on the elicitation of emotions, whereas the latter refer to the experience of emotions.

This definition presupposes a specific view of emotions, which should be stated explicitly first. I will then briefly discuss functions of emotions and in particular the interpersonal functions of emotions. Following this, I will address the question of whether emotions are social constructs before going on to discuss some empirical data illustrating situational influences on the elicitation and experience of emotion.

1.1. What is an Emotion?

What is an emotion and how do emotions differ from other affects such as moods, affective disorders, or personality traits such as anxiousness? In everyday language, emotions simply refer to feeling states such as happiness, excitement, love, fear, or hate. In fact, many people consider love and hate as the best examples for emotions (Fehr & Russell, 1984). In the present context, the term emotion will be reserved for short term, intentional states. Emotions are intentional states because they – unlike moods – have an evaluated object. Thus, we are not angry or surprised in general – but we are angry or surprised at something – the object of the emotion. This notion is central to appraisal theories of emotion (e.g., Frijda, 1986; Scherer 1986).

The notion that emotions are linked to evaluations is old and can be traced back as far as Aristotle. Spinoza (1675) also referred to emotions as "accompanied by an idea of an external cause." Appraisal theories of emotion are usually traced back to Magda Arnold who defined emotion as: "... a felt tendency toward an object judged suitable and away from an object judged unsuitable, reinforced by specific bodily changes according to the type of emotion" (Arnold & Gasson, 1954, p 294).

Appraisal theories view emotions as short term reactions to an internal and external event that is evaluated along a series of features such as pleasantness, goal conduciveness, ability to cope, justice, etc. (e.g., Scherer, 1986; Frijda, 1986). A distinctive feeling state as well as bodily changes, expressive behavior, and actions tend to characterize emotions, but none of these is separately necessary or sufficient for a state to be considered an emotion.

1.2 Functions of Emotions

In considering the situational influences on emotions it is important to first consider the impact of emotions on the situation, that is, the intra- and interpersonal functions of emotions. Intrapersonal functions refer to the functions of emotions for the individual that experiences them. In Western society emotions are often viewed as disruptive or destructive. The notion that emotions pervert reason and should be tightly controlled goes back to Plato and has been sustained though the centuries in Western philosophy and religion. From this view, emotions are described as irrational, disruptive and to be distrusted as the following citation by William Penn (founder of Pennsylvania) illustrates, "Passion is a sort of fever in the mind, which ever leaves us weaker than it found us (William Penn, 1644-1718)." On the other hand, evolutionary points of view underline the importance of emotions as means to fast action and as plans for action. In this sense the short-circuiting of cognitive processes in favor of rapid action, for example, by initiating a flight reaction in response to a perceived threat, is seen as adaptive - even when the perceived threat may turn out on occasion to be one's own shadow.

Nico Frijda (1986) describes emotions as states of action readiness, that is, motivational states that engender a specific form of action. As such they are not so much interruptions of rational behavior but in fact part of it as they coordinate and provide goals. From a biological point of view emotions allow the recruitment of behaviors that are low in an organism's response hierarchy as they activate the physiological support systems necessary to deal with the situation at hand.

That emotions play an important role for both rational thinking and the ability to get along successfully with one's fellow human beings is illustrated by research by Damasio (1994). Patients with damage to the lower middle part of the frontal lobes, who show emotional blunting but no impairment of their intellectual abilities, tend have massive problems in planning their life – they make bad decisions and can dither endlessly over trivial problems. Interestingly, they seem particularly oblivious to the impact of their actions with view to notions of politeness and sociability. According to Damasio, emotions can be conceptualized as somatic markers that help us to sort out options that result in social punishment and to strife towards solutions that are emotionally attractive. This elimination process allows us to reduce decision-making problems to manageable proportions. Further, emotions allow us to set priorities and therefore to be able to neglect certain problems (e.g., should I put my pencil into the left or the right drawer) in favor of other more relevant ones.

In the present context, the interpersonal functions of emotions, which are linked to the communication of intentions and the establishment and maintenance of social networks, are of particular interest. When we consider so-called social emotions such as regret, shame, guilt etc. but also anger and contempt we note that these emotions tend to reinforce social rules. Showing anger at a social transgression can serve as punishment while the display of shame signals acceptance of a moral code. Thus, emotions can help to regulate social norms. In this vein, Turner (1997) asserts that emotions evolved in part to provide the means for effective sanctioning and the enforcement of moral codes within groups of hominids. His basic argument is that the communication of emotions in humans was a necessary prerequisite for social bonding among, what he considers essentially asocial hominids. This notion refers to emotions such as anger, shame, and guilt that help enforce the social rules necessary for smooth interactions but also to emotions such as sadness that may recruit succor.

Emotion displays also serve to regulate power relationships. Ethologists have noted the importance of anger displays in primates for the negotiation of hierarchies. Averill (1997) posits that emotions follow rules that help establish emotion roles. He likens emotion roles to other social roles in that they confer privileges (e.g., we excuse actions that stem from extreme emotions), are subject to restrictions (e.g., notions of "excessive" grief), come with obligations (e.g., the angry person should show a desire to redress the wrong - otherwise the anger seems suspect), and have entry requirements. In the case of anger, he notes that both social recognition and legitimacy are required. Thus, persons in higher authority have more "right" to be angry than those with less authority. Conversely, anger displays are perceived as dominant (Keating, 1985; Knudson, 1996).

The importance of emotions for social relationships is also obvious from research by Rimé *et al.* (1991) who found that the vast majority of all emotional experiences are told to others – usually within hours of the experience. Specifically, Rimé and colleagues asked people how often they had told others about an emotional event that had happened to them. These events include relatively trivial things like an exam or a dispute with someone as well as serious events such as the death of a partner. Over many studies they consistently found that between 80% and 90% of the participants retold the event at least several times. This was the case for both men and women and for individuals from different age ranges. They also found that people who have been told about an emotional event tend to retell this event to others at very high rates. According to Rimé, this social sharing of emotions serves to define and elaborate social relationships. The telling allows the individual who experienced the event to situate the event and to affirm the appropriateness of the emotional reaction in its social context. In sum, emotions have an important influence on social relationships and help regulate interactions on various levels. Yet, the social context in which an event takes place also influences the emotions experienced. Before going on to discuss these influences in detail, I need to situate the emotion concept I am working with more precisely in its theoretical context.

1.3. Are Emotions Biologically Based or Socially Constructed?

In the present context, emotions are viewed as having a biological basis. That is, emotions are seen as having an evolutionary history and as grounded in the biological functions of the brain. This is not to deny that human society has a profound influence on the emotions process. In fact, this influence is the topic of this paper. Yet, this position can be contrasted by other points of view. Social constructivist views of emotions tend to deny emotions their biological basis and to view them as purely social constructions. In this context, the social pre- and proscriptions and the rights, obligations, and conventions associated with the experience of emotions are seen as their primary features. To illustrate this point, I have listed here what Harré (1986) considers the five basic features of emotions: (1) the repertoire of language games available in a culture; (2) the moral order within which the moral appraisals which control both the meaning and the occasioned use of emotional terminologies are themselves meaningful; (3) the social function (acts) which particular emotion displays and emotion talk perform in the dramaturgically shaped episodes of this or that culture; (4) the narrative forms that the unfolding of the situations revealed in 1, 2 and 3 above realize; (5) the systems of rules by which these complicated forms of social action within which the emotional qualifications of actions and actors are maintained, changed, critically accounted and taught.

According to this view, members of different social groups differ fundamentally in their evaluation of emotion eliciting events and their consequences and these differences then lead to fundamental differences in emotional reactions (e.g., Kemper, 1990; Wierzbicka, 1995), and it is maintained that emotion words and their connotations dictate the way events are seen and reacted to. In fact, there is some support for the view that basic aspects of emotion eliciting events differ across cultures (e.g. Wierzbicka, 1995). In particular, research focussing on the language of emotions notes that some emotion words can be found only in some cultures. A good example is the German word "Schadenfreude" that denotes the amusement taken from the misfortunes of others. Other examples are the Ifaluk word "song" that denotes a type of anger that may lead to suicide. Further, it has been noted that emotions can become obsolete. For example, the term "accidie" refers to the emotion associated with the loss of intrinsic motivation towards one's religious duties. These observations have been interpreted as showing that emotions are actually socially constructed phenomena. On the other hand, one may argue that the lack of a word does not deny the existence of the state it refers to nor even its importance. For example, the English language does not have a word for "Schadenfreude" but it would be wrong to deny that the emotion plays an important role in the American cultural context as the genre of slapstick movies is based on this emotion. Also, while there is no simple English equivalent to "song" the notion of righteous anger that may even lead someone to suicide by, for example, hunger strike or self-immolation is certainly available.

Frijda, Markam, Sato and Wiers (1995) studied the relationship between emotion words and emotions and note that once one leaves the surface level of description and considers more fundamental aspects such as underlying appraisal categories, fairly stable cross-cultural emotion patterns can be found for many emotions. Other evidence suggests that the antecedents of emotion events are largely similar across a large number of cultures (Mauro, Sato, & Tucker, 1992; Matsumoto, Kudoh, Scherer, & Wallbott, 1988). For instance, Scherer (1997) based on a questionnaire study conducted in 37 countries on all continents (except Antarctica), notes how similar many appraisal patterns are - yet, he also notes interesting differences, especially as regards the attribution of agency.

In my opinion, the somewhat conflicting findings on this issue can be resolved by assuming that the basic emotion process is biologically grounded and universal, but that the type of events attended to, the appraisal of these events, and the relevant norms for behavior may vary as a function of culture, gender, relative power status, as well as the relationship between the interaction partners (see also Mesquita, Frijda, & Scherer, 1997; Frijda, Markam, Sato, & Wiers, 1995). Yet it is important to remember, as Frijda and Zammuner (1994) note, the way we label an emotion can have an important impact on how we cope with it. If, for example, we label homesickness as a physical illness (as is done in the Tahitian culture) we may in fact face this phenomenon differently than when we label it an emotional reaction (sadness at being separated from those close to us). However, in both cases the state of the individual is due to a perceived loss that can not be readily remedied, that is, the appraisal pattern is the same but both the label given by the person who experiences the state and the option that can be considered for coping can be quite different.

Cultural norms may lead us to attempts at controlling or even suppressing certain emotions because we fear negative consequences. As Harré and Parrot (1996) put it: (emotions) ... are acts embedded in patterns of acts; their display is subject to rules and conventions; they are embedded in culturally specific moral orders and normative systems that allow for assessments of the correctness and impropriety of emotions. The emotions that are subject to these attempts can vary from culture to culture. As a consequence, certain emotions may be less frequently observed in certain cultures. For example, it has been noted that the North American culture calls for control of negative emotions and rewards the display of positive emotions.

In sum, cultural and social influences on emotions are quite central to any consideration of emotions, especially emotions in social context. These influences may be relevant to different aspects of the emotion process that I want to discuss in the following and illustrate with empirical evidence.

2. Levels of Social Influences on the Elicitation and Experience of Emotions

In the present context, I will focus on the elicitation and experience of emotions, that is, the process that leads from the perception of an emotional stimulus to the expression and labeling of the resulting emotional state. Yet, as the social consequences of experiencing and displaying emotions play an important role for the socialization of emotions, it is obvious that the same emotion norms that influence the first process also influence the decoding of emotion displays.

A given emotion-eliciting situation has to be perceived by the observer. Whereas some emotion eliciting situations are quite simple in structure (e.g., encountering a bear in the woods), many social situations are complex and allow for the possibility that interactants focus on different elements of the situation, both when appraising the situation and when attributing emotional states to others via perspective taking. For example, it has been noted that girls in general are more sensitive to social context than boys (Merrum-Terwogt & Olthof, 1989). Similarly, Suh *et al.* (1998) discuss evidence that members of collectivistic cultures are more likely to attend to external socially shared elements of an emotion stimulus whereas members of individualistic cultures pay more attention to internal cues.

Differences in emphasis regarding aspects of a situation essentially change the database for the emotional appraisal process. Whether internal cues are ignored or attended to may lead to differences in appraisal. For example, Markus and Kitayama (1994) interpret findings by Levenson *et al.* (1992) that West Sumatrans show the physiological changes congruent with a facial feedback effect but -- unlike Americans -- do not report feeling the corresponding emotion. This appears due to the fact that for the former a situation where they are alone does not qualify as an emotional situation and the internal cues alone are not attended to.

Differences in the perception of a situation may also lead to differences in perspective taking, that is, the appraisal of a situation in terms of another person's emotional state. Hochschild (1983) describes an incident where a passenger threw a cup of coffee at a flight attendant and notes the difference in perception between the airline and the flight attendant. The described situation can be considered as threatening to self-esteem when one is attending to aspects of the situation relevant to the self but as simply tedious when attention is shifted away from the self. In Hochschild's example, the airline focussed on the situation as an example of public contract work (and thus not a possible source of anger) whereas the flight attendant focussed on the personal insult (and therefore felt angry).

Secondly, the actual appraisal process may differ as a function of situational emotion norms. In this context, Mesquita, Frijda, and Scherer (1997) suggest that cultural differences can be understood as differences in the practice, or propensity, to use certain appraisal dimensions. As mentioned above, Scherer (1997), based on data from 37 countries on different continents, concluded that although a sizable degree of similarities in appraisal profiles was found across countries, there were also some differences. He notes that those differences were most notable as related to notions of Agency (did the other person do this on purpose) and Justice. Similarly, Frijda and Mesquita studied appraisal processes in Dutch, Turkish, and Surinamese individuals who all live in the Netherlands. Their results point to a number of differences, including the attribution of Agency. For example, Dutch subjects attributed less agency to others in anger eliciting situations (inconsiderate behavior of others). Further, Frijda *et al.* (1995) found strong similarities in the use of appraisal dimensions between participants from the Netherlands, Indonesia, and Japan, but also note that Japanese, more often than Dutch or Indonesian individuals, appraised situations as immutable and inevitable and reported action tendencies of apathy and "desire to depend upon someone else."

Yet, culture is not the only source of social context influences on appraisal, gender and power have also been linked to differences in emotional experience and expression (see Henley, 1995). The influence of power on appraisal processes is illustrated by Cantin and Hess (1998) who randomly assigned high or low levels of decisional power to participants in a study designed to induce anger in a dyadic setting in which a confederate provoked the participant. The results show that decisional power was a better predictor of anger behavior than individual differences such as anger proneness, dominance, empathy, or endorsement of display rules. Specifically, low power participants appraised the confederate (who always behaved in the same rude manner) as more pleasant and reacted with less anger than high power participants. Power also influenced strategy choice in managing the conflict with high-power participants showing more assertive conflict resolution strategies. They were also judged by outside observers as being more dominant and less affiliative.

Differences in emotion-antecedent appraisals for members of difference groups entrain likely differences in emotional behavior. For example, we may expect that the Japanese participants in the study reported above, would also display more apathy and in the Cantin and Hess study such differences were in fact observed. However, even in situations were appraisal patterns do not differ, behavioral reactions may – either because of constraints placed by display rules or because of the anticipated consequences of emotion displays.

In this context, it has been suggested that the presence of an audience in and of itself is determining for emotion displays. In fact, Fridlund (1991) suggests that emotion displays are not part of an emotion process but rather social signals. Thus, the actual or implicit audience determines emotional expressivity to a larger degree than underlying emotional states. As support for this notion, findings by Fernandez-Dols and Ruiz-Belda regarding the seemingly emotion incongruent facial displays of sports medal winners as well as a study by Fridlund (1991) who varied the sociality of an emotional situation and found monotonic increases in expressivity as a function of sociality, can be summoned. In fact, Fridlund considers the notion that emotion displays are somehow caused by emotions and then modified by social rules as romanticist. Yet, the notion that emotional expressions are social signal is not inconsistent with the notion of emotional expressions as symptom, as is illustrated by findings by Hess, Banse, and Kappas (1995).

To turn back to social emotion rules, these rules, often referred to as display rules (Ekman & Friesen, 1969) are pre- and proscriptive norms for emotion displays that are pervasive and socialized early in life. Gallois (1994), in a recent review on emotion communication in interpersonal situations, concludes that not only are interpersonal situations involving verbal and nonverbal emotional communication, such as self-disclosure or conflict, highly rule-governed but the rules are perceived as normative for the interactions. Consequently, even minor violations of emotion norms can create substantial problems for the interaction process. Three types of emotion rules can be distinguished.

First, normative rules exist regarding the intensity of emotion displays or the overall emotionality of a group. For example, women are generally expected to be more expressive than are men, and Asian-Americans describe themselves as less emotionally expressive than other ethnic groups (Gross & John, 1995) and are perceived as less expressive by others as illustrated, for example, by Herrera, Bourgeois, and Hess (1998). Second, display rules pro- and pre-scribe specific emotion displays for specific situations (e.g., solemn appearance at a funeral), as well as specific modes of expressive behavior (e.g., physically aggressive behavior in women). Finally, display rules regulate the level of emotionality for different types of social relationships. For example, Aune, Buller, and Aune (1996) describe how rules for the expression of positive and negative emotions change over the course of the development of romantic relationships and note sex differences for the display of emotions as relationships progress from dating to more developed relationships.

We have already discussed the influence of power on appraisal processes. Averill (1997) notes that social norms for anger generally allow more latitude in anger expression to individuals in positions of high power. Similar observations are made by Henley (1977) to account for the fact that women tend to show less anger and smile more. In fact, smiling can be interpreted as a sign of appeasement. From an evolutionary perspective smiling derives from the silent bared-teeth display in primates, which is often used as a sign of appeasement. An interesting illustration of the influence of social context in a wider sense on emotion displays has been provided by Preuschoft and VanHoof (1997). They studied the use of the silent bared-teeth display in primates. As stated above, this display is mainly as a gesture of appeasement but also occurs in contexts other than appeasement. Preuschoft and VanHoof studied the variety of uses, other than appeasement, as a function of the power asymmetry in various species of primates and suggest that primates whose social groups tend to be less hierarchical tend to use the silent bared-teeth display in a wider variety of situations.

Finally, social rules on emotion may influence how an emotion is labeled which in turn influences how the event is coped with. We already noted the tendency in the Tahitian culture to label homesickness not in terms of an emotion such as sadness but rather in terms of a physical illness. In an interesting essay on the social control of negative emotions, Janet Landman (1996) notes that current popular wisdom denigrates regret as a useless waste of time and exhorts people to either suppress regret entirely or at least to re-label it in terms of a valuable lesson learned. She also notes that regret has a number of positive sides that are usually ignored, for example, feeling regret allows us to reaffirm that we have values even if we did once not live up to them - something that in the long run may be more valuable than the emotional equilibrium bought by labeling regret as some irrational fit of the blues.

In sum, an important function of emotions is their influence on social relationships. Conversely, the social context in which an event takes place also influences the experience of emotions. Appraisal theories of emotion provide a useful framework for conceptualizing these influences.

Acknowledgments

The present paper is based in part on research funded by the Conseil de Recherche en Sciences Humaines.

References

Arnold, M. B. and J. A. Gasson (1954) "Feelings and emotions as dynamic factors in personality integration", in: *The Human Person*, M. B. Arnold and J. A. Gasson, eds, New York: Ronald, pp. 294-313.

Cantin, V. and U. Hess (1998) *Angry People versus Angry Situations: The Influence of Dispositional and Situational Variables on the Experience and Expression of Anger*, Manuscript submitted for publication.

Damasio, A. R. (1994) *Descartes' Error*, New York: Putnam.

Fehr, B. and J. A. Russell (1984) "Concept of emotion viewed from a prototype perspective", *Journal of Experimental Psychology: General*, 113:464-486.

Frijda, N. (1986) *The Emotions*, Cambridge, UK: Cambridge University Press.

Frijda, N., S. Markam, K. Sato and R. Wiers (1995) "Emotions and emotion words", in: *Everyday Conceptions of Emotion*, J.A. Russel, J.-M. Fernández-Dols, A. S. R. Manstead and J. C. Wellenkamp, eds, Dordrecht: Kluwer, pp. 121-143

Henley, N. M. (1995) "Body politics revisited: What do we know today?", in: *Gender, Power, and Communication in Human Relationships*, P. J. Kalbfleisch and M. J. Cody, eds, Hillsdale, NJ: Lawrence Erlbaum, pp. 27-61.

Keating, C. F. (1985) "Human dominance signals: The primate in us", in: *Power, Dominance, and Nonverbal Communication*, S. L. E. and J. F. Dovidio, eds, New York: Springer Verlag, pp. 89-108.

Kemper, T.D. (1990) "Social relations and emotions. A structural approach", in: *Research Agendas in the Sociology of Emotion*, T. D. Kemper, ed., Albany, NY: State University of New York Press, pp. 207-237.

Knudson, B. (1996) "Facial expressions of emotion influence interpersonal trait inferences", *Journal of Nonverbal Behavior* 20:165-182.

Matsumoto, D., T. Kudoh, K. R. Scherer and H. Wallbott (1988) "Antecedents of and reactions to emotions in the United States and Japan", *Journal of Cross-Cultural Psychology* 19:267-286.

Mauro, R., K. Sato and J. Tucker (1992) "The role of appraisal in human emotions: A cross-cultural study", *Journal of Personality and Social Psychology* 62:301-317.

Mesquita, B., N. H. Frijda and K. R. Scherer (1997) "Culture and emotion", in: *Handbook of Cross-Cultural Psychology, Vol. 2, Basic Processes and Human Development*, J. W. Berry, ed., Boston, MA: Allyn and Bacon, pp. 254-297.

Rimé, B., B. Mesquita, P. Philippot and S. Boca (1991) "Beyond the emotional event: Six studies on the social sharing of emotions", *Cognition and Emotion* 5:435-465.

Scherer, K. R. (1986) "Vocal affect expression: A review and a model for future research", *Psychological Bulletin* 99:143-165.

Spinoza. B. (1675) "The ethics" (R. H. M. Elwes, Trans.), New York: Dover (current edition, 1955).

Suh, E., E. Diener, S. Oishi and H. C. Triandis (1998) "The shifting basis of life satisfaction judgments across cultures - emotions versus norms", *Journal of Personality and Social Psychology* 74:482-493.

Turner, J. H. (1997) "The evolution of emotions: The nonverbal basis of human social organization", in: *Nonverbal Communication: Where Nature Meets Culture*, U. Segerstråle and P. Molár, eds, Mahwah, NJ: Lawrence Erlbaum Associates, pp. 211-223

Wierzbicka, A. (1995) "The relevance of language to the study of emotions", *Psychological Inquiry* 6:248-252.

THE COMMUNICATION OF EMOTION

URSULA HESS
Department of Psychology, University of Quebec at Montreal, P.O. Box 8888, Station A, Montreal, Quebec, H3C 3P8, Canada

ABSTRACT

The ability to well communicate emotions is important for both the encoder, who would like to be understood, and the decoder, who strives to understand. The present paper focuses on the communication of emotions via facial expressions. In this context, the notion that emotional expressions also communicate information about the encoder's view of the world, their perception of emotion antecedent events, as well as their behavioral intentions and even aspects of their personality such as affiliation and dominance are discussed. Specifically, I focus on the role of social emotion norms, which guide both the overt expression of emotions and the attribution of emotions to others based on nonverbal behaviors. These norms vary with such social aspects of the interaction partners as their gender, power, and culture and influence not only the encoding but also the decoding of emotional expressions. Evidence from our laboratory regarding the communication of emotions between members of different social groups is discussed to illustrate this point.

1. Introduction

Knowing what other people feel is an important element of everyday social interactions. We all like to know whether someone is happy about a gift or angry about a careless remark. The ability to well communicate emotions is relevant for both the encoder, who would like to be understood, and the decoder, who strives to understand. But how do we know what others feel? What cues can we use? Research in the communication of emotion has frequently focussed on nonverbal aspects of emotion communication such as facial and vocal expressions or posture and gestures. Further, research on the use of cues in decoding tasks suggests that observers in general focus most on facial expressions.

A first question to be addressed in the context is the question of what it is that these nonverbal indices communicate. Two general points of view can found in the literature. According to the first, which goes back to antiquity and more recently to Darwin (1872), emotion displays are innate symptoms of an underlying feeling state. Studies showing that specific facial displays can be linked to self-reports of specific affective states (e.g., Cacioppo, Petty, Losch, & Kim, 1986, Pope & Smith, 1994; Rosenberg & Ekman, 1994) lend support to this notion. Further, findings that chimpanzees react differentially to different human expressions (Itakura, 1994) and that human children's ability to interpret monkey vocalizations of aggression, fear, dominance, positive emotions, and submission develops simultaneously with their ability to interpret human emotional behavior

(Linnankoski, Laaski, & Leinonen, 1994) suggests similarities between the expressions of human and non-human primates and thus evolutionary continuity. Finally, research by Ekman *et al.* (1987) supports the notion that expressions can be universally recognized and thus are presumably universally shown.

According to the second point of view (see Fridlund, 1991) emotion displays serve purely communicative functions and do not actually provide valid information regarding underlying emotional states. A number of findings can be summoned in support of this position too. In a now classic study, Kraut and Johnston (1979) found that bowlers were more likely to smile when looking at others than when facing the pins after having scored. Also, fans at a hockey game were more likely to smile when interacting with others than while watching the game. More recently, Fernandez-Dols and Ruiz-Belda (1995) showed that people who just won a medal tend to show facial expressions other than those commonly associated with happiness even though they tend to report having been happy. In addition, Fridlund (1991) demonstrated that individuals are more expressive of positive emotions in the (real or implicit) presence of a friend than when alone. All these findings seem to suggest that smiling is not necessarily a sign of a pleasurable experience but rather a social signal.

Yet, apart from the fact that these studies by and large did not attempt to distinguish between different forms of smiles, this dichotomy is an oversimplification. It has long been recognized in the field of emotion research that emotional expressions can be both symptom and signal and that in some instances they will be only a signal ("fake" expressions) while in others they will be mainly symptom (crying when reading a sad story alone). In fact, facial emotion displays are inherently polyvalent, and may also be used as emblems (Ekman, 1979), as interjections (Motley, 1993), and as signals to indicate that we understand a speaker's state (Bavelas, Black, Lemery, & Mullett, 1986). From the perspective of appraisal theories of emotion, which posit that emotions are based on an appraisal of the stimulus according to a number of appraisal dimensions such as pleasantness, goal condusiveness, coping potential, etc., emotions are per se not only based on information processing but also provide information regarding the situation. From this perspective, Frijda and Tcherkassof (1997) describe emotional facial expressions as modes of action readiness. States of action readiness refer to the individual's intentions with regard to their environment (e.g., I wanted to approach). Hence, when we see an anger display we can conclude that the expressor evaluates the eliciting event as a slight or insult and that s/he feels that the situation can be redressed.

Thus, the role of facial expressions as interactional signals does not negate their role as symptoms of emotional states and emotion displays can be considered to serve several related purposes of which the communication of emotional states is only one. This notion can be conceptualized using a model by Bühler (1934). He distinguished three functions of a message: the symbolic, the symptomatic, and the appeal function. The first refers to the sign content of the message and conveys information directed at the interaction partner. The second, the symptomatic function, corresponds to a readout of the individual's state. The third function regards the possible action of the interaction partner. This model (see Figure 1) applies equally well to emotion displays.

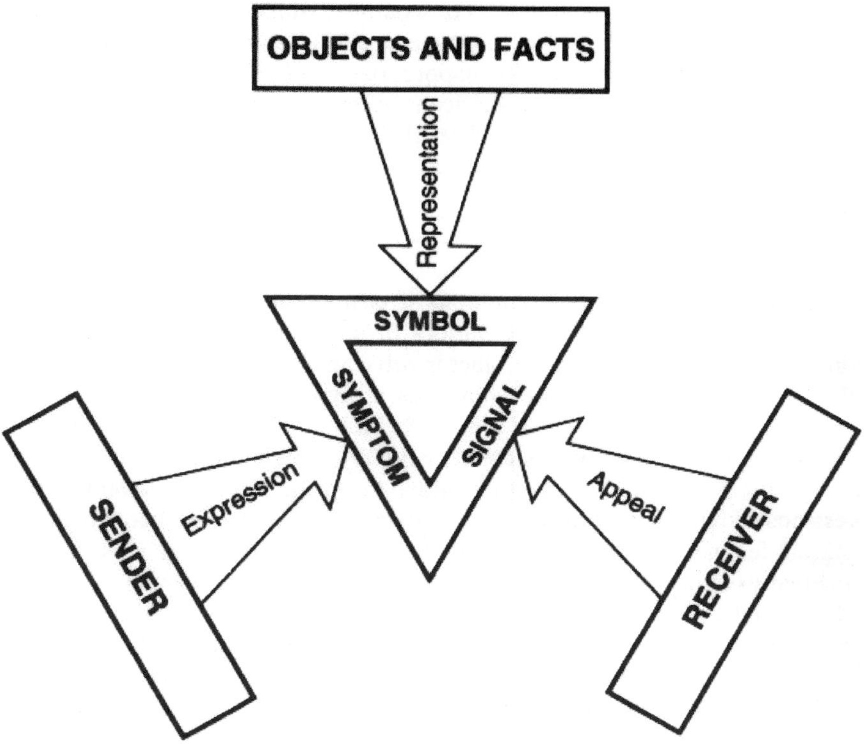

Figure 1. Bühler's Organon Model

For example, an expression of sadness signals that the sender experienced an irreversible loss (Lazarus, 1991). It also suggests a specific internal state of the sender, characterized by a specific subjective feeling state, as well as by a number of physiological and behavioral concomitants; and finally, it may serve an appeal function by motivating the observer to help or to comfort.

This point is illustrated by a study that manipulated both the intensity of the emotional stimulus and the sociality of the situation in terms of the presence or absence of a friend or stranger. Hess, Kappas, and Banse (1995) could show that for friends the facial expressive behavior shown while watching comedy routines varied both with the intensity of the stimulus and the sociality of the situation.

For the condition involving strangers the situation was more complex pointing to the importance of the relationship between encoder and audience (see also Jacobs, Fischer, & Manstead, 1997)

In fact, social aspects of the situation such as the gender, the ethnicity or the relative power of the interaction partners influence the exact mode of expression and the likely behavioral intentions. For example, women are known to be more likely to cry when angry. Also, Japanese consider the expression of anger towards relatives or colleagues as inappropriate and may therefore attempt to suppress its expression, whereas Americans accept and endorse the expression of anger towards close others. Finally, Averill (1997) points out that the appropriate expression of anger requires power to redress the wrong. In the following these points will be discussed in more detail.

2. Encoding of Emotions

Regarding the encoding of emotions it is important to note that emotional facial expressions are shown in a variety of situations and under a variety of conditions many of which do not in fact involve an emotional state. For example, emotional facial expressions are shown as a sign of empathy in the form of mimicry (e.g., Bavelas, Black, Lemery, & Mullet, 1986). Further, emotional facial expressions are shown as conversational emblems and in form of over-learned responses such as the greeting smile. Other examples are emotional facial expressions shown to conform to social demands, which may be incongruent with the current feeling state of the expressor (expression of pleasure at a disappointing gift). Finally, emotional expressions may be shown to mislead others explicitly.

The social context of an interaction can have a number of influences not only on the emotion displays shown but also both on the elicitation of emotions and on the appraisal process itself (see Hess, this volume). In this context, I want to only briefly mention the first influence, that is, display rules. Display rules for the expression of emotions have been studied in a variety of contexts, for example, Aune, Aune, and Buller (1996) studied the development of display rules in romantic relationships and found a curvilinear pattern for the appropriateness of displaying negative emotions across the stages of the relationship. Zeman and Garber (1996) studied display rules for children in grades 1,3, and 5 as a function of the interaction partner. They found changes as a function of age, with younger children reporting more sadness and anger, as well as sex, with girls reporting to be more expressive of sadness and pain, but most interestingly they could show that children report different rules for interactions with peers than for interactions with their parents. Specifically, the children reported to control all emotions more when in the presence of peers and cited fear of negative interpersonal consequences as their reason.

Yet, the largest body of literature on display rules is concerned with sex differences as well as differences among cultures. In general, findings converge on the notion that women are more expressive of emotions in general (e.g., Fischer, 1993) with the exception of outward aggression in anger, and that members of collectivist cultures differ from members of individualistic cultures in the endorsement of the display of certain emotions in certain contexts (Matsumoto, 1996).

Two related concepts in this context are feeling rules (Hochschild, 1979) and naïve emotion theories. Whereas display rules refer only to the appropriateness of the display, feeling rules also refer to the appropriateness of the emotion. That is, these rules can be paraphrased in terms of "I should be happier about ..." or "I really should not get angry at ... for doing this." Naïve emotion theories are really theories of mind, that is, theories that account for the types of situations that elicit specific emotions. These theories may then influence the recognition of emotions not only in others but also in oneself (see e.g., Suh, Diener, Oishi, & Triandis, 1998).

3. Decoding of Emotional Expressions

Research on the decoding of emotional expressions has a long history (see Ekman, 1973). In general, for at least some emotions such as happiness, anger, sadness, fear, disgust, contempt, and surprise, high levels of recognition accuracy are observed (e.g., Fridlund, Ekman, & Oster, 1987). Most of this research has been conducted in a context where social influence variables were controlled and where the basic stimulus material was of considerable clarity. Yet, social context is most likely to influence the decoding of expressions that are somewhat ambiguous. Lets consider the sources of information that decoders have at their disposal. First, there is the actual display, for example, in the case of happiness the upturned corners of the mouth and the wrinkles around the eyes. A first means to decode the expression is therefore pattern matching, that is, the observer notes the upturned corners of the mouth and the wrinkles, labels them as a smile and then makes the link between smiles and happiness. However, not all happiness displays are as simple, for example, some people actually show down-turned corners of the mouth at midranges of activity of the *Zygomaticus major* - the muscle that turns the corners of the mouth up. Another sources of information, may be prior knowledge of the encoder. Together with knowledge regarding the situation this allows the decoder to take the perspective of the encoder and to assess whether the situation was in fact one that is likely to elicit happiness in the encoder. In fact, friends and marital partners often - but not always - are better at decoding each other's expressions. The exceptions tend to occur in the early stages of the relationship - before enough knowledge is available as well as in later stages of the relationship. Colvin, Vogt, and Ickes (1997) in summarizing the literature on this issue note that after a while friends and partners may in fact just assume to know what the other feels without actually looking at the person. Further, research by Noller and her colleagues shows that partners in distressed marriages may in fact get very bad at decoding their facial expressions (e.g., Noller and Ruzzene, 1991). Two explanations are advanced. First, it is possible that the arousal that accompanies certain exchanges in these marriages makes decoding more difficult; second it is possible that judgments are based on theory-driven uncharitable notions about the partner's likely motives rather than on the actual expressive behavior.

However, even when the interaction partners do not know each other, context knowledge, more specifically, cultural and gender specific emotion norms can be employed to predict likely reactions. Thus, emotion norms may lead members of a culture to ascribe less intensity to displays of culturally disapproved emotion displays and to lead members of the same culture to ascribe different levels of intensity to similar emotion displays shown by men and women. For example, Matsumoto and Assar (1992) found that bilingual speakers of Hindi and English decode emotional facial expressions differently when the judgments are made in Hindi as opposed to English, suggesting that linguistic emotion categories focus the individual's attention on aspects of meaning of the environment and thus sensitize the individual for them in line with cultural prescriptions linked to the use of the emotion (see also, Frijda, Markam, Sato, & Wiers, 1995). Also, Hess, Blairy, and Kleck (1997) using photographs of facial expressions by men and women found that different levels of intensity were ascribed to similar emotion displays shown by men and women.

The use of emotion norms in the decoding of emotion displays may often be useful when both interaction partners share the same norms. Yet, norm based decoding biases may lead to failures to recognize certain displays correctly, especially those of medium to low intensity, with negative consequences for the efficacy of emotion communication between members of different social groups. This may be of even more concern in the multicultural context of today's world.

This process is illustrated by data from a series of studies by Hess, Senécal, Herrera, Kirouac, Philippot, and Kleck (1998). In the first study, we asked 544 participants from both a rural and an urban area in Quebec to rate the likely emotional reactions by male and female protagonists for a series of vignettes using an emotion profile. In general, men and women agreed on the patterns of expected emotional reactions of the protagonist. Specifically, for all negative emotion situations, male protagonists were expected to react with more anger, happiness, and serenity (or not differently from women), whereas female protagonists in the same situation were expected to react with more sadness, fear, disgust, contempt, shame, and guilt, for almost all emotion situations. For the happiness situation no differences emerged. Thus, men and women share the beliefs regarding men's and women's likely reactions in a variety of emotion eliciting situations.

In a second study, we asked participants from the same population, that is, college students in Quebec, to rate the likelihood with which they themselves would react with each of the emotions in the emotion profile when confronted with the situations in the vignettes and the results were largely **congruent**. We also asked them about specific behaviors they would show in these situations, and again the results were largely congruent. Specifically, women reported more often that they would cry or isolate themselves in negative emotional situations whereas men were more likely to report to want to hit, criticize, or insult as well as to laugh and smile in the same situations.

In sum, we note than men and women are expected to react differently in general and report different emotional reactions and behaviors for themselves. In particular, women are expected to be more expressive of sadness, and men more expressive of anger.

Having established these "norms" for the participants' population, we showed to a different group a series of drawings of emotional facial expressions by men and women (Senécal, Hess, & Kleck, 1996). The expressions were identical, only the face outlines differed. As expected, the expression was rated as indicating more anger for women than for men. Conversely, men's expressions of lower intensity happiness were rated as happier than women's expressions. This finding suggests that participants used the social norms of women being less expressive of anger to adjust their estimate of a women's likely emotional state when showing an intense expression of anger upwards. In different words, they rated the expression as if taking into account the expected lower level of display intensity and concluded that a women *showing* this much anger must be angrier than a men showing the same level of anger. Similarly, women are expected to smile more frequently, even in situations where there is no emotion eliciting stimulus, thus a low intensity smile shown by a women is less informative of her emotional state than a low intensity smile in a men. No differences where found for disgust expressions and women were expected to be sadder when showing lower intensity expressions of sadness. As these expressions were quite close to a neutral expression, this may suggest that women who do not smile, are expected to be sad, given that sadness was expected for all negative emotion situations in the first two studies. In sum, these studies provide evidence that emotion norms can influence the decoding of emotional facial expressions -- at least in situations where no previous knowledge of the expressor is available. Yet, such situations occur frequently in our everyday lives when we interact with clerks in stores, receptionists and many others whom we have not met before and may not meet again.

The influence of beliefs regarding the likely expressions of men and women as well as of members of different ethnic groups is also visible when it comes to the attribution of behavioral intentions based on emotional facial expressions. Hess, Blairy, and Kleck (1998) studied this issue. They found influences of gender and ethnicity on the attribution of affiliation and dominance. These factors modulated the more important influence of emotional expressions. In general, expressions of happiness and anger were rated as high dominant and expressions of sadness and fear as low dominant. In contrast, anger expressions were rated as low affiliative. Regarding dominance ratings as a function of expressions of happiness and anger in particular, we note that both ethnicity and gender modulate the effect of emotion expression. For example, only Japanese men are rated as low dominant when showing happiness and only Japanese women are rated as low dominant when showing anger. These findings can be explained using a path analytical model including gender, ethnicity, the intensity of the expression, the rated intensity of the expression and the raters' belief that a member of the specific social group in question would show the expression. The model shown in Figure 2 illustrates well that the effects of gender and ethnicity are mediated by the expected likelihood of the expression.

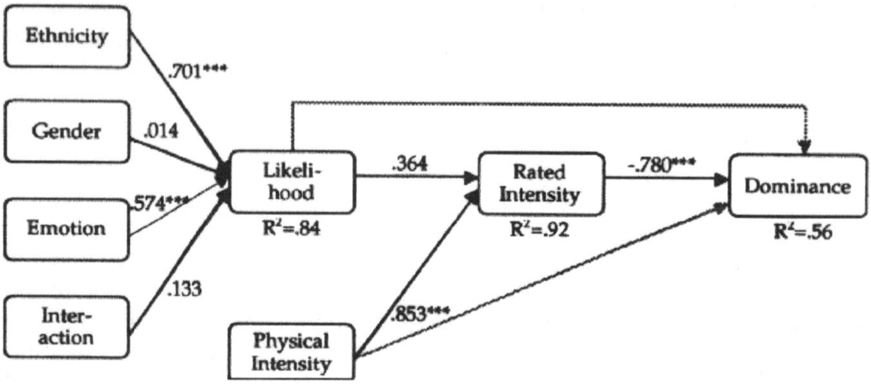

Figure 2. Model for the causal paths between Sex and Ethnicity of the stimulus person, their rated likelihood of showing the expression, the rated intensity of the expression, and dominance and affiliation ratings.

Specifically, only Ethnicity and Emotion predict Likelihood. Caucasians are rated as more likely to show the emotions and happiness is rated as more likely than anger. In the second step, Likelihood and the physical intensity of the stimulus were used to predict the rated intensity of the stimulus. Although, emotion (beta = .29, t = 2.90, p = .016) and the 3-way interaction between Gender, Emotion and Ethnicity (beta = .26, t = 2.55, p = .029) predict intensity when Likelihood is not included as a predictor, neither variable directly predicts Intensity when Likelihood is included in the equation. We can therefore conclude that the influence of these variables on Intensity is mediated by Likelihood. Finally, Intensity significantly predicts Dominance ratings. However, direct contributions by Likelihood as well as by Physical intensity of the stimulus remain, suggesting that the Dominance ratings are not fully mediated by the Intensity ratings. Specifically, we note that the more likely an expression, the more dominant is the expressor perceived. Further, more intense expressions lead to higher ratings of dominance. However, the rated intensity of the expressions is negatively related to perceived dominance. This may be explained when considering the indirect effect of emotion on intensity. In this data set, the emotional expressions of anger were rated as less intense in general but lead to higher ratings of dominance. Thus for dominance, both the physical intensity of the expression and the rated likelihood that the expression is shown by a member of a specific social group, influence dominance ratings.

4. The Active Decoder

So far we have considered the observer from a fairly passive perspective. That is, we have only considered observers as decoders and not as potential interaction partners. Yet, in everyday interactions listeners in a conversation have a much more active role, providing feedback on both their interest and their understanding. In the context of the communication of emotions this implies that observers in a real interaction need to provide some sort of feedback regarding their understanding of the emotional message. Or, to say it differently, they need to signal empathy.

In recent years, two processes associated with the decoding of emotional expressions and the process of empathy have been studied in more detail, emotional contagion and mimicry – the imitation of the emotional behavior of others. The literature on this issue generally concludes that mimicry is a common response to observed emotional expressions in both infants and adults (see Hess, Philippot, & Blairy, in press). Also, evidence for contagion effects is frequently reported. In the clinical literature mimicry has been associated with a better understanding of the patient and some approaches actively encourage therapists to mimic their patients as a means to enhance their empathy. This notion goes back to Lipps (1907) who proposed that imitation is a possible means to the understanding of other selves and was taken up by Freud (1921).

In a recent review, Hess, Blairy and Philippot (in press) conclude that mimicry does not in fact enhance decoding accuracy. However, mimicry seems to play an important role in the establishment of mutual liking and rapport in an interaction.

Conversely, there is evidence that mimicry may depend on the attitudes that the observer holds towards the expressor. That is, mimicry occurs mainly in situations where observers already have positive or at least neutral attitudes toward the target and where they perceive some similarity between themselves and the target. Thus, mimicry may not so much "create" rapport and mutual liking as reinforce it in situations where the basis for such feelings are already present.

Two recent studies (Hess, 1998; Philippot & Yabar, 1998) show that individuals who hold negative attitudes towards members of a different ethnic group tend to not react with mimicry to the facial displays of ethnic out-group members. For example, Herrera, Bourgeois, and Hess (1998), asked French Canadian raters to decode the emotional facial expressions of Japanese and

Caucasian expressors. The expressions were taken from the JACFEE (Matsumoto & Ekman, 1988) and are equivalent in terms of the facial displays. The results show that French Canadian raters mimic expressions of happiness and anger when shown by Caucasian actors but not when shown by Japanese actors. In fact, the facial displays of the French Canadian decoders while ratings the Japanese expressions depended on their racial attitude. The less positive their attitude towards Asians in general the more they tended to smile (as assessed by *Orbicularis Oculi* activity) at low dominance expressions (sadness and fear) and to frown (as assessed by *Corrugator Supercilii* activity) at high dominance expressions (happiness and anger) by the Japanese actors -- a counter mimicry effect. These findings regarding counter mimicry effects suggest that individuals who hold negative attitudes toward others tend to reinforce this negative relational feeling by showing non-matching behaviors which are likely to be perceived as lack of empathy or rapport. As the feeling of being understood by the interaction partner is an important aspect of comfortable interactions, subtle non voluntary nonverbal signals of non understanding are likely to create uncomfortable interactions that the interaction partners may want to avoid in the future. Thus, stereotypes and attitudes that conceive of others as dissimilar and lacking in understanding may reinforce themselves in a subtle way that may be difficult to compensate for.

In summary, the communication of emotions is influenced by a number of social context variables such as sex and cultural background of the interaction partners. This influences pertain not only to the encoding of emotion displays but also to the decoding of these displays. Further, the facial affect displays of the decoder - which when congruent tend to signal empathy - are as well influenced by the relationship between encoder and decoder.

References

Aune, K. S., D. B. Buller and R. K. Aune (1996) "Display rule development in romantic relationships: Emotion management and perceived appropriateness of emotions across relationship stages", *Human Communication Research* 23:115-145.

Averill, J. R. (1997) "The emotions: An integrative approach", in: *Handbook of Personality Psychology*, R. Hogan, J. A. Johnson and S. R. Briggs, eds, San Diego, CA: Academic Press, pp. 513-541.

Bavelas, J. B., A. Black, C. R. Lemery and J. Mullett (1986) "'I show how you feel': Motor mimicry as a communicative act", *Journal of Personality and Social Psychology* 50:322-329.

Bühler, K. (1934) *Sprachtheorie*, Jena: Fischer.

Cacioppo, J. T., R. E. Petty, M. E. Losch and H. S. Kim (1986) "Electromyographic activity over facial muscle regions can discriminate the valence and intensity of affective reactions", *Journal of Personality and Social Psychology* 50:260-268.

Colvin, C. R., D. Vogt and W. Ickes (1997) "Why do friends understand each other better than strangers do?", in: *Empathic Accuracy*, W. Ickes, ed., New York, NY: Guilford Press, pp. 168-193.

Darwin, C. (1872/1965) *The Expression of the Emotions in Man and Animals*, Chicago: The University of Chicago Press (Originally published, 1872).

Ekman, P. (1973) *Darwin and Facial Expression: A Century of Research in Review*, New York: Academic Press.

Ekman, P. (1979) "About brows: Emotional and conversational signs", in: *Human Ethology*, K. F. M. von Cranach, W. Lepenies and D. Ploog, ed., Cambridge: Cambridge University Press, pp. 169-249.

Ekman, P., W. V. Friesen, M. O'Sullivan, A. Chan, I. Diacoyanni-Tarlatzis, K. Heider, R. Krause, W. A. LeCompte, T. Pitcairn, P. E. Ricci-Bitti, K. Scherer, M. Tomita and A. Tzavaras (1987) "Universals and cultural differences in the judgements of facial expressions of emotion", *Journal of Personality and Social Psychology* 53:712-717.

Fernandez-Dols, J.-M. and M.-A. Ruiz-Belda (1995) "Are smiles a sign of happiness? Gold medal winners at the Olympic Games", *Journal of Personality and Social Psychology* 69:1113-1119.

Fischer, A. (1993) "Sex differences in emotionality: Fact or Stereotype?", *Feminism and Psychology* 3:303-318.

Freud, S. (1955) "Group psychology and the analysis of the ego", in: *The Standard Edition of the Complete Psychological Works of Sigmund Freud (Vol. 18)*, J. Strachey, ed. and trans., London: Hograth Press (Original work published 1921).

Fridlund, A. J. (1991) "The sociality of solitary smiling: Potentiation by an implicit audience", *Journal of Personality and Social Psychology* 60:229-240.

Fridlund, A. J., P. Ekman and H. Oster (1987) "Facial expressions of emotion: Review of the literature, 1970-1983", in: *Nonverbal Behavior and Communication*, A. W. Siegman and S. Feldstein, eds, Hillsdale, NJ: Lawrence Erlbaum Associates, pp. 143-224.

Frijda, N., S. Markam, K. Sato and R. Wiers (1995) "Emotions and emotion words", in: *Everyday Conceptions of Emotion*, J. A. Russel, J.-M. Fernández-Dols, A. S. R. Manstead and J.C. Wellenkamp, eds, Dordrecht: Kluwer, pp. 121-143.

Frijda, N. H. and A. Tcherkassof (1997) "Facial expressions as modes of action readiness", in: *The Psychology of Facial Expression*, J. A. Russel and J. M. Fernandez-Dols, eds, New York: Cambridge University Press, pp. 78-102.

Herrera, P., P. Bourgois and U. Hess (1998) "Counter mimicry effects as a function of racial attitudes", *Psychophysiology* 35.

Hess, U. (1998) "Mimicry: Fact and Fiction", Paper presented at the the *Xth Conference of the International Society for Research on Emotion*, Würzburg, Germany, August 4-8.

Hess, U., S. Blairy and R. E. Kleck (1997) "The relationship between the intensity of emotional facial expressions and observers' decoding", *Journal of Nonverbal Behavior* 21:241-257.

Hess, U., S. Blairy and R. E. Kleck (1998) "The influence of expression intensity, gender, and ethnicity on judgments of dominance and affiliation", Manuscript submitted for publication.

Hess, U., S. Blairy and P. Philippot (1998) "Facial reactions to emotional facial expressions: Affect or cognition?", *Cognition and Emotion* 12:509-532.

Hess, U., S. Blairy and P. Philippot (in press) "Facial Mimicry", in: *The Social Context of Nonverbal Behavior*, P. Philippot, R. Feldman and E. Coats, eds, Cambridge University Press.

Hess, U., A. Kappas and R. Banse, (1995) "The intensity of facial expressions is determined by underlying affective state and social situation", *Journal of Personality and Social Psychology* 69:280-288.

Hess, U., S. Senécal, P. Herrera, K. Kirouac, P. Philippot and R. E. Kleck (1998) "Emotional expressivity in men and women: general and personal norms", Manuscript submitted for publication.

Hochschild, A. R. (1979) "Emotion work, feeling rules, and social structure", *American Journal of Sociology* 85:551-573.

Itakura, S. (1994) "Differentiated responses to different human conditions in chimpanzees", *Perceptual and Motor Skills* 79:1288-1290

Jakobs, E., A. H. Fischer and A. S. R Manstead (1997) "Emotional experience as a function of social context: The role of the other", *Journal of Nonverbal Behavior* 21:103-130.

Kraut, R. E. and R. E. Johnston (1979) "Social and emotional messages of smiling: An ethological approach", *Journal of Personality and Social Psychology* 42:853-863.

Lazarus, R. R. (1991) *Emotion and Adaptation*, New York: Oxford University Press.

Linnankoski, I., M. L Laasko and L. Leinonen (1994) "Recognition of emotions in macaque vocalizations by children and adults", *Language and Communication* 14:183-192.

Lipps, T. (1907) "Das Wissen von fremden Ichen", in: *Psychologische Untersuchungen, Band 1*, T. Lipps, ed., Leipzig: Engelmann, pp. 694-722.

Matsumoto, D. (1996) *Culture and Psychology*, Pacific Grove, CA: Brooks/Cole Publishing Company.

Matsumoto, D. and M. Assar (1992) "The effects of language on judgments of universal facial expressions of emotion", *Journal of Nonverbal Behavior* 16:85-99.

Matsumoto, D. and P. Ekman (1988) *Japanese and Caucasian Facial Expressions of Emotion (JACFEE) and Neutral Faces (JACNeuf)*, San Francisco: San Francisco State University.

Motley, M. T. (1993) "Facial affect and verbal context in conversation: Facial expression as interjection", *Human Communication Research* 20:3-40.

Noller, P. and M. Ruzzene (1991) "Communication in marriage: The influence of affect and cognition", in: *Cognition in Close Relationships*, G. J. O. Fletcher and F. D. Fincham, eds, Hillsdale, NJ: Lawrence Erlbaum Associates, pp. 203-233.

Philippot, P. and Y. Yabar (1998) "Friends or Foes? Facial Reactions to In- and Out-Group Members", Paper presented at the the *Xth Conference of the International Society for Research on Emotion*, Würzburg, Germany, August 4-8.

Pope, L. K. and C. A. Smith (1994) "On the distinct meanings of smiles and frowns", *Cognition and Emotion* 8:65-72.

Rosenberg, E. L. and P. Ekman (1994) "Coherence between expressive and experiential systems in emotion", *Cognition and Emotion* 8:201-229.

Senécal, S., U. Hess and R. E. Kleck (1996) "The influence of gender stereotypes on the decoding of emotional facial expressions", Poster presented at the *26th International Congress of Psychology*, Montreal, 16-21 August.

Suh, E., E. Diener, S. Oishi and H. C. Triandis (1998) "The shifting basis of life satisfaction judgments across cultures - emotions versus norms", *Journal of Personality and Social Psychology* 74:482-493.

Zeman, J. and J. Garber (1996) "Display rules for anger, sadness, and pain: It depends on who is watching", *Child Development* 67:957-973.

THE PERCEPTION OF HUMOR

WILLIBALD RUCH
Department of Physiological Psychology, Heinrich-Heine-University of Düsseldorf, Universitätsstraße 1, 40225 Düsseldorf, Germany

ABSTRACT

The chapter describes the humor response; i.e., the perception that something is funny as a distinct qualia. Cognitive models of humor are presented that define the minimal conditions for humor. The relevance of detection of incongruity and its resolution, but also the limitations of this two-stage model are discussed. Techniques for the experimental variation of key ingredients of humor are presented and a major paradigm, the weight-judging task, is introduced. The ascribed role of surprise as a mediator between perceiving and enjoying incongruity is examined. It is argued that the perception of humor is highly individualistic, and the presented summary of findings based on the 3 WD humor test suggests that humor appreciation is primarily affected by personality traits that refer to the individual's general tendency to seek out or avoid stimulus uncertainty. Due to its complexity and central role in human life, the study of humor is recommended for the study of consciousness.

1. Introduction

The perception that something is "funny" is quite a distinct quality in the flowing stream of consciousness (Dennett, 1991). Most experiences we have share the common denominator that we take the information we receive as "serious" and we prepare and carry out appropriate reactions. With a few exceptions, such as when we suspect somebody is lying, or when a news item is biased or too unlikely to be true, we do not doubt the truth value of the communication. Other than these cases we assume cooperation in the communication on the side of the sender. If the information we receive is seen to be new, we want to store it permanently for further use. The information might be important and require an immediate and appropriate response. This is different when we perceive humor: we know this is play, a play with ideas. There is no need to upgrade our knowledge system as the information we received only has an "as if"-truth, it is playing with sense and nonsense. If we are a frequent joke teller, we might be interested in adding the complete joke to our repertoire or adopt its technique, but even if a humorous message touches areas we are serious about or it matches our attitudes we want to keep it separate from what we believe to be literally true. Humor might involve content of high personal significance and its activation during serious interaction would be echoed by an orchestrated set of responses; however, as long as we take the message humorously, we are not required to perform them: we can be passively amused or reciprocate by continuing the humorous interaction.

Although the "humor response" (McGhee 1971 coined this expression denoting the perception that something is funny) is a unique experiential quality, it is not a primary quality of one single object which we perceive directly but involves a comparison. When we say that something is funny we have experienced an incongruity between objects, between elements of an object or between an event and an expectation. This contrast is also reflected in the fact that the terms (e.g., funny, comical) we use to denote the perceived properties of stimuli causing us to engage in such playful processing of incongruity have also second meanings referring to the unusual (e.g. "peculiar," "strange," or "odd") as well as to the suspicious ("There was something funny about these extra charges"). Incongruity often comes unexpectedly and the perception of humor may involve surprise; indeed, there is overlap between surprise and humor as regards both the eliciting stimuli and the information processes preceding these qualia (Meyer, Reisenzein & Schützwohl, 1997).

2. What is Humor?

There is not yet an agreed-upon terminology in humor research. Rather there are different and conflicting terminological systems and two should be mentioned as they assign the key term "humor" different roles. One historical nomenclature stems from the field of aesthetics where *the comic* (or: *the funny*) — defined as the faculty able to make one laugh or to amuse — is distinguished from other aesthetic qualities, such as beauty, harmony, or the tragic. Humor is simply *one* element of the comic — as are wit, fun, nonsense, sarcasm, ridicule, satire, or irony — and basically denotes a smiling attitude toward life and its imperfections: an understanding of the incongruities of existence. Humor in this narrow sense was seen to be based on a sympathetic heart, not on a superior spirit (like wit), moral sense or even haughtiness (like mock/ridicule), or vitality/high spirits (like fun). The other major terminological system, largely endorsed by current Anglo-American writers, uses humor as the umbrella-term for *all* phenomena of this field. Thus, humor replaced *the comic* and is treated as a neutral term; i.e., not restricted to positive occasions for laughter.

Schmidt-Hidding (1963), endorsing the former terminology, described eight styles of the comic/funny according to seven distinguishing features (such as major intention/goal of the comic message, the ideal audience, subject of the comic, attitude of the agent, linguistic features) each of which he summarized from the literature. Based on a factor analysis of these descriptions (Ruch, 1997), one can assume that the field of the comic is at least two-dimensional, with the eight styles forming an arc of about 110° between the factors denoting the more benevolent ("laughing with") and skeptical ("laughing at") comic styles (see Figure 1). Based on the correlation of personality scales with these two factors, one can conclude that the comic styles associated with the first factor are more likely to be performed by individuals prone to positive affect and cheerfulness, whereas the "laughing at"-styles are more typical for individuals scoring high in scales of negative affect and bad mood. Rotating the axes by 90° yields a coordinate system that separates the mental/cognitive from the affective/motivational elements. Thus, the capacity for perceiving incongruity and creating a funny effect might be blended with both benevolence and malevolence; i.e., it may serve to portray human weaknesses in general in an accepting way or to poke fun at the weaknesses of a particular person.

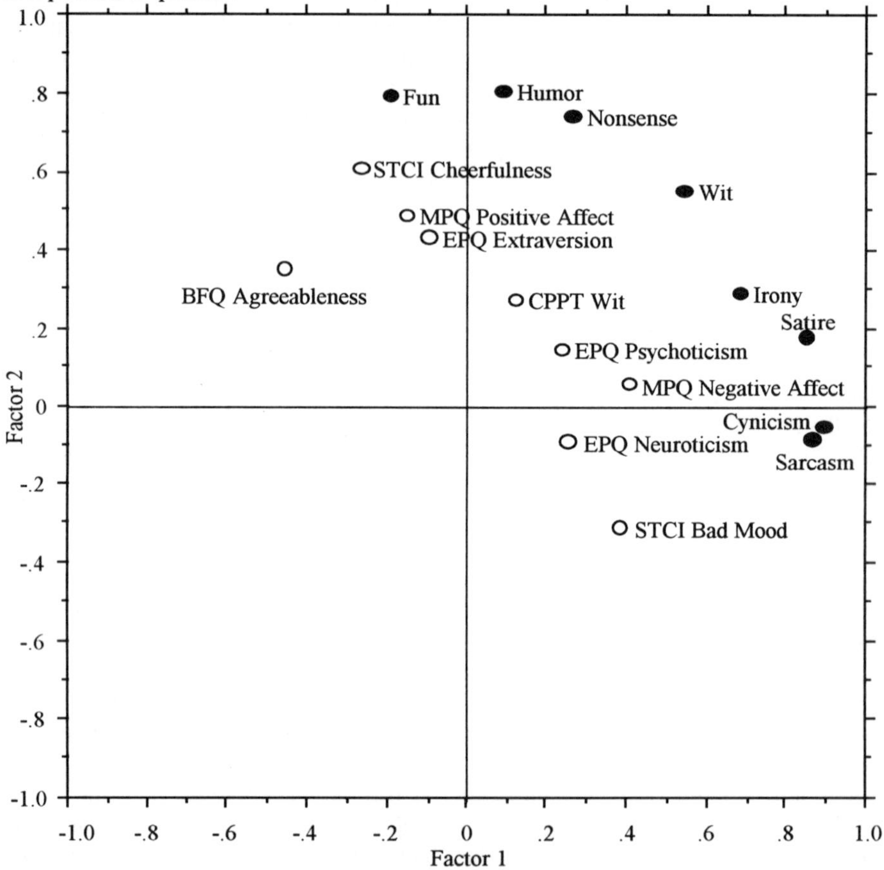

Figure 1. A two-dimensional space of comic styles and the location of personality scales (N = 106).

Similarly, a factor analysis of a comprehensive list of German humor-related nouns (e.g., joker, wit, cynic) resulted in a highly similar two-dimensional space (Ruch, 1997). As also antonyms (i.e., nouns related to humorlessness) were included, the circumplex structure found was more elaborated covering an arc of about 280°. Again, one possible position of the axes separated affective (good vs. bad humoredness) from mental (sensibleness vs. nonsense) elements.

Thus humor (in the broad sense, as I will use it from here on) should be seen as a multifaceted phenomenon hardly to be accounted for by any single theory or to be examined by a single paradigm. Despite the fact that humor might be blended with different motives and may involve almost all topics of relevance to humans, many humor researchers believe in a cognitive nucleus of the funny experience suggesting that the mental dimension described above is the more basic one. They developed models that describe minimal conditions that are necessary to elicit a humor response; i.e., the perception that something is funny. In my further treatment of the perception of humor I will restrict myself to this cognitive dimension.

3. Theories of Humor Appreciation

Numerous theories have been proposed to explain the perceived funniness of humor, with cognitive approaches being the most prominent together with arousal and superiority theories (for a review, see Keith-Spiegel, 1972). Cognitive theories typically analyze the structural properties of humorous stimuli or the way they are processed; sometimes these two levels are also mixed up.

Perhaps beginning with Aristotle, incongruity was considered to be a necessary condition for humor. In this tradition, humor involves the bringing together of two normally disparate ideas, concepts, or situations in a surprising or unexpected manner. Koestler (1964) coined the term "bisociation" to refer to the juxtaposition of two normally incongruous frames of reference, or the discovery of various similarities or analogies implicit in concepts normally considered remote from each other. Although there is general agreement that incongruity is a necessary condition for humor, it was argued that it is not a sufficient one. Sheer incongruity may also lead to puzzlement and even to aversive reactions. To account for this, such variables as the resolution of the incongruity (Suls, 1972), the acceptance of unresolvable incongruity, or the "safeness" of the context in which the incongruity is processed (Rothbart, 1976) have been proposed. Rothbart and Pien (1977) suggested to distinguish between possible and impossible incongruities and between complete and incomplete resolutions. They argue that only possible incongruities can be resolved completely while for an impossible incongruity only a partial resolution is possible and a residue of incongruity is left.

A precise description of what makes a text funny is provided by linguists. Raskin (1985) presented in detail the first formal semantic theory of jokes which — due to its reliance on the concept of "script" (a structured chunk of information about lexemes and/or parts of the world) — became known as the Semantic Script Theory of Humor (SSTH). The SSTH can be summarized as two necessary and sufficient conditions. A text is funny if and only if both of the two conditions obtain: (i) the text is compatible, fully or in part, with two distinct scripts; and (ii) the two distinct scripts are opposite (i.e., the negation of each other, if only for the purpose of a given text), following a list of basic oppositions, such as real/unreal, possible/impossible, etc. For example, Raskin's prototypical joke ("Is the doctor at home?" the patient asked in his bronchial whisper. "No," the doctor's young and pretty wife whispered in reply. "Come right in.") is compatible with the two scripts "doctor" and "lover" and the scripts are opposite on the sex vs. non-sex basis (for an elaborated interpretation see Raskin, 1985). A revision of the SSTH led to the General Theory of Verbal Humor (GTVH, see, for example, Attardo, 1993) which introduced, besides scripts, five other "knowledge resources", that must be tapped into when generating a joke, namely script opposition, logical mechanism, target, narrative strategy, language, and situation.

How are humorous texts processed? Models typically described two distinct albeit different stages or recursive processes. For Kant laughter is "... an affection arising from the sudden transformation of a strained expectation into nothing". In other words, that which is originally perceived in one (often serious) sense is suddenly viewed from a totally different (usually implausible or ludicrous) perspective. Eysenck (1942; p. 307) suggested that "laughter results from the sudden, insightful integration of contradictory or incongruous ideas, attitudes, or sentiments which are experienced objectively." Based on prior work, Suls (1972) presented a two-stage model of humor appreciation that emphasized the understanding aspect. According to this model, the perceiver must proceed through these two stages to find a joke or cartoon funny. In the first stage, "... the perceiver finds his expectation about the text disconfirmed by the ending of the joke ... In other words, the recipient encounters an incongruity — the punchline. In the second stage, the perceiver engages in a form of problem solving to find a cognitive rule which makes the punchline follow from the main part of the joke and reconciles the incongruous parts." (p. 82).

In the doctor's wife joke above, the ending ("come right in") is incongruous, as it does not readily follow the prior "no" and it is not supplemented by a statement to the effect that the patient was welcome to wait for the doctor's return. Thus it does not make sense for her to invite him in. However, the hints *young* and *pretty* help the recipient to reinterpret the text along the lines that not the doctors' patient, but his wife's lover is knocking on the door, and suddenly the ending (including the wife's unexplained whispering) makes sense and follows from the joke body.

Neurological evidence suggests that the two brain hemispheres might be differently involved in the two stages, with the left hemisphere "setting up" the joke and the right "getting it" (see McGhee, 1983). Studying event-related brain potentials in response to jokes (where the last word, containing the incongruity, was presented after the joke body to be used as a trigger point), Derks, Gillikin, Bartolome and Bogart (1997) found that jokes that did not elicit laughter showed no evidence of a N400 while those that elicited laughter showed a negative wave at about 400ms presumably representing the incongruity-triggered disruption and possible extension of the categorization process. However, the activity was present in both hemispheres of the cortex.

According to the model there are two possible outcomes of the second stage: laughter (if the rule is found) or puzzlement (if it is not found). While the latter is plausible, the former has been doubted. Why should the resolution immediately lead to laugher? Having borrowed the flow chart of a problem solving computer program, this model can not go much beyond seeing humor as being a problem solving activity. While the model described the comprehension part well, it does not explain appreciation. It is likely that the cognitive processes continue after resolving the incongruity. Unlike after real problem solving, the recipient is aware that the fit of the solution is an "as if"-fit. Already Lipps (1898) noted that what makes sense for a moment is subsequently abandoned as not really making sense. At a meta-level we experience that we have been fooled; our ability to make sense, to solve problems, has been misused. Thus, in particular for the impossible incongruities and their partial resolution, the two-step (i.e., step I: detection of incongruity or violation of a build-up expectation; step II: resolution of incongruity) model needs to be expanded to include a third stage of detecting that what makes sense is actually nonsense[1]. This third stage then allows to distinguish between joke processing and mere problem solving. If the processes indeed would end with the resolution of the incongruity, we would not be able to distinguish whether we just resolved a problem (as in riddles) or whether we processed humor. We would believe in the outcome of the problem solving activity — that it has truth-value. Some authors postulated even further oscillations between the two interpretations of the text or two perspectives involved; like playing with sense and nonsense (for conflict or ambivalence theories, see Keith-Spiegel, 1972).

The problem solving aspect might also be peripheral, as we might respond more to the connotative elements involved; e.g., in the joke above some might experience a rapid succession of one's sympathy for a patient in pain and one's feelings towards adultery. Furthermore, Derks, Staley and Haselton (1998) rightfully raised the question whether joke comprehension is so demanding that it has a problem solving quality. Based on their results they suggest that perceiving humor is more an automated expert-like behavior. However, the "mastery" studies show inverted-u functions between children's development, complexity of jokes and appreciation (McGhee, 1979) and intelligence predicts humor (Ruch, 1992).

So far little research attention has been paid to the temporal characteristics of the perception of humor; for example, wit is quick, in jokes there is a sudden manifestation of the incongruous, while in humorous stories there might be a gradual realization of the incongruous. Thus, the perception of humor may differ in intensity, duration and its form over time. This, in turn, suggests variations in the techniques and material. Finally, humor may involve different modes; for example, it

---

[1] Thus, while both Suls and Lipps propose two-stage models, these models are different as the incongruity-resolution model covers stages one and two and Lipps' model refers to stages two and three.

can be verbal (e.g., jokes), graphical (cartoons, caricatures), acoustical (funny music), or behavioral (e.g., pantomime), again making matters very complex. So far, the scope of most theories is limited to the analysis of jokes and cartoons.

3.1. Traditional and Nontraditional Vehicles for Transporting Funny Incongruity

The empirical test of the validity of the theoretical models is contingent on the variation of the key ingredients (e.g., degree of incongruity, resolution, salience of contents); however they cannot be varied independently of each other by manipulating a joke or cartoon. For example, making the punch line more incongruous simultaneously means changing its content or other properties. One way out is, for example, to leave the jokes intact, but undertake a differential priming of the two meanings of a key word in a joke (Wilson, 1979), or a priming of the theme (Goldstein, Suls & Anthony, 1972) or even the structure (Derks & Arora, 1993) of the jokes to follow.

Another possibility is the use of artificial humor stimuli. This may take, for example, the form of sequences of words deviating from proper grammatical sequences (Ehrenstein & Ertel, 1978), adjective-noun pairs varying in semantic distance (Godkewitsch, 1974), a domains-interaction approach (Hillson & Martin, 1994) or computer-drawn caricatures with various degrees of exaggeration (Rhodes, Brennan & Carey, 1987). Such studies typically demonstrate the importance of an intermediate degree of incongruity.

A sophisticated way to vary the degree of incongruity is to implement it in a psychophysical task. This became known as the weight-judging paradigm (WJP) in humor (see Deckers, 1993). Participants are instructed to make a series of judgments between a pair of weights. The first weight of the pair remains the same and is referred to as the standard (S) (see Figure 2). The second weight of the pair is the comparison (C). An experiment might involve 6 to 12 comparisons presented by means of a "lazy Susan"-type of circular tray. The subjects lifts the standard and then one of the comparisons and must subsequently indicate whether C is heavier or lighter than S. The weights vary in heaviness with half of them being lighter and half of them being heavier than the standard. The cylinders containing the weights are visually indistinguishable. The expected weight is the mean of the weights lifted so far. Incongruity is operationally defined as the difference between the weight of the critical comparison (CC) and the expectation (i.e., the standard).

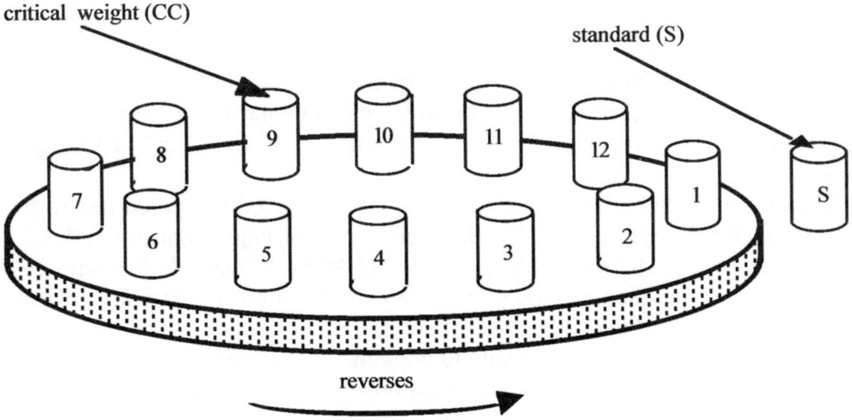

Figure 2. Arrangement of cylinders in a weight-judging task.

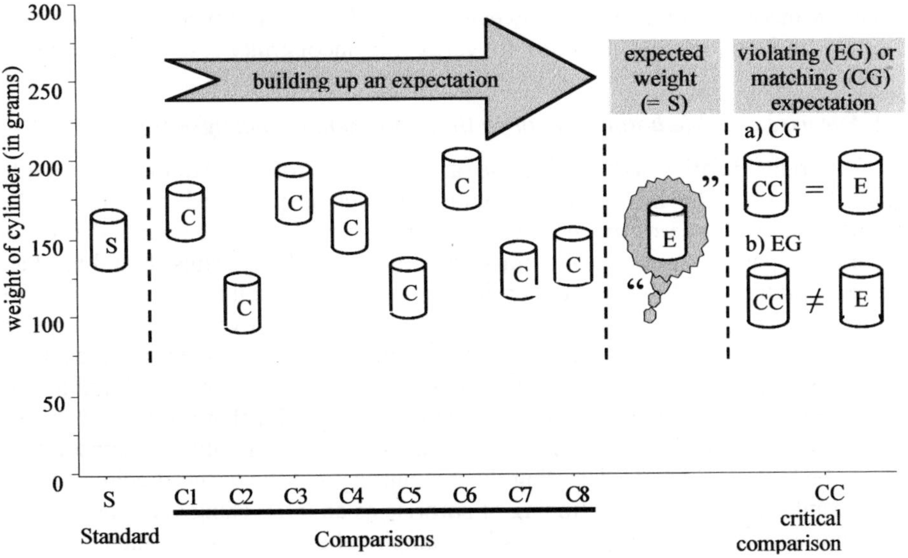

Figure 3. The two stages in the set-up of the weight-judging task: building up an expectation about the weight of the comparison and violating this expectation to different degrees.

For example, in the study by Ruch, Köhler and van Thriel (1999) the weight of the eight comparisons varied between 110 and 140, and 160 and 190 grams (in intervals of 10 grams) with the standard being 150 grams (Figure 3). These eight trials are used to build up an expectation about the weights. In the control group the critical comparison was 150 grams yielding the "expected" outcome; this is equivalent to a congruous ending of a story (or a joke where the incongruity is not perceived by the recipient). In the experimental groups, the CC was either 600, 1000, or 1600 grams creating an incongruity by violating the expectations; these differences from expectation (CC - S) being 450, 850, and 1450 grams, respectively.

As in prior studies, the degree of incongruity affected the perceived surprise, funniness, and amusement, but also the facial signs of enjoyment; i.e., AU12 and AU6 in the *Facial Action Coding System* (FACS; Ekman & Friesen, 1978). However, in WJP-studies no downward turn of the curve of intensity of amusement can be found for higher degrees of incongruity.

While these methods seem to validly implement incongruity into nontraditional humor stimuli and circumvent problems emerging with the manipulation of traditional humor, so far little is known about how they overlap among each other and with traditional humor stimuli. For example, much to their surprise, Ruch and Köhler (1998) report that enjoyment induced by the WJP is correlated with finding incongruity-resolution humor funny and nonsense humor (humor with left over traces of incongruity) aversive; however, the WJP was introduced as a paradigm for the study of incongruity, not to verify the incongruity-resolution model of humor.

3.2. Surprise as a Mediating Factor in the Enjoyment of Incongruity?

Different theories assign surprise a prominent role in humor (see Keith-Spiegel, 1972). The incongruity-resolution model as proposed by Suls (1972) suggests that "degree of incongruity is directly related to the amount of surprise experienced, and the amount of surprise that the punch line creates should produce a corresponding need to solve the problem. ... The prediction here is that the more surprising the punch line, the more one should want to overcome the surprise. When the problem is solved, the recipient should experience greater appreciation". (Suls, 1972; p. 91). This leads to the prediction that the perception of humor will be blended with the subjective experience of surprise. Furthermore, as there are physiological and behavioral indicators of surprise (or the orienting response), they should be located between detecting the incongruity and the onset of enjoyment.

Despite the frequent mentioning of surprise as an intervening variable involved in humor appreciation, so far there is little empirical evidence for its manifestation. Most of the evidence available stems from verbal ratings of surprise. These data, however, show that subjects report elevated degrees of surprise and the expected positive correlation between surprisingness and funniness (Deckers, 1983, Ruch et al., 1999). In a pilot study using a weight-judging task, six out of 24 amused participants displayed facial signs of surprise before a smile. As the observed raising of the eyebrows (AU1 and AU2 in the FACS) was of low intensity, in the subsequent study we utilized facial electromyography (EMG) to be able to detect even slightest contractions of the frontalis muscle[2] in response to incongruities transported via a WJP, 20 jokes presented auditorily, and a "jack-in-the-box"-gag. While surprise and amusement were the most intensive feelings induced, and the amount of felt surprise and amusement were positively correlated, frontalis actions were infrequent and only rarely preceded a smile or laugh in response to jokes (4.11%) and when lifting the incongruous weight (17.85%) in the defined time slot[3]. Also, there were frontalis actions concordant with the incongruity (and felt surprise) but not followed by a joint contraction of the zygomatic major and orbicularis oculi muscles (and felt amusement). Nevertheless, patterns like the one displayed in Figure 4 also emerged.

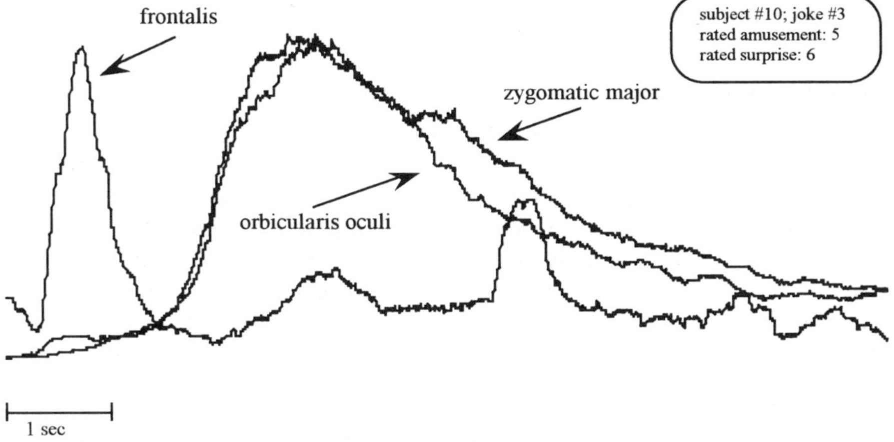

Figure 4. EMG-recordings from the frontalis (surprise) and orbicularis oculi and zygomatic major muscles (enjoyment) of a research participant reporting to be amused and surprised by an auditorily presented joke.

In such cases a facial indicator of surprise preceded the enjoyment display supporting the view that surprise occurs between the detection of the incongruity and the release of the positive affect. Figure 4 shows a steep onset in frontalis activity before the ballistic and highly coordinated contraction of the zygomatic and orbicularis oculi muscles. Note that while the surprise response has returned to baseline after one second, the intensity of the amusement expression is still increasing.

The rare occurrence of surprise casts doubt on the assumption that humor generally induces a sequence of emotions, namely surprise first (as a function of incon-

[2]Facial-EMG was recorded from the zygomatic, corrugator, frontalis, and obicularis oculi muscle regions using the recommended electrode placements (Fridlund & Cacioppo, 1986). The digital data collection was accomplished under the control of an Apple Macintosh Quadra 950, an electromyogram amplifier module (EMG 100 A, BIOPAC Systems, Inc.), and a data collection software (AcqKnowledge 2.1).

[3]While the "Jack-in-the-box" - gag most reliably induced frontalis activity (56.41%), it may be that some of them were overlaid by a startle response.

gruity) and enjoyment second (as a function of its resolution). Thus, facial criteria do not support the equating of the type of surprise that is posited as an intervening variable in models of humor appreciation with the emotion of surprise as it is understood by emotion theory. Also, it might be that the element of suddenness is missing in the jokes and the WJP (but not in the jack-in-the-box gag). These findings also draw attention to the above mentioned different time course of incongruity that is neglected most of the time. In some humor there is a sudden manifestation of the incongruous, while in humorous stories there might be a gradual realization of the incongruous. Also, we have to consider that the experienced recipient is anticipating an incongruous ending (even if not anticipating the exact punch line) which should reduce the amount of surprise.

4. Individual Differences

Humor is in the eye of the beholder and thus the identification of those variables that affect the perception of humor is necessary. Why is it that somebody finds a joke absolutely hilarious, the next considers it boring and still another one embarrassing? Many studies have set out to investigate the questions of "what is funny to whom and why" and enriched our understanding of both humor and personality. In fact, the study of humor appreciation has even been seen to be a means to understand the individual, and it has been suggested that humor inventories might be used as an "objective" test of personality (Cattell & Tollefson, 1966).

4.1. A Taxonomy of Jokes and Cartoons

What aspects are reflected in individual differences in the perception of humor? Naturally, as content and structure have to be distinguished as two different sources of pleasure in humor, one would assume that both are also pivotal in producing individual differences. While intuitive and rational taxonomies typically distinguish only between content classes, factor analytic studies show that structural properties of jokes and cartoons are at least as important as their content, with two factors consistently appearing: namely, incongruity-resolution (INC-RES) humor and nonsense (NON) humor. Jokes and cartoons of these factors have different content (e.g., themes, targets) but are similar with respect to the structural properties and the way they are processed.

Jokes and cartoons of the INC-RES humor category are characterized by punch lines in which the surprising incongruity can be completely resolved. The common element in this type of humor is that the recipient first discovers an incongruity which is then fully resolvable upon consideration of information available elsewhere in the joke or cartoon. There is a certain projective element in these jokes as essential things are not spelled out and have to be supplemented by the recipient; often resolving the incongruity requires attributing motives and traits (e.g., stingy, mean, stupid, absent-minded) to the characters depicted in the jokes. Although individuals might differ with respect to how they perceive and/or resolve the incongruity, they have the sense of having "gotten the point" or understood the joke once resolution information has been identified. At the time this factor was first extracted, it seemed that the two-stage structure in the process of perceiving and understanding humor described by Suls (1972) is a model that fits well to these jokes and cartoons, and hence incongruity-resolution humor was considered to be an appropriate label for that factor.

Nonsense humor also has a surprising or incongruous punch line, however, "... the punch line may 1) provide no resolution at all, 2) provide a partial resolution (leaving an essential part of the incongruity unresolved), or 3) actually create new absurdities or incongruities" (McGhee, Ruch & Hehl 1990; p. 124). In nonsense humor the resolution information gives the appearance of making sense out of incongruities without actually doing so. The recipient's ability to make sense or to solve problems is exploited; after detecting the incongruity he is misled to resolve it, only to later discover that what made sense for a moment is not really making sense. Rothbart and Pien's (1977) impossible incongruities that allow only for partial resolutions are characteristic of the nonsense factor, while their possible incongruities allowing for complete resolutions are more prevalent in INC-RES humor.

While both the incongruity-resolution and the nonsense structure can be the basis for harmless as well as tendentious content, only few contents seem to be salient enough to form independent factors. The pool of jokes and cartoons we analyzed contained different content areas (including aggression), but only sexual humor formed a robust factor overpowering the structure variance. These factors were first extracted in studies of Austrian samples and later replicated in Belgium, England, France, Germany, Israel, Italy, and Turkey (for a review of the development and validation of the taxonomy see Ruch, 1992; Ruch & Hehl, 1998). The 3 WD humor test (Ruch, 1995) was constructed to assess funniness and aversiveness of these three types of humor.

4.2. Humor Appreciation and Personality

The rationale for the prediction of personality correlates of humor appreciation was primarily based on the fact that the two humor structures mainly differ with respect to the degree of resolution obtained: in incongruity-resolution humor a complete resolution of the incongruity is possible while there are residual traces of incongruity in nonsense humor. Thus, in INC-RES the resolution of incongruity contributes to appreciation whereas in NON appreciation is based on the existence of residual incongruity. This consideration and evidence from other sources (see Ruch, 1992) led to the hypotheses that appreciation of the incongruity-resolution structure is a manifestation of a broader need of individuals for contact with structured, stable, unambiguous forms of stimulation, whereas appreciation of the nonsense structure in humor reflects a generalized need for uncertain, unpredictable, and ambiguous stimuli.

According to Wilson's (1973) dynamic theory of conservatism this trait reflects a *generalized fear of both stimulus and response uncertainty*. This should lead more conservative individuals to show greater avoidance and dislike of novel, complex, unfamiliar, incongruous events and to prefer and seek out stimuli which are simpler, more familiar and congruent. This hypothesis was validated for visual art, poetry, and music. Not surprisingly, then, the hypotheses that conservative persons find incongruity-resolution humor more funny and nonsense humor more aversive than liberals were substantiated in several countries, with conservatism being the most potent predictor of humor appreciation found so far. While conservatism does not predict the *seeking* of stimulus uncertainty, the trait of sensation seeking, and in particular the component of experience seeking, does. Experience seeking involves the seeking of stimulation through the mind and the senses, through art, travel, even psychedelic drugs, music, and the wish to live in an unconventional style, and there is evidence that it is closely related to the novelty and complexity dimensions of stimuli (Zuckerman 1994). Therefore it was hypothesized and substantiated in several countries that experience seeking is positively related to appreciation of nonsense humor (for details see Ruch 1992).

The hypothesis that incongruity-resolution humor is preferred by individuals who generally dislike stimulus uncertainty and nonsense humor is appealing to those generally enjoying or searching for uncertainty was also substantiated in the field of aesthetics (Ruch & Hehl, 1998). For example, while appreciation of INC-RES correlated with liking of simple and representational paintings, and simple line drawings (such as a triangle, square, or cross), appreciation of nonsense correlated positively with liking complex and fantastic paintings (e.g., by Dali), liking of complexity and asymmetry in freehand drawings and polygons, and also with producing complexity in black/white patterns and enjoying and enhancing visual incongruity when wearing prism glasses which distort the visual field.

5. Closing Remarks

The study of humor provides an excellent area of consciousness study in as much as it refers to a very complex experience. Having a cognitive nucleus, the humor experience is fed by emotional and motivational factors (e.g., desires, needs, unconscious processes). Even adverse or tragic events can be seen from a humorous perspective as well, thereby blending the perception of humor with qualia of negative valence. Little is know about the conditions how and to what extent humor neutralizes such antagonistic states. As a typical human experience, humor warrants further inquiry.

Acknowledgments

The preparation of this chapter was facilitated by a Heisenberg grant (Ru 480/5-2) from the German Research Council (DFG) awarded to the author and the presented data stem from a DFG-grant (Ru 480/1-2). Special thanks to Lambert Deckers, Pete Derks, and Rod Martin for comments on the chapter.

References

Attardo, S. (1993) *Linguistic Theories of Humor*, Berlin: Mouton de Gruyter.

Cattell, R.B. and D.L. Tollefson (1966) *The Handbook for the IPAT Humor Test of Personality*, Champaign, IL: IPAT.

Deckers, L.H. (1993) "On the validity of a weight-judging paradigm for the study of humor", *Humor* 6:43-56.

Dennett, D.C. (1991) *Consciousness Explained*, Boston: Little, Brown and Company.

Derks, P. and S. Arora (1993) "Sex and salience in the appreciation of humor", *Humor* 6:57-69.

Derks, P., L.S. Gillikin, D.S. Bartolome and E.H. Bogart (1997) "Laughter and electroencephalographic activity", *Humor* 10:283-298.

Derks, P., R.E. Staley and M.G. Haselton (1998) "'Sense' of humor: Perception, intelligence, or expertise?", in: *The Sense of Humor*, W. Ruch, ed., Berlin: Mouton de Gruyter, pp. 143-158.

Ehrenstein, W.H. and S. Ertel (1978) "On the genesis of the funniness judgement", *Psychologische Beitr.* 20:360-374 (in German).

Ekman, P. and W.V. Friesen (1978) *Facial Action Coding System (FACS)*, Palo Alto, CA: Consulting Psychologists Press.

Eysenck, H.-J. (1942) "The appreciation of humour: An experimental and theoretical study", *Brit. J. Psychol.* 32:295-309.

Fridlund, A.J. and J.T. Cacioppo (1986) "Guidelines for human electromyographic research", *Psychophysiology* 23:567-589.

Godkewitsch, M. (1974) "Correlates of humor: Verbal and nonverbal aesthetic reactions as functions of semantic distance within adjective-noun pairs", in: *Studies in the New Experimental Aesthetics*, D.E. Berlyne, ed., Washington, DC: Hemisphere, pp. 279-304.

Goldstein, J.H., J.M. Suls and S. Anthony (1972) "Enjoyment of specific types of humor content: Motivation or salience? ", in: *The Psychology of Humor*, J.H. Goldstein and P.E. McGhee, eds, New York: Academic Press, pp. 159-171.

Hillson, T.R. and R.A. Martin (1994) "What's so funny about that?: The domains-interaction approach as a model of incongruity and resolution in humor", *Motiv. Emotion* 18:1-29.

Keith-Spiegel, P. (1972) "Early conceptions of humor: Varieties and issues", in: *The Psychology of Humor*, J.H. Goldstein and P.E. McGhee, eds, New York: Academic Press, pp. 4-39.

Koestler, A. (1964) *The Act of Creation*, London: Hutchinson.

Lipps, T. (1898) "Komik und Humor. Eine psychologisch-ästhetische Untersuchung", in: *Beiträge zur Ästhetik*, T. Lipps and R.M. Werner, eds, Leipzig: Leopold Voss.

McGhee, P.E. (1971) "Development of the humor response: A review of the literature", *Psych. Bull.*, 76:328-348.

McGhee, P.E. (1979) *Humor: Its Origin and Development*, San Francisco: Freeman.

McGhee, P.E. (1983) "The role of arousal and hemispheric lateralization in humor", in: *Handbook of Humor Research*, P.E. McGhee and J.H. Goldstein, eds, New York: Springer, pp. 13-37.

McGhee, P.E., W. Ruch and F.-J. Hehl (1990) "A personality-based model of humor development during adulthood", *Humor* 3:119-146.

Meyer, W.-U., R. Reisenzein and A. Schützwohl (1997) "Towards a process analysis of emotions: The case of surprise", *Motiv. Emotion* 21:251-274.

Raskin, V. (1985) *Semantic Mechanisms in Humor*, Dordrecht: Reidel.

Rhodes, G., S. Brennan and S. Carey (1987) "Identification and ratings of caricatures: Implications for mental representations of faces", *Cog. Psychol.* 19:473-497.

Rothbart, M.K. (1976) "Incongruity, problem-solving and laughter", in: *Humour and Laughter: Theory, Research and Applications*, A.J. Chapman and H.C. Foot, eds, London: Wiley, pp. 37-54.

Rothbart, M.K. and D. Pien (1977) "Elephants and marshmallows: A theoretical synthesis of incongruity-resolution and arousal theories of humour", in: *It's a Funny Thing, Humour*, A.J. Chapman and H.C. Foot, eds, Oxford: Pergamon Press, pp. 37-40.

Ruch, W. (1992) "Assessment of appreciation of humor: Studies with the 3 WD humor test", in: *Advances in Personality Assessment* (Vol. 9), J.N. Butcher and C.D. Spielberger, eds, Hillsdale, NJ: Erlbaum, pp. 27-75.

Ruch, W. (1995) *Humor-Test 3 WD*, Unpublished manuscript, University of Düsseldorf, Düsseldorf, Germany.

Ruch, W. (1997) "Issues in the definition and assessment of humor--and the sense thereof", *Proced. of the 8th ISHS Conference*, Edmond, University of Central Oklahoma, 40.

Ruch, W. and F.-J. Hehl (1998) "A two-mode model of humor appreciation: Its relation to aesthetic appreciation and simplicity-complexity of personality", in: *The Sense of Humor*, W. Ruch, ed., Berlin: Mouton de Gruyter, pp. 109-142.

Ruch, W. and G. Köhler (1998) "A temperament approach to humor", in: *The Sense of Humor*, W. Ruch, ed., Berlin: Mouton de Gruyter, pp. 203-230.

Ruch, W., G. Köhler and C. van Thriel (1999) "Surprise: an emotional response to humor?", Manuscript in preparation.

Schmidt-Hidding, W., ed. (1963) *Europäische Schlüsselwörter. Band I: Humor und Witz*, München: Huber.

Shultz, T.R. (1976) "A cognitive-development analysis of humour", in: *Humour and Laughter: Theory, Research and Applications*, A.J. Chapman and H.C. Foot, eds, London: Wiley, pp. 11-36.

Suls, J.M. (1972) "A two-stage model for the appreciation of jokes and cartoons: An information-processing analysis", in: *The Psychology of Humor*, J.H. Goldstein and P.E. McGhee, eds, New York: Academic Press, pp. 81-100.

Wilson, Ch.P. (1979) *Jokes: Form, Content, Use and Function*, London: Academic Press.

Wilson, G.D. (1973) "A dynamic theory of conservatism", in: *The Psychology of Conservatism*, G.D. Wilson, ed., London: Academic Press, pp. 257-266.

Zuckerman, M. (1994) *Behavioral Expressions and Biosocial Bases of Sensation Seeking*, Cambridge: Cambridge University Press.

THE EXPRESSIVE PATTERN OF LAUGHTER

WILLIBALD RUCH* and PAUL EKMAN°
*Department of Physiological Psychology, Heinrich-Heine-Universität Düsseldorf, Universitätsstraße 1, 40225 Düsseldorf, Germany
°Human Interaction Laboratory, University of California as San Francisco, 401 Parnassus Avenue, San Francisco 94143, USA

ABSTRACT

Laughter as a vocal expressive-communicative signal is one of the least understood and most frequently overlooked human behaviors. The chapter provides an overview of what we know about laughter in terms of respiration, vocalization, facial action, and body movement and attempts to illustrate the mechanisms of laughter and to define its elements. The importance of discriminating between spontaneous and contrived laughter is pointed out and it is argued that unrestrained spontaneous laughter involves inarticulate vocalization. It is argued that we need research integrating the different systems in laughter including

1. Introduction

Laughter is a conspicuous but frequently overlooked human phenomenon. In ontogenetic development it emerges, later than smiling, around the fourth month; however, cases of *gelastic epilepsy* (from Greek; *gelos* = laughter) among neonates demonstrate that all structures are there and functional on date of birth. Further evidence for its innateness comes from twin studies as well as from the fact that laughter is easily observable among deaf-blind children (even among deaf-blind thalidomide children, who could not "learn" laughter by touching people's faces).

Man is not the only animal that laughs. Like smiling, laughter has its equivalent in the repertoire of some nonhuman primates. Beginning with Darwin (1872), many writers have been struck by the notable acoustic, orofacial, and contextual similarities between chimpanzee and human laughter. Especially among juvenile chimpanzees a "play-face" with associated vocalization was noted to accompany actions such as play, tickling, or play-biting (Preuschoft, 1995; van Hooff, 1972).

2. Laughter as an Inarticulate Utterance

Laughter is estimated to be about 7 million years old (Niemitz, 1990). There is disagreement on how human speech developed phylogenetically. It could have originated from non-verbal vocal utterances, a prelingual gestural system, or sounds initially used to supplement the facial channel. However, it is safe to assume that laughter –like other utterances such as moan, sigh, cry, groan, etc.–

was there before man developed speech and served as an expressive-communicative social signal.

Considering the stage in the evolution of voice when laughter emerged allows us to make several deductions about the nature of the sound, its generation and the cerebral organization of laughter. If laughter was part of the human vocal repertoire before the speech centers were developed, it is likely that the sound was generated almost exclusively by laryngeal modulations, modified to some degree by supralaryngeal activity but not by articulation. This is because articulation requires voluntary control over the vocal system. Thus the production of speech sounds needs the coordination of respiration, phonation, resonance, and articulation, an analysis of laughter will involve mainly the consideration of the first three. But is it consistent with current knowledge to hypothesize that a laugh-pulse –a vocalization segment initiated by an aspirated "h" type sound followed by the utterance of one of several vowel sounds that is then abruptly terminated– is an inarticulate sound? Indeed, the /h/, an voiceless fricative glottal sound, is the *only* consonant produced at the level of the larynx; but as laughter is not "speech" we should not expect phonemes of a language to be uttered. Likewise, the assumption of resting articulators would suggest that the laughter "vowel" (but also the sound for other prelingual utterances, such as moaning or utterances of astonishment) is the central vowel schwa, or /ǝ/ (as in *a*bout). For the production of this "neutral vowel" one needs to open the mouth and lower the jaw, but keep all other articulators passive. We know from everyday experience that there is a tremendous variety in the quality of the laughing sounds, so obviously deviations from the schwa occur. This does not contradict the hypothesis of an inarticulate vocalization *per se* but puts forward the task of defining the deviations from the resting articulation position that are due to altered emotional states and separate them from those supralaryngeal conditions modulating the sound that are due to voluntary actions.

A second factor that we need to take into account is that there are different ways to generate laughing sounds and it is possible to intentionally modulate the emotion-driven spontaneous laughter. As the cerebral organization of vocal behavior progressed phylogenetically after these early facio-vocal signals became hard-wired, higher centers obtained control over the laughter "timer". Thus, in addition to laughing spontaneously (emotional laughter), we can laugh voluntarily or on command (contrived or faked laughter), and we can even speak or sing the laughing sounds, which typically are phonetically represented as "ha ha ha." These forms of utterances differ in degree of volitional control and –inversely– emotionality, and imply different neural pathways and systems involved (for reviews of neuronal control of vocalization see, for example, Jürgens, 1998; Ploog, 1986; for reviews or models of laughter: Arroyo *et al.*, 1993; Fried *et al.*, 1998). The distinction made between spontaneous and voluntary laughter is consistent with clinical observations; certain patients with degenerative brain disorders may be unable to move their abdomen voluntarily, but demonstrate vigorous expiratory movement of the abdomen when laughing spontaneously (Bright *et al.*, 1986). Finally, it should be noted that there are also *voluntary* attempts to regulate *spontaneous* laughter; for sake of brevity the combination of voluntary and spontaneous in regulated laughter is only mentioned here.

In *spontaneous* laughter we are following an impulse, an urge to laugh without restraining ourselves. There is no attempt to suppress the response or exert any control over its expression; the laughing person has been described as abandoning himself or herself to the body response (Plessner, 1941). The involuntary aspect can also be seen in the fact that during laughter our self-awareness and self-attention is diminished. Trying to direct attention during a laughter episode stops or reduces laughter. During laughter the state of consciousness is altered; as Hall and Allin (1897; p. 8-9) put it: "[t]he objective world has vanished and is forgotten, the proprieties and even the presence of others are lost, and the soul is all eye and ear to the one laughable object. Care, trouble, and even physical pain are forgotten, and the mind, as it were, falls back through unnumbered millennia and catches a glimpse of that primeval paradise where joy was intense and supreme." While descriptions of emotional experience are rather sparse and reports of subjects mainly refer to the awareness of the impulse to laugh, spontaneous laughter is clearly enjoyable.

In *voluntary* laughter we may want to produce a sound pattern like that of natural laughter. A typical everyday situation demanding faked laughter is when we want to signal somebody that we enjoyed a humorous message and join the laughter of others but actually do not feel any enjoyment. While voluntary laughter may pass over into involuntary laughter, we can not voluntarily produce emotional laughter. Interestingly, research has shown that contrived and spontaneous laughter within a person are strikingly similar with respect to the respirational pattern (suggesting a "laughter signature"); however, it is doubtful whether this is also the case for supralaryngeal systems. The many ways in which they differ still need to be described; this will be especially important when studying individual differences as laughter gets socialized during ontogenetic development. Research has made use of the fact that we can voluntarily alter key features, and thus studies of laughter at different vowels, pitch, and voice quality exist (Citardi *et al.*, 1996; Habermann, 1955). Laughing on command is embarrassing to some, and thus the results obtained for contrived laughter are of limited value for describing spontaneous laughter as it may merge voluntary movements with unintended affective states.

Speaking or singing "ha ha" allows modulating the sound in many ways (e.g., stretching the length of vowels, emphasizing particular syllables), just as with other spoken words. As with voluntary laughter we *intend* to produce a particular sound pattern and we have even more control over the outcome; the major difference being that phonation is not based on forced breathing but on well dosed air supply resulting in less tracheal resonance, breathiness, and aspiration. Spoken laugh sounds may be devoid of emotional intonation; thus there will be less variability in melody, and the utterance will be more clearly articulated and often match the phonemes of a language. Emotional laughter and speech are rather independent; in fact, laughter and speech may co-occur, thereby lengthening the duration of laugh-pulses (Nwokah *et al.*, in press). It remains to be verified that damage of the higher speech centers has little impact on the expression of tickle-induced emotional laughter.

3. Description of Laughter

Laughter is not a term used consistently, nor is it precisely defined in research articles and even encyclopedias. In everyday life a smiling face is often referred to as "laughter" although the vocal element is entirely missing. Research studies usually focus on one or two systems of laughter rather than the integration of all the components involved. For example, a study of the acoustics of laughter typically restricts the investigation to the phonatory system and may arrive at the conclusion that the duration of laughter is, say, below two seconds. Sound production is contingent on air flow and the deviation from resting breathing exceeds the rhythmic forced exhalations underlying laughter by far; therefore duration estimates of eight seconds are not uncommon in respiration studies (e.g., Habermann, 1955). In studies of primate laughter the face gets most attention and also studies of joy in humans typically focus on the face, or even on the emotion-specific actions neglecting the mouth opening altogether. The analysis of videotaped laughter suggests a mean duration of five seconds while the study of facial muscle contractions exceeds this value due to the offset period where facial changes are barely visible (Ruch, 1990).

3.1. Laughter Segmentation

Laughter bout. The term *laughter bout* is used here to refer to the whole behavioral-acoustic event, including the respiratory, vocal, and facial and skeletomuscular elements. Prototypically, a laughter bout may be subdivided into an *onset* (i.e., the pre-vocal facial part which in explosive laughter is very short and steep); an *apex* (i.e., the period where vocalization or forced exhalation occurs), which –in sustained laughter– might be interrupted by inhalations; and an *offset* (i.e., a post-vocalization part; usually a long-lasting smile fading out smoothly).

Laugh cycle and laugh-pulse. The laughter vocalization period is composed of *laugh cycles*, i.e., repetitive *laugh-pulses* (Moore & von Leden, 1958) interspersed with pauses. There is laughter with only one or two pulses ("exclamation" laughter, "chuckle"; Nwokah *et al.*, 1993), but studies typically report a mode of four pulses in a laugh cycle (Provine & Yong, 1991; Rothgänger *et al.*, 1998). The upper number of pulses in a laugh cycle is limited by the lung volume, and different studies give numbers between 9 and 12 (Boeke, 1899; Provine & Yong, 1991); a laughter *episode* –two or more laughter bouts separated by inspirations– will have more.

Table 1 presents an attempt to integrate empirical findings from studies of different systems involved in laughter. Obviously the same phenomenon was studied and converging information about the rhythmic pattern of laugh-pulses can be extracted from studies of air flow and pressure, respirational and laryngeal muscles, acoustics, and the oscillations in the pressure a finger puts on a plate. During intense laughter, the laugh-pulse would be extractable from different parts of the body, as the massive respiration movements cause vibrations of the trunk and limbs to occur which may be detected by an acceleration sensor on the body surface. While so far there is no convincing evidence for Darwin's (1872) claim that the muscles of the limbs are thrown into rapid vibratory movements at the same time as the respiratory muscles, initial support can be found in a sample figure provided by Santibañez and Bloch (1986). In this illustration of the EMG-recordings from the brachioradialis (forearm muscle which bends the elbow) and rectus femoris (anterior muscle of the upper thigh) muscles, an increase in amplitude can be found, and the rhythmic pattern seems to match the one of the abdominal muscles.

There are roughly five pulses per second, but the interpulse interval varies as a function of the position of the pulse in a sequence (Boeke, 1899; Provine & Yong, 1991; Rothgänger *et al.*, 1998). Further segmentation of the laugh-pulse can be obtained in the different systems involved in laughter. Boeke, recording his own laughter on an Edison Sonograph, already discovered the existence of pauses between the laughter sound pulses whose length exceeded the duration of the ha-utterances by a factor of two. Due to the dynamics of respiration, the duration of sound pulses decreases sequentially from an earlier to later position, and the duration of the interpulse pause increases. Differences in operational definitions of pulse and interpulse pause add further variations to the estimations of length (see Table 1).

Acoustic segmentation of the laugh-pulse. Acoustic analyses of laughter suggest a distinction between the vowel-like *utterance nucleus* and the preceding aspirated "h"-type sound initiating the sound pulse. As the glottis is open during the interpulse pauses, aspiration continues. Thus, not surprisingly, Provine and Yong (1991) report that laughter played backwards still sounds like "ha ha".

Vibratory movements in a laugh-pulse. At the level of *laryngeal movements* a laugh-pulse can be further split up into the number and duration of *vibratory cycles of the vocal cords*, and even further into their *contour*; i.e., the *phases* of when vocal cords are opening, closing, or closed. Using ultra high speed motion picture photography (4000 exposures a second), Moore and van Leden (1958) found in their analysis of the vibration of the cords during a man's laughter a range of 5 to 15 cycles of opening and closing in laugh-pulses. Accordingly, the duration of laugh-pulses varied from 30 to 100 ms. Each vibratory cycle interrupts the air column ascending from the lungs and the rate of these cycles is the basis for the fundamental frequency of the sound. Further segmentation may be achieved by analyzing the time when vocal cords are opening, closing, or closed, and this vibratory curve contour reflects the dynamics of respiration as it undergoes a

Table 1. The laugh-pulse as studied in respiration, acoustics, and body movement

Author(s)	variable studied	number of pulses	duration (in ms) IPI	pulse	IPP
Respiration					
Bloch, Lemeignan and Aguilera (1991)	rectus abdominis muscle (EMG)	4.17/s	240		
Hoit, Plassman, Lansing and Hixon (1988)	rectus abdominis muscle (EMG)	5-6/s			
Air pressure					
Agostoni, Sant'Ambrogio and Portillo Carrasco (1960)	gastric air pressure (swallowed balloon)	5-6 cycles/s			
	thoracic air pressure (swallowed balloon)	5-6 cycles/s			
Schroetter (1925)	exhaled air in closed system	up to 5/s			
Laryngeal activity					
Moore and van Leden (1958)	vocal fold vibrations (ultra high speed camera)			30-100	
Luschei, Ramig, Baker and Smith (1997)	thyroarytenoid muscle (EMG)	5 Hz			
	cricothyroid muscle (EMG)	5 Hz			
	posterior cricoarytenoid muscle (EMG)	5 Hz			
Acoustics					
Boeke (1899)	"ha"-sounds/utterances	4.12/s	243	60	183
Bickley and Hunnicutt (1992)	laughter syllables		204	97	167
	laughter syllables		224	68	156
Nwokah, Davies, Islam, Hsu and Fogel (1993)	laugh events		200		
Nwokah, Hsu, Davies and Fogel (1998)	events			110	
	events			130	
Mowrer, LaPointe and Case (1987)	laugh bursts	5.55/s	180		
Provine and Yong (1991)	laughter notes	4.76/s	210	75	135
Rothgänger, Hauser, Cappellini and Guidotti (1998)	plosives	4.67/s	213	126	87
Body vibrations (sentics)					
Clynes (1980)	transient finger pressure (sentograph)	5.03/s	199		

Notes. Authors typically gave either frequency per second or the duration of utterances, or other information which was then transformed into one or both parameters by the author of the present chapter.
IPI = interpulse interval; IPP = interpulse pause

progressive change even within one laugh-pulse. These modifications in the vibratory pattern codetermine how laughter sounds and may be a key factor in distinguishing types of laughter.

3.2. Laughter Respiration

A respiration cycle consists of inspiration, inspiration pause, expiration, and expiration pause. No matter where in a respiration cycle a person is, laughter typically begins with an initial forced exhalation, followed by a more or less sustained sequence of repeated expirations of high frequency and low amplitude, which may or may not be phonated as "ha ha ha;" i.e., the laugh cycles. While in the case of sustained laughter the expiratory phases will be interrupted by inhalations, there is no evidence for Darwin's (1872; p. 199) assertion that "... [t]he sound of laughter is produced by a deep inspiration followed by short, interrupted, spasmodic contraction of the chest, and especially of the diaphragm." No inspiration preceding the laugh is necessary as laughter is produced at a low lung volume (Bright *et al.*, 1986). Normally the laugh cycles are initiated around functional residual capacity (FRC; i.e., at the lung volume after a normal expiration) and terminate close to residual volume (i.e., the air volume remaining in the lung after maximal expiration), or sometimes even exceed the level of maximal voluntary exhalation (Bright *et al.*, 1986; Lloyd, 1938). Thus, most likely the initial forced exhalation is expelling the tidal volume, and the following sequence of laugh-pulses is based on the expiratory reserve volume. The increase in depth of respiration –the amplitude during laughter may be up to 2.5 times higher than during resting respiration– is therefore due to the stronger expiration; inspiration may add to the amplitude in case of laugher episodes, where single deep inhalations intersperse the expiratory sequences.

The rhythmic laughter respiration pattern is produced by saccadic contractions of auxiliary expiration muscles; i.e., muscles that are typically passive during normal expiration, such as the diaphragm (Agostoni *et al.*, 1960), the abdominal (rectus abdominis; Hoit *et al.*, 1988; Santibañez & Bloch, 1986) and the rib cage muscles (triangularis sterni; De Troyer *et al.*, 1987). Of the three muscles mentioned, only the diaphragm is involved in resting breathing in humans; its contraction causes inspiration. The role of the diaphragm is not entirely clear, however, as the discharges in the EMG recordings (albeit parallel to the air pressure) may be indicating reflex contractions due to the passive distention occurring as a function of the violent contractions of the ribcage and abdominal muscles. The triangularis sterni is passive during quiet breathing but involved in different active respiration maneuvers; i.e., respiration below FRC. It contributes to the deflation of the rib cage during active expiration such as in coughing and its neural activation is largely coupled with that of the abdominals. The relative contribution of rib cage and abdomen to the volume may vary even within one laughter cycle (Bright *et al.*, 1986; Habermann, 1955).

The respiratory muscles function in concert with the larynx; while without any closing of the glottis there may be single or a few forced exhalations, the adduction (closing) prevents the air to be exhaled too quickly, and allows the building up and maintaining of subglottal air pressure. The initial forced exhalation increases the transdiaphragmatic air pressure by about 5440 Pa (Agostoni *et al.*, 1960) to about 6120 Pa (Schroetter, 1925); this pressure plateau is maintained and forms the basis for the sustained period of phonation of the laugh utterances. The heightened pressure makes the air stream up the airways through the larynx where the rhythmic closing and opening of the glottis interrupts the air stream. These vibrations are carried through the vocal tract whose shape amplifies or dampens certain frequency spectra, and finally the air escapes through the mouth or the nose.

3.3. Phonation

The acoustic sequence of laughter pulses are produced by a series of rapid, continuous, stereotypic laryngeal adjustments, which are separated into four stages: interpulse pause, adduction (closing) of the arytenoid cartilages, vibration of vocal chords, and abduction (opening) of the arytenoid cartilages (Moore & von Leden, 1958). As confirmed by laryngeal EMG during laughter the thyroarytenoid and –to a lesser extent– the cricothyroid muscles are involved in the closing of the glottis and the opening is achieved by the posterior cricoarytenoid muscle (Luschei *et al.*, 1997). Both groups contract in the same pace (see Table 1), but with a time lag.

The *interpulse pause* is an instant of quiet between the audible moments of laughter. The arytenoid cartilages rest open allowing the breath stream to flow unimpeded through the larynx, and the vocal cords remain motionless. Some of the aspirate sound is produced at this time but becomes more audible as the vocal cords approach each other. The fact that the "h" sound is an eddy-current phenomenon created during the closing stage of the valve-like movements of the vocal folds is in line with the assumption of the absence of cortically controlled articulation. Occasionally it was argued that glottal stops occur between the laugh-pulses; however, both videoendoscopy and photoglottography show that the glottis is widely open at the end of a laugh-pulse (Citardi *et al.*, 1996; Moore & van Leden, 1958) and the acoustic analyses reported quiet aspirations in the periods between voicing (e.g. Luschei *et al.*, 1997).

During the *adduction stage* the arytenoid cartilages carry the cords toward each other and when the glottal space has been narrowed to a small slit, the cords begin to vibrate. As *vibration* occurs not only when the arytenoids are fully approximated but start at the end of the adduction and continue when abduction starts again there are transitions between an initial swing of the flaccid cords and the full vibrations and again towards the end of the vibratory movements. Moore and von Leden (1958) report that it takes about 10 ms to accomplish the transition from quiescence to full vibration, and that changes in vibratory curve contours and in the length of cycles occur as the laugh-pulse progresses. During the *adduction phase* of the laugh-pulse the arytenoids and vocal cords move into their rest positions. There is a progressive diminution of the mesial excursion of the cords and the slowing of the cord movement as the interarytenoid space enlarges; for example, in the sample laugh-pulse provided by Moore and von Leden (1958) there is a progressive drop of about 30 cycles per second.

Thus, whereas during singing we try to keep the fundamental frequency constant for each note, due to the dynamics of air flow there is tremendous variation even *within* a single laugh-pulse. Acoustic analyses also demonstrate the changes in fundamental frequency *between* laugh-pulses; typically there is a progressive decrease in pitch and intensity (loudness) of laughter at later pulses of a cycle. Boeke noted already in 1899 that the melodic variation in laughter is higher than in speech.

3.4. Supralyrangeal Modulation

The buzzing sound produced in the glottis is carried into the resonance tract whose form codetermines the sound of laughter. While the laryngeal and respiratory movements during laughter appear to be highly stereotyped, the acoustic output is quite variable. There is little systematic research on the activity of the 10 or so movable articulators and thus at present a comprehensive evaluation whether the activity is compatible with the hypothesis of an inarticulate sound is not possible.

While one could predict that in courteous laughter without much emotional involvement the vowel uttered is schwa-like, in emotional laughter some supralaryngeal conditions modulating the sound occur, partly molded by the emotional state. Despite the assumption that there is no cortically controlled articulation in laughter, qualitative differences might be related to emotional states in various ways. Variations in pitch can be obtained by increasing or reducing the length of the vocal tract, for example, by lifting or lowering the larynx, or by protruding or retracting the lips. Indeed, both Habermann (1955) and Citardi *et al.* (1996) observed movement of the larynx in the superior–inferior direction during *voluntary* laughter. A further determinant of pitch, the lengthening and tension of the vocal cords, is susceptible to emotional arousal (contraction of the vocalis muscles) as is the general degree of tension vs. relaxation in the laryngeal area. Width or narrowness of the pharynx affects the voice quality, and it has been argued that in positive states there is a widening of the throat (as in taking up food) producing a-type sounds while in disgust there is a narrowing of the throat (as in expelling bad food) compatible with the sound during contemptuous or scornful laughter.

Of the major articulators in speech, the tongue (involved in producing high and low, and front and back vowels) is likely to be in a resting central position during joyful laughter, but the jaw and lips are not. The act of opening the mouth and the degree of aperture of the mouth (i.e., lowering of the jaw) affects the sound; this action is coupled with respiration (allows the escape of air). Habermann (1955) reported nasals occurring; this seems to be likely only for mild laughter, when the mouth is not opened, the soft palate lowered, and the air escapes through the nose. Last but not least, two important articulators, the lips and the cheeks, are typically not in a resting position as they are part of the emotion-driven facial actions. For example, the facial display of enjoyment involves the retraction of the lip corners and the cheeks are lifted upwards (see description of the specific muscles in section 2.5 below); the contraction of these two pairs of muscles changes the form of the mouth aperture and tenses the skin of the upper ending of the vocal tract thereby affecting the sound at the onset and apex of these actions. As Tartter (1980) demonstrated, the same spoken message sounds different when the sender is smiling or showing a resting face; listeners can reliably infer smiling from the voice. Intense smiling is also incompatible with the utterance of vowels which require a protruding of the lips, such as an "o" or "u," making these sounds unlikely to occur during joyful laughter. Taken together, these considerations would predict that laughter vowels would deviate from being a neutral vowel and the reason for a contrary finding is due to other constraints during emotional laughter.

Chimpanzee laughter. As the laryngeal apparatus and vocal tract of humans and primates are different in many ways (see Lieberman, 1975) one should not expect the similarities between primate and human laughter to be very strong. Chimpanzee laughter has been described as a soft repetitive guttural sound of low intensity panting noises during exhalation that roughly sounds like human laughter. The sound is based on rhythmic breathing resulting in staccato vocalizations. Spectro graphic analyses are sparse; Berntson *et al.* (1989) report that while generally "noisy," the laughter recordings from a juvenile male during active play evidenced temporal structure, with periodic louder voiced and voiceless components. They provide a sample figure with four distinct syllable-like louder egressive voiced segments. These segments were characterized by some low fundamental frequency voicing, with intermittent noise. Provine (1996) concludes that the major difference is that in human laughter *several* laugh-pulses may occur on one expiration, whereas chimpanzee laughter is produced during *each* brief expiration and inspiration. He notes that "... [t]he sounds of chimpanzee and human laughter are sufficiently different that without viewing the characteristic "play face" and source of stimulation (such as play and tickle), naive human beings may be unable to identify the chimpanzee vocalization as laughter" (Provine, 1996; p. 40).

3.5. Facial Actions

If the theory applies that sounds were initially added to the facial display to supplement or underscore a message then the emotion-specific facial configuration molded and also limited the sound utterance. Thus, the identification of the facial actions in spontaneous laughter is important. As laughter is commonly associated with the emotion of joy, the facial configuration named (Ekman *et al.*, 1990) the *Duchenne display* (to honor Duchenne who first described how this pattern distinguished enjoyment smiles from other kinds of smiling) might serve as a first starting point against which, for descriptive purposes, deviations may be judged. The *Duchenne display* refers to the joint contraction of the zygomatic major and orbicularis oculi muscles (pulling the lip corners backwards and upwards and raising the cheeks causing eye wrinkles, respectively). While laughter has been described as occurring during different emotional states, so far it is not clear whether or not these laughs are also morphologically different. Furthermore, those voluntary facial actions aimed at modulating the intensity, duration or quality of the sound emitted, or to suppressing laughter altogether are not completely identified. Another starting point may be the facial display during primate laughter. This is circular, in part, as the human facial actions have been the basis for homologizing the primate displays.

Primate laughter. The facial features of primates have been described by salient features, rather than their muscular underpinning. The play face (or relaxed open-mouth display) in chimpanzees has been described as the jaws being widely open, the mouth corners normal or slightly retracted, the upper lip covering the teeth, lower lips loose, and the lower teeth exposed. This description seems to involve the action of the zygomatic major, but not the orbicularis oculi muscles, although it is developed in chimpanzees. However, most of this work (e.g., Chevalier-Skolnikoff, 1973; van Hooff, 1972) was done before the rediscovery of Duchenne, and therefore the actions of this muscle were perhaps not specifically studied.

Human laughter. The facial expression of laughter has been given some attention by researchers during the past and present century. A summary of hypotheses put forward and the results of some studies of laughter using facial-EMG (surface and needle electrodes) or the *Facial Action Coding System* (Ekman & Friesen, 1978) are presented in Table 2. It should be noted, however, that early descriptions had to be based on real time observations, or on inferences from knowledge about facial musculature since no recording tools were available. As intense laughter involves movement of head and trunk, facial measurement continues to be a problem.

Table 2 confirms that the two muscles forming the *Duchenne smile* have been found to be involved in laughter as well, explaining the smooth transitions between smiling and laughter in both the onset and offset of a laughter bout. Sumitsuji (pers. comm.) observed that there is little innervation of the orbicularis oculi muscle during *voluntary* laughter; thus the eye region might serve as a marker for distinguishing emotional from voluntary laughter, much as it does for distinguishing enjoyment from non enjoyment smiling. Interestingly, in two forms of pathological laughter (epileptic laughter and pseudo bulbar palsy; representing excitation or the loss of inhibition, respectively), the activated motor pattern includes the contraction of the orbicularis oculi muscle. Furthermore, laughter includes the relaxation of some muscles (masseter, pterygoids) allowing for a lowering of the lower jaw so that the air is expelled through the mouth.

Table 2. Hypotheses and empirical findings regarding the involvement of facial muscles in laughter

	upper face muscles						*mid-face muscles*						*lower face muscles*					
Author(s)	1	2	3	4	5	6	7	8	9	10	11	12	13	14	15	16	17	18
Hypotheses: laughter																		
Bloch et al. (1987)					X	X	X											X
Darwin (1872)					X	X	X		X	X?								
Dearborn (1897)			X	X		X	X	X	X	X	X		X			X		
Hecker (1873)							X		X	X								
Heller (1902)					X	X	X						X					
Hjortsjö (1970)	X	X			X	X	X	X		X		X	X					X
Piderit (1867)							X	X	X				X		X			
Poeck and Pilleri (1963)					X		X				X			X				
Raulin (1900)					X?		X	X					X					
Hypotheses: strong laughter																		
Darwin (1872)			X		X	X	X		X	X?								
Heller (1902)	X	X	X	X	X	X	X			X			X				X	
Piderit (1867)			X				X	X	X				X		X			
Results: FACS-studies																		
Grammer (1990)							X											
Keltner and Bonanno (1997)					X		X										–	
Ruch (1990, 1994)					X	X	X										–	
Ruch (1997)					X		X										–	
Results: EMG-studies																		
Gallo and Palla (1995)																	X	
Santibañez et al. (1986)														X			X	
Sumitsuji (1967)	–	–	–		X		X		X				X	X	X	X		
Tanaka (1976)	–	–	–		X		X									X		
Tanaka et al. (1991)	–	–	–		X		X		X				X			X		
Supressed laughter																		
Tanaka (1976)	X	X			X		X											
Pathological laughter																		
Tanaka et al. (1991)	X	X	X		X		X		X				X			X		
Yamada et al. (1994)			X															

Notes. 1=frontalis, pars medialis-raises the inner brows, producing horizontal furrows in the medial region of the forehead; 2=frontalis, pars lateralis-raises the outer brows, producing horizontal furrows in the lateral regions of the forehead; 3=corrugator supercilii-draws the brows together and downward, producing vertical furrows between the brows; 4=pyramidalis/procerus/depressor supercilii-pulls the medial part of the brows downward and may wrinkle the skin over the bridge of the nose; 5=orbicularis oculi, pars orbitalis-tightens the skin surrounding the eye causing crows-feet wrinkles; 6=orbicularis oculi, pars palpebralis-tightens the skin surrounding the eye causing the lower eyelid to raise; 7=zygomaticus major-pulls the lip comers up and back; 8=zygomaticus minor-deepens nasolabial furrow; 9=levator labii superioris/levator labii superioris, caput infraorbitalis-raises the upper lip and flares the nostrils, exposing the canine teeth; 10=levator labii superioris alaeque nasi-raises the center of the upper lip and flares the nostrils; 11=nasalis, pars alaris bzw. dilatores nasi-dilates nostrils; 12=buccinator-compresses and tightens the cheek, forming a dimple; 13=risorius-stretches lip corners straight to the side; 14=orbicularis oris-tightens, compresses, protrudes, or inverts the lips; 15=depressor anguli oris-pulls the lip comers downward; 16=mentalis-raises the chin and protrudes the lower lip; 17=masseter-adducts the lower jaw; 18=caninus/levator anguli oris-pulls the lip corner up sharply and puffs up cheeks. A "X" includes action of that muscle, a "–" means lowering of muscle tone. Authors either gave names or picture of muscles or a description of facial appearance from which the muscle action was then inferred utilizing the information provided by the *Facial Action Coding System* (Ekman & Friesen, 1978). An "?" was added when the muscle is not entirely clear from the description given.

However, the number of postulated and confirmed muscles involved in the laughter facial expression are far more than those usually identified. So how do we make sense of the true complexity of the neuromuscular patterns of laughter given the limited and contradictory information currently available? First, several of the additional muscles may be involved in auxiliary movements, perhaps of secondary importance; this might be the case for the muscles located around the mouth, such as the levator labii superioris, depressor anguli oris, risorius, or orbicularis oris. They are involved in the radial opening of the lips aimed at widening the mouth to let the air stream out more easily. Second, some of the hypotheses relate to the *intensity* of laughter. It has been suggested that strong laughter also involves the muscles in the upper face, such as the corrugator supercilii and the frontalis muscles. Moreover, claims have been made that with increasing intensity of laughter almost *all* muscles get involved and tensed (Heller, 1902) and that the most intense form of laughter is not well distinguishable from the facial display of crying (Darwin, 1872; Piderit, 1867). There is no support for this from the existing studies of healthy adults; on the contrary, Tanaka and Sumitsuji (1991) report a lowering of muscle tone in the upper face during laughter (as has been found for the corrugator supercilii in smiling). However, in pathological laughter –usually of high intensity– the frontalis and corrugator muscles are active (Tanaka & Sumitsuji, 1991). This would seem to support an intensity hypothesis; however, as patients do not enjoy laughter fits, the additional movements might be of voluntary nature reflecting efforts to control or suppress laughter rather than being part of an innate response pattern.

Third, some muscles may be active only part of the time and the course of contraction of muscles over time can be different for different muscles. For example, the masseter (chewing muscle) whose relaxation initially helps to lower the jaw, was found to have elevated activity during laughter (Gallo & Palla, 1995; Santibañez & Bloch, 1986). Unfortunately, studies often report only intensity of a muscle contraction without considering its distribution over time. Fourth, one has to take into account that researchers might have erred in transforming their observations into the list of muscles due to the lack of profound knowledge of the effects of muscle contraction on the appearance on the surface at that time. This might have resulted in identifying the wrong muscle (e.g., the early anatomists' labeling of the "risorius" seems to imply that they considered it to be the *major* laughing muscle), or too many muscles (e.g., the strong action of the zygomatic major is able to cause changes –like opening the nostrils or raising the upper lips– that can also be attributed to separate muscles). Fifth, in previous studies, the type of laughter often was not controlled for and this might account for some of the additional muscle actions observed. Instead of pure enjoyment there might have been voluntary laughter, a blend of enjoyment with other emotions (e.g., surprise should add the elevation of the eyebrows caused by the frontalis muscles), movements associated with regulating the intensity of laughter, attempts to suppress laughter, or even accidental movements. Of course, a significant yet unexplored remaining cause for these surplus facial actions might be that laughter during different emotional states (e.g., scorn, nervousness, embarrassment) is morphologically different.

The information in Table 2 is not yet complete, as not all facial muscles have been studied so far. It seems advisable to supplement future facial EMG-studies by videotapes of face and body to control for the nature of movement. Furthermore, it seems essential to vary and control for the quality of laughter, guarantee for different intensity levels, and also examine the course of muscle contractions over time rather than studying sheer intensity at apex. We also need to start separating voluntary and spontaneous actions based on the existing experience (Ekman, 1997).

Further facial changes. Several authors described a brightening of the eyes (even among chimpanzees); the sparkling quality was explained either by enhanced lacrimation or an enhanced tenseness owing to the contraction of the muscles around the eye (Darwin, 1872), or that the eyeballs get filled with blood or other fluids (due to the enhanced circulation; Piderit, 1867). Furthermore, there is the observation that a laughter episode is frequently terminated by a closing of the lids.

3.6. Body Movement

While smiling is purely facial, at higher intensity levels laughter involves the whole body; however, changes in posture and body movements have received least attention in the study of laughter. Darwin writes (1872; p. 206-207) "During excessive laughter the whole body is often thrown backward and shakes, or is almost convulsed." Indeed, one can expect actions associated with the respiratory movements; for example, the backward tilt of the head facilitates the forced exhalations, and the forced inhalation interrupting two laughter cycles will raise and straighten the trunk. Furthermore, as noted above, the massive respiratory movements underlying the laughter pulse may cause the observed shaking of the shoulders and vibrations detectable on the trunk but also extremities.

There are also actions not coupled with respiration but their exact list and the conditions of their emergence are not known. As laboratory studies rarely induce violent laughter one has to rely on observations and self-reports such as the one by Hall and Allin who issued an 11 item questionnaire on tickling, fun, wit, humor, and laughing as a supplement to *The American Journal of Psychology*. In a subsequent report, they present a qualitative analyses based on responses of 3000 people (Hall & Allin, 1897). Regarding body movements they summarize (p. 5) "... [i]n the height of the laugh ... some plant the elbows on the knees; others rock violently sideways, or more often back and forth; the hands are thrown into the air or clapped on the thighs; ... the limbs jerk; the foot is stamped; the fists pound; ... waves of nervous tremor pass over the body; ... the hand is placed over the eyes, mouth, or both; ... little children jump up and down, lie on the floor and roll all over the room; some swing the hands in the air; the breast heaves up and down; some turn around on the heel from left to right ... the head shakes from side to side; ... others always hold the sides with both bands; others roll the head, features often twitch or tremble convulsively." Obviously, research is needed sorting accidental actions, attempts to regulate laughter and genuine elements of a laughter pattern.

4. Concluding remarks

The present review makes clear that the we need multi-level studies of laughter, as different systems work together in the generation of the expressive pattern and there are multiple dependencies among respiration, facial action, acoustics, and body movements. The review was restricted to some basic elements (leaving out issues such as neurohormonal effects of laughter, or social factors in laughter), since these provide the basis for tackling the yet unanswered questions of how to detect whether laughter is faked or felt, how to distinguish among different types of emotional laughter on a morphological basis, and what is the relationship between smiling and laughter. The list of unanswered questions and unsolved problems seems endless, but the study of laughter is a worthwhile subject as it is a window to ancient affective experience; a prototype of the prelingual utterance of joy.

Acknowledgments

The preparation of this chapter was facilitated by a Heisenberg grant (Ru 480/1-2) from the German Research Council (DFG) awarded to the first author. Thanks to Eva Nwokah and Rod Martin for comments on an earlier version of the chapter.

References

Agostoni, E., G. Sant'Ambrogio and H. del Portillo Carrasco (1960) "Elettromiografia del diaframma e pressione transdiaframmatica durante la tosso, lo sternuto ed il riso", *Accademia Nazionale dei Lincei, Roma* 28:493-496.

Arroyo, S., R. Lesser, B. Gordon, S. Uematsu, J. Hart, P. Schwerdt, K. Andreasson and R. Fisher (1993) "Mirth, laughter and gelastic seizures", *Brain* 116:757-780.

Berntson, G.G., S.T. Boysen, H.R. Bauer and M.S. Torello (1989) "Conspecific screams and laughter: Cardic and behavioral reactions of infant chimpanzees", *Devel. Psychobiol.* 22:771-787.

Bickley, C. and S. Hunnicutt (1992) "Acoustic analysis of laughter", *Proc. Int. Conf. Spoken Language Process* 2:927-930.

Bloch, S., M. Lemeignan and N. Aguilera (1991) "Specific respiratory patterns distinguish among human basic emotions", *Int. J. Psychophysiol.* 11:141-154.

Bloch, S., P. Orthous and G. Santibañez (1987) "Effector patterns of basic emotions: A psychophysiological method for training actors", *J. Soc. Biol. Struct.* 10:1-19.

Boeke, W. (1899) "Mikroskopische Phonogrammstudien", *Pflüg. Arch.* 76:497-516.

Bright, K.E., T.J. Hixon and J.D. Hoit (1986) "Respiration as a laughing matter", in: *WHIMSY IV*, D.L.F. Nilsen, ed., Tempe: Arizona State University, Department of English, pp. 147-148.

Chevalier-Skolnikoff, S. (1973) "Facial expression of emotion in nonhuman primates", in: *Darwin and Facial Expression*, P. Ekman, ed., New York: Academic Press, pp. 11-89.

Citardi, M.J., E. Yanagisawa and J. Estill (1996) "Videoendoscopic analysis of laryngeal function during laughter", *Ann. Ot. Rhin. Lar.* 105:545-549.

Clynes, M. (1980) "The communication of emotion: Theory of sentics", in: *Emotion. Theory, Research, and Experience* (Vol. 1, *Theories of Emotion*), R. Plutchik and H. Kellerman, eds, New York: Academic Press, pp. 271-301.

Darwin, Ch. (1872) *The Expression of the Emotions in Man and Animals*, London: Murray [3rd edition edited by Paul Ekman, New York: Oxford University Press, 1998].

Dearborn, G.V.N. (1897) "The emotion of joy", *Psychol. Rev., Mon. Suppl.*, 2.

De Troyer, A., V. Ninane, J.J. Gilmartin, C. Lemerre and M. Estenne (1987) "Triangularis sterni muscle use in supine humans", *J. Appl. Physiol.* 62:919-925.

Ekman, P. (1997) "What have we learned by measuring facial behavior", in: *What the Face Reveals*, P. Ekman and E.L Rosenberg, eds, Oxford: Oxford University Press, pp. 469-485.

Ekman, P. and W.V. Friesen (1978) *The Facial Action Coding System (FACS)*, Palo Alto, CA: Consulting Psychologists Press.

Ekman, P., R.J. Davidson and W.V. Friesen (1990) "The Duchenne smile: Emotional expression and brain physiology II", *J. Personal. Soc. Psychol.* 58:342-353.

Fried, I., Ch.L. Wilson, K.A. MacDonald and E.J. Behnke (1998) "Electric current stimulates laughter", *Nature* 391:650.

Gallo, L.M. and S. Palla (1995) "Activity recognition in long-term electromyograms", *J. Oral. Rehabil.* 22:455-462.

Grammer, K. (1990) "Stangers meet: Laughter and nonverbal signs of interest in opposite-sex encounters", *J. Nonverbal Behav.* 14:209-236.

Habermann, G. (1955) *Physiologie und Phonetik des lauthaften Lachens*, Leipzig, Germany: Barth.

Hall, G.S. and A. Allin (1897) "The psychology of tickling, laughing, and the comic", *Am. J. Psychol.* 9:1-41.

Hecker, E. (1873) *Die Physiologie und Psychologie des Lachens und des Komischen*, Berlin: Dümmler.

Heller, H.V. (1902) *Grundformen der Mimik des Antlitzes*, Vienna, Austria: Schroll.
Hjortsjö, C.H. (1970) *Man's Face and Mimic Language*, Lund: Studentlitteratur.
Hoit, J.D., B.L. Plassman, R.W. Lansing and T.J. Hixon (1988) "Abdominal muscle activity during speech production", *J. Appl. Physiology* 65:2656-2664.
Jürgens, U. (1998) "Neuronal control of mammalian vocalization, with special reference to the squirrel monkey", *Nat. Wiss.* 85:376-388.
Keltner, D. and G.A. Bonanno (1997) "A study of laughter and dissociation: Distinct correlates of laughter and smiling during bereavement", *J. Personal. Soc. Psychol.* 45:1313-1324.
Lieberman, P. (1975) *On the Origins of Language*, New York: Macmillan.
Lloyd, E.L. (1938) "The respiratory mechanism in laughter", *J. Gen. Psychol.* 19: 79-189.
Luschei, E.S., L. Ramig, K. Baker and M. Smith (1997) *Coordination of the laryngeal and respiratory systems during spontaneous laughter*, Unpublished Manuscript, Department of Speech Pathology, University of Iowa, Iowa City, USA.
Moore, P. and H. von Leden (1958) "Dynamic variations of the vibratory pattern in the normal larynx", *Fol. Phoniat.* 10:205-238.
Mowrer, D.E., L.L. LaPointe and J. Case (1987) "Analysis of five acoustic correlates of laughter", *J. Nonverbal Behav.* 11:191-199.
Niemitz, C. (1990) "Visuelle Zeichen, Sprache und Gehirn in der Evolution des Menschen—Eine Entgegnung auf McFarland", *Z. Sem.* 12:323-336.
Nwokah, E., H. Hsu, P. Davies and A. Fogel (in press) "The integration of laughter and speech in vocal communication: A dynamic systems perspective", *JSLHR.*
Nwokah, E., P. Davies, A. Islam, H. Hsu and A. Fogel (1993) "Vocal affect in three-year-olds: a quantitative acoustic analysis of child laughter", *J. Acoust. Soc. Amer.* 94:3067-3090.
Piderit, T. (1867) *Mimik und Physiognomie*, Detmold, Germany: Meyer.
Plessner, H. (1941) *Lachen und Weinen*, Bern, Switzerland: Francke.
Ploog, D. (1986) "Biological foundations of the vocal expressions of emotion", in: *Emotion: Theory, Research, and Experience* (Vol. 3), R. Plutchik and H. Kellerman, eds, New York: Academic Press, pp. 173-197.
Poeck, K. and G. Pilleri (1963) "Pathologisches Lachen und Weinen", *Schw. Arch. Neurol. Neurochirur. Psychiat*, 92:323-370.
Preuschoft, S. (1995) *'Laughter' and 'smiling' in Macaques--An Evolutionary Perspective*, Utrecht: University of Utrecht.
Provine, R.R. (1996) "Laughter", *Am. Sci.* 84:38-45.
Provine, R.R. and Y.L. Yong (1991) "Laughter: A stereotyped human vocalization", *Ethology* 89:115-124.
Raulin, J.M. (1900) *Le rire et les exhilarants*, Paris: Baillière.

Rothgänger, H., G. Hauser, A. Cappellini and A. Guidotti (1998) "Analysis of laughter and speech sounds in Italian and German students", *Nat. Wiss.* 85:394-402.

Ruch, W. (1990) *Die Emotion Erheiterung*, Unpublished Habilitation thesis, University of Düsseldorf, Department of Psychology, Düsseldorf, Germany.

Ruch, W. (1993) "Exhilaration and humor", in: *Handbook of Emotions,* M. Lewis and J.M. Haviland, eds, New York: Guilford Publications, pp. 605-616.

Ruch, W. (1994) "Extraversion, alcohol, and enjoyment", *Pers. Individ. Diff.* 16:89-102.

Ruch, W. (1997) "State and trait cheerfulness and the induction of exhilaration: A FACS study", *Europ. Psychol.* 2:328-341.

Santibañez, G. and S. Bloch (1986) "A qualitative analysis of emotional effector patterns and their feedback", *Pav. J. Biol. Sci.* 21:108-116.

Schroetter, H. (1925) "Zur Kenntnis des Energieverbrauches bei emotiven Äußerungen des Seelenlebens", *Monatsschr. Ohrenheilkd. Laryngo-Rhinol.* 59:82-108.

Sumitsuji, N. (1967) "Electromyographic studies on the facial expression", *Psych. Neurol. Jap.* 69:1101-1119.

Tanaka, M. (1976) "Psychophysiological study on the emotional reaction induced by television and movies in schizophrenia", *Med. J. Os. Univ.* 28: 205-216.

Tanaka, M. and N. Sumitsuji (1991) "Electromyographic study of facial expressions during pathological laughing and crying", *Electromyography and Clinical Neurophysiology* 31:399-406.

Tartter, V.C. (1980) "Happy talk: Perceptual and acoustic effects of smiling on speech", *Percept. Psychophys.* 27:24-27.

Van Hooff, J.A.R.A.M. (1972) "A comparative approach to the phylogeny of laughter and smiling", in: *Non-verbal Communication,* R.A. Hinde, ed., Cambridge: Cambridge University Press, pp. 209-241.

Yamada, N., H. Tanabe, H. Kazui, M. Ikeda, M. Hashimoto, S. Sakamoto, J. Sawada and J. Shiraishi (1994) "A case of lobar atrophy with stereotypic unusual laughing", *Brain and Nerve* 46:781-786.

AFFECT BALANCE AND TOTAL AFFECT VERSUS POSITIVE AND NEGATIVE AFFECT AS FUNDAMENTAL MEASURES OF EMOTIONAL EXPERIENCE: SIMPLE STRUCTURE IS NOT ALWAYS SO SIMPLE

MICHAEL W. GILLESPIE

Department of Sociology, University of Alberta, Edmonton, Alberta T6G 2H4, Canada

ABSTRACT

This paper compares two ways of representing the two-dimensional structure that emerges from a factor analysis of self-report inventories of emotional experience. The first and dominant representation is produced by rotating the two factors to simple structure. The result is two (apparently) unidimensional, relatively independent factors termed positive and negative affect. The second is the less commonly used unrotated structure that treats each emotion as a complex measure of two dimensions: valence--a bipolar continuum with positive emotions loading at one end and negative emotions loading at the other—and arousal, a unipolar continuum on which both positive and negative emotions load positively. Since rotation changes the pattern of correlations between the factors and any criterion variable used to test its validity, judgment of the relatively validity of the two structures requires the comparison of the two sets of correlations generated by the two structures. To this end, I report the results of two structural equation models of survey data from two different data sources: emotional responses to Carter and Reagan in the 1980 American Election Survey and three surveys of quality of life that use the Bradburn affect-balance scale. The results, I argue, provide compelling evidence for the validity of the unrotated solution.

1. Introduction

1.1. The two-dimensional structure of self-report inventories of emotional experience

This paper addresses the two-dimensional structure of self-report inventories of emotional experience.[1] The need for two dimensions arises in factor analyses of

---

[1]By "emotional inventory," I mean a set of questions in a survey (interview or self-administered interview) that asks the respondent whether he or she has experienced a set of emotions within some specified time. The time unit, usually the "past week" or "past few weeks," can vary between the immediate ("right now") and the extended ("the past year." The response categories can vary between dichotomous ("yes" or "no") to a variety of graded sets (e.g., "never," "hardly ever," "sometimes," "frequently," "all the time"). Most inventories do not specify an object of the emtions, but some, for example, the emotional response to political figures, do. The structure described above generally occurs, regardless of the difference in the time reference, the response categories used, or whether or not an object of emotion is specified.

responses to these inventories due to the relatively small magnitude of the negatively signed correlations between positive and negative emotions. Whether they use factor analysis or not, researchers from a variety of disciplines typically respond to these correlations by rotating the two-dimensional solution to simple structure (Bradburn, 1969; Watson & Tellegen, 1985). This solution divides reports of emotional experience into two (apparently) unidimensional, relatively independent subsets of positive and negative affect.

Table 1. Correlations of Emotional Responses to Reagan in the 1980 Election Campaign with One Another and With Selected Criterion Variables

	Variable											
	h	p	s	a	f	d	u	ft	pq	np	talk	n-int
h	1.0											
p	.48	1.0										
s	.31	.28	1.0									
a	-.21	-.17	-.08	1.0								
f	-.26	-.20	-.14	.28	1.0							
d	-.31	-.25	-.13	.46	.33	1.0						
u	-.24	-.26	-.11	.29	.55	.38	1.0					
ft	.61	.53	.28	-.39	-.43	-.46	-.46	1.0				
pq	.61	.55	.29	-.34	-.38	-.43	-.43	.76	1.0			
nq	-.49	-.42	-.25	.28	.34	.40	.36	-.61	-.62	1.0		
talk	.07	.05	.09	.12	.16	.12	.16	-.01	.01	-.01	1.0	
n-int	-.12	-.13	-.11	-.08	-.06	-.05	-.07	-.02	-.08	.09	-.28	1.0

*Source: 1980 American Election Survey (Miller, 1983). Listwise deletion of missing cases (n = 1,134). The emotions are: hope (h), pride (p), sympathy (s), anger (a), fear (f), disgust (d), and unease (u). The criterion variables are the Reagan feeling thermometer (ft), a scale of positive leadership qualities attributed to Reagan (pq), a scale of negative leadership qualities attributed to Reagan (nq), talk about the campaign (talk), and interest in the campaign coded to reflect no interest (n-int).

The correlations in Table 1 illustrate the pattern that gives rise to the two-dimensional structure. The data on which these correlations are based come from the 1980 American National Election Survey conducted by the University of Michigan's Institue for Survey Research (Miller, 1982). Table 1 contains the correlations between seven emotional responses to Reagan plus a set of criterion variables described in the next section. Three emotions—hope, pride, and sympathy—are positive. Four—anger, fear, disgust, uneasy—are negative. Looking at just the correlations between the emotions, we see moderate, positive correlations between emotions with the same value (positive with positive, or negative with negative) accompanied by smaller negative correlations between emotions with the opposite valence (positive with negative). The fact that positive and negative emotions form two relatively independent clusters seems to justify their separation on *prima facie* grounds.

A less obvious way of representing these correlations, however, uses an unrotated structure that treats each item (emotion) as measuring two dimensions. One dimension is a bipolar continuum with positive emotions loading at one end and negative emotions loading at the other. The second dimension is a unipolar continuum on which both positive and negative emotions load positively. Emotions with the same "sign" cluster together because of their consistent loadings on both dimensions, while the inconsistent loadings of oppositely signed emotions produce oppositely-signed correlations that cancel one another out.

The first unrotated dimension clearly lends itself to being interpreted as a measure of emotional valence, the extent to which the person's emotional experience is positive or negative, "good" or "bad," "pleasant or unpleasant." Interpretation of the second unrotated dimension is less straightforward. While some methodologically minded investigators interpret it as representing an artificial acquiescent response set (Green *et al.*, 1993), others, particularly those who view emotion from a circumplex perspective, interpret it as a measure of emotional arousal (Russell & Pratt, 1980; Fisher, 1997). The compatability of the two unrotated dimensions with valence and arousal would appear to give it a theoretical advantage over the rotated solution that divides emotion into positive and negative affect. Both valence and arousal were seen as fundamental dimensions of emotional experience in the early scientific psychology of Wilhelm Wundt (1906). This theory received empirical support from the pioneering work of Osgood and his associates on both the dimensions of semantic space (Osgood & Tannenbaum,1957) and the dimensions of facial expression of emotion (Osgood, 1966). Moreover, valence and arousal continue to appear in the current work of Shaver and his associates on the meaning of emotion words (Shaver *et al.*, 1987) and in the work of Peter Lang and his associates on the dimensions that underlie affective picture processing (Bradley *et al.*, 1993; Lang, 1984).[2]

---

[2] The concept of emotional arousal, in particular, has played a significant if somewhat controversial role in psychological theories of emotion. The concept figured prominently in theories of emotion developed in the middle of the twentieth century due to the discovery by physiologists in the 1950's of the reticular activating formation (Morruzi & Mougan, 1949). Probably the most famous of these theories was that of Schacter and Singer (1962), (but see Duffy [1957]). It treats emotion as a combination of physiological arousal plus cognitive cues. The concept of arousal fell somewhat into disfavor due to difficulties in demonstrating the role of the reticular activating formation in emotion (Lacey, 1967), problems in replicating the Schacter and Singer experiment (Marshall, *et al.*, 1979; Maslach, 1969, but see Schachter & Singer, 1979), and the rise of discrete emotion theories that regard as simplistic the attempt to reduce emotion to valence and arousal (Ekman, 1993; Izard, 1991; Tomkins, 1991). Nevertheless, arousal continues to play a role in theories of emotion to date for a variety of reasons in addition to the work referred to above. First, recent work by Thayer (1989) that uses both self-report and physiological measures of arousal successfully addresses the earlier problems raised by Lacey. Second, the work of Schacter and Singer spawned a "misattribution of arousal paradigm," and, recently, Sinclair and his associates, working in this paradigm, successfully "replicated" the Schachter and Singer paradigm using a design superior to that of both Schacter and Singer and the two unsuccessful attempts at replication (Sinclair, *et al.*, 1994). Finally, discrete emotion theorists, particularly Izard (1991), have come to see discrete and dimensional approaches to emotion as complementary rather than competitive. The issue that

Despite the compatability of the unrotated solution with theories of emotion, the division of emotional inventories into measures of positive and negative affect continues as the dominant position. Although interesting theories have emerged to justify this position (Cacioppo *et al.*, 1999; Marcus, 1988; Watson *et al.*, 1999), the basis for this dominance seems to be empirical. For example, Watson (1988) uses the finding that self-reported stress correlates positively with negative affect but is independent of positive affect as support for keeping positive and negative affect separate. All this finding does, however, is support the need for two dimensions to represent measures of affect. It begs the question of whether the unrotated or rotated solution is the best way to represent this structure.

The reason for this indeterminancy stems from an easily demonstrated but apparently unappreciated result of factor rotation. As shown in below, rotation changes the pattern of correlations between the factors and any criterion variable used to test the validity of the structure. In terms of the unrotated structure, Watson's result translates into a finding of negative and positive correlations of self-reported stress with valence and arousal, respectively. In other words, self-reported stress "behaves" empirically like any other report of negative effect, so that there is no reason to privilege this finding as evidence for the rotated solution. Moreover, the same argument can be applied to the array of empirical evidence recently marshaled in support of the unrotated solution (Cacioppo *et al.*, 1999; Watson *et al.*, 1999). A balanced approach to the evidence requires that the results be interpreted in terms of both solutions. Although the issue is in principal unresolvable, one can argue for a provisional position by comparing the interpretability of the two sets of results on grounds of parsimony, theoretical reasonablness, ease of interpretation, and other criteria.

To illustrate this strategy, I compare in sub-section 1.2 rotated and unrotated factor plots based on a factor analysis of the correlations reported in Table 1. In addition to the seven emotion items, the analysis includes five criterion variables: three that I interpret as relatively pure (unidimensional) measures of valence and two as less pure measures of arousal. Using Lisrel notation (Jöreskog & Sörbom, 1988a), I demonstrate in sub-section 1.3 the reason for the different behavior of these criterion variables in the two plots. Sectons 2 and 3 apply this strategy to two sets of results. The first set is based on an analysis of the American election survey data that adds variables on Carter to the variables on Reagan. The set second is based on three surveys in which the emotion inventory is used as a measure of quality of life.

---

remains and which defines the focus of this paper, however, is how best to use reports of discrete emotions to measure emotion dimensions.

1.2. Rotated and Unrotated Factor Plots of the Emotions Aroused by Reagan Candidacy

Figure 1 contains the the two-dimensional rotated factor plot from a factor analysis of the correlations in Table 1. In addition to the seven emotions, the matrix includes five criterion variables: (a) the Reagan "feeling thermometer," a global measure of emotional response to Reagan that asks respondents to place themselves on a scale from 0 to 100 according to how warm or cold they feel towards Reagan, (b and c) appraisal of the Reagan's leadership ability by means of a six-item Likert scale of positive leadership qualities (pq) and a three-item Likert scale negative qualities (nq), and (d and e) two measures of political motivation: whether the respondent talks about the campaign (talk) and the respondent's interest in the campaign coded so that a high score represents no interest (n-int).[3] Looking at this plot, we see that the three positive emotions--sympathy (s), pride (p), and hope (h)--define the horizontal axis, positive affect, and that the four negative emotions—disgust (d), uneasy (u), fear (f), and anger (a) define the vertical axis, negative affect, albeit not as well.

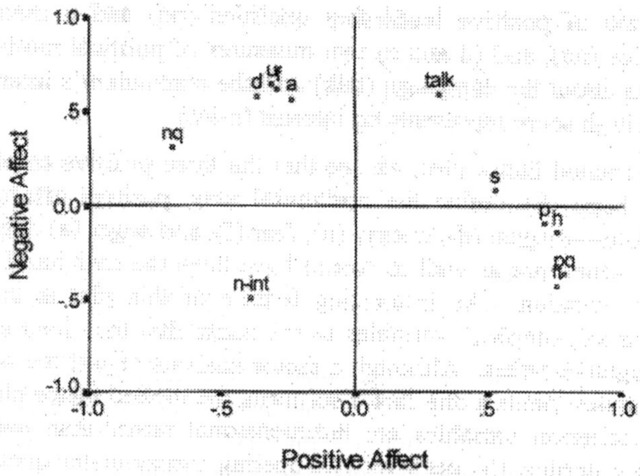

Figure 1. Rotated Factor Plot of Subjects' Emotional Response to Reagan, Appraisal of Reagan's Leadership Qualities, and Interest in the Campaign. Source: the University of Michigan 1980 American Election Survey (n = 1,134). See Table 1 for abbreviations legend.

---

[3] Despite the inclusion of these additional variables, the two-dimensional solution contintues to be the optimum solution using the usual criterion of extracting only factors with eigenvalues greater than one. To facilitate comparison of this factor plot with the unrotated one (see Figure 2), I used varimax rather than oblique rotation. In any event, oblique rotation (using "direct oblimin") yields only a trivial correlation between the two factors.

An interesting feature of this plot is that all five criterion variables appear as "complex" variables in the sense that they load significantly on both positive and negative affect. In the case of the feeling thermometer measure, this bidimensionality raises the possibility of two unidimensional feeling thermometer questins--one that asks how warm the respondent feels about Reagan, the other that asks how cold the respondent feels about Reagan. In the case of the talk and interest variables, this bidimensionality also raises the possibility of breaking these questions down into two sets of unidimensional items: interest and talk about the positive feelings aroused by Reagan and interest and talk about the negative feelings aroused by Reagan. Finally, a factor analysis that combines the respondent's appraisals of *both* positive and negative leadership qualities yields a single factor (Abelson *et al.*, 1982). The factor plot above, however, suggests that this factor represents a complex variable. Again, this bidimensionality raises the possibility of asking two questions about the feelings that these appraisals raise: one question about the positive feelings raised by the appraisal or Reagan and one about the negative feelings raised by this same appraisal.

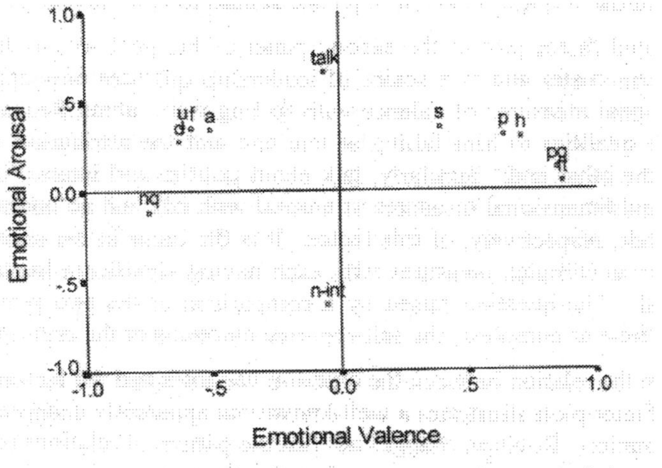

Figure 2. Unrotated Factor Plot of Subjects' Emotional Response to Reagan, Appraisal of Reagan's Leadership Qualities, and Interest in the Campaign. Source: the University of Michigan 1980 American Election Survey (n = 1,134). See Table 1 for abbreviations legend.

The unrotated factor plot in Figure 2 avoids these problems of interpretation. Looking at this plot, we see that the emotions now become complex variables, loading on both dimensions, while the criterion variables appear as more or less unidimensional markers of the two dimensions. The feeling thermometer and the respondent's attribution of positive qualities to Reagan's of leadership, for example, define the positive end of the horizontal axis, while the attribution of negative qualities to Reagan defines the negative end. Similarly, talk about the campaign and low interest in the campaign, falling at opposite ends of the vertical axis, define this factor. Using these criterion variables to interpret the two dimension, one can argue that the horizontal dimension represents the *valence* of the respondent's feelings about Reagan, a continuum that stretches from negative to positive. While offering a less straightforward interpretation, the second factor appears to represent the degree to which the respondent is activated or aroused by the campaign. The loadings of the seven emotions on the two unrotated axes, of course, support these interpretations. The three positive emotions load positively on the horizontal (valence) dimension, while the four negative emotions load negatively. All seven emotions, on the other hand, load positively on the second dimension (arousal).

1.3. The Algebra of Factor Rotation

The shift in the relation between the criterion variables and the dimensions in the rotated and unrotated factor plots illustrates the consequence of rotation described in the previous section. Rotation changes not just the pattern of relations (or loadings) between the factors and their putative measures but also the pattern of correlations between the factors and *any* variable, whether intended as a measure or not. This feature is particularly easy to demonstrate in the case of a confirmatory factor analysis that restricts the loadings to values of 1, 0, or –1. Eqs. 1 - 7 use LISREL notation to make this point (Jöreskog & Sörbom, 1988a).

Consider, first, the unrotated solution specified in Eq. 1 (see Figure 2). In this equation "y" is a 7 x 1 column vector of the three positive and four negative emotions, "Λ_y" is a 7 x 2 matrix of factor loadings, "η" is a 2 x 1 column vector of factors--valence (η_v) and arousal (η_a)--and "ε" is a 7 x 1 column vector of measurement error terms.

$$(1) \quad y = \Lambda_y \eta + \varepsilon$$

Eq. 2 presents the structure of the rotated solution (See Figure l) as a linear transformation of the unrotated structure. As eq. 2 shows, the 7x1 column vectors of the emotions and measurement error terms remain unchanged. Post-mulitiplying the 7x2 matrix of factor loadings by the transformation matrix (T), however, rotates the loadings from complex to simple structure. In addition, post-multiplying the loading matrix requires pre-multiplication of the 2x1 column vector of unrotated factors by the *inverse* of transformation matrix (T^{-1}).

$$(2) \quad y = \Lambda_y (T T^{-1}) \eta + \varepsilon = (\Lambda_y T)(T^{-1} \eta) + \varepsilon = \Lambda_y^* \eta^* + \varepsilon$$

The need to pre-multiply the column vector of factors by the inverse of the transformation matrix follows, of course, from the elementary mathematical principle that an operation on one part of an equation requires a compensating operation somewhere else in order to preserve the equality. Cliff (1987, p. 320) refers to this requirement as "balancing the books" in his discussion of factor rotation.

A consequence of this transformation is that, if valence and arousal are fundamental, positive and negative affect represent artifical linear combinations of these basic dimensions. Eqs. 3 and 5 present these relations. As shown by eq. 3 and eq. 4, respectively, positive affect equals arousal plus valence, while negative affect equals arousal minus valence.[4]

(3) $\eta_{\mathrm{p}}^{*} = \eta_{\mathrm{v}} + \eta_{\mathrm{a}}$

(4) $\eta_{\mathrm{n}}^{*} = -\eta_{\mathrm{v}} + \eta_{\mathrm{a}}$

The key implication of these relations is that the covariance of any criterion variable, x, with positive and negative affect, respectively, will be an artifical linear combination of the covariances of the criterion variable with valence and arousal. Eqs. 5 and 6 present this result formally. The covariance of x with positive affect equals sum of the covariances of x with valence and arousal, while the covariance of x with negative affect equals the covariance between x and arousal minus the covariance between x and valence.

(5) $\sigma_{\mathrm{x},\eta_{\mathrm{p}}*} = \sigma_{\mathrm{x},\eta_1} + \sigma_{\mathrm{x},\eta_2}$

(6) $\sigma_{\mathrm{x},\eta_{\mathrm{n}}*} = -\sigma_{\mathrm{x},\eta_1} + \sigma_{\mathrm{x},\eta_2}$

These results show why self-reported stress correlates with negative affect but not positive affect. The positive covariance of self-reported stress with arousal augments its positive covariance with "negative" valence but cancels out its negative covariance with valence. These results also explain the behavior of these criterion variables in the factor plots in Figures 1 and 2. As suggested by the unrotated factor plot, three of the criterion variables—the Reagan feeling thermometer, appraisal of Reagan's positive qualities, and appraisal of Reagan's negative qualities—covary with valence but not arousal. In terms of eqs. 5 and 6, these results translate into positive covariances of the Reagan feeling thermometer and the attribution of positive qualities with positive affect and into negative

[4] It is important to remember that the question of which dimensions are fundamental is one of substance, not mathematics. If one regards positive and negative affect as fundamental, one could reverse the transformation process and express valence and arousal as artifical combinations of positive and negative affect. Eq. 3 would be rewritten as: $\eta_{\mathrm{v}} = .5 \times \left(\eta_{\mathrm{p}}^{*} - \eta_{\mathrm{n}}^{*}\right)$, and Eq. 4 would be rewritten as: $\eta_{\mathrm{a}} = .5 \times \left(\eta_{\mathrm{p}}^{*} + \eta_{\mathrm{n}}^{*}\right)$. In other words, valence equals positive affect minus negative affect, while arousal equals positive affect plus negative affect.

covariances with negative affect. In the case of the attribution of negative leadership qualities, the result is a negative covariance with positive affect and a positive covariance with negative affect. The second factor plot shows that the two measures of campaign activation—talk about politics and (no) interest in the campaign—covary with arousal but not with valence. In terms of eqs. 5 and 6, these results translate into a positive covariance between talk about politics with both positive and negative affect and a negative covariance between no interest in politics and these two variables.

Eq. 7 extends the results of Eqs. 5 and 6 to the analysis of the emotional responses to both candidates in the 1980 election. It consists of the 4x4 variance-covariance matrix of the positive and negative affect dimensions for both Reagan and Carter. The first two rows and columns correspond to the positive and negative affect dimensions, while the third and fourth rows and columns correspond to the positive and negative affect dimensions for Carter. Eq. 7 presents these variances and covariances as linear combinations of the variances and covariances for the valence and arousal dimensions for the two candidates. To simplify the presentation, however, I assumed that the valence - arousal covariances both within and between candidates are zero. Consequently, eq. 7 only approximates the true variance-covariance matrix.

$$\underset{4\times 4}{\Psi^*} \cong \left(\begin{array}{cc:cc} \sigma_1^2+\sigma_2^2 & & & \\ -\sigma_1^2+\sigma_2^2 & \sigma_1^2+\sigma_2^2 & & \\ \hdashline \sigma_{31}+\sigma_{42} & -\sigma_{31}+\sigma_{42} & \sigma_{31}^2+\sigma_{42}^2 & \\ -\sigma_{31}+\sigma_{42} & \sigma_{31}+\sigma_{42} & -\sigma_{31}^2+\sigma_{42}^2 & \sigma_{31}^2+\sigma_{42}^2 \end{array}\right)$$

(7)

The 2x2 sub-matrix in the lower left submatrix of eq. 7 is particularly relevant to this discussion. It contains the covariances of Reagan positive and negative affect with Carter positive and negative affect. The simplifying assumptions described in the previous paragraph imply equalities between the two main diagonal elements and between the two off-diagonal elements. That is, the covariance between Reagan and Carter positive affect equals the covariance between Reagan and Carter negative affect. Both equal the covariance between Reagan and Carter valence (σ_{31}) plus the covariance between Reagan and Carter arousal (σ_{42}). Similarly, the covariance between Reagan positive affect and Carter negative affect equal the covariance between Reagan negative affect and Carter positive affect. Both equal the covariance between Reagan and Carter arousal minus the covariance between Reagan and Carter valence. It seems likely that the covariance between Reagan and Carter valence is negative (since liking one candidate should correlate with disliking the other) and that the covariance between the Reagan and Carter arousal is positive (since one is more likely to be aroused by the pair of candidates rather than just one or the other). If these hypotheses are correct, the two covariances in the main diagonal will cancel one another, while they will augment one another in the off-diagonal entries.

2. Two Structural Equation Models of the Emotional Responses to Reagan and Carter during the 1980 Presidential Campaign

Table 2 contains the loadings of the seven emotions aroused by Reagan and those aroused by Carter on two sets of four factors in a confirmatory factor analysis of the reactions of respondents in the 1980 American election survey to the candidacies of Reagan and Carter. The first model treats the each emotion as a measure of valence and arousal; the second divides the emotions into measures of positive and negative affect. Both models use very simple loadings of either 1, -1, or 0. The item reliabilities and goodness of fit statistics apply to both models. Although the chi-square values (LR) are high relative to the degree of freedom, both goodness of fit indexes suggest reasonable fit. Because the emotion items are dichotomous, the data used in the analyses consist of tetrachoric, polychoric, and biserial correlations (Jöreskog & Sörbom, 1988b).[5]

2.1. Comparing the Correlations of Selected Criterion Variables with Valence and Arousal with the Correlations of the Criterion Variables with Positive and Negative Affect

Table 3 contains the correlations between several criterion variables and the two sets of latent variables defined in Table 2. The criterion variables represent relatively pure measures of either valence or arousal/activation. As in the exploratory factor analysis described earlier, the former consists of candidate feeling thermometers and the appraisal of the candidate's qualities as a leader. The latter consist of talk about politics and interest in the election (now coded so that a high score represents strong interest) and a new variable, whether the respondent cares who wins the election.

---

[5] The use of tetrachoric, polychoric, and biserial, and polyserial correlations represent estimates of the correlations between the indicators that would have been obtained if the dichotomous and ordinal variables were continuous. (A tetrachoric correlation is based on the relation between two dichotomous variables. A polychoric correlation is based on the relation between two ordinal variables. A biserial correlation is based on the relation between a dichotomous variable and a continous variable. A polyserial correlation is based on the relation between an ordinal variables and a continuous variable.

Table 2. Reliabilities and Loadings of Emotion Items for Reagan and Carter on Unrotated and Rotated Latent Affect Variables

	Model							
	Unrotated				Rotated			
Item	Rvalnc	Rarousl	Cvalnc	Carousl	Rpaffct	Rnaffct	Cpaffct	Cnaffct
rhope	1.00	1.00	-	-	1.00		-	-
rproud	1.00	1.00	-	-	1.00		-	-
rsymp	1.00	1.00	-	-	1.00		-	-
rangry	-1.00	1.00	-	-	-	1.00	-	-
rfear	-1.00	1.00	-	-	-	1.00	-	-
rdisgust	-1.00	1.00	-	-	-	1.00	-	-
runeasy	-1.00	1.00	-	-	-	1.00	-	-
chope	-	-	1.00	1.00	-	-	1.00	-
cproud	-	-	1.00	1.00	-	-	1.00	-
csymp	-	-	1.00	1.00	-	-	1.00	-
cangry	-	-	-1.00	1.00	-	-	-	1.00
cfear	-	-	-1.00	1.00	-	-	-	1.00
cdisgust	-	-	-1.00	1.00	-	-	-	1.00
cuneasy	-	-	-1.00	1.00	-	-	-	1.00

LR = 669.14, df = 239, GFI = .979, AGFI = .976; Analysis based on tetrachoric, polychoric, and biserial correlations.

The results for the unrotated model lend themselves to a relatively parsimonious interpretation. The two candidate valence factors correlate substantially and in the predicted direction with criterion variables that measure the respondent's liking of the candidate and attribution of positive and negative qualities. Naturally, the correlations are higher between the candidate valence factors and the criterion variables that pertain to the same candidate. With two exceptions, the candidate valence factors are independent of the arousal criterion variables, and the magnitude of the two correlations is relatively small in the case of these exceptions. Both arousal factors correlate significantly and positively with all three criterion variables that measure campaign activation/arousal, and, with one exception, all are independent of the valence criterion variables.

The correlations between the criterion variables and positive and negative affect, in the right hand side of Table 2 illustrate the points made the discussion of the algebra of factor rotation. These correlations appear to be an artifact of the correlations of the criterion variables with valence and arousal. The correlations of positive affect with the criterion variables that represent pure measures of valence mirror the correlations of negative affect with these variables. The correlation of a criterion variable with positive affect equals its correlation with negative affect in magnitude, but the two differ in sign. This mirroring occurs because the criterion variables covary with the valence component of positive and negative affect but are independent of the arousal component. All three criterion measures of arousal correlate positive with both positive and negative affect. These correlations occur because these criterion variables covary positively with the arousal component of positive and negative affect but are independent of valence.

I also decided to use the respondent's party identification and vote in the 1980 election as additional criterion variables for judging the relative validity of the unrotated and rotated solutions. A problem with these variables is that both consist of a set of partially-ordered nominal categories. A correspondence analysis of the cross-classification of these variables, however, yields two readily interpretable dimensions for each variable (Weller & Romney, 1990). The first and strongest dimension corresponds to the valence of party identification and vote. In the assignment of scores on this dimension to the categories of party identification (PIDval) strong Republicans and strong Democrats define the negative and positive extremes, respectively, "apolitical" respondents and "pure independents" occupy the middle, and "weak identifiers" and "independent leaners" lie between the middle and the two extremes. Respondents who voted for Reagan or Carter with a strong preference define the two extremes of this dimension when applied to the respondent's vote (VTval). Respondents who neither voted for nor preferred a candidate fall in the middle. Those who voted for a candidate with weak support or else who either strongly or weakly preferred a candidate fall between the middle and one of the two extremes.

Table 3. Correlations of Unrotated and Rotated Latent Affect Variables for Carter and Reagan with Selected Criterion Variables

	Model							
	Unrotated				Rotated			
Criterion	Rvalnc	Rarousl	Cvalnc	Carousl	Rpaffct	Rnaffct	Cpaffct	Cnaffct
Rft	.815**	-.016	-.355**	.030	.721**	-.705**	-.289**	.327**
Rpq	.772**	.040	-.419**	.050	.710**	-.642**	-.329**	.393**
Rnp	-.650**	.026	.358**	-.054	-.568**	.569**	.276**	-.342**
Cft	-.427**	-.076	.848**	.013	-.413**	.329**	.723**	-.736**
Cpq	-.338**	-.016	.757**	.084	-.310**	.281**	.681**	-.620**
Cnq	.194**	.100*	-.590**	-.018	.223**	-.119**	-.508**	.508**
Talk	-.047	.468**	-.115**	.432**	.191**	.263**	.120**	.325**
e-intrst	.068**	.360**	-.020	.274**	.239**	.113**	.120**	.160**
Carewin	.009	.216**	.003	.194**	.115**	.095**	.100**	.099**

*p < .05; **p < .01

The second dimension appears to measure the extent of the respondent's arousal or activation by the campaign. Strong identifiers and voters with a strong preference define the high ends of this dimesion for party identification (PIDact) and vote (VTact), respectively, while apoliticals and non-voters with no preference define the low end.

Table 4 contains the correlations between these criterion variables and the two sets of affect latent variables. With some exceptions, the pattern of correlations resembles the pattern observed in Table 3. The Reagan and Carter valence factors correlate substantially in the predicted direction with the valence components of party identification and vote. Both the Reagan and Carter arousal factors also correlate positively and significantly with the arousal components of party identification and vote, although the correlations are not strong. In addition, there are small but significant, positive correlations between Reagan valence and the activation components of both party identifcation and vote. Finally, Reagan arousal correlates negatively with the valence components of party identification and vote, although only the latter is significant. These exceptions to the pattern of correlations in Table 3 suggest that respondents with positive feelings toward Reagan were more active in the campaign.

Table 4. Correlations of Unrotated and Rotated Latent Affect Variables for Carter and Reagan with Party Identification and Vote in the November Election

	Model							
	Unrotated				Rotated			
Criterion	Rvalnc	Rarousl	Cvalnc	Carousl	Rpaffct	Rnaffct	Cpaffct	Cnaffct
PIDval	-.467**	-.077	.420**	-.024	-.455**	.362**	.342**	-.380**
PIDact	.172**	.160**	.034	.111**	.234**	-.071**	.027	.087**
VTval	-.587**	-.102*	.580**	-.052	-.576**	.454**	.463**	-.535**
VTact	.213**	.154**	.006	.138**	.267**	-.109**	.074**	.067*

*p < .05; **p < .01

The correlations of negative affect with the valence components of party identification and vote mirror the correlations of positive affect with these variables. Similar to the corresponding correlations in Table 3, this pattern is best seen as an artifact of the correlations between the valence components of party identification and vote with the valence component of the affect variables and their relative independence of the arousal. With respect to the activation components of party identification and vote, their correlations with positive and negative affect are generally positive though small and, in one case not significant. More noteworthy are the two negative, significant correlations between the activation components and Reagan negative affect. This departure from the pattern observed in Table 3, however, reflects the positive correlations of Reagan valence with the activation components of party identification and vote described in the previous paragraph. Again, these correlations in are best seen as an artifact of the corresponding correlation in the left half of this table.

2.2. Comparing the Correlations between Reagan and Carter Valence and Arousal with the Correlations between Reagan and Carter Positive and Negative Affect

An advantage of expanding the analysis to include the emotional reactions to both candidates is that the emotions aroused by one candidate can serve as criterion variables to judge the validity of the structure used to model the correlations between the emotions aroused by the other candidate. Table 5 contains the relevant correlations for the two sets of factors. The upper left-hand submatrix contains the correlations of the valence and arousal factors, while the lower right-hand submatrix contains the correlations of the positive and negative affect.[6]

Table 5. Correlations of Unrotated and Rotated Latent Variables for Reagan and Carter with One Another

	Model							
	Unrotated				Rotated			
Variable	Rvalnc	Rarousl	Cvalnc	Carousl	Rpaffct	Rnaffct	Cpaffct	Cnaffct
Rvalnc	1				-	-	-	-
Rarousl	.041	1			-	-	-	-
Cvalnc	.406**	.008	1		-	-	-	-
Carousl	.107*	.872**	.052	1	-	-	-	-
Rpaffct	-	-	-	-	1			
Rnaffct	-	-	-	-	.479**	1		
Cpaffct	-	-	-	-	.130*	.593**	1	
Cnaffct	-	-	-	-	.455**	.047	.528**	1

$^*p < .05$; $^{**}p < .01$

The comparison of the two sets of correlations provides perhaps the most compelling evidence for the superiority of the unrotated solution. The pattern of correlations in the upper left-hand submatrix is simple and easily interpreted. There are two substantial correlations. A moderate negative correlation between Reagan valence and Carter valence (r = -.406), and a strong positive correlation between Reagan arousal and Carter arousal (r = .872). All the remaining correlations, including the the significant negative correlation between Reagan valence and Carter arousal, are small by comparison. Valence and arousal appear to represent distinct, relatively independent factors both within and between candidates. Respondents who like one candidate tend to dislike the other, and respondents are typically aroused by the campaign or contest between Reagan and Carter rather than by one candidate or the other. This interpretation, of course, is consist with the positive correlations of both Reagan and Carter arousal with the criterion measures of arousal.

The correlations for the rotated factors in the submatrix in the lower-right triangle of Table 5 comprise both sensible and puzzling results. First, within candidates, there are substantial negative correlations between positive and

---

[6] It is, of course, logically possible to correlate the valence and arousal factors for one candidate with negative and positive affect factors of the other. Since adopting one structure (unrotated) for one candidate and the other (rotated) for the other candidate makes no sense, however, the cells in Table 5 that would contain these correlations are empty.

negative affect (r = -.470, -.528). Respondents who feel positive about a candidate tend not to feel negative about him, and vice versa. On the other hand, the cross-candidate correlations seem puzzling. There are substantial positive cross-candidate, cross-affect correlations (r = .455, .593), but the negative cross-candidate, same-affect correlations are small, and one is not significant (r = -.130, -047). It makes sense that feeling positive feelings about one candidate should go with negative feelings about the other. Why, then, are the presence of one set of feelings about a candidate not associated with the absence of these feelings for the other? While one should not discount the possibility of other interpretations, the most plausible answer is that all the correlations in the lower-left triangle of Table 5 are an artifact of the correlations between valence and arousal in the upper-right triangle.

2.3. Discussion of the Results of Analyzing the Emotional Response to Reagan and Carter

A person's report of his or her emotional response to an object represents a complex process. Part of the report is drawn from a small set of discrete emotions that have evolved due to the way they function to make good things seem better and bad things seem worse (Tomkins, 1991). In addition, a large part of the report consists of the many nuances that emotion words attempt to express. These nuances vary across cultures and within cultures wax and wane in currency as cultural fashions shift (Lutz, 1986; Shaver *et al.*, 1987).[7] Behind these sources lie least two dimensions. This two-dimensional structure suggests that the person's emotional response consists of a two-part process. The way researchers conceptualize this process, however, differs. The dominant theory argues that separate positive and negative activation systems comprise the two parts and that an object can activate one, the other, or both systems (Cacioppo & Bernston, 1994). In contrast, I argue that the two parts consist, first, of non-specific emotional response termed "arousal" or "activation" and, second, a valence component that attaches value to this emotional response.[8] My position resembles

---

[7] Consider, for example, the conversational shift from "you're welcome" to "no problem" as a response to to "Thank you." The possible changes in the emotional meaning that accompanies this shift seems a topic worthy of study.

[8] In terms of Tomkins (1991) theory of discrete emotions, surprise comes closest to providing a concrete example of undifferentiated arousal. Because surprise is typically followed quickly by a various emotions, some researchers seem reluctant to include surprise as an emotion (Shaver, et. al., 1987). The value attached to undifferentiated arousal probably comes from a variety of sources—the nuances of emotion words (Shacter and Singer, 1962; Sinclair, et. al., 1994), one or more discrete emotions (Tomkins, 1991), and positive and neural substrates described by Gray (1987) and Cacioppo and Berntson (1994). I do not pretend to understand how arousal and valence conjoin to produce an emotional experience, never mind the *report* of an emotional experience. In particular, I want to avoid the impression that a two-*part* process necessarily entails a two-*step* process.

earlier conceptions of emotion, particularly Schacter and Singer (1962), but also is consistent with the work of Lang and his associates on affective picture processing (Bradely *et al.*, 1994). In this work students' physical and physiological responses to slides--smiling, frowning, heart rate, and electrical skin conductance--are correlated with previously obtained ratings of each slide's valence and arousal. Indeed, their results and mine could be viewed as homologous in the sense that the candidacies of Reagan and Carter correspond to the slides and the self-reported emotions correspond to the valence and arousal ratings.

While the difference between the two conceptions might seem a matter of taste, a consideration of the phenomena of emotional experience and the ease of interpreting the correlations between the responses to Reagan and Carter should underscore the significance of the difference and and suggest that the choice of valence and arousal as fundamental dimensions of emotional experience is the correct one. With respect to the experience of emotion, it is hard to see how the feeling thermometer questions or the valence and arousal ratings obtained by Lang and his associates could work if positive and negative affect represented separate activation systems. In the case of the countless voter surveys that have used the feeling thermometer question, for example, someone—an interviewer, respondent, or, indeed, an analyst--would have felt the need for two questions (or sets of questions): one about "cold" feelings, the other about "warm" feelings.[9] Moreover, why, if there are two activation systems, does attribution of positive and negative qualities to the candidate yield a single factor rather than two.[10] For those who discount the phenomena of experience as a basis for making this choice, consider the correlations between the emotional responses to Reagan and Carter. Interpreting emotional experience as valence and arousal yields an extremely simple, easily interpreted result. Respondents who value one candidate tend to devalue the other, and respondents tend to respond (or not respond) to both Reagan and Carter, yoked together by virtue of their contest for the presidency, rather than to one candidate or the other. What, in contrast, in the two-activation system would produce substantial positive cross-candidate, cross-sytem correlations but weak negative or absent cross-candidate, same-system correlations?

---

[9] Marcus (1988) does raise this possibility that the feeling thermometer is two-dimensional in discussing the results of his analysis of respondents' reactions to Reagan and Mondale in the 1984 election survey. Rather than a felt need for a warm and cold question, however, he is motivated by a small difference between the magnitude of the positive and negative correlations of the Reagan feeling thermometer with positive and negative Reagan affect.

[10] Abelson, et. al. (1982) raise this same point in their analysis of the 1980 election data.

3. Valence and Arousal versus Positive and Negative Affect as Psychological Consequences of Stress

The preceeding sections focus on respondents' reports of their emotional reactions to specific objects. In subsequent sections, I examine the more typical situation in which data on emotion inventories is collected: quality of life surveys that specify no object of emotion.[11] In such surveys the emotion inventories appear to be used as measures of psychological well-being along with a variety of other measures, most notably self-reported happiness (Bradburn, 1969) and life satisfaction (Campbell & Converse, 1978). In analyzing data from these surveys, I move from a focus on the global and vague concept of quality of life to a more specific, albeit equally vague, concept of stress. I use life event inventories as the source of stress (the stressors) and responses to emotion inventories as a measure of the psychological consequences of the stress produced by the life events.

3.1. Two Models of the Relation between Stress and Affect

A hallmark of the literature of on social stress is that the consequences of stress are limited to negative life events such as the death of a spouse (Hobfall and Spielberger, 1987).[12] This position runs counter to earlier conceptions of the stress process that view both positive and negative life events as sources of stress (Holmes & Rahe, 1967; Selye, 1956). The reason for this shift in perspective is empirical. The results of countless studies suggest that negative life events lower scores on measures of psychological well-being. In contrast, positive life events, such as a job promomotion, appear to have no effect, either positive or negative, on well-being.

An important feature of this research is that the reports of negative emotion dominate scales of psychological well-being—for example, such as the DSM measures of depression. When researcher analyze the reports of negative and positive emotion separately, they find the consequences of negative life events are

---

[11] Because of the absence of information on the objects of emotion, some researchers might regard the emotion inventories as reports of "mood" rather than "emotion" (e.g., Frijda, 1994). While not disputing the need to attend to the differences among the variety of affects, I ignore this an similar distinctions on the grounds that most data on emotion inventories would not permit the study of these subtle conceptual differences.

[12] The same event, of course, can be viewed positively or negatively depending on the context of the event (Wheaton, 1994) and the way the event is resolved (Turner and Avison, 1992). I use the terms "positive" and "negative" fairly loosely, therefore, to refer to the way an event would be perceived typically or on average by the population studied. I term as "neutral" events that are ambiguous or for which good and bad consequences tend to balance out. Although one can categorize some events on a priori grounds—e.g., the death of a spouse versus an increase in one's finances--I use the data on the relation between the event and reports of emotional experience to determine its valence. Although reliance on the data to determine the status of an event introduces a certain degree of circularity, I break out of this circle by deciding when one of many failed predictions represents not the invalidity of a decision about the valence of a life event but a general pattern of no support for the theory.

limited to an increase in negative affect. Negative life events neither increase nor decrease positive affect. Moreover, the relation between positive life events and positive affect remains a mystery because those studies that examine the effects of both positive and negative life events either do not use reports of positive affect as part of their measure of well-being or do not separate positive and negative affect when both are present.

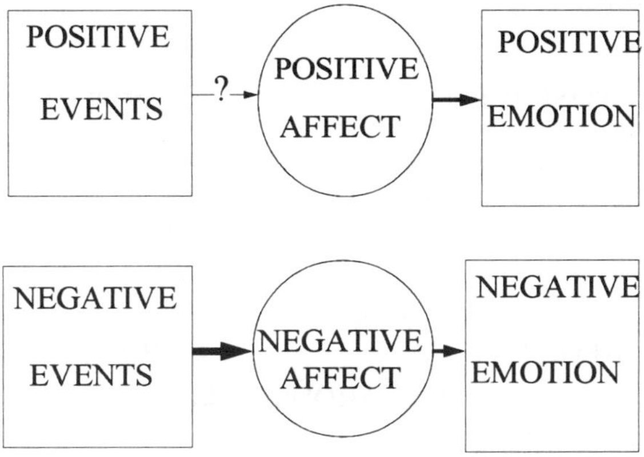

Figure 3. A Model of the Effects of Positive and Negative Life Events on Positive and Negative Affect

Figure 3 diagrams the model of the relation between life events and the responses to emotion inventories that has developed out of this research. The boxes contain the observed variables reported by the respondents in quality of life surveys: positive life events, negative life events, positive emotions, and negative emotions. The circles refer to latent variables, factors or constructs, that the researcher believes mediate the effects of the life events on the reports of emotional experience. The arrows in the diagram refer to the causal effects that the researcher believes link the observed variables and constructs. Unless a negative sign is specified, the reader should assume that the effect is positive. The model Figure 3 treats positive and negative emotions as measures of positive and negative affect, respectively. The only effect of life events on emotion that is specified is the positive effect of negative events on positive emotion through the effect of negative events on negative affect. The model is incomplete, however, because of the paucity of research on the relation between positive events and positive emotion. Consequently, I attach a to the arrow to leave open the possibility of a positive, negative, or no affect. This model, of course, relies on the rotated solution to represent the two-dimensional structure of emotion inventories.

Figure 4 contains a theoretical model of the relation between life events and the emotions reported in quality of life surveys. Based on the unrotated solution, it specifies valence and arousal as latent variables that mediate life events and reported emotion. It is a more complex model than the one specified in Figure 3. The reason for this complexity is that the life events affect both valence and arousal. In this respect, they resemble the slides studied by Lang and his associates. According to this theory behind the model in Figure 4, each life event, depending on its valence, will increase, decrease, or, in the case of "neutral" events, have no effect on valence. Both positive and negative, on the other hand, will increase arousal.[13]

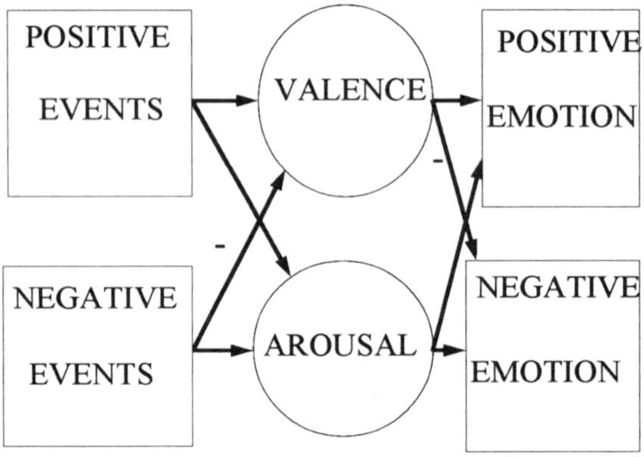

Figure 4. A Model of the Effects of Positive and Negative Life Events on Valence and Arousal.

Using the results from eqs. 5 and 6, one can derive implications about the relations between positive and negative life events and positive and negative emotion. These implications provide alternative explanations of the research results described above. In addition, they supply a hypothesis about the relation between positive life events and positive emotion. First, negative events increase negative emotion because they reduce valence and increase arousal. These two effects combine to produce a strong positive relation between negative life events and negative emotion. Second, positive events have a weak or non-existent net effect on negative emotion because because their positive effect on valence decreases negative emotion, but their positive effect on arousal increases it. These two effects tend to cancel one another out. Third, negative events have a weak or non-existent effect on positive emotion because their negative effect on valence reduces positive emotion, but their positive effect on arousal increases it. Again, these two effects tend to cancel one another out. Finally, positive events should increase positive emotion because they increase valence and increase arousal.

[13]To simplify the presentation, the models in Figures 3 and 4 do not include neutral events. Neutral events, by definition, should have no effect on the valence component of reported emotion. On the other hand, they will increase arousal. (See the previous note [12] for a discussion of neutral events.)

These two effects should combine to produce a strong positive relation between positive events and positive emotion. This last relation represents the hypothesis to be tested in the next section.

3.2. Results of an Analysis of the Effects of Life Events on Reported Emotion

To compare the results generated by the two models, I estimated two structural equation models using data from the 1978 American Quality of Life Survey (Campbell *et al.*, 1978). These data consist of 3,536 face to face interviews of a representative sample of the continental United States conducted by the University of Michigan's Survey Research Center. I used the ten-item Bradburn affect – balance scale to measure valence and arousal and positive and negative affect (Bradburn, 1969). The scale contains five positve emotions and five negative emotions. Table 6 contains thereliabilities and two sets of factor loadings for each item for the two models. The first model treats each item as a two-dimensional measure of valence and arousal, while the second model divides the items into two sets of unidimensional measures of positive and negative affect (Compare these loadings with those in Table 2.).

Table 7 contains estimates of the effects of three life events indexes of positive, negative, and "neutral" or ambiguous events, plus the respondent's age and sex, on the latent affect variables as defined by the unrotated and rotated models. The life events questions ask whether a particular event, such as death of the spouse occurred within the five year period prior to the interview. The respondent's sex is a dummy variable with men coded 1 and women coded zero. Although information on the number of events is available, I used each event coded as yes (1) or no (0) to construct the indexes.[14]

---

[14] The three sets of events are selected from a substantially larger inventory that consists of thirty-one events. In analysis of the full set, negative events dominate the inventory, but a large number appear to be neutral or ambiguous due to the absence of significant effects on valence. Nearly all of the events increase arousal, however. I refer to each set of life events as an index because I constrained the effects of the life events within a set to equal one another. This is equivalent to a simple sum of the number of events in a set that the respondent experienced. An index differs from a "scale" also based on the sum of the items in that its specification places no constraints on the relations between the events; instead, the constraints are limited to their effects. In a scale, on the other hand, the relations among the items are at the heart of judging the scale's reliability and validity.

Table 6. Loadings of the Ten Bradburn Items on Unrotated and Rotated Factors

		Model			
		Unrotated		Rotated	
Item	Reliability	Valence	Arousal	Positive affect	Negative affect
Excited	.242	1	1	1	0
Restless	.237	-1	1	0	1
Proud	.258	1	1	1	0
Lonely	.351	-1	1	0	1
Pleased	.368	1	1	1	0
Bored	.305	-1	1	0	1
On Top of the World	.194	1	1	1	0
Depressed	.310	-1	1	0	1
Things going My Way	.243	1	1	1	0
Upset	.288	-1	1	0	1

Source: The Quality of American Life. 1978 (Campbell and Converse, 1978). LR Chi-square = 1540.48, n = 3,536, df = 197, GFI = .965, AGFI = .942.

Consistent with the predictions based on the model in Figure 4, the results on the left-hand side of Table 7 show that the three-item indexes of positive, negative, and neutral events increase, decrease, or have no (net) effect on valence. All three indexes, however, have significant positive effects on arousal. Age has significant positive and negative effects on valence and arousal, respectively, although the age's effect on arousal is substantially stronger than its effect on valence. With respect to sex, there is no difference between men and women's valence, but men score slightly lower than women on arousal.

The results for the rotated model on the right hand side of Table 7 show, consistent with the literature, that negative life events increase negative affect. Contrary to the literature, negative events significantly reduce positive affect, but the coefficient is small. Consistent with the prediction derived from the model in Figure 4, positive events significantly and substantially increase positive affect. Neutral events also have significant positive effects on both positive and negative affect, although the coefficients are relatively small. While these last results may seem difficult to interpret from the perspective of the rotated model in Figure 3, they follow from the prediction of the unrotated model in Figure 4 that all events increase arousal. Indeed, all the relations on the right-hand side of Table 7 are best viewed as artifacts of the effects of age and sex on valence and arousal.

Table 7. Effects of Three Life Events[a] Indexes on Unrotated and Rotated Latent Affect Variables, Controlling for Age and Sex

	Model			
	Unrotated		Rotated	
Event Index	Valence	Arousal	Positive Affect	Negative Affect
Positive Events • *Job promotion* • *Received an honor* • *Improved finances*	.161*	.178*	.256*	-.012
Neutral Events • *Retired from a job* • *Child left home* • *Changed jobs*	.006	.095*	.070*	.051*
Negative Events • *Death of a spouse* • *Troubles with spouse* • *Injury or illness* • *Trouble finding a job*	-.238*	.158*	-.092*	.266*
Age	.081*	-.269*	-.115*	-.217*
DSex (1 = men)		-.111*	-.051*	-.087*

[a]The survey asked for events that occurred within the past five years rather than the more typical period of one year.

*Coefficients are partial, standardized regression coefficients. Starred coefficients (*) are statistically significant ($p < .01$).

3.3. Results of an Analysis of the Effects of Change on Reports of Emotion

The theoretical basis for the positive effect of life events on emotional arousal in Figure 4 is that the *change* engendered by life events, whether positive or negative, increases arousal. This position, of course, is consistent with Selye's (1956) application of his concept of physiological stress to the social life of humans. It also is consistent the attempt made by Holmes and Rahe (1967) to use magnitude estimation methods to assign weights to different life events based on the perception of the amount of change produced by the life event.[15] This conception has fallen into disfavor due largely to the apparent finding that the effect of life events is limited to the effect of negative events on negative affect. Nevertheless, the concept of stress as change continues to lurk in contemporary formulations, albeit one informed more by stress as a mechanical rather than physiological phenomenon (Wheaton, 1994).

[15] The idea of stress as a non-specific physiological response clearly resembles the idea of undifferentiated arousal as a component of emotional response. It should be noted, however, that Selye's physiological research focused on the generalization across different *noxious* stimuli. The generalization of stress to include positive as well as negative events seems to have occurred in Holmes and Rahe's application of Selye's ideas in their development of a stress index.

To test the validity of this interepretation, I analyzed the effects on the Bradburn scale of respondents' reports of the amount and desirability of change in eight life domains during the year preceeding the interview. The data come from the 1978 and 1979 Edmonton Area Surveys. They consist of 345 and 346 face-to-face interviews, respectively, conducted by the University of Alberta's Population Research Laboratory. The domains include work, finances, housing, schooling, health, family life, friendships, and personal habits. Although the domains lend themselves theoretically to the index construction, the correlations across domains for both the amount and desirability of change are sufficiently strong to permit the the construction of two scales: the amount of change and the desirability of change (including no change).[16]

Table 8. Effects of Amount and Desirability of Perceived Change in Eight Areas of Life[a] on Unrotated and Rotated Latent Affect Variables, Controlling for Age and Sex

	Model			
	Unrotated		Rotated	
Change Scale	Valence	Arousal	Positive Affect	Negative Affect
Amount of Change	-.059	.281*	.276*	.274*
Desirability of Change	.408*	-.016	.242*	-.192*
Age	.209*	-.150*	.069	-.222*
DSex (1 = men)	.078	-.073	.011	-.092*

[a]The areas of change consist of (a) "In the work that you do or in the situation at work," (b) "In your economic situation, (including finances, obligations, standard of living)," (c) "In your place of residence or in your housing conditions," (d) "In your educational situations," (d) "In your own or your family's health," (e) "In your family life," (f) "In your friendships and social life," and (g) "In your personal habits." The amount of change was measured on a four-point scale (1 = no change, ... 4 = many changes); desirability of change (or no change) is measured on a seven-point scale (1 = very undesirable ... 7 = very desirable).
*Coefficients are partial, standardized regression coefficients. Starred coefficients are statistically significant ($p < .01$).

Table 8 contains estimates of the effects of these two scales, plus the respondent's age and sex, on the latent affect variables generated by unrotated and rotated models of the ten emotions in the Bradburn affect-balance scale. The results for the unrotated model consist of a simple and easily interpreted pattern of effects. Consistent with this model, the amount of change positively affects arousal but not valence, while the desirability of change positively affects valence but not arousal. From the perspective of the rotated model, these findings translate, first, into positive effects of the amount of change on both positive and negative and positive affect, and, second, into positive and negative effects of the desirability of change on positive and negative affect, respectively. I regard these

[16] See note 14 for a discussion of the difference between an index and a scale.

results, however, as an artifact of the effects of the amount and desirability of change on valence and arousal. The effects of age and sex largely replicate the results reported in Table 7, except that the effect of sex is no longer statistically significant. An interesting finding, not reported in Table 8, is a strong indirect effect of age on arousal via the strong, negative impact of age on the amount of change ($\beta = -.594$).

3.4. The Results of An Analysis of the Correlations between Husbands' and Wives' Emotional Experience

The final data set I examine comes from the 1980 Edmonton Area Couple Survey. These data consist of face-to-face interviews conducted separately with the husband and wife for a total of 178 couples (Northcott & Kinzel, 1980). Table 9 contains the cross-spousal correlations between factors based on the husband's emotional experience and factors based on his wife's emotional experience. I use a shortened version of the Bradburn affect-balance scale to generate these factors since the 1980 survey deleted five of the items from usual ten-item scale.[17]

Table 9. Correlations between Husbands' and Wives' Affect for Unrotated and Rotated Models*

	Model			
	Unrotated		Rotated	
Husband Variables / Wife Variables	H-Valence	H-Arousal	H–Positive Affect	H–Negative Affect
W-Valence	.376**	-.236	-	-
W-Arousal	-.294**	.333**	-	-
W–Positive Affect	-	-	.138	-.103
W–Negative Affect	-	-	-.189	.510**

*Source: 1980 Edmonton Area Couple Survey (n = 178). Likelihood-ratio chi-square, degrees of freedom, p-value, and Goodness of fit index are: 41.93, 35, .195, .958.

**Correlation statistically significant ($p < .05$, two-tail test).

Unlike the comparable correlations between the reactions Reagan and Carter presented in Table 5, the unrotated model yields a more complex picture than the rotated one. The "same" cross-spousal variables correlate positively, while the "different" cross-spousal variables correlate negatively. That is, husband's valence and arousal correlate positively with wife's valence and arousal (r = .376, .333), while husband's valence and arousal correlate negatively with wife's arousal and valence (r = -.294, -.236), although the latter correlation does not quite attain statistical significance. In the case of the right-hand side of Table 9, on the other hand, the only appreciable correlation is the strong positive correlation between husband's and wife's negative affect (r = .510).

---

[17] The deleted items are restless, proud, on top of the world, things going my, and upset at a criticism. (See Table 6 for the full set of ten items.)

Despite the apparent complexity of the unrotated solution, I find these results easier to interpret. The cross spousal, same-variable correlations possibly arise from mutual influence processes whereby one spouse's valence positively affects the other's valence, and one spouse's arousal positively affects the other's arousal. Although different causal processes could underlie the negative correlation between one spouse's valence and the other's arousal, one possibility is that the spouse's valence behaves like a life even by affecting both the valence and arousal of the other spouse. According to this interpretation, the low valence of one spouse acts like a negative life event by reducing the other spouse's valence and increasing his or her arousal. Although Gottman (1994) has noted the strong correlation between spouses' negative affect coupled with the absence of a correlation between their positive affect, an interpretation of these results from the perspective of positive and negative affect seems elusive. Instead, all four correlations on the right-hand side of Table 9 are better interpreted as an artifact of the correlations on the left-hand side.

3.5. Discussion of the Results of Using Emotional Experience to Measure the Psychological Consequences of Stress

The direction that the research on the stress has taken seems to reflect a morbid preoccupation of psychology and sociology with the misfortunes that are part of the human condition. This preoccupation goes well with the finding that negative events make us feel bad, but positive events have no effect one way or the other. This result, as I argue above, however, is an artifact of using emotion inventories dominated by negative emotions as the criterion variable to measure the psychological consequences of stress. (The reliance on negative emotion as a criterion variable, of course, further illustrates the disciplines' preoccupation with the morbid.) Apart from the skewed image of the human condition conveyed by this focus, the failure to attend to the positive, as well as to the negative, produces misleading results about the effect of life events on emotion. It has led to the conflation of valence with arousal in the conception of both the impact of life events and the structure of emotional experience. Consider, for example, the direction research on the effects of life events might have taken had researchers focused exclusively on measures of positive emotion. As suggested by the results in Table 7, a "Pollyanna" picture may have emerged in which positive events increase positive affect, but negative events have no effect one way or the other. Such a perspective would be equally unbalanced, although in the opposite direction from the current focus. In order to separate the effects of valence and arousal, researchers need to attend to both positive and negative emotion and to treat each emotion as a measure of both valence and arousal.

4. General Summary and Discussion

4.1. Summary

This paper has attempted to uncover the fundamental dimensions that underlie self-reports of emotional experience. In contrast to the dominant practice of separating positive and negative emotions into two apparently unidimensional scales of positive and negative affect, I propose treating each emotion as a bidimensional measure of valence and arousal. Formally, the two positions differ on the decision whether to rotate the two dimensional structure that underlies reports of emotional experience to simple structure. The separation of emotional experience into positive and negative affect rests on the decision to rotate; the treatment of each report of emotional experience as measuring both valence and arousal rests on the unrotated solution. This formal difference between the two positions poses an apparently subtle problem for attempts to validate one position or the other through the use of correlations between the two factors and various criterion variables. As the section on the algebra of rotation shows, rotation automatically changes the correlations between the factors and the criterion variables. A balanced approach, therefore, requires a comparison of the two sets of correlations based on rotated and unrotated structures.

Using results from the analysis of a variety of data sets, I suggest that the correlations of a variety of criterion variables with the two unrotated factors, interpreted as valence and arousal, are fundamental, while the correlations of these criterion variables with positive and negative affect are artifacts. In the case of data on the candidacies of Reagan and Carter, the pattern of correlations seems more parsimonious. The first unrotated factor, valence, correlates substantially in the predicted direction with criterion variables easily interpreted as pure measures of valence: candidate feeling thermometers, the appraisal of qualities that indicate either good or bad leadership, the respondent's position on a continuum of party identification that ranges from strong Republican to strong Democrat, and whether the respondent voted for Reagan or Carter. The second unrotated factor, arousal, correlates significantly, if less substantially, with criterion variables readily interpreted as measures of arousal: talk about politics, interest in the campaign, whether the respondent cares who wins the election, the strength of the respondent's identification with a political party, and whether the respondent voted in the 1980 election.[18] As evidence of the discriminant validity of the unrotated

[18] The lower correlations could reflect the fact that arousal is the second component in a principal components analysis of the emotion items. On the other hand, the lower correlations also could occur because these criterion variables combine in varying degrees arousal with "potency" or, less abstractly in the context of the campaign, the respondent's commitment to act politically. The existence of a third dimension, of course, is consistent with the work of Osgood and Tannenbaum (1957) on the semantic differential and subsequent research by Heise and his associates on affect control theory (Heise, 1988). It also is consistent with the results of a multidimensional scaling of the meaning of emotion words conducted

factors, the correlation of valence and arousal with arousal and valence criterion variables, respectively, are invariably small and usually not significant. Finally, perhaps the best evidence for the construct validity of the unrotated solution is the pattern of cross-candidate correlations that consist of a moderate negative correlation between Reagan and Carter valence and a strong, positive correlation between Reagan and Carter arousal. By comparison, the correlations between these criterion variables and the rotated factors, positive and negative affect, are often puzzling and, therefore, seem best interpreted as an artifact of their correlations with valence and arousal.

In the case of the data from quality of life surveys, the unrotated factors correlate in a theoretically expected way with indexes that essentially count the number of positive, negative, and neutral life events reported by the respondent. The pattern of relations is more complex than the results of the analyses of the emotional reactions to presidential candidates. This complexity occurs, however, because, according to the theory diagrammed in Figure 4, life events affect both valence and arousal. Consequently, the experience of a life event increases, decreases, or has no effect on emotional valence, depending on whether the event is positive, negative, or neutral. All events, however, increase arousal. These results account for puzzling finding in stress research that has led to a focus on the effect of negative life events on negative emotions. As I argue above, had researchers used measures of positive emotions as a measure of psychological well-being, the focus of stress research would be on the effect of positive events on positive affect. Both of these emphases, real and hypothetical, however, reflect a reliance on correlations that are an artifact of the effects of positive, negative, and neutral life events on valence and arousal.

The theory that underlies these relations is based on early conceptions of stress as change. Consistent with this conception, I present evidence that the self-reported desirability of the change in eight life domains of the respondent's life correlates positively with emotional valence but is independent of arousal, while the amount of change correlates positively with arousal but is independent of valence. These correlations more closely resemble those presented in the analysis of the election campaign data because the criterion variables—the desirability of respondent's experience of change and the amount of change experienced—represent relatively pure measures of valence and arousal. Returning to complexity, I present the results of an analysis of couple data that reveal positive correlations between husband and wife valence and arousal and negative correlations between one spouse's valence and the other spouse's arousal. Interpreting these correlations, I argue that the affect of one spouse operates like a

---

by Shaver and his associates (Shaver, et. al., 1987). The two dimensions yielded in their first analysis are easily interpreted as valence and arousal. In a second three-dimension solution, they find that the second dimension splits into two that resemble the activation (arousal) and potency dimensions of the semantic differential.

life event in its effect on the emotions of the other spouse in the sense that the valence and arousal components of one spouse's emotions affect the valence and arousal components of the other spouse's emotions.

4.2. Discussion

In the summary of a recent debate on the merits of the unrotated and rotated solutions in representing the structure of emotions, Diener's (1999) suggestions for future research argues for moving beyond the correlational analysis of self-report data to the use of experimental designs and data on a variety of criterion variables including presumably the physiological and neurological measures that Lang and Cacioppo work with in their respective paradigms (Bradley *et al*., 1994; Cacioppo *et al*., 1999). While I endorse this suggestion (who would not!), I maintain that my correlational analyses of self-report data go a long way toward resolving the debate in favor of valence and arousal, rather than positive and negative affect, as fundamental dimensions of emotional experience. Moreover, proponents of the rotated solution who believe that they can salvage their perspective with criterion variables judged superior to the self-report criterion variables used in this paper should heed the lessons of the section on the algebra of rotation, particularly Eqs. 5 and 6. It presents the formal, substance-free, relation between the two sets of factors. The message of this section is that one can transform the correlations between *any* criterion variable and positive and negative affect into different correlations between the criterion variable and valence and arousal. It does not matter whether the variable is a cause or effect of emotion, whether the criterion variable occurs naturally or is experimentally induced, whether the criterion variable is a self-report measure or is obtained through by observing the person's behavior, a physiological state, or neural substrate.

Even though the correlations of a criterion variable with positive and negative affect appear to provide *prima faciae* support for the rotated solution, the transformed correlations of a criterion variable with valence and arousal could lend themselves to an interpretation that is equally plausible and possibly superior. The formal structure of the unrotated solution also can shed light on otherwise puzzling findings that do not involve the correlations between emotion inventories and criterion variables. For example, Diener (1999) describes a finding of strong bipolarity of positive and negative affect in the case of "intense emotional episodes" in contrast to the usual finding of relative independence of positive and negative affect in the case of "typical low-intensity mood experience." Diener attributes these contrasting findings to the different structure of emotional experience during a momentary episode versus a trait-like aggregation of experience over a series of episodes. Without discounting the importance of the trait-state distinction, another explanation is that the variance of arousal variance will be attenuated, by definition, in intense emotional episodes, particularly when compared to typical situation in which emotion inventories are administered. As both the upper-left and lower-right triangles of equation of the matrix in eq. 7 show, the valence of the person's experience will dominate his or report in such episodes, and, therefore, the correlation between positive and negative affect will approach –1.[19]

Readers who discount the results of correlational analyses in favor of the gold standard of experimental data should recall the analysis of the life events data. The results of this analysis raise the possibility that experimental manipulations could affect both valence and arousal. Researchers who restrict their analysis to the effect of a manipulation on either positive or negative affect run the risk of confounding the effect of the manipulation on valence with its effect on arousal..

Although the analysis of emotion inventories comprises the *raison d'etre* for this paper, future work on valence and arousal as fundamental dimensions of emotional experience should incorporate other measures of emotion. In addition to physiological and neurological variables, the self-report measures should include bipolar items that resemble those used in the semantic differential (Osgood & Tannenbaum, 1957) or self-administed manikin developed by Lang and his associates (Bradley *et al.*, 1994).

The use of such items could accomplish at least three things. First, it would help address an apparent shortcoming of using emotion inventories to measure valence and arousal. The problem is that this method precludes the occurrence of certain logically possible combinations of valence and arousal—for example, either maximum or minimum valence scores in combination with either maximum or minimum arousal scores. This problem may be more apparent than real since circumplex theories of emotion imply that these combinations are either non-existent or rare. Nevertheless, the use of experimentally independent, (relatively) pure measures of valence and arousal will permit a test of this implication rather than producing it as an artifact of the measurement procedure. Second, the use of pure measures of valence and arousal would permit the investigation of more complex relations between valence and arousal. Although the variables are linearly independent, there is the possibility of a quadratic relation such that valence

---

19 The upper-left triangle of the matrix in Eq. 7 consists of the variance-covariance matrix for Reagan positive and negative affect, while the lower-left triangle consists of the variance-covariance matrix for Carter positive and negative affect. Both matrices express these quantities as a linear combination of the variances in valence and arousal. Recall that Eq. 7 simplifies this expression by assuming that the covariance between variance and arousal is zero. With this simplification, the positive affect and negative affect variances equal one another, and the covariance between positive and negative affect equals the variance in arousal minus the variance in valence. As the arousal variance approaches zero, the covariance between positive and negative affect approaches the negative value of the positive and negative affect variances, and, therefore, the correlation between positive and negative affect will approach –1. Naturally, relaxing the assumption of zero covariance between valence and arousal will complicate the picture, but I doubt that the valence-arousal covariance will be large enough to alter it appreciably.

increases then declines as arousal increases.[20] Third, the use of bipolar measures of affect permit the investigation of three and possibly more dimensions of affect.[21]

I closing, I shall use the "searchlight hypothesis" as developed by Crick (1984) and elaborated by Baars (1988) to attempt a deeper understanding of valence and arousal. The idea of a searchlight is that consciousness involves an attentional process that illuminates a conceptual or semantic space. In applying this hypothesis to valence and arousal, I view valence as the illumination process and arousal as partly a byproduct of the attentional process much as heat is a byproduct of a searchlight's illumination. Adopting Tomkins' (1991) notion of play to describe the loose relations between the modules of biological systems, I suggest that a non-optimal level of urgency is aroused in the attention process.

The idea of arousal as a byproduct of attention speaks both to the analysis of emotional reactions to presidential candidates and to the use of arousal to measure the psychological consequences of stress. The idea of arousal as a byproduct arose in the former endeavor when I attempted to use the two candidate arousal factors and the respondent's interest in the election as measures of a more general factor that mediates the effect of the strength of party identification of the strength of the respondent's vote. Unfortunately for this hypothesis, I found no support for the existence of this general factor. Instead, interest in the campaign serves as the mediating variable, and the relations of the two factors to strength of party identification and strength of vote are no longer significant when interest in the campaign is controlled (Gillespie, 1994). In terms of the searchlight hypothesis,

---

20 This quadratic function would account for the "positivity offset" and "negativity bias" discussed extensively by Cacioppo, et. al. (1999). In terms of valence and arousal, positivity offset represents the positive effect of valence on arousal as valence increases from a (hypothetical) baseline to some optimal, moderate level. The negativity bias represents the negative effect of valence on arousal as arousal increases from moderate to high levels. This relation, of course, resembles the J-curve between stress and learning first observed at the beginning of the twentieth century.

21 Based on the analysis of emotion words conducted by Heise and his associates (1988), Osgood and Tannenbaum (1957), and Shaver, et. al. (1987), the dimension of "potency" is an obvious candidate for a third emotion dimension. Despite the theoretical and empirical rationale for this dimension, uncovering its existence has proved elusive. The self-administered manikin developed by Lang and his associates, for example, contains a bipolar scale designed to measure this dimension: a series of otherwise identical manikins that very in size from small to large (Bradley, et. al., 1994). They dropped this dimension in their research program, however, because of its high correlation with valence. Lutz's multidimensional scaling of Ifaluk emotion words, on the other hand, yielded a two-dimensional solution in which potency, rather than arousal, was the second dimension after valence (Lutz, 1986). Social relations on the Ifaluk atoll, however, were strongly hierarchical, and many of emotion words pertained to feelings of deference, obligations, and rights manifested in these relations. One reason for the failure of potency to emerge in factor analyses of emotional experience is the paucity of items that capture these sentiments. (It may be that expressions of feelings about one's power vis a vis that of others may be more taboo than the expression of feelings about sex, at least in industrial and post-industrial societies.) As suggested by a consideration of criterion measures of arousal in the analysis of emotional reactions to the Reagan and Carter candidacies, the potency dimension also could represent a willingness or commitment to act, and these sentiments also are absent from emotion inventories, again, possibly because due to the difficulty of expressing them in a form to which subjects can easily respond.

interest in the campaign is analogous to the optimum level of urgency required by the attentional process, while the two arousal factors, after controlling for interest, represent the difference between optimum and non-optimum levels of urgency and are analogous to the heat produced by a searchlight.[22]

Arousal figures as a byproduct in the study of the stress process because of the fact that is part of the response to stressful situations. In terms of Selye's (1956) theory stress, arousal supplies the urgency needed by the attention process to change things that make us feel bad and reproduce things that make us feel good. His concept of "eustress" attempts to make the point that a certain level of arousal is necessary if we are to function. Arousal feels stressful when it becomes excessive. To return to the searchlight hypothesis, stress might be thought of as too much consciousness. To move beyond the realm of speculation, however, requires studies of emotion that treat valence and arousal as possible independent variables and investigate their effects, if any, on mental and physical health.

References

Abelson, R. P., D. R. Kinder, M. D. Peters and S. T. Fiske (1982) "Affective and Semantic Components in Political Personal Perception", *Journal of Personality and Social Psychology* 42:619-30.

Baars, B. (1988) *A Cognitive Theory of Consciousness*, Cambridge: Cambridge University Press.

Bradburn, N. M. (1969) *The Structure of Psychological Well-Being*, Chicago: Aldine.

Bradley, M. M., M. K. Greenwald and A. O. Hamm. (1993) "Affective Picture Processing", in: *The Structure of Emotion*, N. Birbaumer and A. Öhman eds, Seattle: Hogrefe & Huber,

Cacioppo, J. T. and G. G. Berntson (1994) "Relationship between Attitudes and Evaluative Space: A Critical Review with Emphasis on the Separability of Positive and Negative Substrates", *Psychological Bulletin* 115:401-423.

Cacioppo, J. T., W. L. Gardner and G. G. Bernston (1999) "The Affect System Has Parallel and Integrative Processing Components: Form Follows Function", *Journal of Personality and Social Psychology* 76:839-855.

Campbell, A. and P. E. Converse (1980) *Quality of American Life, 1978* (machine-readable data file), Conducted by the Institute for Social Research, the Univeristy of Michigan, ISPSR ed., Ann Arbor, Mich.: Inter-University Consortium for Political and Social Research.

Cliff, N. (1987) *Analyzing Multivariate Data*, San Diego: Haarcourt Brace Jovanovich.

---

22 Interestingly, talk about politics is the criterion measure of arousal that correlates the strongest with Reagan and Carter arousal. The image of "heated" discussions of politics is, of course, consistent with my equation of arousal with heat.

Crick, F. (1984) "Function of the Thalamic Reticular Complex: The Searchlight Hypothesis", *Proceedings of the National Academy of Sciences* 83:4586-4590.

Diener, E. (1999) "Introduction to the Special Section the structure of Emotion", *Journal of Personality and Social Psychology* 76:803-804.

Duffy, E. (1957) "The Psychological Significance of the Concept of 'Arousal' or 'Activation'", *The Psychological Review* 64:265-275.

Fisher, G. A. (1997) "Theoretical and Methodological Elaborations of the Circumplex Model of Personality Traits and Emotions", in: *Circumplex Models of Personality and Emotions*, R. Plutchik and H. R. Conte, eds, Washington, DC: American Psychological Association, pp. 245-270.

Frijda, N. H. (1994) "Varieties of Affect: Emotions and Episodes, Mooods, and Sentiments", in: *The Nature of Emotion*, P. Ekman and R. J. Davidson, eds, New York: Oxford University Press, pp. 59-67.

Gillespie, M. W. (1994) "Separating Valence from Arousal in the Emotional Response of Voters to Presidential Candidates; the 1980, 1984, and 1988 Presidential Elections", Paper presented to the *1994 Annual Meetings of the American Political Science Association*, New York, NY.

Gottman, J. M. (1994) *What Predicts Divorce? The Relationship Between Marital Processes and Marital Outcomes*, Hillsdale, N.J.: Lawrence Erlbaum Associates.

Gray, J. A. (1987) *The Psychology of Fear and Stress* (2nd ed), Cambridge: Cambridge University Press.

Green, D. P., S. Goldman and P. Salovey (1993) "Measurement Error Masks Bipolarity in Affect Ratings", *Journal of Personality and Social Psychology* 76: 856-867.

Heise, D. R. (1988) "Affect Control Theory: Concepts and Model", in: *Analyzing Social Interaction: Advances in Affect Control Theory*, L. Smith-Lovin and D. R. Heise, eds, New York: Gordon and Breach Science Publishers.

Hobfoll, S. E. and C. D. Spielberger (1994) "Family Stress: Integrating Theory and Measurement", *Journal of Family Psychology* 6:99-112.

Holmes, T. H. and R. H. Rahe (1967) "The Social Readjustment Rating Scale", *Journal of Psychosomatic Research* 11:213-218.

Jöreskog, K. and D. Sörbom (1988a) *LISREL 7*, Mooresville, IN: Scientific Software, Inc.

Jöreskog, K. and D. Sörbom (1988b) *PRELIS*, Mooresville, IN: Scientific Software, Inc.

Lacey, J. (1967) "Somatic Response Patterning and Stress: Some Revisions of Activation Theory", in: *Psychological Stress*, M. H. Appley and R. Trumbull, eds, New York: Appleton-Century Crofts.

Lang, P. J. (1984) "Cognition in Emotion: Concept and Action", in: *Emotions, Cognition, and Behavior*, C. E. Izard, J. Kagan and R. B. Zajonc, eds, Cambridge: Cambridge University Press.

Lutz, C. (1986) "The Domain of Emotion Words on Ifaluk", in: *The Social Construction of Emotions*, R. Harré, ed., Oxford: Basil Blackwell.

Marcus, G. E. 1988 "The Structure of Emotional Response: 1984 Presidential Candidates", *American Political Science Review* 82:732-61.

Marshall, G. D. and P. G. Zimbardo (1979) "Affective Consequences of Inadequately Explained Physiological Arousal", *Journal of Personality and Social Psychology* 37:970-988.

Maslach, C. (1979) "Negative Emotional Biasing of Unexplained Arousal", *Journal of Personality and Social Psychology* 37:953-969.

Morruzi, G. and H. W. Magoun (1949) "Brain Stem Recticular Formation and Activation of the EEG", *Electroencephalography and Clinical Neurophysiology* 1:455-473.

Northcott, H. C. and C. Kinzel (1980) *Edmonton Area Series Report No. 14*, Edmonton, Alberta: University of Alberta Population Research Labortory.

Osgood, C. (1966) "Dimensionality of the Semantic Space for the Communication of Emotion via Facial Expression", *Scandinavian Journal of Psychology* 7:1-30

Osgood, C., G. Suci and P. Tannenbaum (1957) *The Measurement of Meaning*. Urbana, IL: University of Illinois.

Russell, J. A. and G. Pratt (1980) "A Description of the Affective Quality Attributed to Environments", *Journal of Personality and Social Psychology* 38:311-322.

Schacter, S. and S. Jerome (1962) "Cognitive, Social and Physiological Determinants of Emotional State", *Psychological Review* 69:379-399.

Schacter, S. and J. Singer (1979) "Comments on the Maslach and Marshall-Zimbardo Experiments", *Journal of Personality and Social Psychology* 37:989-995.

Selye, H. (1956) *The Stress of Life*, New York: McGraw-Hill.

Shaver, P., J. Schwartz, D. Kirson and C. O'Connor (1987) "Emotion Knowledge: Further Exploration of a Prototype Approach", *Journal of Personality and Social Psychology* 52:1061-1086.

Thayer, R. E. (1989) *The Biopsychology of Mood and Arousal*, New York: Oxford University Press.

Tomkins, S. S. (1991) *Affect Imagery Consciousness, Volume III: The Negative Affects, Anger and Fear*, New York: Springer Publishing Company.

Turner, R. J. and W. R. Avison (1992) "Innovations in the Measurement of Life Stress: Crisis Theory and the Significance of Event Resolution", *Journal of Health & Social Behavior* 33:36-50

Watson, D. (1988) "Intraindividual and Interindividual Analyses of Positive and Negative Affect: Their Relation to Health Complaints, Perceived Stress, and Daily Activities", *Journal of Personality and Social Psychology* 54:1020-1030.

Watson, D. and A. Tellegen. (1985) "Toward a Consensual Structure of Mood", *Psychological Bulletin* 98:219-235.

Watson, D., D. Weise, J. Vaidya and A. Tellegen (1999), "The Two General Activation Systems of Affect: Structural Findings, Evolutionary Considerations, and Psychobiological Evidence", *Journal of Personality and Social Psychology* 76:820-838.

Weller, S. C. and A. K. Romney (1990) *Metric Scaling: Correspondence Analysis*, Newbury Part, CA: Sage.

Wheaton, B. (1994) "Sampling the Stress Universe", in: *Stress and Mental Health: Contemporary Isses and Prospects for the Future*, W. R. Avison and I. Godlib, eds, New York: Plenum Press, pp. 77-114.

Wundt, W. (1906) *Outlines of Psychology*, (C. H. Judd trans.), New York: Stechert.

THE CONTRIBUTION OF THE FACE IN THE DEVELOPMENT ON EMOTION AND SELF

JONATHAN COLE
Clinical Neurological Sciences, University of Southampton, Southampton General Hospital, Southampton 5016 6YD United Kingdom

ABSTRACT

It would appear that emotions, (inner brain states), must be expressed as feelings not only in the somatic body, via autonomic change and possibly facial expression and gesture, but also communicated to others for their full effect. Channels for this communication include gesture, vocalization and the focus of the present work, facial expression. This has developed further levels of complexity in man, compared with non-human primates, with the development of subtler emotional states as man's ancestors developed and regulated more complex social groups. Such regulation required knowledge not just of others' likely behavior but their thoughts and emotions. This development of a theory of mind and interpersonal relatedness may have required the development of an area of the somatic self which was visible - the face. The development of such "theories" of mind, (and the emergence of consciousness of others), may therefore have paralleled the development of a mobile emotionally eloquent face. Such models are considered in parallel with biographical narrative accounts of the affects that clinical problems with the expression of emotion on the face, or its interpretation, have on the development of self and social esteem. These cases include those with congenital and acquired blindness, and those with Mobius Syndrome, (a congenital absence in the movements of facial expression). Any consideration of the qualia of emotion and of consciousness must, I believe, include some subjective account of what it is like to live without normal emotion expression on the somatic self, or to live without the ability to perceive this in others. Those with facial differences describe a loss of social relatedness leading to profound social isolation and to an impoverished sense of self. But there are additional and more specific problems. Some with Mobius Syndrome described difficulties in experiencing emotions not expressed on the face, and talked of "thinking emotions" rather than experiencing them. Without facial expression they also have some problems calibrating and controlling emotion. They may, therefore, lack an aspect of the qualia of emotions (they don't have the qualitative experience, they don't know what it feels like) and so they think it or conceptualize it. If that is so, then for those of us without visibly different faces facial expressions, (our own and others'), and the to and fro facial exchanges which are conversations may be an element in the way that we generate emotional qualia, or experience the emotional expression of others.

1. Introduction

"The horror of being faceless, of *forgetting one's own appearance, of having no face. The face is the mirror image of the self" 'John Hull, blinded in his forties. (Cole, 1997)*

"People, they are their faces. I would like to be my face"' James, with Moebius Syndrome, in his fifties.(Cole, 1997)

Emotions, (inner brain states), may be expressed as feelings within the somatic body, via autonomic change and possibly facial expression and gesture, but also need to be communicated to others for their full effect. Channels for this communication include gesture, vocalization and the focus of the present work, facial expression. Facial mobility may have developed in man, compared with non-human primates, with the development of social intelligence and subtler emotional states as man's ancestors regulated more complex social groups. Such regulation required knowledge not just of others' likely behavior, but of their thoughts and emotions. This, in turn, may have required the development of a visible area of the self - the face. The emergence of self, and consciousness of others, may therefore have paralleled the development of a mobile emotionally eloquent face.

Though such models are speculative, there is little doubt that for most people sense of self includes an embodied component. When describing ourselves we describe physical characteristics - height, hair color, sex, eye color, rather than our intellectual or emotional makeup. Wittgenstein's dictum, "the human body is the best picture of the human soul," is relevant here. But I would like to go further: the body part we feel most embodied in is our face, even though it is difficult to describe and so forms little part of how we describe people.

The face is an unique identifier, allowing observable differentiation from others, (identical twins excepted). It also shows our approximate age, our gender and something of our health. Through it we can reveal or attempt to conceal our emotions. It also, perhaps most importantly, through mutual gaze and facial conversations, allows the sharing of emotions and the development of relationships between people. To learn something of the importance of the face on our experience of, and sense of, self I suggest that we, in addition to adopting an experimental approach, should seek to enter the experiences of others for whom there is a facial problem.

1.1 Mature Onset Blindness. Loss of Sight, Loss of Face, Loss of Self?

I began by talking with people who had lost perception of the face of others entirely in adult life, the mature onset blind. John Hull went completely blind in his forties. He wrote a diary covering the first few years after this and the profound mental adjustments he was forced to make, (Hull, 1992). In my later conversation with him I asked him about his later experiences of coming to terms with the loss of the visible face.

Initially on going blind he needed to remember a visible face in order to perceive a person and so tried to remember the faces of his loved ones. Without these, he felt isolated and detached. Just as when sighted people, who shut their eyes, still envisage the world through a visual memory, so John retained a visual memory of others for some time. He became most depressed not when he went blind but as these visual - mainly facial - images faded and he finally left the visual world, nearly two years after the loss of sight itself.

He was also tortured by a loss of self identity resulting from his blindness: "To what extent is the loss of the image of the face connected with loss of the image of the self?' (Hull, 1992) He became aware of facial actions in a way we usually are not. 'Nearly every time I smile, I am aware of it ...aware of the muscular effort: not that my smiles have become more forced.. but it has become a more or less conscious effort. It must be because there is no reinforcement... no returning smile.. like sending off dead letters... I can feel myself stopping smiling ...must ask someone if it is true." (Hull, 1992) Slowly he reclaimed some perception of family and friends, in an affective dimension, which depended on mainly sound and voice. Crucially though, he was far more passive, and depended on people disclosing themselves through speech. There were, for instance, no covert looks at a person when they were off-guard.

On meeting him I began by asking how he constructed identity and character without the visible face?' "I no longer have a visual image. Everything is in the voice and it doesn't occur to me to construct a physique through the voice. I know what my closest friends are thinking and feeling because its all in the voice - but they have to speak. There is a big problem with the child and the face. Its hard to tell moods. If my thirteen year old is taciturn a glance at his face would tell me how he was. At an emotional level anger, impatience and such emotions are more easily expressed in the voice than thoughtfulness or sadness. It is very difficult to detect sadness. My emotional range is narrowed."

I suggested that normally one never consciously constructs character and feelings towards someone - it just happens. I asked if once blind he had to become more intellectual and if these constructions required effort and thought. "Marilyn [his wife] and I have sweated blood on this one because it was so difficult for the sighted person to enter into that world. A year or two ago we had a visitor when Joshua was about three; when the friend had gone Marilyn said, apropos of nothing, 'What comes into your mind when I say Joshua?' I said, 'Well, Joshua.' And she said, 'What, what exactly is it.' 'Well, its the memory running through my hand, the feeling, kicking, laughing body, and the things Joshua and I have done together.' 'Yes, but what of Joshua himself?' 'If you mean what does he look like - nothing.' 'I can't bear that.' 'Darling, if we are going to say do we share the same Joshua we might as well say, Do we share the same world?' and in saying that there is a deep and important sense in which we do not. We do not share the same world." (Cole, 1997)

Trying to explore the interaction between thought and facial action I asked about his use of face since going blind. "I am sometimes fearful that my face is becoming less expressive but Marilyn tells me this is not the case. I often feel that

I'm thought to be too serious. It's hard for a blind person to have fun because so much is visual and shared, especially in the family, people making funny faces, teasing each other and I can be out of it. Laughing together is one of the best things. Not knowing about other person's tears is worse, tears are silent. It's perhaps more important to know about tears than about smiles. I think there's no doubt that the loss of the face is a profound loss. A deeply dehumanizing loss."

I.2. Moebius Syndrome: "Living in My Head"

Next I went to talk with someone who might be considered to live without face, certainly without a mobile expressive one. Those with Moebius syndrome are born without the ability to move any of the muscles of facial expression, or to move their eyes laterally. James was in his fifties. He described feelings of low self worth, of isolation, even in company, and of loneliness which might not be surprising. Worse, though, he seemed to have lost sight of the fact that many of these feelings were related to his facial problems, thinking instead that was just the way he was and that there was no way to reach others. "I have a notion which has stayed with me over much of my life - that it is possible to live in your head, entirely in my head."

He also described never having experienced happiness or sadness or anger through facial expression. This reduced embodiment seemed to have led to a dissociation from his feelings which had a more thinking aspect and were subsequently made affective intellectually. When he met his wife, "Initially I thought I was in love with her rather than feeling it: It was some time later when I realized that I really felt in love." This dissociation re-emerged later: "I sort of think happy or I think sad, not really saying or recognizing actually feeling happy or feeling sad. Perhaps I have had a difficulty in recognizing that which I'm putting a name to is not a thought at all but it is a feeling, maybe I have to intellectualize mood. Of course since I have never been able to move the face, I've never associated movement of the face with feeling of an emotion." (Cole, 1997).

There was also a terror that emotions might run out of control. It seems that we need to express powerful feelings in order to calibrate them. "I have a fear of being out of control with emotions, feeling something that I can't manage. I have also found it very difficult to communicate feelings throughout my life, though I am getting better at it now. I don't really know how I communicate happiness or sadness. Some people cry when they're sad. I don't think I cry. I am afraid of such feelings."

"People, they are their faces. I think now I would like to be my face. Me is something I see in the mirror. I have always had a difficulty with mirrors and photographs. I don't like being confronted by me. I don't want to but I have to look in a mirror, and apart from shaving I never do. I may have turned my back on my face" (Cole, 1997).

I suggested that to become a whole person he had to look in the mirror and say, "This is my face but I am not this: I exist behind and yet beyond it." Many with Moebius have successfully expanded and accentuated their use of voice intonation and gesture to overcome their lack of facial expressiveness. "Yes, I think you're right. If you say where does 'me' now reside, I think I am slowly coming out of my head a bit. I am not sure I can locate where I am but I don't think I am entirely in my head or even my mind. I have an expression of living 'a life of the mind,' but I do accept that the mind is not easily able to communicate its thoughts or even its feelings."

2. Conclusion

For Wittgenstein the face was the interlocutor between the self and the world, and facial action and feeling were intimately linked, "The content of an emotion - here one imagines something like a picture. The human face might be called such a picture... We describe a face immediately as sad... even when we are unable to give any other description of the features... We see emotion." (Wittengenstein, 1980). Such themes also occupied Merleau-Ponty. He considered the relations between emotion and consciousness and the importance of our social existence for the full experience of selfhood; "What is it to be moved, what is the meaning of emotion? Can one conceive of a consciousness which is incapable of emotion?... the intellectual elaboration of our experience of the world is constantly supported by the affective elaboration of our inter-human relations." (Merleau-Ponty, 1964). Merleau-Ponty was in no doubt as to the importance of the face in this: "I exist in the facial expression of others, as they do in me."

The face, therefore, is not only a visible expression of emotion, but to some extent the self is constituted in the face and is developed and experienced in the interactions between faces. The narratives of loss from those with facial problems may allow some insights into the consequences of this and reflect on the normal intimate relations between face, emotion and self.

References

Cole, J.(1997) *About Face,* Cambridge, MA: MIT Press

Hull, J. (1992) *Touching the Rock*, New York: Random House.

Merleau-Ponty, M. (1964) *The Primacy of Perception,* Evanston, IL: Northwestern University Press.

Wittgenstein, L. (1980) *Remarks on the Philosophy of Psychology,* Chicago: University of Chicago Press.

EMOTIONS AND LEARNING IN A DEVELOPING ROBOT

R. MANZOTTI, G. METTA and G. SANDINI
LJRA~Lab, DIST, University of Genova, Via Opera Pia 13, 16145, Genova, Italy

ABSTRACT

The role of emotions has been underestimated in the field of robotics. We claim that emotions are relevant for the building of purposeful artificial systems from at least two perspectives: a cognitive and a phenomenological one. The cognitive aspect is relevant for at least two reasons. First, emotions could be the basis for binding between internal values and different external situations (the somatic marker theory). Second emotions could play a crucial role, during development, both for taking difficult decisions whose effects are not immediately verifiable and for the creation of more complex behavioral functions. Thus emotions can be seen, from a cognitive point of view, as a reinforcement stimulus and in this respect, they can be modeled in an artificial being. Inasmuch, emotions can be seen as a medium for linking rewards and values to external situations. From the phenomenological perspective, we accept the division between feelings and emotions. Emotions are, in James' words, the body theatre in which several emotions are represented and feelings are the mental phenomenological perception of them. We could say that feelings are the qualia of the body events we could call emotions. We are using this model of emotions in the development of our project: Babybot. We stress the importance of emotions during learning and development as endogenous teaching devices.

1. Emotions and Qualia

1.1. Cognitive and Phenomenological Side

Traditionally emotions have not been considered in the development of robots. The main reason has been the confusion between their cognitive and phenomenological aspect. It is important to clarify the difference.

Speaking from a purely cognitive perspective, we can see emotions as simple devices assigning to a particular situation a global value with no conscious analysis of the details. Thus, emotions can be seen as reward variables able to represent wide collections of external situations. For example, if animals (or particular classes of them) have no conscious experience they could still have unconscious emotions.

Phenomenological emotions, or feelings[1], can be seen as the conscious perception of cognitive unconscious emotions (Damasio 1994). They require a conscious being, whose existence and properties are still far from being obvious (Dennett 1991). Without entering into the consciousness debate, we only state that we accept the existence of consciousness as a real phenomenon waiting for a

[1] Some refer to the dualism of unconscious emotion versus conscious emotion as emotion versus feeling, others as mood versus emotion. We prefer the former choice.

proper answer (Chalmers, 1996; Searle, 1992). Given this position, feelings have to be explained like subjective entities. What we can hope to give in this context is just a cognitive framework for their phenomenological counterpart.

1.2. James'(1890) Theatre

Emotions are useful tools to represent possible future reward or punishment. The idea is that they can be embodied not just in the neural structure of the brain but that they could be physically part of the body itself. The body would be the theatre in which the emotions are represented (unconsciously) and a specific situation can be associated to a particular body response. In this way the body becomes, through emotions, a processing element of the cognitive architecture of an active being. This structure has several advantages among which the fact that the system is thus able to develop a sort of subjective personality through an explicit, conscious representation of the body. This representation is embedded in the body itself, which after all, is its own best representation (Brooks, 1991). For example if I have eaten too much it is probable that my mood will change and that, therefore, my disposition toward different kinds of physical activities will change too. The blood flux toward my stomach has modified my body variables and I will represent my future emotions in a different way. Indirectly, the body has processed the information concerning my eating and, through following emotional responses, will intervene on the way the brain will make future decisions.

More complex emotional responses can be obtained as a result of neural structure explicitly devoted to the activation of a particular emotional response. They can be activated by specific (and unconscious) stimuli regarding various relevant aspects of the environment (visual expressions, dangerous situations, phylogenetically-selected stimuli, etc.). Among these dedicated neural structures we can cite the amygdala, the cingulate cortex and the thalamus (Adolphs, *et al.*, 1998; Le Doux, 1996; Morris, *et al.*, 1998).

1.3. Feelings as Qualia of What?

Emotions are thus a specific body response to a particular kind of stimulus. This response can be different according to the global state of the body. For example, the sight of a naked person of the opposite sex can elicit different responses in the same person according to its overall conditions. Eventually the brain, through several internal sensory channels (blood pressure, heart rate, breathing rate, etc.), can perceive this body state. This perception involves the existence of the mental object that carries the meaning of the emotion: a feeling or qualia of an emotion. Phenomenologically speaking, qualia can be seen as the carriers of the meaning of perception. When we have a red sensation (we are in such state that it is something to be us seeing something red) we are in some kind of relation with the meaning of «red» (intentional, identity or whatever). Then a particular qualia has a meaning that is in relation with the kind of information it is carrying. We stress the difference we see between the subjective meaning carried by qualia and the objective information carried by the relations between qualia (Shannon & Weaver, 1963). Speaking of feelings, as we have said, they are the conscious perception of emotion (that is a complex body state perceived through several channels). A particular feeling can thus be seen as the qualia of a particular emotion. If an emotion represents something (danger, joy, happiness, whatever), its qualia is the subjective meaning associated with that something. If a red qualia is the meaning of a red patch then a feeling, or qualia of an emotion, is the meaning of something more complex but equally real (Nagel, 1974; Stubenberg. 1998).

2. Emotion and Reinforcement

The role of emotions in learning is related to the pleasure/pain feeling associated to certain emotional states. This feeling can be exploited to guide the learning phase and, perhaps, to achieve more efficient performance. In humans the emotional basis of learning has a cultural as well as an evolutionary side the association between a certain body state and some external event may be unconscious. In our robotic experiment the implementation of emotional states and feelings is still rather simple and it is essentially related to the generation of a pleasure/pain sensory feedback to reinforce/inhibit specific sensori-motor behaviors. On the other hand, it is important to stress that the overall system is acting and learning not only on the basis of exteroceptive and proprioceptive sensory data coding physical parameters (such as geometric relationships, speed of motion etc.), but also on internally generated body signals explicitly coding an "emotional" parameter (and not only geometric information about the body status). The overall state of the body (i.e. after having being modified by an emotional response), together with the normal sensations, is eventually perceived by the system which chooses the best action.

3. BabyBot Project

3.1. The Experimental Setup

The overall goal of the babybot project is to study the implementation of artificial autonomous system following a developmental approach or, in other words, to investigate how the development of, for example, sensori-motor coordination in human infants can be modeled in an artificial being (Sandini *et al.* 1997). The specific experiment we describe here involves the learning of arm-reaching control movements toward "good" visual targets (and the inhibition of such reaching toward "bad" targets). In the framework of the present paper, therefore, learning has both a physical and an emotional component. Of the 10 degrees of freedom of babybot we use only 4: two of them controlling the direction of gaze of a color camera and two controlling the motion of the arm (which is, therefore, constraint to move on a plane). The distinction between *good* and *had* objects is based on color. More specifically reaching toward green objects is reinforced and the experimenter punishes reaching toward red ones. To increase the difficulty of an otherwise simple associative task, visual processing, which is based on a space-variant color camera (Sandini & Tagliasco, 1980), provides, besides color information, also a set of non-relevant signals such as target position and size (computed in the image plane).

3.2. Learning Motor Control and Emotional Value

The motor control strategy adopted is based on the so-called "force fields" approach first proposed by Mussa-Ivaldi, *et al.* (1993). Arm reaching, as described in more details in (Sandini *et.al.*, 1997), is controlled by a direct mapping between gaze angles and arms angles built, during the learning phase, through the interaction between the robot and the environment. A supervised learning technique has been used in this case. It is worth mentioning the fact that as learning progresses the system smoothly switches from a purely, built-in, reflexive behavior to a state where reflexes are used as primitives to perform coordinated reaching.

A neural network carries on the higher level decision. It receives sensory and reinforcement inputs and generates the decision (i.e. reaching/no reaching) as output. The network learns to distinguish between bad and good objects on the basis of the reinforcement signal given by the teacher at the end of each motor action. It is worth noting that the relevant information (color) is embedded into a set of non-relevant sensory signals. Despite this fact, the neural network is able to learn which subset is relevant for the task at hand.

Figure 1. shows the experimental setup described in section 3.1, along with some exemplar reaching trajectories recorded at the end of the training phase. In summary the system exhibits an increase in success rate (the percentage of correct decisions) from *50%* (random decision) to about 90% in about 40 trials and simultaneously the reaching accuracy increases to about 1 cm. from an initial situation where motor reflexes are just driving the end-effector within the field of view (approximately 10-15 cm. in our case).

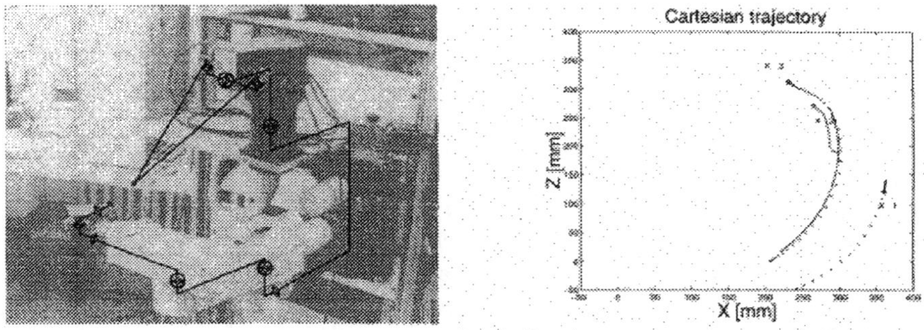

Fig. 1. The robot (left figure), and trajectories of the hand (right figure).

Acknowledgements

Research activity described here is supported by the Italian Ministry of Research and University (MURST grant No.9709112600) and by the EU TMR grants VIRGO (FMRX- CT96- 0049) and SMART-Il.

References

Adolphs, R., D. Tranel and A. Damasio (1998) "The human amygdala in social judgement", *Nature* 393:470-474.

Brentano, F., (1973) *La Psicologia dal Punto di Vista Empirico*, (translated by G. Gurisatti, original: F. Brentano (1873), *Psychologie vom empirisehen Standpunkt*), Roma: Laterza.

Brooks, R. (1991) "Intelligence without representation", *Artificial Intelligence* 47:139-159

Chalmers, D. (1996) *The Conscious Mind. In Search of a Fundamental Theory*, New York: Oxford University Press.

Damasio, A.R. (1994) *Descartes Error: Emotion, Reason and the Human Brain*, New York: G.P. Putnam's Sons.

Dennett, D. (1991) *Consciousness Explained*, Boston: Little, Brown and Company.

Hyman, S.E. (1998) "A new image for fear and emotion", *Nature* 393:417-419.

James, W. (1890) *The Principles of Psychology*, NewYork: Dover.

LeDoux, J.E. (1996) "In search of an emotional system in the brain: Leaping from fear to emotion and consciousness", in *The Cognitive Neurosciences*, M. Gazanniga, ed., Cambridge, MA: MIT Press, pp. 1049-1061.

Morris, J.S., A. Öhman and R. J. Dolan (1998) "Conscious and unconscious emotional learning in the human amygdala", *Nature* 393:467-470.

Murphy, S.T. and R.B. Zajonc (1993) "Affect, cognition and awareness: Affective priming with optimal and suboptimal stimulus exposures", *Journal of Personality and Social Psychology* 64:723-739.

Mussa-Ivaldi, F.A., S.F. Giszter and E. Bizzi (1993) "Convergent force fields organized in the frog's spinal cord", *The Journal of Neuroscience* 13:467-491.

Nagel, T. (1974) "What is it like to be a Bat?", *The Philosophical Review* 83:435-450.

Sandini, G. and V. Tagliasco (1980) "An anthropomorphic retina-like structure for scene analysis", in: *CGJP*, 14 No. 3:365-372.

Sandini, G., G. Metta and J. Konczak (1997) "Human sensori-motor development and artificial systems", in: *AJR&JHAS* '97, Japan.

Searle, J. (1992), *The Rediscovery of the Mind*, Cambridge, MA: MIT Press.

Shannon, C. and W. Weaver (1963) *The Mathematical Theory of Communication*, Urbana, IL: University of Illinois Press.

Stubenberg, L. (1998) *Consciousness and Qualia*, Amsterdam: John Benjamins.

THE MENTAL REPRESENTATION OF ROMANTIC JEALOUSY: A BLENDED EMOTION (AND MORE)

DON J. SHARPSTEEN
Department of Psychology, University of Missouri at Rolla, Rolla, MO, 65401, USA

ABSTRACT

Some jealousy researchers have argued that the term "jealousy" is simply a label for the anger, sadness, and/or fear produced by an appraisal of threat to one's relationship from a romantic rival. I argue here that "jealousy" refers to a particular organization of feelings, thoughts, and behaviors that is distinct from other emotions. In the present study, subjects were able to distinguish changes in the intensity of a fictitious person's jealousy from changes in his or her anger, sadness, and fear. This happened despite the fact that the statements used to characterize the person's emotional state were all prototypic jealousy features. Although anger, sadness, and fear commonly occur during jealousy episodes, they appear not to be the essence of what people understand to be jealousy.

1. Introduction

Some jealousy researchers (e.g., Hupka, 1984) have argued that the term "jealousy" is simply a label for whatever emotion events follow appraisal of threat to one's relationship from a romantic rival. That is, jealousy is the anger, sadness, and/or fear produced by that situation. If so, then the intensity of a person's jealousy should vary with the intensity of these emotions. Further, in the context of a jealousy episode, people should not be able to distinguish an emotion of jealousy from the emotions of anger, sadness, and fear (in other words, jealousy is not itself an emotion).

But Sharpsteen (1991, 1993) has argued that the term "jealousy " refers to a particular organization of events, feelings, thoughts, and behaviors. If that is the case, then people should be able to judge the intensity of jealousy independently of the intensity of anger, sadness, and fear and distinguish jealousy from its emotional concomitants.

The organization of characteristics that represents jealousy should be a part of people's jealousy *prototypes* (mental representations of jealousy episodes). Based on people's descriptions of typical jealousy episodes, Sharpsteen (1993) compiled a list of 86 prototypic jealousy features. If typical jealousy episodes comprise emotions such as anger, sadness, and fear, then the features of people's jealousy prototypes should be organized in a way that allows them to discriminate among these constituents of their jealousy experiences.

Sharpsteen and Kirkpatrick (1997) asked subjects to sort 86 cards, each with a prototypic jealousy feature typed on it, into piles representing the emotions involved in jealousy. A cluster analysis of the resulting co-occurrence matrix produced four distinct clusters, one clearly representing anger, another fear, and another sadness. The fourth they labeled "idealized jealousy" because it, uniquely, included features that cognitive-motivational models of emotion posit as constituents of any emotion episode--antecedents and appraisals; cognitive, afective, and behavioral responses; and coping strategies and consequences. This organization of features, they suggest, serves as a cognitive template for identifying an emotion event as jealousy, as opposed to simple anger, sadness, or fear.

The goal in the present study was to show that people can distinguish jealousy from anger, sadness, and fear despite the fact that each is represented in their jealousy prototype.

2. Method

2.1 Subjects

Ninety-six subjects (48 men, 48 women), drawn from Introductory Psychology classes, volunteered to participate in exchange for course credit.

2.2 Materials and Procedure

Each subject (48 men and 48 women) received a booklet containing ten separate questionnaires ostensibly filled out by ten jealous people both before and after discussing their jealousy with their partners. These people had supposedly rated the applicability of various thoughts, feelings, and behaviors to their emotional state. The items on these sham questionnaires were three features from the "idealized jealousy" cluster randomly mixed with three features from either the anger, sadness, or fear clusters. Features were selected randomly with the one constraint that they had been judged relatively central to the concept of jealousy in a previous study (see Sharpsteen, 1993). Additionally, two sets of features from each cluster were used. Thus there were 12 versions of sham questionnaires, each containing one of the two sets of jealousy features and one of the two sets of features from just one of the other emotion clusters. For any given subject, each of the ten before-and-after questionnaires he or she received had the same six features.

Sham ratings of the applicability of these features to the jealous person's emotional experience were distributed so that, from "before" questionnaires to "after" questionnaires, average ratings for jealousy features either increased or decreased *or* ratings for the features from the other emotion cluster either increased or decreased *or* average ratings for both sets of features remained the same (i.e., there were five ways in which "after" ratings might differ from "before" ratings). Also, there were two levels of average "before" ratings (either 35 or 65 on a 100-point scale) and these average ratings were always the same for both sets of features. Where average ratings increased or decreased, the change was always 20 points. Thus there were ten "before-and-after" questionnaires in all: two levels of average ratings on the "before" questionnaires and five ways in which those average ratings might have changed (or not) on the "after" questionnaires.

For each sham questionnaire, subjects were asked to examine the applicability ratings on the "before" questionnaire and "after" questionnaire and then rate the intensity of each jealous person's jealousy, anger, sadness, and fear/worry on a 1-to-7 scale.

3. Results

For each of the three groups of subjects (those receiving anger, sadness, or fear features in addition to the jealousy features), MANOVAs and planned contrasts were performed on "change" scores ("after" minus "before" ratings) for each of the four emotions. As shown in Table 1 (which presents means collapsed across feature sets and initial sham-rating levels; key findings are in bold print) change scores for jealousy were significantly larger (in the appropriate direction) when sham ratings for jealousy features changed than when sham ratings for the other emotion's features changed. Similarly, change scores for the other emotion were significantly larger when sham ratings for its features changed than when ratings for jealousy features changed. Thus the key prediction was supported: Subjects were able to distinguish jealousy from its concomitant emotions, despite the fact that the statements representing all four emotions were prototypic jealousy features.

Table 1. Mean "Change Scores" for Ratings of Emotions for Features Included in Questionnaire

	Jealousy Lowered	Other Emotion Lowered	No Change	Other Emotion Raised	Jealousy Raised
Jealousy with Anger	**-1.63**	-.39	.09	.70	**1.22**
	-.36	**-1.09**	.03	**1.61**	.45
Jealousy with Sadness	**-1.14**	-.22	.19	.47	**1.27**
	-.19	-1.67	.27	1.41	.31
Jealousy with Fear	**-1.25**	-.27	.13	.47	**1.47**
	-.63	**-.98**	.30	**1.41**	.97

Mean change scores for emotions not represented in a given before-and-after questionnaire (e.g., ratings of anger or sadness for subjects presented with jealousy and fear features) are presented in Table 2. By and large, subjects apparently inferred that levels of emotions for which they had no explicit information would be consistent with the overall intensity of the jealousy experience. Ratings of the jealous person's anger generally changed with changes in the applicability of jealousy, sadness, and fear features. Sadness ratings changed with changes in the applicability of anger and fear features, and increased with increased applicability of jealousy features, but did not decrease with decreased applicability of jealousy features. This might suggest that subjects expect sadness to linger after jealousy has diminished. Fear, however, varied with the applicability of jealousy features but not with the applicability of anger or sadness features.

Table 2. Mean "Change Scores" (as Absolute Values) for Ratings of Emotions Not Included in Questionnaire

		Jealousy Lowered	Other Emotion Lowered	No Change	Other Emotion Raised	Jealousy Raised
Emotion Rated:	Feature Type Included:					
Anger	Sadness	.75	.27	.14	.73	1.20
	Fear	.56	.70	.13	.73	.89
Sadness	Anger	.39	.56	.06	.66	.92
	Fear	.42	1.11	.06	1.06	.63
Fear	Anger	1.00	.39	.28	.58	1.16
	Sadness	1.02	.42	.17	.44	1.44

4. Discussion

In terms of its mental representation, jealousy's features include feelings, thoughts, and behaviors typically associated with anger, sadness, and fear (i.e., it is a blended emotion). But, as shown here, the organization of prototypic jealousy features describes something more than those concomitant emotions and that "something" was apparently regarded as jealousy by subjects in the present study.

Subjects clearly distinguished jealousy features from anger, sadness, and fear features. Changes in the intensity of jealousy increased more with changes in the sham ratings assigned to features from the "idealized jealousy" cluster than with changes in the sham ratings for other features included in the questionnaire (and, generally, did not change at all in the latter condition). Thus, when given explicit information about anger, sadness, or fear, subjects made distinctions between those emotions and jealousy. When not given explicit information, however, subjects tended to infer that changes in jealousy implied changes in anger, sadness, and fear. In other words, their implicit theories about jealousy represent these emotions as linked, but distinguishable.

One implication of these findings is that jealousy might be considered a candidate for a basic emotion, rather than being a blend of basic emotions. Indeed, evolutionary accounts of jealousy suggest that it could have evolved independently of other emotions. If so, jealousy's functions must go beyond those of anger, sadness, and fear (or else its evolution, or theirs, would have been unnecessary). Second, characterizations of emotions (and not just jealousy) as "blended" have apparently been too simplistic and ambiguous. If jealousy is simply a blend of more basic emotions, and defined by its context, then how is it that subjects make distinctions between jealousy and other emotions when that context is a given (as it was in this study)? People apparently know about, and can make inferences about, *patterns* of events (i.e., idealized jealousy) that define an emotion episode as jealousy (even without seeing all the features that form that pattern). That is, jealousy episodes possess a reliable structure beyond the situation that provokes them and the emotions that occur during them. Certainly, this pattern could involve elements of anger, sadness, and fear, but it is also something more.

Acknowledgements

This work was supported, in part, by the Arts and Sciences Training and Education Fund of the University of Missouri at Rolla.

References

Hupka, R. B. (1984) "Jealousy: Compound emotion or label for a particular situation?", *Motivation and Emotion* 8:141-155.

Sharpsteen, D.J. (1991) "The organization of jealousy knowledge: Romantic jealousy as a blended emotion", in: *The Psychology of Jealousy and Envy*, P. Salovey, ed., New York: Guilford Press, pp. 31-51.

Sharpsteen, D. J. (1993) "Romantic jealousy as an emotion concept: A prototype analysis", *Journal of Social and Personal Relationships* 12: 69-82.

Sharpsteen, D. J. and L.A. Kirkpatrick (1997) "Romantic jealousy and adult romantic attachment", *Journal of Personality and Social Psychology* 22: 627-640.

ALEXITHYMIA AND THE BIOCYBERNETICS OF SHAME

WILLIAM FRAWLEY° and RAOUL N. SMITH*
°*Dept. of Linguistics/Cognitive Science, University of Delaware, Newark, DE 19716 USA*
**College of Computer Science, Northeastern University, Boston, MA 02115 USA*

ABSTRACT
Alexithymia is a cognitive-affective disturbance of the experience and expression of emotions. The cognitive and behavioral effects of alexithymia are strikingly similar to disruptions of consciousness reported in the cognitive science literature. Just as blindsight and prosopagnosia may allow covert object and face recognition without report to awareness or other domains, so alexithymia appears to disconnect emotional knowledge and the rest of mental life. This construal of alexithymia in information-processing terms motivates our subsequent analysis: a preliminary relational network of *shame*. This in turn allows us to make interesting conclusions about the nature and representation of emotions and the use of such representations to determine interventions in affective disorders.

1. Introduction

Alexithymia is a cognitive-affective disturbance of the experience and expression of emotions. It is characterized by the inability to recognize and verbally describe one's own emotions; markedly reduced symbolic thought restricting expression of attitudes, feelings, wishes, and drives; inability to use feelings as signs of emotional problems; utilitarian thought concerned with the minutiae of external events; reduced recall of (often colorless, bizarre) dreams; difficulty in discriminating between emotional states and bodily sensation; stiff posture; lack of affective facial expressions; and impaired capacity for empathy. Alexithymia was first observed among patients who expressed classical psychosomatic disorders -- so called infantile personalities (Ruesch, 1948) and emotional illiterates (Freedman & Sweet, 1954) -- and later observed in post-traumatic stress (Krystal, 1968), alcoholism, and drug addiction (Rybakowski *et al*., 1988). Taylor *et al.* (1997) survey the syndrome.

The cognitive and behavioral effects of alexithymia are strikingly similar to disruptions of consciousness reported in the cognitive science literature (Lane *et al*., 1997; Frawley & Smith, 2000). Just as blindsight and prosopagnosia may allow covert object and face recognition without report to awareness or other domains (Farah, 1995), so alexithymia appears to disconnect emotional knowledge and the rest of mental life, as if there were a bug in the interface between affect and cognition. Alexithymics seem to have lost the "feel of their feelings" and look very much like the theoretical zombies of the absent qualia argument. Seen as a dissociation rather than a loss (thanks to Cal Izard for this idea), alexithymia's cognitive and behavioral consequences become clearer. If affective knowledge indexes memory, frames problem-solving, tags plans, signals error, guides repair and focuses attention (Oatley & Larocque, 1995), then its dissociation from other informational domains might naturally result in utilitarian problem-solving, face processing that is informationally sound but lacks affect, emotions that are experienced but defectively expressed, and disruptions of motor output. Moreover, if a principal cognitive function of emotions is to signal goodness of fit between inside and outside (Damasio, 1994), then alexithymia might be understood broadly as a problem of mind/world regulation.

As a dissociation, alexithymia's well known resistance to therapeutic intervention is also clarified. Dynamic, conversational psychotherapies require the exploration and articulation of feelings, fantasies, and somatic locales of affective disturbance. Because alexithymics superficially answer the questions that drive psychotherapy, or even fail to understand the questions themselves, they are considered poor candidates for psychotherapy. Alexithymics may not articulate their feelings well, but still experience the feelings via covert recognition: the dissociations of alexithymia thus suggest interventions built on reconstructing the links and tags between affective and cognitive, rather than restoring the emotional knowledge itself (cf. Lesser's (1985) instructional, reality-based approach).

The advantage of construing alexithymia in informational terms is that the benefits of cognitive science as a whole accrue to the task of explaining emotions and their disruptions. Metatheoretically, functionalism and levels of explanation can help guide inquiry. Emotions are one thing that brains-bodies do; affective knowledge falls out of the wetware computations. We might best explain emotions by aligning the three standard levels of accounts in cognitive science -- (1) material/ physical/ hardware, (2) knowledge/ software /algorithm, (3) output/ phenomenal/ ecological -- rather than by reducing emotions to the physical or eliminating them altogether as empty folk psychological terms.

Alexithymia might be the functional outcrop of varied material cause -- multiply realized -- much as some of the classic symptoms of visual agnosia (like failure at object integration) surface in both acquired disorders (head trauma) and genetic ones with no obvious neural correlate (Williams syndrome and Turner syndrome). This would accord with the remarkably similar characteristics seen across cases of different, or even unknown, physical cause and effect -- drug abuse, alcoholism, and incest produce alexithymia as much as do lesions in anterior cingulate gyrus (see Damasio, 1994). This embrace of functionalism might in turn push for detailed studies of the material correlates for the disorder since there might be many-to-one or one-to-many relations across the hardware-software-output triad. Picard (1997) makes a similar point in explaining the difficulties in juggling the biological and computational problems in arousal. There is no single measure of arousal, and multiple hardware systems contribute to unitary arousal at the phenomenal and knowledge levels.

This metatheoretical stance would support Izard's (1993) view that cognitive processes (appraisal, attribution, and construal) are neither the sole, nor the primary, elicitors of emotions. Izard's four interacting classes of emotion activators -- Neural, Sensorimotor, Motivational, Cognitive -- might call for subdivision and alignment. Subcellular information processing, for example, might determine emotion thresholds, manifested at another level as proneness to experience emotions and therefore characteristic mood. But this does not mean that, say, disgust is explained by locating a peptide or even a dedicated neural assembly in which the peptide is expressed. As Pert (1993) says, emotion, like mind generally, is physically distributed and emerges from an interactive "psychosomatic communication network" (p. 178), also involving the organism in the world.

In the end, detailed functional studies of alexithymia would more finely differentiate the syndrome. Some versions of alexithymia might result from quite different material losses. Just as we now know that autism and Williams syndrome come in at least two kinds so alexithymia might be unpacked as a cover term for differential convergence and divergences across the physical, knowledge, and ecological levels.

The cognitive science of alexithymia also turns our attention to the computational-representational structures that comprise affective knowledge. Important basic work has already been done in this regard, but we think that much of it needs augmentation both in quantity and kind to capture the full connectivity of emotions. In Figures 1 and 2, we present portions of the knowledge structures that comprise shame. Fuller affective representations using these relational structures as a base would allow us to: (1) incorporate a wide variety of information into the knowledge structure, including developmental and clinical observations that might serve as locales for links between the cognitive and the affective, predictable sites for breakdown, and the basis for interventions; (2) split the core from the periphery in affective knowledge and drive inquiry into universal vs. culture-specific and innate vs. learned emotion knowledge; (3) yield a computationally tractable knowledge base to support implementations to test emotion theory and drive discovery and applications for specific kinds of therapeutic interventions.

Here we look at one emotion, shame, to underscore the benefits of a rich representational system. (The exposition below should be read with the relational representations of Figures 1 and 2 at hand.) Why shame? It has a determinate conceptual structure; it lies at the basis of a number of other syndromes and so has wide effects (Bradshaw 1988 calls it "the master emotion": it plays an important role, for example, in the progression of alcoholism).

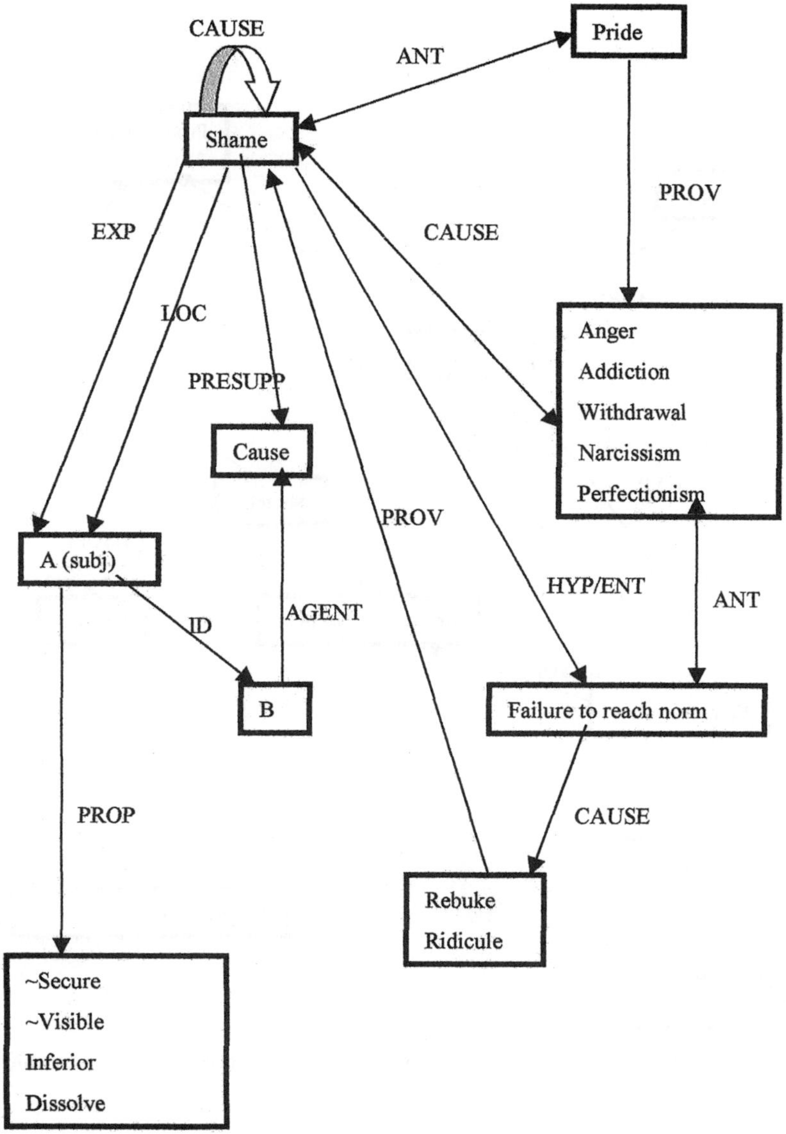

Figure 1. A partial network of Shame, focusing on cause

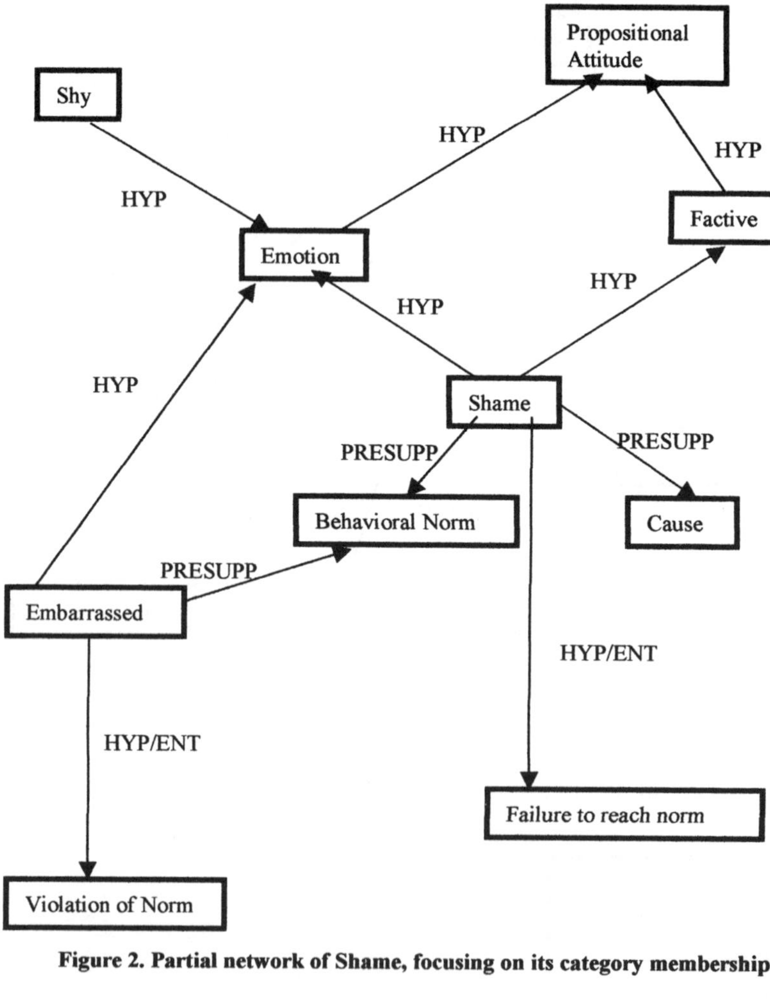

Figure 2. Partial network of Shame, focusing on its category membership

Shame has the form:

A_1(subject) Shame$_{e1}$ B(Agent) doing C
Roles: A Experiencer, B Agent (of), C Reason/Actional Source
Degrees of Separation: Typically A = B, but if not, A is ashamed at/of B's C as if B were in *A's normative world* or is ashamed *for* B in *B's normative world*

Shame is technically unaccusative (the subject (A) is a logical undergoer, e marks its logical position and A and e are coindexed) which affects the concept's pattern of information flow. As an entirely internal experience (agent-focused cognitive unit: Ortony, Clore & Collins, 1988), it patterns like other self-centered emotions (satisfied and relieved vs. admire, gloat, and love). Unaccusative emotions contrast with transitives – (compare. envy vs. admire: A1 envy$_{e1}$ B vs. A admires B -- and may have causative derivatives depending on the separability of agent and subject and the independence of the action of the agent -- B shames A into X only if A and B are different), but transitives have no derivative middle because the objects are not substantially affected: A pities B , but not B pities easily.

Shame, like all emotions, is a propositional attitude and (apparently like all other emotions) factive (presupposes its complement: the shame-inducing act, C, holds regardless of whether shaming itself is true). Factivity makes emotions a coherent class of mental predications, and distinct from, say, belief or desire, which are either neutral with respect to their propositional content or inherit the truth properties of that content. These facts in turn would suggest that the proper way to analyze emotions is via theory of mind and the recent debates about the linguistic and situational origins of mental content and the transitions from desire to belief (Carruthers & Smith, 1996; de Villiers & de Villiers, in press).

Shame presupposes some social-behavioral norm as a backdrop but entails a failure to reach this norm. Consequently shame can be triggered by almost any breakages in interpersonal relations (Kaufman, 1996). Shame and embarrassment are thus neither synonyms nor antonyms, but mere overlaps since the latter presupposes the norm, but entails its violation.

When we examine the finer relational structure of shame -- roughly all lexical inferences -- we see some remarkable facts. Shame enters into two kinds of meta-relations: one self-reflexive (shame calls itself) and the other reciprocal with respect to anger (shame calls anger which then induces shame) (relationally, shame META shame (self-reflexive) and shame META anger (reciprocal)). These metarelations make shame vicious (and hence difficult to treat): shame makes you ashamed and it can then induce anger, which then returns the favor by invoking shame, which makes you more angry, which then makes you ashamed at your anger (Scheff, 1987). This is quite unlike the informational structure surrounding pity, pride, or gratification. Pride does not make you proud, and even though it can make you brazen, that does not make you proud.

The unaccusative structure of shame locates it entirely within the subject (relationally, shame LOC subject). Shame's lack of overt manifestation for its truth makes it hard to ascribe (Oatley & Larocque, 1995) and subject to intervention strategies that force the sufferer to focus on externals only (Kaufman, 1996). Shame's association with the discovery of hidden shortcomings differentiates it from embarrassment, which requires some overt violation and the revelation of empirical evidence. Shame is therefore more like shyness, disappointment and relief, and unlike guilt, which is tied to overt acts of reparation (Kaufman, 1996).

Properties of the subject further refine the locative nature of shame (relationally, shame SUBJ PROP inferior insecure, dissolved, falling short, invisible...). These properties account for difficulty of therapy (how can you stabilize a subject that is made dissolute and insecure by the very object of therapy?) and the connection of shame with retreat (as an entirely internal invisible state retreating from discovery, shame is very much a cover up, unlike embarrassment or guilt.).

Finally, shame can be the background condition for entire scripts and scenarios (collections of knowledge structures) (relationally, shame FRAME [X]). This is widely reported in the clinical literature, where shame can be the ideology for entire families. Shame's broad effects link it with fear and embarrassment, but contrast with being happy for or gratified (a happy for/gratified family?). One wonders if negative emotions are the major background conditions for others, as if their values set the initial weights or determine all adjustment values in a neural net of other emotion states.

A number of insights emerge from this relational analysis of the affective knowledge surrounding shame. One is that we can generate the standard flow model of affective knowledge from the bottom up by converting the concepts and relations to procedural questions and then grouping them by informational domain. Elicitors would then correlate in part with causes and presuppositions; appraisals would correlate in part with category membership and related notions; expression would correlate partly with consequences and results; and so on. Indeed, this way of seeing emotions would yield a method of doing comparative emotion analysis since we could take a single emotion concept in several different languages, align the analyses according to the information-flow model of emotions (generated from relational structures), and determine the overlaps and divergences. All these findings are rendered in Figure 3, where we have converted the relational analysis of English *shame* to the information-flow model of emotions and then presented for comparison renderings of the shame concept in Danish and Javanese. The reader is invited to inspect the convergences and divergences for him-/herself.

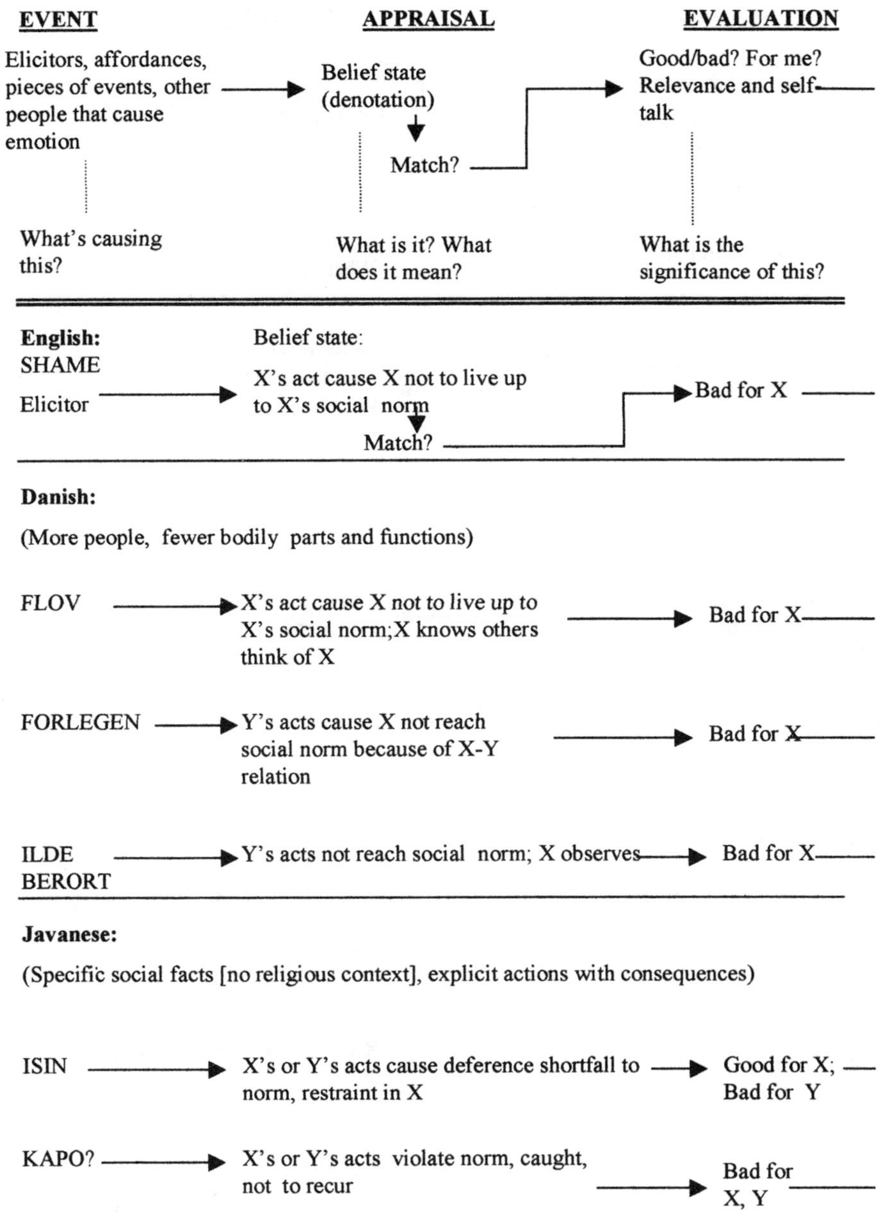

Figure 3. Information flow in emotion and cross-language comparison

READINESS		EXPRESSION
→ Plans and programs for action: Cognitive-verbal, somatic, expressive, motor What should I do?	→	Execution: Think/speak, feel, display, move What do I THEN do?
→ Prepare for inaction, invisibility, covering up; overreactions and excessive behavior	→	Hide, lower eyes, feel sick, disappear, be reticent, be perfectionist
→ Want others not to think of X	→	Avoid, exhibit physical symptoms
→ Escape		
→ Sympathize		
→ Cover, disappear, reduce	→	X avoid others' eyes, blush, lower head
→	→	No blush

The relational analysis also makes intensity and familiarity, often represented as distinct values, emergent properties of the overall connectivity of emotional knowledge rather than independently manipulated parameters. Shame might increase in intensity from the positive or negative values of its relata rather than from a separate intensity variable: its intensity might increase from an increase in anger but from a decrease in expectations about the presupposed norm.

We conclude with one theoretical and one practical observation. If emotions are computable, can a machine feel, and how would we know so? Churchland (1996) argues that consciousness is our autoconnection to our wetware, not a separate substance. Emotions might then be the feel of our autoconnections -- the goodness of fit mechanisms that the body/brain does Since these might be multiply realized, there is no reason why a machine might not sense its own goodness of fit, or lose that feel in alexithymia.

Treating alexithymia would be a great service to health care delivery because the syndrome may be one of the largest indirect causes of its increased cost. Alexithymics often express their psychic problems via physical complaints, which their physicians take as suggesting some organic pathology necessitating extensive texts, none of which discloses an organic problem. The kinds of emotion representations we have developed for shame could underlie computerized emotion tutors that could be one of a set of network services for aiding recovery from the dissociation of the cognitive and affective.

References

Bradshaw, J. (1988) *Healing the Shame that Binds You*, Deerfield Beach, FL: Health Communications, Inc.

Carruthers, P. and P. Smith (1996) *Theories of Theories of Mind*, Cambridge, UK: Cambridge University Press.

Churchland. P. (1996) *The Engine of Reason, the Seat of the Soul*, Cambridge, MA: MIT Press.

Damasio, A. (1994) *Descartes' Error*, New York: Avon.

de Villiers, J. and P. de Villiers (in press) "Linguistic determinism and the understanding of false beliefs", in: *Children's Reasoning and the Mind*, P. Mitchell and K. Riggs, eds, Hove: Psychology.

Farah, M. (1995) *Visual Agnosia*, Cambridge, MA: MIT Press.

Frawley, W. and R.N. Smith (2000) "A processing theory of alexithymia", Manuscript under review.

Freedman, M. and B. Sweet (1954) "Some specific features of group psychotherapy and their implications for selection of patients", *International Journal of Group Psychotherapy* 4:355-36.

Izard, C.E. (1993) "Four systems for emotion activation: Cognitive and noncognitive processes", *Psychological Review* 100:68-90.

Kaufman, G. (1996) *The Psychology of Shame*, Berlin: Springer.

Krystal, H. (1968) *Massive Psychic Trauma*, New York: International Universities Press.

Lane, R.D., G. Ahern, G.E. Schwartz and A.W. Kaszniak (1997) "Is alexithymia the emotional equivalent of blindsight?", *Biological Psychiatry* 42:834-44.

Lesser, I. (1985) "Current concepts in psychiatry", *The New England Journal of Medicine* 312:690-692.

Oatley, K. and L. Larocque (1995) "Everyday concepts of emotions following every-other-day errors in joint plans", in: *Everyday Conceptions of Emotion*, J. Russell *et al.*, eds, Dordrecht: Kluwer, pp. 145-65.

Ortony, A., G. Clore and A. Collins (1988) *The Cognitive Structure of Emotions*, Cambridge, UK: Cambridge University Press.

Pert, C. (1993) "Interview", in: *Healing and the Mind*, B. Moyers, ed., NY: Doubleday.

Picard, R. (1997) *Affective Computing*, Cambridge, MA: MIT Press.

Ruesch, J. (1948) "The infantile personality", *Psychosomatic Medicine* 10:134-144.

Rybakowski, J. *et al.* (1988) "High prevalence of alexithymia in male patients with alcohol dependence", *Drug and Alcohol Dependence* 21:133-136.

Scheff, T. (1987) "The shame-rage spiral", in: *The Role of Shame in Symptom Formation*, H. Block Lewis, ed., Hillsdale, NJ: Lawrence Erlbaum, pp. 109-149.

Taylor, G. *et al.* (1997) *Disorders of Affect Regulation: Alexithymia in Medical and Psychiatric Illness*, Cambridge, UK: Cambridge University Press.

THE FUNCTION OF EMOTIONAL EXPERIENCE IN DECISION-MAKING, PROBLEM SOLVING AND CREATIVE ACTIVITY

LIS NIELSEN

Department of Psychology, University of Arizona, Tucson, Arizona 85721, U.S.A.

ABSTRACT

The evidence that the conscious experience of emotion plays a functional role in higher cognition is reviewed. A variety of functions of emotion have been posited in the literature. These include: emotional experience as a motivator of cognitive processing, emotion as a mechanism for attentional bias, and emotion as a determinant of distinct processing styles or modes. Similarities and differences in these proposals are examined. Some implications for research and application are suggested. Problem solving and decision-making can be highly emotional processes. Gearing up to tackle difficult tasks, coping with the frustrations of failure, balancing competing attractive or unattractive options - each of these experiences belies the notion that problem solving is a purely rational process. Feelings, sometimes very strong emotional feelings, frequently come into play. Are these emotional experiences mere epiphenomena to the cognitive functions they accompany? In what follows I review evidence to the contrary, evidence which suggests that the experience of emotion itself may play a functional role in directing the way we solve problems and make decisions in a variety of domains.

1. Conscious Emotion as a Motivating Force

The subjective reports of creative artists and scientists suggest that emotion is a primary motivating force in their work. These individuals report a rich interplay of cognition and emotion in the creative process. Feelings are the pushes and pulls that propel one forward, draw one along, or hold one back on the path to a problem solution (May, 1980; Nielsen, 1998). There is a particular feeling associated with being caught up in the flow of these exploratory and creative processes (Csikzentmihalyi, 1990). And like the processes leading up to them, creative breakthroughs or insights are reported as being highly emotional experiences. Are there systematic relationships between experienced cognition and emotion in problem solving? Do emotions support or drive creative problem solving in particular ways? While phenomenological reports do not reveal the mechanisms of emotional motivation, they do paint a picture of an inextricable link between emotional experience and the creative drive. In her recent autobiography Walking in the Shade, the author Doris Lessing writes:

> *The Golden Notebook* was written at high pressure -- pressure from within, which brings me to another murky area. Sometimes the emotional pressures that fuel a novel are very far from its subject matter. ...*The Golden Notebook*'s fuel was feelings of loss, change:

> that I had been dragged to my emotional limits by Jack and then Clancey -- rather, I had been dragged by my emotional needs, which really had nothing to do with them as individuals. ... Peter, the third and last child, was growing up. Loss, departures, the ending of dramas begun long ago, the need for drawing lines -- *finis*. All this dynamic energy went into *The Golden Notebook*: emotional energy, which is so much stronger than we think ... And this besides having to acknowledge that what is so often called 'intellectual' is in fact emotional. What is more violently emotional and passionate -- and poisonous -- than a room full of intellectuals in ideological debate? ...But I slide past this dangerous area, holding my breath (1997, pp. 337-338).

Conflict can both motivate and impede progress on a creative work. The Danish poet Sören Ulrik Thomsen (1994) describes the experience of tension created by the juxtaposition of an intense desire to always be writing and the strong aversion to embarking on a new writing task. The aversion is to the difficult process of mental reframing - the destruction of current representations of the problem to allow for the emergence of a creative product. Often the behavior of the artist caught up in the throes of creation seems to mirror this process of destruction and recreation. Thomsen writes:

> I have written and written until I could write no more, fallen asleep and awakened after a brief time with new words in my head, arisen without washing, donning some of the dirty clothes that lay in heaps on the floor... I have eaten just enough to keep me vertical, yet smoked constantly. And when the ashtray overflowed its banks, I didn't empty it but continued to scrape the pipe out so that a pyramid of ash finally grew on the desk, until it became so great that it had to be pushed over on the floor with the rest of the debris... It was not that I was so absorbed in the process to notice things about me, on the contrary, I was embarrassingly aware of it all. When I mention this it is because I see this behavior as a code for the strong imaginary drive behind the artistic creative process, namely that it is bound up with destruction... There is a breaking down here that occurs synchronously and proportional to the building up of the work (1995, p. 8, author's translation).

Some investigators have proposed (see Section 3) that positive emotion fosters creative thinking. Yet both the self-reports of artists and recent work on expert problem solving paint a different picture. The evidence suggests that what distinguishes creative from mundane problem solvers is a willingness to go to the extents their feelings carry them - to sometimes push the limits of those feelings and risk emotional pain in pursuit of the creative solution. This ability may be rooted in their greater capacity for tolerating the emotional tension and anxiety that necessarily arises when one seeks to press the boundaries of one's competence in genuine acts of creation (May, 1980; Nielsen, 1998). Indeed, expert writers typically frame problems in more difficult ways in order to challenge their skill; they are willing to take more risks, make more mistakes, agonize more, and generally spend more time on the problem at hand (Scardamalia & Berieter, 1991). As Thomsen writes, "Only those who are prepared to risk miserable defeat will be given the chance to reach extraordinary heights and produce true works of art" (1994, p. 46, author's translation).

Are these feelings of tension and aversion and of being dragged to one's emotional limits playing a causal role in the creative process, or are they merely results or epiphenomenal markers of other processes associated with creative activity? Artists seem to know no other way that it could be - but does that imply causality? And how might the causal mechanisms operate?

2. Emotion as a General Biasing Mechanism

There is converging evidence that emotion functions as a general biasing mechanism for attention and cognitive processing. The basic idea is that affective systems provide inputs to decision-making or attentional mechanisms, and these inputs serve to bias ongoing cognition and behavioral responses. Among the earliest to propose such a link between emotion and cognition was the cognitive psychologist Herbert Simon. Simon (1967) argued that any information processing system that has multiple, complex, and potentially competing goals that need to be accomplished in real time would require some mechanism to organize behavior in terms of current goal priority. Simon proposed that emotion functions as an "interrupt" mechanism that shifts the cognitive system from one goal to the other.

Evolutionary psychologists are motivated by similar concerns about how organisms manage the competing demands posed by complex environments. Emotions evolved as adaptations for dealing quickly and effectively with frequently encountered environmental challenges. In light of the efficiency of emotional responses and the flexibility they afford the organism, Tooby and Cosmides argue, "[T]he entire structure of attention, from perceptual systems to the contents of high-level reasoning processes, should be regulated by emotional state" (1990, p. 413). According to Tooby (1985), the subjective experience of emotion is an internal, attention-focusing signal that targets all available resources to the problem at hand and motivates the appropriate action tendencies.

Lang *et al.* (1997) provide experimental evidence that supports the evolutionary view. Their studies of affective modulation of the eye-blink startle response demonstrate emotional influences on attention and behavioral responses. Lang *et al.* (1997) found that the priming of negative emotional state amplifies our natural withdrawal responses, while positive emotional set engages our attention and triggers approach behaviors. There is now considerable evidence that a number of innate, adaptively functional, affective modulators of attention exist, upon which further attentional biases can be constructed through learning (See Öhman & Soares, 1994, 1998; LeDoux, 1996).

Recent work has examined how these affective mechanisms function in higher domains of social and cognitive decision-making. Damasio and colleagues (1994; Bechara *et al.*, 1997; Adolphs *et al.*, 1996) argue that feedback from the autonomic systems that mediate emotional arousal helps to bias and resolve decisions. They have found that brain-injured subjects with deficits in these arousal-mediating systems have difficulty making decisions under conditions of uncertainty. Earlier results of Bruchlinski and Tichomirow (1975) similarly suggest that emotional arousal guides problem solving, functioning as a decision making heuristic. Both groups of investigators argue that the function of arousal is to bias some decisions over others by focusing attention and narrowing down the area of a problem space or decision tree to be explored.

Of these investigators, only Bruchlinski and Tichomirow (1975) discuss the links between these automatic appraisals or attentional biases and the conscious experience of emotion. In studies of complex problem solving, they found that changes in physiological arousal and concomitant emotional experience precede and prime insights and cognitive breakthroughs. Bechara *et al.* (1997) offer supporting evidence that the affective biasing of decisions takes place prior to conscious awareness of decision principles, though they remain silent on the role of conscious emotional processes in the stages leading up to this awareness.

Other writers have suggested the arousal that comes after an insight might itself play a functional role. Baddeley (1976) suggests that this Aha! experience might be a signal that helps to focus attention on important information. However, Weisberg (1995) has argued that this experience is merely a reaction of surprise to the suddenness with which an insight appears.

In Clore and Parrott's (1991) "affect as information" model, emotional experience - like other types of experience (bodily or cognitive) - provides information to decision-making. Emotional experience is the output of our appraisal system, indicating the significance that events hold for us. While we may be mistaken about the causes of our experience, if these experiences are deemed relevant to the decisions that need to be made, they will influence those decisions.

Must affect be conscious to influence cognition? Clore and Parrott (1991) argue that it is by virtue of being felt that affect modulates our judgment and decision processes, though they admit that we are often unaware of relying on our experience in this way. Subjective reports of creative artists also suggest that felt affect has a strong motivating force, in line with Tooby's (1985) evolutionary view. On the other hand, Simon (1967) speaks only of central nervous system processes whose primary role is to interrupt cognitive processing (after which they may or may not lead to consciously felt emotion). Similar roles for unconscious affective processes have been suggested by others (See Lang *et al.*, 1997; LeDoux, 1996; Öhman & Soares, 1994, 1998). Damasio (1994) suggests that the biasing properties of arousal systems may operate cortically in an 'as-if' fashion, bypassing bodily circuits - the question remains whether these alleged cortical states give rise to the experience of affect or whether embodiment is required for affect to function in this way. Further work is needed to determine whether the affective biasing of cognitive processes is always experientially mediated.

3. Emotion as a Shaper of Processing Style

As the startle modulation studies of Lang and colleagues (1997) suggest, specific emotions function to bias cognition and behavior in characteristic ways. Much of the work that looks at the differential effects of emotion types on cognition has focused on how experienced emotion impacts on cognitive processes, implying that the experience itself might be doing a portion of the functional work. In this section I will look at the differential changes in cognitive processing that occur as the result of either positive or negative emotional state.

3.1. The Role of Positive Emotions

By inducing positive affect through a variety of means (giving subjects candy, showing happy films, providing success feedback, etc.), Alice Isen and colleagues (Isen, 1993) have studied the effects of positive emotion on many aspects of decision-making. They found that positive affect promotes creativity in problem solving, success in negotiation, and thoroughness and efficiency in decision processes. Positive affect also encourages working beyond the problem and promotes more flexible thinking (i.e., it enables people to see more similarities and differences between stimuli) (Isen, 1993). Finally, Isen (1993) claims that in addition to fostering creative thinking, positive affect increases the aversion to risk and desire to maintain the positive state.

Drawing on research by Isen and others, Fredrickson (1998) proposes a "broaden and build" model of positive emotions. On this model, the function of positive emotion is to broaden the "momentary thought-action repertoire" enabling the building up of intellectual, social, and physical resources, through exploratory thinking, integration of knowledge, pleasant social interaction, and play.

It is interesting to contrast these claims with the point raised earlier regarding the motivating and dominant role of negative affect in creative activity. Recall that expert writers purposely frame problems in challenging, difficult ways to arrive at more creative solutions (Scardamalia & Berieter, 1991). Is it possible to reconcile these two views? Some of the controversy may be rooted in how creative activity is defined. For Isen, creative thinking means finding better solutions to verbal puzzles or coming up with more remote associations to words. In contrast, the tasks set by expert writers or scientists are more highly constrained and involve a larger investment of the self. This suggests that contextual constraints may make an important difference in the two cases. Fredrickson's arguments also speak to this concern. Positive emotions might facilitate creativity in social and recreational contexts, while creativity in more artistic or intellectual pursuits may require more endurance and involve more pain.

It is true that some aspects of creative activity are experienced as highly positive. Without the balance of positive flow states and strong desires to produce or solve problems, it is uncertain whether the creative tension would be tolerated over extended periods. Nevertheless, it appears that negative emotion may do some of the important functional work. Isen's work also suggests that positive states play a causal role in the creative process. More work is needed to differentiate the conditions under which negative or positive affect fosters creative activity.

3.2. The Role of Negative Emotions

Like positive emotions, negative emotions have also been proposed to shift cognition and behavior into particular modes. Much of the emphasis on negative emotions emphasizes their adaptive role in quickly shifting behavior to cope with threats to physical or emotional integrity. Negative emotions like fear, anger, sadness, grief, tend to narrow (as opposed to broaden) the focus of attention, enabling problem-focused coping to proceed (Fredrickson, 1998; Lang *et al.*, 1997). Depending on the particular negative emotion, different processing modes may be engaged. According to Oatley and Johnson-Laird (1995), anger encourages additional effort or aggression, while fear encourages vigilance and quick planning. Sadness tends to slow down action and prompt the search for alternatives to the current situation (Oatley & Johnson-Laird, 1995). In general, negative emotions encourage more self-focus and a more narrow attentional focus on the eliciting stimulus, which can be contrasted with the breadth of focus allowed by positive emotions.

In a study of the relation between experienced emotion and problem solving, Nielsen (in preparation) found that insights into the solution to a difficult logical problem are more frequently preceded by negative emotional experiences than by positive experiences. Nielsen suggests two possible explanations for this pattern. First, harking back to the role of negative emotion in creativity, negative experiences may be markers of the effort involved in the difficult restructuring required for insights to occur. Or secondly, they may be the outcome of an appraisal that one's current tactic is flawed, and, following Lang *et al.* (1997) a signal that prompts a withdrawal from and abandonment of erroneous problem conceptions (Nielsen, in preparation).

In an applied setting, Greenberg and colleagues (1993) discuss the function of experienced negative emotion in facilitating the therapeutic process. They propose that experiencing intense negative emotion in an accepting therapeutic environment allows its full expression while simultaneously facilitating acceptance of the emotion on the part of the client. The negative experience may then give way to calm, allowing the client to take a more active role in dealing with the core problem from which the negative emotion arises. Here, the experience of the negative affect plays an important functional role - the client must both live and move through the negative emotion in order to achieve a more positive state (Greenberg et al., 1993).

4. Implications and Directions for Future Research

As we engage in the solving of problems, both creative and mundane, our emotional experiences may be important in determining whether and how we arrive at solutions. Emotional experience can bias our attention and narrow or widen its scope. Specific types of emotional experiences can differentially influence the flexibility of our thinking and decision-making ability. Likewise, our capacity to tolerate particular emotional states may influence our success at solving more challenging problems.

The evidence presented here suggests that emotional experience plays a functional role in higher cognitive processes, though more evidence is needed to demonstrate causality and to identify the mediating mechanisms. Whether affect plays a causal role or merely serves as a marker of modes or shifts in cognitive processing, it will be important to discover to what extent an understanding of the role of emotional experience can be exploited to improve people's problem solving potential. An increased awareness of the emotional phenomenology of problem solving and decision-making might enable one to choose to follow or not to follow one's feelings, instead of being automatically directed by them. Individual differences in abilities to experience, tolerate, and regulate emotion may turn out to importantly influence the extent to which this is possible.

Understanding the mechanisms that mediate these functional influences of emotion on higher cognition is a key area of current research, both at the level of the neural systems that mediate the effects of emotional responses on cognitive processing, and at the level of phenomenal experience, where the record of those emotional experiences is to be found. As we come to understand how these systems interact, applications of this understanding to pedagogical, social, and clinical domains ought to be considered.

References

Adolphs R., D. Tranel, A. Bechara, H. Damasio and A.R. Damasio (1996) "Neuropsychological approaches to reasoning and decision-making", in: *Neurobiology of Decision-Making*, A.R. Damasio *et al.*, eds, Berlin: Springer-Verlag.

Baddeley, A.D. (1976) *The Psychology of Memory*, New York: Basic Books.

Bechara, A., H. Damasio, D. Tranel and A.R. Damasio (1997) "Deciding advantageously before knowing the advantageous strategy", *Science* 275:1293 -1295.

Bruschlinski, A.W. and O.K. Tichomirow (1975) *Zur Psychologie des Denkens*, Berlin: VEB Deutscher Verlag der Wissenschaften.

Clore, G. and W.G. Parrott (1991) "Moods and their vicissitudes: Thoughts and feelings as information", in: *Emotion and Social Judgments*, J.P. Forgas, ed., Oxford: Pergamon Press.

Csikszentmihalyi, M. (1990) *Flow: The Psychology of Optimal Experience*, New York: Harper Perennial.

Damasio, A.R. (1994) *Descartes' Error: Emotion, Reason and the Human Brain*, New York: G.P. Putnam's Sons.

Fredrickson, B.L. (1998) "What good are positive emotions?", *Review of General Psychology* 2:300-319.

Isen, A.M. (1993) "Positive affect and decision-making", in: *Handbook of Emotions*, M. Lewis and J.M. Haviland, eds, New York: The Guilford Press, pp. 261-278.

Lang, P.J., M.M. Bradley and B.N. Cuthbert (1997) "Motivated attention: Affect, activation and action", in: *Attention and Orienting: Sensory and Motivational Processes*, P.J. Lang, R.F. Simons and M.T. Balaban, eds, Hillsdale, NJ: Lawrence Erlbaum Associates.

LeDoux, J. (1996) *The Emotional Brain: The Mysterious Underpinnings of Emotional Life*, New York: Simon and Schuster.

Lessing, D. (1997) *Walking in the Shade*, London: Harper Collins.

Greenberg, L.S., L.N. Rice and R. Elliot (1993) *Facilitating Emotional Change: The Moment by Moment Process*, New York: Guilford Press.

May, R. (1980) *The Courage to Create*, New York: Bantam Books.

Nielsen, L. (1998) "Modeling creativity: Taking the evidence seriously", in: *Toward a Science of Consciousness II*, S.R. Hameroff, A.W. Kaszniak and A.C. Scott, eds, Cambridge, MA: The MIT Press.

Nielsen, L. (in preparation) "Relationships between cognition and emotion in problem solving: The phenomenology and physiology of insights".

Oatley, K. and P.N. Johnson-Laird (1995) "The communicative theory of emotions: Empirical tests, mental models, and implications for social interaction", in: *Goals and Affect*, L.L. Martin and A. Tesser, eds, Hillsdale, NJ: Erlbaum.

Öhman, A. and J.F. Soares (1994) "'Unconscious anxiety': Phobic responses to masked stimuli", *Journal of Abnormal Psychology* 103:231-240.

Öhman, A. and J.F. Soares (1998) "Emotional conditioning to masked stimuli: Expectancies for aversive outcomes following non-recognized fear-relevant stimuli", *Journal of Experimental Psychology: General* 127:69-82.

Scardamalia, M. and C. Berieter (1991) "Literate expertise", in: *Toward a General Theory of Expertise: Prospects and Limits*, K.A. Ericsson and J. Smith, eds, Cambridge, UK: Cambridge University Press.

Simon, H.A. (1967/1979) "Motivational and emotional controls of cognition", in: *Models of Thought*, H.A. Simon, ed., New Haven: Yale University Press.

Thomsen, S.U. (1994,1995) "A dance on words: reflections on the artistic creative process, parts 1 and 2", *Kritik* 110:31-51, and 116:1-21.

Tooby, J. (1985) "The emergence of evolutionary psychology", in: *Emerging Syntheses in Science, 1985, Proceedings of the Founding Workshops of the Santa Fe Institute*, Santa Fe, NM: The Santa Fe Institute.

Tooby, J. and L. Cosmides (1990) "The past explains the present", *Ethology and Sociobiology* 11:375-424.

Weisberg, R.W. (1995) "Prolegomena to theories of insight in problem solving: A taxomomy of problems", in: *The Nature of Insight*, R.J. Sternberg and J.E. Davidson, eds, Cambridge, MA: The MIT Press.

CONCLUSION

SOME FUTURE DIRECTIONS IN THE STUDY OF EMOTION AND CONSCIOUSNESS

ALFRED W. KASZNIAK
Center for Consciousness Studies, Departments of Psychology, Neurology & Psychiatry, University of Arizona, 1503 E. University, Tucson, Arizona 85721, U.S.A.

ABSTRACT

The papers in this volume demonstrate the significant progress that has been made over the past decades toward understanding psychological, biological, and social aspects of emotion. The present paper focuses on two areas of recent research that appear to hold particular promise for providing future answers to questions about emotion and consciousness: (1) the interrelationship of emotion, consciousness, and embodied cognition, and (2) the life-span development of expressive, physiological, and experiential aspects of emotion. Recent research in these areas is selectively reviewed, with suggestions for future research directions.

1. Introduction

As illustrated by the diverse collection of papers in this volume, empirical research and scholarship on emotion and on the interrelationships of emotion and consciousness studies is vigorous and thriving. Remarkable progress has been made over the past few decades toward understanding psychological, biological, and social aspects of emotion and in extrapolating the clinical implications of this understanding for the assessment and treatment of emotional disorders (Davidson, 2000; Lewis & Haviland-Jones, 2000). Relevant to the focus of the present volume, this progress has included a sharpening of our conceptual models for the various components of emotional qualia (see Clore & Ortony, 2000), and the development of experimental methodologies for dissociating conscious and nonconscious aspects of emotion (see Öhman, Flykt, & Lundqvist, 2000). Research reviewed by Nielsen (this volume) clearly points to an important contribution of emotion in decision-making, problem solving, and creative activity. The growing body of emotion research has also motivated recent theoretical formulations (e.g., Damasio, 1999; Watt, this volume) which posit aspects of emotion to play a possibly necessary role in the genesis of consciousness itself. The goal of the present paper is to explore promising areas for future research on questions relevant to understanding emotion and consciousness. Although this exploration will necessarily be selective and biased by the author's particular interests, it will hopefully provide a sense of the current excitement about new developments and possibilities for continuing investigation. Although several other areas of important recent research could also be described, the present focus will be upon the following two: (1) studies concerning the interrelationship of emotion, consciousness, and embodied cognition, and (2) research on the life-span development of expressive, physiological, and experiential aspects of emotion.

2. Emotion, Consciousness, and Embodied Cognition

There is a growing convergence between work on emotion-consciousness interrelationships and recent developments in cognitive science that emphasize the embodiment of cognition (e.g., Bermúdez, Marcel, & Eilan, 1995; Clark, 1997; Damasio, 1999; Varela, 2000; Varela, Thompson, & Rosch, 1991). This perspective views "...our bodies both as physical structures and as lived, experiential structures - in short, as both 'outer' and 'inner,' biological and phenomenological." (Varela, Thompson, & Rosch, 1991, p. xv). Emotion, by this view, is intrinsic to embodiment, in the sense that emotions are generally characterized by physiological changes that prepare the organism for possible action (Frijda, 1986; Lazarus, 1991; Levenson, 1994). As described by Varela (2000):

> ...one of the most striking discoveries over the last few years is the understanding that affect or emotion is at the very foundation of what we do every day as coping with the world: that reason or reasoning is almost like the icing on the cake... Mind is fundamentally something that arises out of the affective tonality, which is embedded in the body. (p. 78)

Several authors (Ashby, Isen, and Turken, 1999; Bower, 1981; Christianson, 1992; Clore, Schwarz, & Conway, 1994; Fredrickson, 1998; Isen, 2000; Lane, Nadel, Allen, & Kaszniak, 2000; Niedenthal & Kitayama, 1994; Nielsen, this volume; Palfai & Salovey, 1992; Schwarz, 2000) have reviewed evidence that emotion motivates aspects of cognitive activity, biases attentional and perceptual processes, affects memory encoding and retrieval, and shapes the manner or style in which cognitive processes are applied to problem-solving, reasoning, and decision processes. Emotion appears to play a particularly important role in social cognition and judgment (Bodenhausen, 1993; Forgas, 1995; Schwarz & Clore, 1996; Zajonc, 1998). Promising new research is being directed toward answering questions about the psychological and neurobiological processes that may explain these effects of emotion on cognition.

2.1 Psychological Models of Emotional Influences on Cognition

Much of the current research interest in emotion-cognition relationships can be traced to experiments conducted during the 1970s and early 1980s (see Bower, 1981; Kelley, 1986), showing that information is more likely to be recalled from memory if it is congruent rather than incongruent with an individual's current emotional experience. During this period, other investigators (e.g., Johnson & Tversky, 1983) demonstrated current mood to similarly influence cognitive estimations of the likelihood of positive versus negative outcomes and events. Subsequent studies showed that current affect (as either assessed by measures of present mood or manipulated by watching films or exposure to examination stress) influences choices and judgments about unrelated topics (see Schwarz & Clore, 1996).

As reviewed by Lerner and Keltner (2000), studies of *incidental* affective influences on cognition (those in which an emotional experience that should be irrelevant influences judgements and choices) have suggested that both direct and indirect mechanisms mediate the observed effects. Indirect influences, typically explained within associative network models (e.g., Wright & Bower, 1992), are assumed to operate when mood-congruent information is selectively retrieved from memory and then used in unrelated judgments. Direct influences have been explained within the affect-as-information model (Clore, 1992; Schwarz & Clore, 1983). Within this model, individuals are posited to use their current emotional experience in a heuristic manner in making judgments. Interestingly, this heuristic use of present feelings in judgment appears to occur only when the experienced emotion is seen as relevant to the object of judgment. When feelings are attributed to a source irrelevant to the judgment (e.g., a bad day at work or the weather), then the emotional experience is apparently no longer considered as informative for the choice or judgment at hand and hence does not influence it (Clore, 1992; Schwarz & Clore, 1983). According to Clore and Ortony (2000), this may explain the more pervasive effects of mood, in comparison to specific emotions, on judgment. Unlike specific emotions, which are generally focused upon a causal object (e.g., fear of some particular event or situation), the situational causes of moods are typically not salient. Consequently, mood-based feelings can easily be "...misattributed to whatever stimulus is being processed at the time" (Clore & Ortony, 2000, p. 46), and thus be taken as informative for judgments about the stimulus.

Lerner and Keltner (2000) note that most such claims about the mechanisms mediating effects of emotion on judgment share the common feature of basing predictions on the valence of the experienced emotion or mood. That is, positive emotion or mood leads to more positive judgments and negative emotion or mood leads to more negative judgments. However, Lerner and Keltner (2000) have empirically demonstrated that emotions of the same valence can have different effects on judgment. Persons assessed (by self-report questionnaires) to be characterized by the baseline or dispositional emotion of fearfulness were found to make more pessimistic judgments of future events, while angry persons made more optimistic judgments. Lerner and Keltner explain these differential effects of fear and anger (both negatively valenced emotions) by positing that "...each emotion is defined by a tendency to perceive new events and objects in ways that are consistent with the original cognitive-appraisal dimensions of the emotion." (p. 473) Explaining their appraisal-tendency model, Lerner and Keltner (2000) point out that the three central appraisal themes of fear, which are uncertainty, unpleasantness, and situational control (see Lazarus, 1991; Smith & Ellsworth, 1985), are conceptually related to risk perception. Given this appraisal structure, fear would be expected to be associated with a tendency to perceive situational control and uncertainty in new situations, with fearful persons consequently perceiving greater risk in these situations. In contrast, the appraisal structure of anger would be associated with a tendency to perceive individual control and certainty in new situations, resulting in the perception of less risk. Future research, exploring the validity of this appraisal-tendency approach with other different negative or positive emotions would appear to be quite promising, given the greater specificity in predictions about emotion-judgment relationships than those provided by valence-based approaches.

Future studies might profitably explore possible mechanisms by which appraisal-tendency effects occur, and examine "...whether dispositional and momentary emotions exert different or similar influences on judgement." (Lerner & Keltner, 2000, p. 486). This later question is of particular interest from the perspective of emotion-consciousness relationships, since momentary emotions are more likely to be consciously associated with situational causes of the emotion, reducing effects on judgments of other events and objects (see Clore & Ortony, 2000). Lerner and Keltner (2000) also note that dispositional emotions might have a greater influence than momentary emotions on judgment and choice because dispositional emotions tend to manifest early in life and function as relatively stable schemas for organizing and interpreting events (Damasio, 1994; Gasper & Clore, 1998).

2.2 Do Negative and Positive Emotions Influence Cognition through Similar Mechanisms?

Much of the research on emotion-cognition relationships, and much of emotion research generally, has focused upon the negative emotions (e.g., fear, anger, sadness, and disgust), with relatively little investigation of positive emotions. Fredrickson (1998) suggests that this marginalization of positive emotions may be due to several factors. First, in comparison to negative emotions, positive emotions are fewer in number and less differentiated from each other in facial expression, autonomic physiology, phenomenal experience, and initiating appraisal processes. Fredrickson (1998) also argues that these asymmetries between positive and negative emotions may reflect their evolutionary origins. Assuming that natural selection operates on emotions only responsive to situations of threat or opportunity, negative emotions would be greater in number and differentiation than positive emotions because there are more different threats than opportunities (Nesse, 1990). Failing to respond appropriately to an opportunity is likely to be much less costly (in terms of future gene replication potential) than failing to respond appropriately to a threat. A second reason proposed by Fredrickson (1998) to explain the neglect of positive emotions is that negative emotions present a greater range of individual and societal problems that have attracted the attention of psychologists. Finally, and perhaps most importantly in its consequences for research on emotion-cognition relationships, general theories of emotion have often been built to fit the specifications of prototypic emotions, and negative emotions have most frequently been taken as the prototypes.

For example, Fredrickson (1998) notes that many current theorists (e.g., Frijda, 1986; Lazarus, 1991; Levenson, 1994) view emotions as characterized by specific action tendencies (urges to act in specific ways). The skeletal muscle and autonomic physiological changes characteristic of particular emotions are seen as being in the service of these action tendencies. For example, fear is accompanied by increased blood flow to the large muscles to facilitate running away (Levenson, 1992). Fredrickson (1998) points out that "...pairing specific emotions with specific action tendencies seems easy enough when working within the subset of negative emotions, but fitting the positive emotions into this purportedly emotion-general model raises problems" (p. 302). This results in theorists proposing rather vague action tendencies for positive emotions. Fredrickson (1998) provides examples of this, which include Frijda's suggestion that joy is associated with a kind of "free activation," which serves as an "...aimless, unasked-for readiness to engage in whatever interaction presents itself" (Frijda, 1986, p. 89). Fredrickson argues that positive emotions do not fit models based upon specific action tendencies because they "...do not typically arise in life-threatening circumstances, and perhaps by consequence, they do not seem to create well-defined urges to pursue a specific course of action" (Fredrickson, 1998, p. 303). As an alternative to theories that attribute nonspecific action tendencies to positive emotions, Fredrickson proposes a new thought-action tendency model. In this model, negative emotions are seen as functioning to narrow the individual's momentary thought-action repertoire through activation of evolutionarily adaptive specific action tendencies (for examples, see LeDoux, this volume; Öhman, 2000). Many positive emotions (e.g., joy, interest, contentment, love), in contrast, are proposed by Fredrickson to broaden the momentary thought-action repertoire, resulting in "...novel, creative, and often unscripted paths of thought and action" (Fredrickson, 1998, p. 304). An increasingly large body of empirical research (see Isen, 2000) relating positive affect and decision making is consistent with Fredrickson's "broaden-and-build" model of how positive emotion influences cognition and behavior.

Continuing research on differences between the cognitive correlates and consequences of positive and negative emotions is thus another area that holds promise for producing new information that would be of both theoretical and practical importance. Practically, a greater emphasis upon the role of positive emotion in both cognitive coping with chronic stress and enhancing health is likely to result in greater progress toward understanding successful outcomes than has the past focus upon negative emotion (Folkman & Moskowitz, 2000; Salovey, Rothman, Detweiler, & Steward, 2000). Increased research on the practical consequences of positive emotion would be consistent with current trends toward the broader development of a "positive psychology" (Seligman & Czikszentmihalyi, 2000).

On the theoretical side, continued research on differences between the cognitive consequences of positive and negative emotions would facilitate the development of more detailed and articulated models of the psychological mechanisms by which emotion and cognition interact. Of particular interest are questions concerning the ways in which consciousness may play a differential role in the mediation of negative versus positive emotion effects on attention, decision, choice, and judgment. For example, the negative emotions of fear and anxiety involve defensive responses (LeDoux, this volume; Öhman, 2000) that have evolved to operate very rapidly and automatically, without any necessary conscious processing of the eliciting stimuli. The well-documented selective attention bias produced by fear and anxiety, which focuses on potentially threatening information (Mathews & MacLeod, 1994; Mogg & Bradley, 1998; Öhman, 2000), appears to be initiated automatically and prior to conscious awareness. It is presently unclear, however, whether the increased flexibility in selection and switching of cognitive perspective that accompanies positive emotion (see Isen, 2000) is also initiated preattentively and nonconsciously. Fredrickson (1998), as noted above, has speculated about the different kinds of natural selective pressures that may have shaped evolution of the cognitive consequences of negative versus positive emotion. Different evolutionary histories could have allowed for the possibility of positive emotion involving more conscious processes in the mediation of its effects on problem solving and decision. This possibility deserves empirical testing, perhaps employing experimental methodologies that have been used to explore the nonconscious activation of decision biasing by fear-eliciting stimuli that have been backward-masked (Öhman, Flykt, & Lundqvist, 2000).

2.3 Neurobiological Models of Emotional Influences on Cognition

The neuroanatomic structures and neurophysiologic processes described by LeDoux (1996; this volume) as subserving fear conditioning in the rat serves as a basis for a neurobiological understanding of how the negative emotion of fear might nonconsciously influence cognition. Direct thalamo-amygdala connections appear to permit rapid detection and integrated response to threatening stimuli for which evolution has developed a specific preparedness. The initial detection and response to such stimuli do not appear to require the necessarily slower participation of the cerebral cortex and its mechanisms of conscious decision-making, although conscious processes can certainly modify the expression and action components of fear responses. Connections from the amygdala to cortical sensory processing areas, hippocampal memory systems, and frontal cortex executive systems (LeDoux, 1996) provide possible mechanisms by which automatic aspects of fear response could influence attention, perception, and decision-making. The neural pathways involved in fear response thus help in understanding how and why negative emotion can have effects upon cognition that are automatic and occur prior to conscious awareness.

Damasio (1994) proposed the "somatic marker hypothesis," in which the amygdala and ventromedial prefrontal (VMF) cortex are seen as components of a neural system involved in the activation of somatic (including emotional) states that influence decision-making. Empirical evidence is now available that supports of this hypothesis. Bechara, Damasio, Damasio, and Lee (1999) studied a group of patients with bilateral amygdala but not VMF damage, and another group with VMF but not amygdala damage. A computer-administered gambling task provided a measure of decision-making. On each trial of this task, the participant selects a single card from one of four decks pictured on the computer screen. After "mouse" clicking to turn each card, the participant receives money, with the amount displayed on the screen. For some cards the participant both wins money and pays a penalty (amount also displayed on the screen). The ultimate future yield of choices from any particular card deck varies because the penalty amounts are higher in the two high-paying decks, leading to a negative balance, and lower in the two low-paying decks, leading to a final gain. Skin conductance responses (SCRs) were recorded during decision-making performance as an index of somatic state activation. Both the amygdala damage and the VMF damage groups were impaired on the gambling task (making fewer advantageous card deck choices). Both groups also failed to show anticipatory SCRs while they pondered risky choices. Healthy individuals show anticipatory SCRs during risk-related decision-making, with functional magnetic resonance imaging (fMRI) demonstrating medial prefrontal and orbitofrontal activation to be associated with both the generation and afferent representation of these SCRs (Critchley, Elliott, Mathias, & Dean, 2000). Despite the similarity of amygdala damage and VMF damage patients in decision choices and absent anticipatory SCRs, Bechara et al. (1999) did find group differences: The VMF damage patients showed SCRs when they received a reward or punishment (money), but the amygdala damage patients did not. Further, in a separate Pavlovian conditioning experiment, the VMF damage patients were able to acquire a conditioned SCR to visual stimuli paired with an aversive loud sound, whereas the amygdala damage patients did not. The authors interpreted these group differences as suggesting different contributions of the amygdala and VMF to decision-making. The decision-making impairment of patients with amygdala damage was seen as the possibly indirect consequence of an inability to evoke the necessary somatic states after winning or losing money. This would preclude the evocation of a somatic state while pondering a decision with future consequences. In contrast, the decision-making impairment of patients with VMF damage was seen as "...related to an inability to integrate effectively all of the somatic effectors such as the hypothalamus and brainstem nuclei" (Bechara et al., 2000, p. 5473). This proposal is consistent with the role of ventromedial frontal structures in emotion posited by other investigators (e.g., Kaszniak, this volume; Kaszniak, Reminger, Rapcsak, & Glisky, 1999; Lane, 2000).

It will be important for future research to more fully address the question of why VMF damage patients fail to trigger emotional signals or somatic states when they ponder decisions. Data have recently been reported (Bechara, Damasio, & Damasio, 2000; Tranel, Bechara, Damasio, & Damasio, 1998) which suggest that VMF damage patients may lack the ability to re-experience the associated emotional state when recalling previous instances of punishment. This inability may possibly result in their lack of anticipatory SCRs in decision-making. Eight VMF damage patients were asked to think about and describe a past situation in which they felt either happiness, sadness, fear, or anger. After a brief description of the situation was obtained, the participant was asked to imagine and re-experience each emotion while physiological activity (SCR, heart rate, respiratory frequency, skin temperature, and facial EMG) was monitored. All the patients were able to retrieve previous emotional experiences, and showed greater physiological activity during imagery of the angry than of neutral situations. However, the patients' ability to re-experience fear was less reliable: several could not do so, and those who could showed a less intense response. Although clearly preliminary, these results do encourage further investigations of the possibility that the decision-making impairment of VMF damage patients may be due to an inability to re-experience the emotion of a previously fearful situation. As Bechara and colleagues (2000) point out, such research would have both theoretical and practical implications. Recent studies using gambling tasks have suggested that impaired decision making may be a core problem in substance abusers (Grant, Contoreggi, & London, 1997; Petry, Bickel, & Arnett, 1998). Further, the "acquired sociopathy" of patients with VMF damage (Blair & Cipolitti, 2000; Eslinger & Damasio, 1985) may shed light on the mechanisms underlying idiopathic sociopathy. In this respect, it is of interest to note that early dysfunction in the prefrontal cortex can be a cause of abnormal social and moral behavior development (Anderson, Bechara, Damasio, Tranel, & Damasio, 1999).

But what about neurobiological mechanisms subserving the documented effects of positive emotion on cognition? A recent neuropsychological theory proposes that positive emotion may influence cognition through somewhat different brain systems and mechanisms than negative emotion. Ashby, Isen, and Turken (1999) have proposed a theory that accounts for many of the effects of positive emotion on cognitive (including memory and problem-solving) task performance by assuming that positive emotion is associated with increased brain dopamine levels. More specifically, the theory posits that positive emotion facilitates creative problem solving, in part, through increased dopamine release in the anterior cingulate cortex, which results in improved cognitive flexibility and switching of cognitive perspective. This proposal is of particular interest in the questions it raises about the role of consciousness in emotion-cognition relationships. If Ashby and colleagues (1999) are correct about the role of increased dopamine release in the anterior cingulate, and if other investigators (e.g., Kaszniak, this volume; Kaszniak et al., 1999; Lane, 2000) are correct in positing a critical role for the anterior cingulate in the conscious experience of emotion, then this would suggest the possibility that the consequences of positive emotion for decision-making may require conscious mediation. Continued empirical investigation designed to test each component of this theoretical prediction would seem to be a very promising area for future research.

Another important question for future research concerns the degree to which attention to conscious emotional experience may be capable of influencing the effects of emotion on cognition. Several recent books (e.g., Brazier, 1998; Kornfield, 2000; Levine, 2000) and articles (e.g., Masters, 2000) have focused upon the relationship between meditation, particularly that of vipassana or mindfulness-based Buddhist practices, and emotion. These practices involve assuming an attitude of "pure observation" which begins by conscious attention to the individual activities of the body, including those of emotion. Mindfulness-based meditation has been held to be capable of achieving greater emotion-cognition integration through careful conscious attention to emotional experience. The construct of mindfulness, defined as "...a state of conscious awareness in which the individual is implicitly aware of the context and content of information" (Langer, 1992, p. 289), has begun to receive attention by psychologists interested in consciousness and cognition.

There is recent behavioral (reaction time) and cerebral electrophysiological (event-related potential) evidence consistent with the interpretation that long-term mindfulness meditation practice results in a broader attentional focus (Schnyer, Kaszniak, Allen & Trujillo, 1999). There is also evidence that mindfulness-based practices, integrated with cognitive therapy procedures (Segal, Williams, &

Teasdale, in press), provide a cost-efficient psychological approach to preventing relapse/recurrence in recovered recurrently depressed patients (Teasdale et al., 2000). In addition, preliminary evidence supports the effectiveness of a mindfulness-based stress reduction program (MBSR; Kabat-Zinn, 1990) in the treatment of generalized anxiety disorder and panic (Kabat-Zinn et al., 1992), and in the self-regulation of chronic pain (Kabat-Zinn, Lipworth, Burney, & Sellers, 1986). In regard to the application of MBSR in chronic pain, it is interesting to note that there is experimental evidence for the conclusion that the experienced negative valence of acute pain (e.g., cold pressor induced) is reduced the more that one attends to painful sensations themselves and the less that attention includes their significance (Ables, Blanchard, & Leventhal, 1983; Dar & Leventhal, 1993).

These scholarly, practice, and research developments encourage future efforts to empirically evaluate the effect of mindfulness-based meditation practice on the various components of emotion response, and on the particulars of emotion-cognition relationships. Such studies would not only shed additional light on the role of consciousness in emotion, but would also be of practical importance, given the possible applications for personal enhancement of life quality and psychotherapeutic intervention in emotional disorder.

3. Life-Span Development of Emotional Expression, Physiology, and Experience[1]

Another particularly active area of recent research concerns the development, from infancy through old age, of the various components of emotion. Although the relevant literature is far too large to accurately summarize here, select aspects will be reviewed and particularly promising areas for future research will be noted.

3.1 Emotional Development in Infancy and Childhood

Research on emotional development in infancy and childhood has increased significantly over recent decades, with several published volumes summarizing the resultant body of information (Fox, 1994; Saarni, 1999; Salovey & Sluyter, 1997; Schore, 1994). Although infants cannot provide verbal report of their emotional experience, emotional response can be reliably coded from facial and other signals (Izard et al., 1995). Summarizing the results of infant studies of emotional development, Lewis (2000) has proposed the following model: At birth, the child shows evidence of a bipolar emotional organization, with the negative pole manifest by general distress behavior (crying, irritability) and the positive pole manifest by interest in the environment and satiation behavior (Steiner, 1979). By

[1] The author wishes to thank Ms. Sheryl Reminger for assistance in preparation of that portion of this section on older age.

3 months, joy (excitement/happiness), surprise, sadness and disgust emerge, followed by anger between 4 and 6 months. Behavior indicative of fearfulness, thought to require additional cognitive development, is seen by 7 to 9 months (Camras, Malatesta, & Izard, 1991). By the end of the second year of life, objective self-awareness (self-referential behavior) emerges, giving rise to what Lewis (2000, p. 277) calls "self-conscious emotions" (embarrassment, empathy, and envy). Between 2 and 3 years of age, children develop the ability to evaluate their behavior against a standard (external or internal), giving rise to "self-conscious evaluative emotions" (Lewis, 2000, p. 278), which include the rudiments of pride, shame, and guilt. Thus, by 3 years of age, the child shows evidence of a highly differentiated emotional life. However, the full development self-conscious evaluative emotions, such as guilt in response to wrongdoing or shame in response to failure, may not manifest until 5 to 7 years of age (Brown & Dunn, 1996; Izard & Ackerman, 1998).

Of particular interest from the perspective of questions concerning emotion-consciousness relationships is a growing body of research on children's awareness of and ability to verbally report their emotion. As noted by several investigators and theorists (e.g., Harris, 2000; Saarni, 1999), the development of emotional self-awareness (extending from early through middle childhood) is critically influenced by, and itself in turn influences, child-caregiver interaction. Awareness of one's own emotions appears to be the most basic skill in the development of what has been termed "emotional competence" (Saarni, 1999) or "emotional intelligence" (Mayer & Salovey, 1997). According to several theorists (e.g., Lewis & Brooks, 1978; Saarni, 1999), the development of a sense of self is a prerequisite for emotional self-awareness. Thus, young infants are thought to have emotional states, but to lack a conscious awareness of these emotional states. In addition, conscious awareness of emotional experience appears dependent upon the ability to differentiate emotional bodily changes from those not associated with emotion (Lewis, 2000), and the cognitive capacity to evaluate context and agency (who or what is the cause of some internal change; see Lewis, 1991). Children as young as 4 years of age may demonstrate an understanding of at least some of the relationship between situational causes and emotion (Strayer, 1986). Additional sophistication in emotional self-awareness may appear as early as 5 to 6 years of age, when children become aware of simultaneous multiple or conflicting emotions (Wintre & Vallance, 1994), although some children may not show this development until late childhood (Donaldson & Westerman, 1986). Because emotional self-awareness appears to facilitate problem-solving (Saarni, 1999), continuing research concerning the biological, psychological, and social contributors to development of this capacity would appear to be particularly fruitful.

3.2 Emiotional Development and Change in Adulthood and Aging

During the early and middle adult years, individual differences in trait emotions or moods appear to be robustly stable (Helson & Klohnen, 1998; Diener & Suh, 1998), and correlated with stable differences in relevant neurochemical systems (Davidson, 1998). In cross-sectional comparisons of different adult age groups, self-report of unpleasant emotions remains stable over successive age cohorts, while positive affect shows some decrease (Diener & Suh, 1998). However, since cross-sectional data trends may reflect age cohort differences rather than true developmental changes, the interpretation of this apparent positive affect decrease with progressively older cohorts must await the availability of longitudinal data. With increasing age in the middle adult years, emotional experience appears to have greater saliency (Carstensen, 1992; Carstensen, Gross, & Fung, 1998; Carstensen & Turk-Charles, 1994), and there may develop a general bias toward emotion-based responses within a variety of situations (Blanchard-Fields, 1994). There is also cross-sectional data that suggests a continuing emotional integration and interaction between emotion and cognition from early through middle adulthood. As Labouvie-Vief (1998) notes, cognition informs and is informed by emotion throughout adult development, and through such mutual exchange, "...affects that otherwise would be experienced as too disruptive or too frightening can be understood as simply one facet of the human condition" (p. 231). Labouvie-Vief (1998, pp. 231-232) also notes that such aspects of emotion-cognition interaction may be associated with "...advanced forms of wisdom and spiritual leadership" in that these involve an "...ability to maintain an attitude of serenity vis-à-vis the most extreme forms of affect, from horror and suffering to mystic ecstasy (Miller & Cook-Greuther, 1990)." Given the relative paucity of available empirical data on the development of wisdom in adulthood (see Baltes & Staudiger, 2000), this would appear to be a particularly important area for continuing emotion-cognition research.

The relationship of aging, beyond the middle adult years, to changes in emotional expression, physiology, and experience has been debated in the research literature (for recent reviews, see Cacioppo, Berntson, Klein, & Poehlmann, 1998; Levenson, 2000). Since older age is known to be associated with changes in daily functioning on many levels (cognitive, motor, etc.; see Kaszniak & Newman, 2000), it has been assumed that some change in emotional functioning should occur with age. Neurobiological research has contributed to the hypothesis of age-related emotional change. For instance, the amygdala, a subcortical structure of central importance in emotion (LeDoux, 1996; this volume), has been reported to be smaller in neuronal volume for older versus younger adults, particularly in the medial, medial central, and cortical nuclei (Herzog & Kemper, 1980). Nonetheless, psychological research to date has not been conclusive about the relationship between aging and emotion.

Levenson, Carstensen, Friesen, and Ekman (1991) studied a group of older individuals who participated in two emotion-eliciting tasks: a directed facial action task, in which muscle-by-muscle instructions were given for constructing facial emotional expressions (without mention of the target emotion), and a "relived emotion task," in which participants were asked to recall a time that strongly provoked a particular emotion. Data was collected on self-reported intensity of felt emotion, observed facial expression (using the Facial Action Coding System; Ekman & Friesen, 1978), and physiological responsiveness. The results from this study were then compared to those obtained from similar investigations with younger adult participants. In the relived-emotions task, no differences were found between younger and older participants in their rated intensity of emotional experiences, or in the percentage of trials showing spontaneous emotional facial expressions. The older and younger groups also did not differ in the frequency with which different emotions produced significant autonomic physiologic differences, regardless of whether the emotion was produced by the directed facial action task or the relived-emotions task. Even though the pattern of autonomic differentiation between emotions was similar for older and younger participants, Levenson et al. (1991) did find the older adults to show a smaller magnitude of cardiovascular reactivity (similar results were obtained by Levenson, Carstensen, & Gottman, 1994). It is not clear whether this reflects an emotion-specific diminution in cardiovascular reactivity or age-related changes in the general reactivity of the cardiovascular system (see Cacioppo, Berntson, Klein, & Poehlmann, 1998; Frolkis, 1997). Levenson et al. (1991) also found that older participants reported experiencing target emotions significantly less often in the directed facial action tasks. Their ability to contract muscles into emotional facial configurations was also rated as lower. This led Levenson et al. (1991) to conclude that aging either leads to a lessening of voluntary muscle control, or that older persons experience less motivation or involvement in tasks of this nature.

A similar result was found in a study conducted by Malatesta, Fiore, and Messina (1987). In that study, older individuals were asked to pose emotional facial expressions. Photographed images of these expressions were then coded for emotional content by younger raters. These raters were shown to be poor at judging what target emotion was being expressed. This suggests that older adults may not be as capable of voluntarily expressing particular emotions. However, the Malatesta et al. (1987) results cannot directly address the question of whether age-related changes occur in spontaneous, nonvoluntary emotional facial expression, which is mediated by different neuronal circuitry than voluntary expression (Matsumoto & Lee, 1993; Rinn, 1984). In addition, the use of young adult raters to assess the facial expressions of older adults complicates interpretation of the findings (i.e., younger persons may have less experience in observing the facial expressions of older versus younger adults).

In contrast, Moreno, Borod, Welkowitz, and Alpert (1990) failed to find effects of age on voluntary emotional expressivity. In this study, women in three age groups were photographed while deliberately posing four emotional expressions plus a neutral expression. Chimeric left- and right-side composites of these photographs were then created and rated for intensity by three college-age raters. Although the results did show greater emotional intensity ratings for left-sided facial expression than for right-sided facial expression (a facial expression asymmetry found consistently across many studies; Borod, Haywood, & Koff, 1997), there was no significant main effect of age group nor any group by hemiface interaction. However, a significant interaction of age group and emotion was found. Post hoc tests for individual emotions revealed a significant effect of age for the expression of disgust, which was expressed more intensely by the older age group than by the other two age groups. The authors concluded that facial expressive capacities do not decline throughout the life span, and that the right cerebral hemisphere is generally more responsible for this continuation of function. However, since Moreno and colleagues studied only posed facial expression, their study could not address the question of whether facial asymmetries are invariant across age groups for spontaneous emotional expression.

Malatesta and colleagues (Malatesta, Izard, Culver, & Nicolich, 1987) failed to find differences in the intensity of nonvoluntary emotional expression across different age groups. In this study, groups of young, middle-aged, and older women were asked to recount emotional experiences relating to specific types of emotion while their facial expressions were videotaped. Another group of participants (representing all three age groups) then viewed the tapes (without sound) and was asked to judge the dominant emotion and the intensity of that particular emotion. No differences were found between the intensity ratings of young, middle-aged, and older participants asked to recount emotional experiences. In addition, no age-related differences in judged facial expressivity were found. However, there was a trend for judges, particularly younger judges, to make more errors when decoding the expressions of older individuals. In addition, the oldest group showed the poorest performance in decoding the emotional signals of others. These results were seen as paralleling an earlier study of emotional experience by Malatesta and Kalnok (1984), which had participants in three different age groups complete questionnaires addressing emotional experiences, emotional expressiveness, emotional quality, and the importance of emotion. Results of this earlier study showed that while older individuals assigned less importance to their everyday emotions than younger individuals, they did not report experiencing greater or lesser degrees of positive or negative emotions, nor did they describe a greater inhibition of emotion communication.

Nonetheless, other researchers report apparent decreases in emotional experience with age. Lawton, Kleban, Rajagopal, and Dean (1992) found evidence for a "diminished emotionality", or a lessening of affective intensity, with age. In their study, participants were asked to rate different dimensions of their personal affective experience. Older participants reported both decreased surgency and decreased physiological responsiveness to emotional stimuli, and their responses also suggested an increased stability of mood. Lawton et al. (1992) interpreted this decreased surgency with age as a result of increased emotional self-regulatory capacities. However, because the data of Lawton and colleagues was based upon self-report, it is difficult to know whether results reflect real age-group differences in emotional experience, or differences in the recall of past emotional experience.

Although previous investigations of the relationship between aging and emotion have thus generated conflicting results, some consistencies are present. First, aging does appear related to the production of voluntary and posed facial expressions, with older age groups showing a decrease in this ability. However, age differences generally do not appear to occur when spontaneous facial expressions are observed (e.g., during participants' recounting of previous emotional experiences). Nonetheless, this lack of age group differences does not unambiguously address the question of possible age group differences in facial expression, since other age-related differences (e.g., the possibility that older persons might differentially recall more intense emotional experiences than younger adults) could also affect observed facial expressions. The use of stimuli developed and standardized for the purpose of generating emotional responses, rather than the use of recalled emotional experience or imagery, could more directly address the question of how emotional expression may be influenced by adult age. In addition, previous studies of emotional expression and aging have primarily relied on judges' ratings of observed facial expressions. This methodology also does not allow for unambiguous interpretation. For example, age-related changes in facial skin may obscure the observability of actual facial muscle activity. Facial expression in response to standardized emotional stimuli may be too subtle to be reliably recorded by observational methods, although detectable via facial electromyography (Cacioppo, Petty, Losch, & Kim, 1986). Also, observer biases, particularly those of observers in different age groups, may confound apparent age-group differences in rated expression. More direct and quantifiable methods could provide less ambiguous information about the nature of age-related changes in emotion.

Using such methods to investigate the relationship between aging, emotional experience, and emotional expression, my colleagues and I (Reminger, Kaszniak, & Dalby, 2000) compared older (mean age = 68.4 years) and younger (mean age = 26.4 years) participants. The methodology employed was similar to that described (see Kaszniak, this volume) for our study of facial expression and emotional experience in Parkinson's disease (Dalby, Reminger, Kaszniak, & Montgomery, in press). Bilateral facial electromyographic (EMG) activity (recorded from zygomatic and corrugator muscle regions) was used as the dependent measures of facial expression, and Self-Assessment Manikin (SAM) ratings of valence and arousal were used as the dependent measures of emotional experience. Emotional states were generated through presentation of pictures (the International Affective Picture System [IAPS] slides; Lang, Bradley, & Cuthbert, 1999) designed and standardized to stimulate specific emotions.

Despite sufficient power to detect possible group differences, neither emotional experience ratings (Figs. 1 & 2) nor Facial EMG patterns (Figs 3 & 4) showed a significant effect of age group. The expected patterns of emotional experience and asymmetrical EMG change were found in response to the emotional stimuli. Greater corrugator muscle activity occurred in response to negative slides and greater zygomatic activity occurred in response to positive slides, with left-sided activity consistently greater than right.

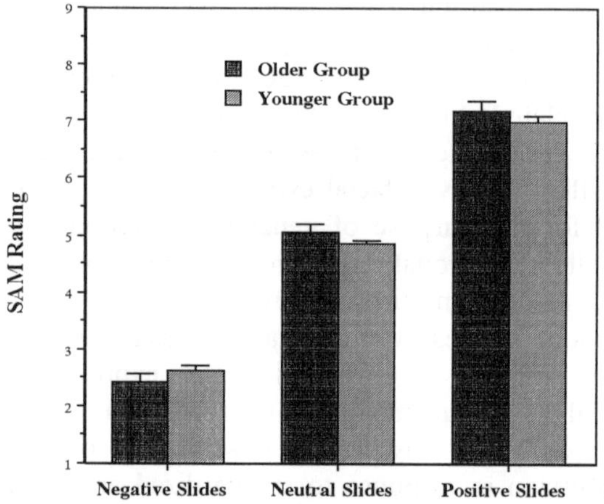

Figure 1. Emotional valence experience ratings of older and younger subjects. SAM = Self-Assessment Manikin (Modified from Reminger, Kaszniak, & Dalby, 2000)

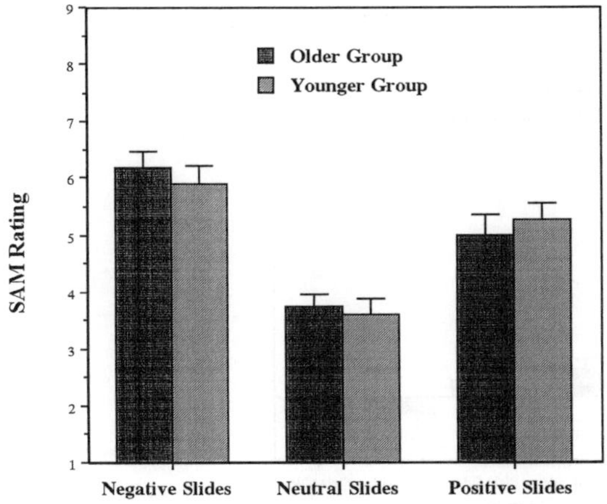

Figure 2. Emotional arousal experience ratings of older and younger subjects. SAM = Self-Assessment Manikin (Modified from Reminger, Kaszniak, & Dalby, 2000)

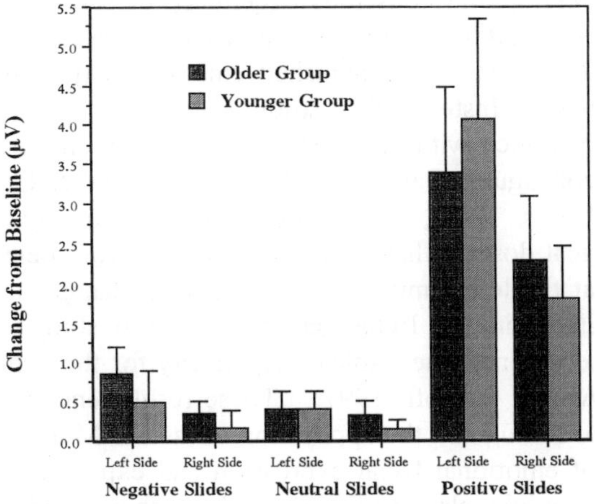

Figure 3. Mean electromyographic zygomatic muscle activity of older and younger subjects in response to emotionally arousing slides. (Modified from Reminger, Kaszniak, & Dalby, 2000)

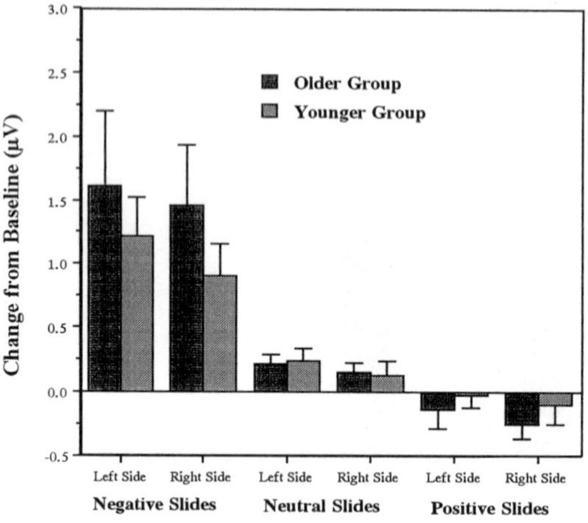

Figure 4. Mean electromyographic corrugator muscle activity of older and younger subjects in response to emotionally arousing slides. (Modified from Reminger, Kaszniak, & Dalby, 2000)

The results of the Reminger et al. (2000) study are thus consistent with the interpretation that facial expression and conscious experience of emotion remain invariant from younger to older adulthood. This is contrary to the results of some previous studies that have suggested both diminished emotional experience and expression in older individuals. Instead, this study suggests a resilience of emotional expression and experience with age. This conclusion is in agreement with that of Carstensen and colleagues (Carstensen, Graff, Levenson, & Gottman, 1996; Carstensen & Turk-Charles, 1998), based on other data, who suggest that emotion may be a psychological domain that is relatively spared from deleterious effects of aging. The fact that the left hemiface showed greater changes in EMG activity than the right hemiface, in the Reminger et al. (2000) study, is also consistent with age invariance in a right hemisphere superiority for the expression of emotion (see Borod, Haywood, & Koff, 1997). These results encourage the further use of IAPS emotional stimuli and the application of facial EMG and SAM ratings in the measurement of emotional facial expression and experience across the life span. Future studies might also profitably apply such methods to the examination of emotion changes in age-related illnesses such as Alzheimer's disease (AD). In AD, the appraisal of emotional stimuli (Albert, Cohen, & Koff, 1991; Allender & Kaszniak, 1989; Lavenu, Pasquier, Lebert, Petit, & Van der Linden, 1999) is impaired and disordered emotional behavor is common (for review, see Duke & Kaszniak, 2000). Current work in my laboratory has been pursuing such application.

3.3 Emotion and Sense of Self across the Life Span

The possibility of stability across the adult life span in the facial expression, conscious experience, and physiologic patterning components of emotion raises interesting questions about whether this emotion stability may be an important component of the constancy in sense of self over a lifetime. Several investigators and theorists (see Bermúdez, Marcel, & Eilan, 1995) have hypothesized bodily experience, of which emotion is a major component, to be at the core of experience of the self. Further, several studies (although containing a number of methodological limitations) have suggested an affective influence on self-focused attention (for review, see Palfai & Salovey, 1992). Studies of mother-infant interactive sequences (e.g., Trevarthen, 1993) demonstrate that infants learn emotion regulation in such early interactions. The quality and timing of these interaction experiences appear to influence the developing patterns of neuronal connectivity in brain regions (particularly the frontal lobes, which show rapid development during the first year of life). These patterns appear to be essential to the future socioemotional development of the child (for review of relevant evidence, see Panksepp, 1998; Schore, 1994; 1996). As noted by Schore (1996, p. 66), "...high levels of positive affect are vitally important for the infant's continuing neurobiological development." The mother-infant interactional patterns also likely establish the basis for intersubjectivity, the development of empathic capacity (Schore, 1996; Trevarthen & Aitken, 1994), and an initial sense of self (Trevarthen, 1993). Once established during the first 2 years of life, the prerational and nonverbal stream of emotion that originally functioned to bind an infant to its parent may continue throughout life to be a "...primary medium of intuitively felt affective-communication between persons" (Schore, 1996, p. 73). As argued by Gaussen (2000), the core of a sense of self may reside in early-established patterns of affect regulation, "...responsible for the maintenance of the sense of continuity of the affect of the individual" (p. 131). Gaussen (2000) goes on to conclude "...patterns of affect regulation integrate a sense of self across (emotional) state transitions, thereby allowing for a continuity of inner experience" (p. 132). This is similar to Damasio's (1994) proposal for the role of emotion in the continuous and consistent reconstruction of the self at each moment of time. If these theorists are correct, then this may provide one answer to the question of why most people experience a continuity in their sense of self, despite many physical and cognitive changes with aging. As Izard and Ackerman (1998) propose:

> The role of emotion feelings in the continuity of self is explained in part by the constancy of the quality of emotion feelings. A happy or sad feeling today is the same in quality as the happy or sad feeling of yesterday and of tomorrow. Continuity of self-identity also is fostered by the stability of the frequency of each emotion feeling. Our characteristic mood of the past is likely to be our characteristic mood of the future. Thus, a certain dependability and regularity in our emotion experiences provides stability in basic aspects of self-identity, self-concept, and personality. (p. 19)

The adaptive advantage conveyed by having stable emotion experiences in the context of a changing cognitive system and other developmental processes may have led to the evolution of life span stability in emotion and sense of self. As suggested by Dougherty, Abe, and Izard (1996), a constancy of a given emotion feeling would appear to guarantee the "reliability of the motivational function of that emotion" (p. 31). Developmental changes in the conscious experience (and other components) of an emotion like fear could result in inappropriate responses (with possible gene survival consequences) to danger.

In future research concerning these intriguing issues in the adult development of emotion, it will be important for longitudinal studies to be conducted, since most of the currently available research is cross-sectional. Longitudinal studies would be capable of disentangling possible adult developmental changes from historical period and cohort effects that can affect age group differences. Longitudinal investigations, as well as additional cross-sectional comparisons of adult age groups, should also be aimed at examining whether relationships between emotion, cognition, consciousness, and sense of self change over the adult life span. Given the role of emotion in cognitive decision-making, and the importance of careful decision-making under conditions of possible risk in older age (e.g., whether to change residence, whether to pursue particular medical interventions or other health care strategies), such research would have considerable practical as well as theoretical importance.

It appears that emotion, consciousness, and sense of self exist in complex interrelationship throughout life. Emotion has been held to be a necessary contributor to a developing sense of self in childhood and its stability throughout adulthood. Emotion may even be necessary for the experience of consciousness itself. Conversely, the development of a sense of self seems necessary for the experience of self-conscious emotions. The continuing effort to disentangle these interrelationships is clearly an important agenda for future research.

References

Ables, T.A., E.B. Blanchard and H. Leventhal (1983) "Cognitive control of pain: Attention to the sensory aspects of the cold pressor stimulus", *Cognitive Therapy and Research* 7:159-177.

Albert, M.S., C. Cohen and E. Koff (1991) "Perception of affect in patients with dementia of Alzheimer type", *Archives of Neurology* 16:791-795.

Allender, J.A. and A.W. Kaszniak (1989) "Processing of emotional cues in patients with dementia of the Alzheimer type", *International Journal of Neuroscience* 46:147-155.

Anderson, S.W., A. Bechara, H. Damasio, D. Tranel and A.R. Damasio (1999) "Impairment of social and moral behavior related to early damage in human prefrontal cortex", *Nature Neuroscience* 2:1032-1037.

Ashby, F.G., A.M. Isen and A.U. Turken (1999) "A neuropsychological theory of positive affect and its influence on cognition", *Psychological Review* 106:529-550.

Baltes, P.B. and U.M. Staudinger (2000) "Wisdom: A metaheuristic (pragmatic) to orchestrate mind and virtue toward excellence", *American Psychologist* 55:122-136.

Bechara, A., H. Damasio and A.R. Damasio (2000) "Emotion, decision making and the orbitofrontal cortex", *Cerebral Cortex* 10:295-307.

Bechara, A., H. Damasio, A.R. Damasio and G.P. Lee (1999) "Different contributions of the human amygdala and ventromedial prefrontal cortex to decision-making", *The Journal of Neuroscience* 19:5473-5481.

Bermúdez, J.L., A.J. Marcel and N. Eilan, eds (1995) *The Body and the Self*, Cambridge, MA: MIT Press.

Blair, R.J.R. and L. Cipolotti (2000) "Impaired social response reversal: A case of 'acquired sociopathy'", *Brain* 123:1122-1141.

Blanchard-Fields, F. (1994) "Age differences in causal attributions from an adult developmental perspective", *Journals of Gerontology: Psychological Sciences* 49:P43-P51.

Bodenhausen, G.V. (1993) "Emotions, arousal, and stereotypic judgments: A heuristic model of affect and stereotyping", in: *Affect, Cognition, and Stereotyping: Interactive Processes in Group Perception*, D.M. Mackie and D.L. Hamilton, eds, San Diego, CA: Academic Press, pp. 13-37.

Borod, J.C., C.S. Haywood and E. Koff (1997) "Neuropsychological aspects of facial asymmetry during emotional expression: A review of the normal adult literature", *Neuropsychology Review* 7:41-60.

Bower, G.H. (1981) "Mood and memory", American *Psychologist* 36:129-148.

Brazier, D. (1998) *The Feeling Buddha: A Buddhist Psychology of Character, Adversity and Passion*, New York: Fromm International.

Brown, J.R. and J. Dunn (1996) "Continuities in emotion understanding from three to six years", *Child Development* 67:789-802.

Cacioppo, J.T., G.G. Berntson, D.J. Klein and K.M. Poehlmann (1998) "Psychophysiology of emotion across the life span", in: *Annual Review of Gerontology and Geriatrics (Volume 17): Focus on Emotion and Adult Development*, K.W. Schaie and M.P. Lawton, eds, New York: Springer, pp. 27-74.

Cacioppo, J.T., R.E. Petty, M.E. Losch and H.S. Kim (1986) "Electromyographic activity over facial muscle regions can differentiate the valence and intensity of affective reactions", Journal *of Personality and Social Psychology* 50:260-268.

Camras, L.A., C. Malatesta and C. Izard (1991) "The development of facial expressions in infancy", in: *Fundamentals of Nonverbal Behavior*, R. Feldman and B. Rime, eds, New York: Cambridge University Press, pp. 73-105.

Carstensen, L.L. (1992) "Social and emotional patterns in adulthood: Support for socioemotional selectivity theory", *Psychology and Aging* 7:331-338.

Carstensen, L.L., J. Graff, R.W. Levenson and J.M. Gottman (1996) "Affect in intimate relationships: The developmental course of marriage", in: *Handbook of Emotion, Adult Development, and Aging*, C. Magai and S.H. McFadden, eds, San Diego, CA: Academic Press, pp. 227-247.

Carstensen, L.L., J.J. Gross and H.H. Fung (1998) "The social context of emotional experience", in: *Annual Review of Gerontology and Geriatrics (Volume 17): Focus on Emotion and Adult Development*, K.W. Schaie and M.P. Lawton, eds, New York: Springer, pp. 325-352.

Carstensen, L.L. and S. Turk-Charles (1994) "The salience of emotion across the adult life span", *Psychology and Aging* 9:259-264.

Carstensen, L.L. and S. Turk-Charles (1998) "Emotion in the second half of life", *Current Directions in Psychological Science* 7:144-149.

Christianson, S.-A., ed. (1992) *The Handbook of Emotion and Memory: Research and Theory*, Hillsdale, New Jersey: Lawrence Erlbaum Associates.

Clark, A. (1997) *Being There: Putting Brain, Body and World Together Again*, Cambridge, MA: MIT Press.

Clore, G.L. (1992) "Cognitive phenomenology: Feelings and the construction of judgment", in *The Construction of Social Judgments*, L.L. Martin and A. Tesser, eds., Hillsdale, New Jersey: Lawrence Erlbaum, pp. 133-163.

Clore, G.L. and A. Ortony (2000) "Cognition in emotion: Always, sometimes, or never?", in: *Cognitive Neuroscience of Emotion*, R.D. Lane, L. Nadel, G.L. Ahern, J.J.B. Allen, A.W. Kaszniak, S.Z. Rapcsak and G.E. Schwartz, eds., New York: Oxford University Press, pp. 24-61.

Clore, G.L., N. Schwarz and M. Conway (1994) "Affective causes and consequences of social information processing", in: *Handbook of Social Cognition. Volume 1, Basic Processes*, R.S. Wyer, Jr. and T.K. Srull, eds, Hillsdale, NJ: Lawrence Erlbaum Associates, pp. 323-417.

Critchley, H.D., R. Elliott, C.J. Mathias and R.J. Dolan (2000) "Neural activity relating to generation and representation of galvanic skin conductance responses: A functional magnetic resonance imaging study", *The Journal of Neuroscience* 20:3033-3040.

Dalby, P.R., S.L. Reminger, A.W. Kaszniak and E.B. Montgomery (in press) "Expressive and subjective features of emotion in idiopathic Parkinson's disease: A test of the facial feedback hypothesis", *Journal of the International Neuropsychological Society*.

Damasio, A.R. (1994) *Descartes' Error: Emotion, Reason and the Human Brain*, New York: G.P. Putnam.

Damasio, A. (1999) *The Feeling of What Happens: Body and Emotion in the Making of Consciousness*, New York: Harcourt Brace & Company.

Dar, R. and H. Leventhal (1993) "Schematic processes in pain perception", *Cognitive Therapy and Research* 17:341-357.

Davidson, R. (1998) "Affective style and affective disorders: Perspectives from neuroscience", *Cognition and Emotion* 12:307-330.

Davidson, R.J., ed. (2000) *Anxiety, Depression, and Emotion*, New York: Oxford University Press.

Diener, E. and M.E. Suh (1998) "Subjective well-being and age: An international analysis", in: *Annual Review of Gerontology and Geriatrics (Volume 17): Focus on Emotion and Adult Development*, K.W. Schaie and M.P. Lawton, eds., New York: Springer, pp. 304-324.

Donaldson, S. and M. Westerman (1986) "Development of children's understanding of ambivalence and causal theories of emotion", *Developmental Psychology* 22:655-662.

Dougherty, L.M., J.A. Abe and C.E. Izard (1996) "Differential emotions theory and emotional development in adulthood and later life", in: *Handbook of Emotion, Adult Development and Aging*, C. Magai and S. McFadden, eds., San Diego, California: Academic Press, pp. 27-41.

Duke, L.M. and A.W. Kaszniak (2000) "Executive control functions in degenerative dementias: A comparative review", *Neuropsychology Review* 10:75-99.

Ekman, P. and W.V. Friesen (1978) *Facial Action Coding System (FACS): A Technique for the Measurement of Facial Action*, Palo Alto, CA: Consulting Psychologists Press.

Eslinger, P.J. and A.R. Damasio (1985) "Severe disturbance of higher cognition after bilateral frontal lobe ablation: patient EVR", *Neurology* 35:1731-1741.

Folkman, S. and J.T. Moskowitz (2000) "Positive affect and the other side of coping", *American Psychologist* 55:647-654.

Forgas, J.P. (1995) "Emotion in social judgments: Review and a new affect infusion model (AIM)", *Psychological Bulletin* 117:39-66.

Fox, N.A., ed (1994) "The development of emotion regulation: Biological and behavioral considerations", *Monographs of the Society for Research in Child Development* 59 (2-3, Serial No. 240).

Fredrickson, B.L. (1998) "What good are positive emotions?", *Review of General Psychology* 2:300-319.

Frijda, N.H. (1986) *The Emotions*, Cambridge, UK: Cambridge University Press.

Frolkis, V.V. (1977) "Aging of the autonomic nervous system", in: *Handbook of the Psychology of Aging*, J.E. Birren and K.W. Schaie, eds., New York: Van Nostrand Reinhold, pp. 177-189.

Gasper, K. and G.L. Clore (1998) "The persistent use of negative affect by anxious individuals to estimate risk", *Journal of Personality and Social Psychology* 74:1350-1363.

Gaussen, T. (2000) "The development of personhood and the brain", in: *The Psychology of Awakening*, G. Watson, S. Batchelor and G. Claxton, eds., York Beach, ME: Samuel Weiser, pp. 123-134.

Grant, S., C. Contoreggi and E.D. London (1997) "Drug abusers show impaired performance on a test of orbitofrontal function", *Society for Neuroscience Abstracts* 23:1943.

Harris, P.L. (2000) "Understanding emotion", in *Handbook of Emotions, Second Edition*, M. Lewis and J.M. Haviland-Jones, eds., New York: Guilford, pp. 281-292.

Helson, R. and E.C. Klohnen (1998) "Affective coloring of personality from young adulthood to midlife", *Personality and Social Psychology Bulletin* 24:241-252.

Herzog, A.G. and T.L. Kemper (1980) "Amygdaloid changes in aging and dementia", *Archives of Neurology* 37:625-629.

Isen, A.M. (2000) "Positive affect and decision making", in: *Handbook of Emotions, Second Edition*, M. Lewis and J.M. Haviland-Jones, eds., New York: Guilford, pp.417-435.

Izard, C.E. and B.P. Ackerman (1998) "Emotions and self-concepts across the life span", in: *Annual Review of Gerontology and Geriatrics (Volume 17): Focus on Emotion and Adult Development*, K.W. Schaie and M.P. Lawton, eds., New York: Springer, pp. 1-26.

Izard, C.E., C.A. Fantauzzo, J.M. Castle, O.M. Haynes, M.F. Rayias and P.H. Putnam (1995) "The ontogeny and significance of infants' facial expressions in the first nine months of life", *Developmental Psychology* 31:997-1013.

Johnson, E. and A. Tversky (1983) "Affect, generalization, and the perception of risk", *Journal of Personality and Social Psychology* 45:20-31.

Kabat-Zinn, J. (1990) *Full Catastrophe Living*, New York: Delta.

Kabat-Zinn, J., L. Lipworth, R. Burney and W. Sellers (1986) "Four-year follow-up of a meditation-based program for the self-regulation of chronic pain: Treatment outcomes and compliance", *Clinical Journal of Pain* 2:159-173.

Kabat-Zinn, J., A.O. Massion, J. Kristeller, L.G. Peterson, K.E. Fletcher, L. Pert, W.R. Lenderking and S.F. Santorelli (1992) "Effectiveness of a meditation-based stress reduction program in the treatment of anxiety disorders", *American Journal of Psychiatry* 149:936-943.

Kaszniak, A.W. and M.C. Newman (2000) "Toward a neuropsychology of cognitive aging", in: *Psychology and the Aging Revolution: How we Adapt to Longer Life*, S.H. Qualls and N. Ables, eds., Washington, DC: American Psychological Association, pp.43-67.

Kaszniak, A.W., S.L. Reminger, S.Z. Rapcsak and E.L. Glisky (1999) "Conscious experience and autonomic response to emotional stimuli following frontal lobe damage", in: *Toward a Science of Consciousness III: The Third Tucson Discussions and Debates*, S.R. Hameroff, A.W. Kaszniak, and D.J. Chalmers, eds, Cambridge, MA: MIT Press, pp.201-213.

Kelley, C.M. (1986) "Depressive mood effects of memory and attention", in: *Handbook for Clinical Memory Assessment of Older Adults*, L.W. Poon, T. Crook, K.L. Davis, E. Eisdorfer, B.J. Gurland, A.W. Kaszniak and L.W. Thompson, eds., Washington, DC: American Psychological Association, pp. 238-243.

Kornfield, J. (2000) *After the Ecstasy, the Laundry: How the Heart Grows Wise on the Spiritual Path*, New York: Bantam Books.

Labouvie-Vief, G. (1998) "Cognitive-emotional integration in adulthood", in: *Annual Review of Gerontology and Geriatrics (Volume 17): Focus on Emotion and Adult Development*, K.W. Schaie and M.P. Lawton, eds., New York: Springer, pp.206-237.

Lane, R.D. (2000) "Neural correlates of conscious emotional experience", in: *Cognitive Neuroscience of Emotion*, R.D. Lane, L. Nadel, G.L. Ahern, J.J.B. Allen, A.W. Kaszniak, S.Z. Rapcsak and G.E. Schwartz, eds, New York: Oxford University Press, pp. 345-370.

Lane, R.D., L. Nadel and A.W. Kaszniak (2000) "The future of emotion research from the perspective of cognitive neuroscience", in: *Cognitive Neuroscience of Emotion*, R.D. Lane, L. Nadel, G.L. Ahern, J.J.B. Allen, A.W. Kaszniak, S.Z. Rapcsak and G.E. Schwartz, eds, New York: Oxford University Press, pp. 345-370.

Lang, P.J., M.M. Bradley and B.N. Cuthbert (1999) *International Affective Picture System (IAPS): Technical Manual and Affective Ratings*, Gainesville, FL: The Center for Research in Psychophysiology, University of Florida.

Langer, E.J. (1992) "Matters of mind: Mindfulness/Mindlessness in perspective", *Consciousness and Cognition* 1:289-305.

Lavenu, I., F. Pasquier, F. Lebert, H. Petit and M. Van der Linden (1999) "Perception of emotion in frontotemporal dementia and Alzheimer disease", *Alzheimer Disease and Associated Disorders* 13:96-101.

Lawton, M.P., M.H. Kleban, D. Rajagopal and J. Dean (1992) "Dimensions of affective experience in three age groups", *Psychology and Aging* 7:171-184.

Lazarus, R.S. (1991) *Emotion and Adaptation*, New York: Oxford University Press.

LeDoux, J. (1996) *The Emotional Brain: The Mysterious Underpinnings of Emotional Life*, New York: Simon and Schuster.

Lerner, J.S. and D. Keltner (2000) "Beyond valence: Toward a model of emotion-specific influences on judgement and choice", *Cognition and Emotion* 14:473-493.

Levenson, R.W. (1992) "Autonomic nervous system differences among emotions", *Psychological Science* 3:23-27.

Levenson, R.W. (1994) "Human emotions: A functional view", in: *The Nature of Emotion: Fundamental Questions*, P. Ekman and R. Davidson, eds., New York: Oxford University Press, pp. 123-126.

Levenson, R.W. (2000) "Expressive, physiological, and subjective changes in emotion across adulthood", in: *Psychology and the Aging Revolution: How we Adapt to Longer Life*, S.H. Qualls and N. Ables, eds., Washington, DC: American Psychological Association, pp. 123-140.

Levenson, R.W., L.L. Carstensen, W.V. Friesen and P. Ekman (1991) "Emotion, physiology, and expression in old age", *Psychology and Aging* 6:28-35.

Levenson, R.W., L.L. Carstensen and J.M. Gorrman (1994) "Influence of age and gender on affect, physiology, and their interrelations: A study of long-term marriages", *Journal of Personality and Social Psychology* 67:56-68.

Levine, M. (2000) *The Positive Psychology of Buddhism and Yoga: Paths to a Mature Happiness*, Mahwah, NJ: Lawrence Erlbaum Associates.

Lewis, M. (1991) "Ways of knowing: Objective self awareness or consciousness", *Developmental Review* 11:231-243.

Lewis, M. (2000) "The emergence of human emotions", in: *Handbook of Emotions, Second Edition*, M. Lewis and J.M. Haviland-Jones, eds., New York: Guilford, pp. 265-280.

Lewis, M. and J. Brooks (1978) "Self-knowledge and emotional development", in: *The Development of Affect*, M. Lewis and L. Rosenblum, eds., New York: Plenum, pp. 205-226.

Lewis, M. and J.M. Haviland-Jones, eds (2000) *Handbook of Emotions, Second Edition*, New York: Guilford.

Malatesta, C. Z., M.J. Fiore and J.J. Messina (1987) "Affect, personality, and facial expressive characteristics of older people", *Psychology and Aging* 2:64-69.

Malatesta, C.Z., C.E. Izard, C. Culver and M. Nicolich, (1987) "Emotion communication skills in young, middle-aged, and older women", *Psychology and Aging* 2:193-203.

Malatesta, C.Z. and M. Kalnok (1984) "Emotional experience in younger and older adults", *Journal of Gerontology* 39:301-308.

Masters, R.A. (2000) "Compassionate wrath: Transpersonal approaches to anger", The *Journal of Transpersonal Psychology* 32:31-51.

Mathews, A. and C. MacLeod (1994) "Cognitive approaches to emotion and emotional disorders", *Annual Review of Psychology* 45:25-50.

Matsumoto, D. and M. Lee (1993) "Consciousness, volition, and the neuropsychology of facial expressions of emotion", *Consciousness and Cognition* 2:237-254.

Mayer, J. and P. Salovey (1997) "What is emotional intelligence?", in: *Emotional Development and Emotional Intelligence*, P. Salovey and D. Sluyter, eds., New York: Basic Books, pp. 3-31.

Miller, M.E. and S.R. Cook-Greuther, eds (1990) *Transcendence and Mature Thought in Adulthood: The Further Reaches of Adult Development*, Lantham, MD: Rowman and Littlefield.

Mogg, K. and B.P. Bradley (1998) "A cognitive-motivational analysis of anxiety", *Behavior Research and Therapy* 36:809-848.

Moreno, C.R., J.C. Borod, J. Welkowitz and M. Alpert (1990) "Lateralization for the expression and perception of facial emotion as a function of age", *Neuropsychologia* 28:199-209.

Nesse, R.M. (1990) "Evolutionary explanations of emotions", *Human Nature* 1:261-289.

Niedenthal, P.M. and S. Kitayama, eds. (1994) *The Heart's Eye: Emotional Influences in Perception and Attention*, New York: Academic Press.

Öhman, A. (2000) "Fear and anxiety: Evolutionary, cognitive, and clinical perspectives", in: *Handbook of Emotions, Second Edition*, M. Lewis and J.M. Haviland-Jones, eds., New York: Guilford, pp.573-593.

Öhman, A., A. Flykt and D. Lundqvist (2000) "Unconscious emotion: Evolutionary perspectives, psychophysiological data and neuropsychological mechanisms", in: *Cognitive Neuroscience of Emotion* , R.D. Lane, L. Nadel, G.L. Ahern, J.J.B. Allen, A.W. Kaszniak, S.Z. Rapcsak and G.E. Schwartz eds, New York: Oxford University Press, pp. 296-327.

Palfai, T.P. and P. Salovey (1992) "The influence of affect on self-focused attention: Conceptual and methodological issues", *Consciousness and Cognition* 1:306-339.

Panksepp, J. (1998) *Affective Neuroscience*, Oxford, UK: Oxford University Press.

Petry, N.M., W.K. Bickel and M. Arnett (1998) "Shortened time horizons and insensitivity to future consequences in heroin addicts", *Addiction* 93:729-738.

Reminger, S.L., A.W. Kaszniak and P.R. Dalby (2000) "Age-invariance in the asymmetry of stimulus-evoked emotional facial muscle activity", *Aging, Neuropsychology, and Cognition* 7:156-168.

Rinn, W.E. (1984) "The neuropsychology of facial expression: A review of the neurological and psychological mechanisms for producing facial expressions", *Psychological Bulletin* 95:52-77.

Saarni, C. (1999) *The Development of Emotional Competence*, New York: Guilford.

Salovey, P., A.J. Rothman, J.B. Detweiler and W.T. Steward (2000) "Emotional states and physical health", *American Psychologist* 55:110-121.

Salovey, P. and D.J. Sluyter, eds (1997) *Emotional Development and Emotional Intelligence: Educational Implications*, New York: Basic Books.

Schnyer, D.M., A.W. Kaszniak, J.J.B. Allen and L. Trujillo, (1999, June) "Assessing attentional changes in long-term mindfulness meditators using a cued attention task and event-related brain potentials", Paper presented at the *Third Annual Meeting of the Association for the Scientific Study of Consciousness*, London, Canada.

Schore, A.N. (1994) *Affect Regulation and the Origin of the Self: The Neurobiology of Emotional Development*, Hillsdale, NJ: Lawrence Erlbaum Associates.

Schore, A.N. (1996) "The experience-dependent maturation of a regulatory system in the orbital prefrontal cortex and the origin of developmental psychopathology", *Developmental Psychopathology* 8:59-87.

Schwarz, N. (2000) "Emotion, cognition, and decision making", *Cognition and Emotion* 14:433-440.

Schwarz, N. and G.L. Clore (1983) "Mood, misattribution and judgments of well-being: Informative and directive functions of affective states", *Journal of Personality and Social Psychology* 45:513-523.

Schwarz, N. and G.L. Clore (1996) "Feelings and phenomenal experiences", in: *Social Psychology: Handbook of Basic Principles*, E.T. Higgins and A. Kruglanski, eds, New York: Guilford, pp. 433-465.

Segal, Z.V., J.M.G. Williams and J.D. Teasdale (in press) *Mindfulness and the Prevention of Depression: A Guide to the Theory and Practice of Mindfulness-Based Cognitive Therapy*, New York: Guilford Press.

Seligman, M.E.P. and M. Csikszentmihalyi (2000) "Positive psychology: An introduction", *American Psychologist* 55:5-14.

Smith, C.A. and P.C. Ellsworth (1985) "Patterns of cognitive appraisal in emotion", *Journal of Personality and Social Psychology* 48:813-838.

Strayer, J. (1986) "Children's attributions regarding the situational determinants of emotion in self and other", *Developmental Psychology* 22:649-654.

Steiner, J.E. (1979) "Human facial expression in response to taste and smell stimulation", *Advances in Child Development and Behavior* 13:237-295.

Teasdale, J.D., Z.V. Segal, J.M.G. Williams, V.A. Ridgeway, J.M. Soulsby and M.A. Lau (2000) "Prevention of relapse/recurrence in major depression by mindfulness-based cognitive therapy", *Journal of Consulting and Clinical Psychology* 68:615-623.

Tranel, D., A. Bechara, H. Damasio and A.R. Damasio (1998) "Neural correlates of emotional imagery", *International Journal of Psychophysiology* 30:107.

Trevarthen, C. (1993) "The self born in intersubjectivity: The psychology of an infant communicating", in: *The Perceived Self: Ecological and Interpersonal Sources of Self-Knowledge*, U. Neisser, ed., New York: Cambridge University Press, pp. 121-173.

Trevarthen, C. and K.J. Aitken (1994) "Brain development, infant communication, and empathy disorders: intrinsic factors in child mental health", *Developmental Psychopathology* 6:597-633.

Varela, F.J. (2000) "Steps to a science of inter-being: Unfolding the dharma implicit in modern cognitive science", in: *The Psychology of Awakening*, G. Watson, S. Batchelor and G. Claxton, eds, York Beach, ME: Samuel Weiser, Inc., pp.71-89

Varela, F.J., E. Thompson and E. Rosch (1991) *The Embodied Mind: Cognitive Science and Human experience*, Cambridge, MA: MIT Press.

Wright, W.F. and G.H. Bower (1992) "Mood effects on subjective probability assessment", *Organizational Behavior and Human Decision Processes* 52:276-291.

Wintre, M. and D. Vallance (1994) "A developmental sequence in the comprehension of emotions: Intensity, multiple emotions, and valence", *Developmental Psychology* 30:509-514.

Zajonc, R. (1998) "Emotions", in: *The Handbook of Social Psychology, Volume 1*, D. Gilbert, S. Fiske and G. Lindzey, eds., New York: Oxford University Press, pp. 591-632.

PARTICIPANTS

Members of the Advisory Board, Lecturers and Members of the Organizing Committee are indicated respectively by three, two and one asterisks

Arturo Aguilar
Universidad Nacional Autonoma de Mexico
Apdo. Postal Cop-084
04331 Mexico, D.F.
Mexico
samayoa@servidor.unam.mx

Torin Alter
Department of Philosophy
The University of Alabama
Box 870218
Tuscaloosa, AL 35487-0218
USA
talter@philos.as.ua.edu

*Nunzia Aprile
Istituto Italiano per gli Studi Filosofici
Via Gennaro Serra 75
I-80132 Napoli
Italy
Phone +39 081 2542086
Fax +39 081 7645058

Aaron Ben-Ze'ev
Center for the Interdisciplinary Research of Emotions
University of Haifa
Haifa 31905
Israel
Phone +972 4 8240622/5
Fax +972 4 8342101
benzeev@research.haifa.ac.il

Phyllis Books
Assisi Foundation
1905 Salem Street
Irving, TX 75061
USA
Phone +1 972 986 2827
Fax +1 972 513 1140
101351.620@compuserve.com

**Margaret Bradley
Department of Psychology
University of Florida
P.O. Box 100165
Gainesville, FL 32610
USA
Phone +1 352 392 2439
Fax +1 352 392 6047
bradley@nervm.nerdc.ufl.edu

Teresa Brennan
Schmidt College of Arts and Letters
Florida Atlantic University
777 Glades Road
Boca Raton, FL 33431
USA
TBrennan@fau.edu

Barbara Cabott
1238 NW Glisan "C"
Portland, OR 97209
USA
Phone +1 503.242.0136
Fax +1 503 827 0361
bcabott@imagine.com

Clotilde Calabi
Dipartimento di Filosofia
Università di Milano
Via Festa del Perdono
I-20122 Milano
Italy

**C. Richard Chapman
Department of Anesthesiology, Box 356540
University of Washington
Seattle, WA 98195-6540
USA
Phone +1 206 685 2082
Fax +1 206 543 2958
crc@u.washington.edu

Jonathan Cole
Department of Clinical Neuroscience
University of Southampton
Southampton General Hospital
Southampton SO16 6YD
United Kingdom
jcole@minstead.demon.co.uk

*Antonio Cotugno
Istituto di Cibernetica, CNR

Via Toiano 6
I-80072 Arco Felice (Napoli)
Italy
Phone +39 81 8534113
Fax +39 81 5267654
neurosistemi@sole.cib.na.cnr.it

William Frawley
Departments of Linguistics/Cognitive Science
University of Delaware
Newark, DE 19716
USA
billf@copland.udel.edu

**Nico Frijda
Department of Psychology
Subdepartment of Psychonomie, Rm. 704
Roetersstraat 15
1018 WB Amsterdam
Netherlands
Phone +31 20 525 6743
pn_frijda@macmail.psy.uva.nl

**Guido Gainotti
Università Cattolica del Sacro Cuore
Facoltà di Medicina e Chirurgia "Agostino Gemelli"
Istituto di Neurologia
Servizio di Neuropsicologia
Largo Agostino Gemelli 8
I-00168 Roma
Italy
Phone +39 06 35501909
Fax +39 06 305 1343
iclnp@rm.unicatt.it

Dario Galati
Dipartimento di Psicologia
Università di Torino
Via G.L. Lagrange 3
I-10123 Torino
Italy
Phone +39 011 549475
Fax +39 011 549653
Galati@psych.unito.it

Shaun Gallagher
Department of Philosophy
Canisius College
Buffalo, NY 14208
USA
Phone +1 716.888.2329
Fax +1 716 888 2525
gallaghr@canisius.edu

Michael Gillespie
Department of Sociology
University of Alberta
Edmonton, Alberta T6G 2H4
Canada
mgillesp@gpu.srv.ualberta.ca

Andrew Gluck
392 Central Park West, apt 8C
New York, NY 10025
USA
Phone +1 212 316 2810
Andy_Gluck@msn.com

**Paul Griffiths
Unit for History and Philosophy of Science
Carslaw Building F07
University of Sydney
Camperdown, NSW 2006
Australia
Phone +61 2 9351 4801
Fax +61 2 9351 4124
paulg@scifac.usyd.edu.au

Valerie Gray Hardcastle
Department of Philosophy
Virginia Tech
Blacksburg, VA 24061-0126
USA
valerie@vt.edu

**Ursula Hess
Department of Psychology
University of Québec at Montréal
P.O. Box 8888, Station "Centre-ville"
Montréal, Qc H3C 3P8
Canada
Phone +1 514 987 3000 x4834
Fax +1 514 987 7953
Hess.Ursula@UQAM.ca

Margarlit Kaplinsky
11 Bergson str.
Ramat-Aviv 69106
Tel-Aviv
Israel
strivlin@shani.net

**Alfred W. Kaszniak
Center for Consciousness Studies
Department of Psychology
Bldg. 68, Room 312
University of Arizona
Tucson, AZ 85721
USA
Phone +1 520 621 5149
Fax +1 520 621 9306
kaszniak@u.arizona.edu

Chantal Kerssens
Erasmus Universiteit Rotterdam
P.O. Box 1738
NL-3000 DR Rotterdam
Netherlands
Phone +31 10 4087805
Fax +31 10 4363981
kerssens@mpp.fgg.eur.nl

Maria Susana Koreck
Instituto "Juan XXIII"
Fitz Roy 291
8000 Bahia Blanca
Argentina
Phone +54 91 528275
Fax +54 91 535311
mskoreck@criba.edu.ar

***,**Richard Lane
Department of Psychiatry
University of Arizona
AZ Health Sciences Center 7401
P.O. Box 245002
Tucson, AZ 85724
USA
Phone +1 520 626 6254
lane@u.arizona.edu

**Peter Lang
Department of Psychology
University of Florida
P.O. Box 100165
Gainesville, FL 32610
USA
Phone +1 352 392 2439
Fax +1 352 392 6047
langlab@nervm.nerdc.ufl.edu

*Jim Laukes
Center for Consciousness Studies
Department of Psychology
Bldg. 68, Room 312
University of Arizona
Tucson, AZ 85721
USA
Phone +1 520 621 9317
Fax +1 520 621 9306
jlaukes@u.arizona.edu

***,**Joseph LeDoux
Center for Neural Science
New York University
4 Washington Pl., Rm 809
New York, NY 10003
USA
Phone +1 212 998 3530
Fax +1 212 995 4011
ledoux@cns.nyu.edu

Pierre Livet
Université d'Aix-Marseille 1
CEPERC and CREA
29 Avenue R. Schuman
Aix en Provence, 13621
France
livet@newsup.univ-mrs.fr

Fulvio Magara
Anatomisches Institut
Universität Zürich
CH-8057 Zürich
Switzerland
magara@anatomie.unizh.ch

Riccardo Manzotti
Laboratorio Integrato di Robotica Avanzata
Dipartimento di Informatica, Sistemistica e Telematica
Università di Genova
Via Opera Pia 13
I-16145 Genova
Italy
Phone +39 010 353 2946
Fax +39 010 353 2154
manzotti@lira.dist.unige.it

Daniel Weiss Miller
106 St. Marks. Ave
Prospect Heights
Brooklyn, NY 11217

USA
Phone +1 718 789 0515
Fax +1 718 789 2550
btrmiller@aol.com

Ken Mogi
Sony Computer Science Laboratory Inc.
Takanawa Muse Bldg.
3-14-13, Higashigotanda
Shinagawa-ku
Tokyo 141-0022
Japan
Phone +81 3 5448 4380
Fax +81 3 5448 4273
kenmogi@csl.sony.co.jp

Carlo Musio
Istituto di Cibernetica, CNR
Via Toiano 6
I-80072 Arco Felice (Napoli)
Italy
Phone +39 081 8534131
Fax +39 081 5267654
carlom@biocib.cib.na.cnr.it

Marco Neppi Modona
Dipartimento di Psicologia
Università di Torino
Via G.L. Lagrange 3
I-10123 Torino
Italy
Phone +39 011 549475/549508
Fax +39 011 549653
neppi@psych.unito.it

Anna Neugebauer
Istituto Internazionale di Genetica e
Biofisica, CNR
Via Marconi 10
I-80125 Napoli
Italy
Phone +39 0 81 7257268
Fax +39 81 5936123
neugebau@iigbna.iigb.na.cnr.it

Mary Newman
Department of Psychology
University of Arizona
Bldg. 68, Room 312
Tucson, AZ 85721
USA
Phone +1 520 621 7447
Fax +1 520 621 9306
newmanm@u.arizona.edu

Lis Nielsen
Department of Psychology
University of Arizona
Tucson, AZ 85721
USA
Phone +1 520 621 7447
Fax +1 520 621 9306
lnielsen@u.arizona.edu

**Arne Öhman
Department of Clinical Neuroscience
Karolinska Institute
Stockholm, S-171 76
Sweden
Phone +11 48 8 729 5065
Fax +11 46 8 307298
arneo@psyk.ks.se

***,**Jaak Panksepp
Department of Psychology
Bowling Green State University
Bowling Green, OH 43403
USA
Phone +1 419 372 2819
jpankse@bgnet.bgsu.edu

Michael Pauen
Hanse-Institute for Advanced Study
Admiralstr. 15
28215 Bremen
Germany
Phone +49 (0)421 3761301
Fax +49(0)421 3761301
pauen@mailer.uni-marburg.de

***Tomas Radil
Czech Academy of Sciences
Institute of Physiology
1083, Videnska
CZ-142 20 Praha 4
Czech Republic
Phone +42 02 4752509
Fax +42 02 4719517
Radil@biomed.cas.cz

Patrice Renaud
Université du Québec à Hull

Pavillon Alexandre Taché
293 Boulevard Alexandre Taché
P.O. Box 1250, Succ. B
Hull, Québec J8X 3X7
Canada
Phone +1 800 567 1283, ext. 1 4412
Fax +1 819 595 2384
patrice_renaud@uqah.uquebec.ca

**Willibald Ruch
Institute for Psysiologische Psychologie
Universität Düsseldorf
Universitätsstr., Geb. 23.03
D-40225 Düsseldorf
Germany
wruch@rz.uni-duesseldorf.de

Silvia Santillo
Istituto di Cibernetica, CNR
Via Toiano 6
I-80072 Arco Felice (Napoli)
Italy
Phone +39 81 8534113
Fax +39 81 5267654
silvia@sole.cib.na.cnr.it

Francis Schwanauer
Department of Philosophy
University of Southern Maine
P.O. Box 9300
Portland, ME
USA
Phone +1 207 780 4243
franz@portland1.usm.maine.edu

Don Sharpsteen
Department of Psychology
University of Missouri at Rolla
1870 Miner Circle
Rolla, MO 65409
USA
djsharp@umr.edu

Raoul N. Smith
Northeastern University
Boston, MA 02115
USA
Phone +1 617 373 5205
Fax +1 617 373 5121
rnsmith@ccs.neu.edu

Karola Stotz
Department of Philosophy
Ghent University
Blandijnberg 2
9000 Ghent
Belgium
Phone +32 9 264 39 69
Fax +32 9 264 41 97
karola.stotz@rug.ac.be

***Cloe Taddei-Ferretti
Istituto di Cibernetica, CNR
Via Toiano 6
I-80072 Arco Felice (Napoli)
Italy
Phone +39 81 8534113
Fax +39 81 5267654
cloe@sole.cib.na.cnr.it

Yoshi Tamori
Kanazawa Institute of Technology
3-1, Yatsukaho, Matto
Ishikawa 924-0838
Japan
Phone +81 76 274 8250 x 2222
Fax +81 76 274 8251
yo@his.kanazawa-it.ac.jp

Settimo Termini
Università di Palermo
Dipartimento di Matematica
Via Archirafi 34
I-90123 Palermo
Italy
Phone +39 091 6040268
Fax +39 091 6165425
termini@dipmat.math.unipa.it

**Douglas F. Watt
Department of Neuropsychology
Quincy Hospital
114 Whitwell Street
Quincy, MA 02169
USA
Phone +1 617 376 5400
Fax +1 617 376 1635
DrWatt@email.msn.com

Philip Wong
Department of Psychology
New School for Social Research

65 Fifth Ave.
New York, NY 10003
USA
Phone +1 212 229 5778
Fax +1 212 989 0846
pswong@newschool.edu

Branka Zei
Hopitaux Universitaires de Genève
Psychiatrie de Liaison
51 Bvd. De la Cluse
1205 Genève
Switzerland
zei-branka@diogenes.hcuge.ch